Applied Mathematical Sciences

Volume 120

The mathematization of all sciences, the fading of traditional scientific boundaries, the impact of computer technology, the growing importance of computer modeling and the necessity of scientific planning all create the need both in education and research for books that are introductory to and abreast of these developments.The purpose of this series is to provide such books, suitable for the user of mathematics, the mathematician interested in applications, and the student scientist. In particular, this series will provide an outlet for topics of immediate interest because of the novelty of its treatment of an application or of mathematics being applied or lying close to applications. These books should be accessible to readers versed in mathematics or science and engineering, and will feature a lively tutorial style, a focus on topics of current interest, and present clear exposition of broad appeal. A compliment to the Applied Mathematical Sciences series is the Texts in Applied Mathematics series, which publishes textbooks suitable for advanced undergraduate and beginning graduate courses.

More information about this series at http://www.springer.com/series/34

Andreas Kirsch

An Introduction
to the Mathematical Theory
of Inverse Problems

Third Edition

 Springer

Andreas Kirsch
Department of Mathematics
Karlsruhe Institute of Technology (KIT)
Karlsruhe, Germany

ISSN 0066-5452   ISSN 2196-968X   (electronic)
Applied Mathematical Sciences
ISBN 978-3-030-63345-5   ISBN 978-3-030-63343-1   (eBook)
https://doi.org/10.1007/978-3-030-63343-1

Mathematics Subject Classification: 45Q05, 47A52, 47J06, 65J20, 65J22, 65L08, 65L09, 65N20, 65N21, 65R30, 65R32, 78A45, 78A46, 81U40, 86A22, 31A25, 34A55, 34B24, 35J05, 35R25, 25R30

This Springer imprint is published by the registered company Springer Nature Switzerland AG
The registered company address is: Gewerbestrasse 11, 6330 Cham, Switzerland

Preface to the Third Edition

The field of inverse problems is growing rapidly, and during the 9 years since the appearance of the second edition of this book many new aspects and subfields have been developed. Since, obviously, not every subject can be treated in a single monograph, the author had to make a decision. As I pointed out already in the preface of the first edition, my intention was—and still is—to introduce the reader to some of the basic principles and developments of this field of inverse problems rather than going too deeply into special topics. As I continued to lecture on inverse problems at the University of Karlsruhe (now Karlsruhe Institute of Technology), new material has been added to the courses and thus also to this new edition because the idea of this book is still to serve as a type of textbook for a course on inverse problems. I have decided to extend this monograph in two directions. For some readers, it was perhaps a little unsatisfactory that only the abstract theory for linear problems was presented but the applications to inverse eigenvalue problems, electrical impedance tomography, and inverse scattering theory are of a nonlinear type. For that reason, and also because the abstract theories for Tikhonov's method and Landweber's iteration for nonlinear equations have come to a certain completion, I included a new chapter (Chapter 4) on these techniques for locally improperly posed nonlinear equations in Hilbert spaces with an outlook into some rather new developments for Banach spaces. The former Chapters 4, 5, and 6 are moved to 5, 6, and 7, respectively. The additional functional analytic tools needed in this new Chapter 4 result in two new sections of Appendix A on convex analysis and weak topologies.

As a second new topic, a separate section (Section 7.6) on interior transmission eigenvalues is included. These eigenvalue problems arise naturally in the study of inverse scattering problems for inhomogeneous media and were introduced already in the former editions of this monograph. Besides their importance in scattering theory, the transmission eigenvalue problem is an interesting subject in itself, mainly because it fails to be self-adjoint. The investigation of the spectrum is a subject of the present research. Special issues of *Inverse Problems* [37] and recent monographs [34, 55] have addressed this topic alreadyf for the study of complex eigenvalues, one is until now restricted to radially symmetric refractive indices which reduces the partial differential equations to ordinary differential equations. Classical tools from complex analysis make it possible to prove the existence of complex eigenvalues (Subsection 7.6.1) and uniqueness for the corresponding inverse spectral problem (Subsection 7.6.3). I think that this analogue to the inverse Sturm–Liouville problem of Chapter 5 is a natural completion of the study of interior transmission eigenvalues.

Finally, a rather large number of mistakes, ambiguities, and misleading formulations has been corrected in every chapter. As major mistakes, the proofs of Theorems 4.22 (a) and 6.30 (d) (referring to the numbering of the second edition) have been corrected. I want to thank all of my colleagues and the readers of the first two editions for the overwhelming positive responses and, last but not least, the publisher for its encouragement for writing this third edition.

Karlsruhe, Germany   Andreas Kirsch
December 2020

Preface to the Second Edition

The first edition of the book appeared 14 years ago. The area of inverse problems is still a growing field of applied mathematics and an attempt at a second edition after such a long time was a difficult task for me. The number of publications on the subjects treated in this book has grown considerably and a new generation of mathematicians, physicists, and engineers has brought new concepts into the field. My philosophy, however, has never been to present a comprehensive book on inverse problems that covers all aspects. My purpose was (as I pointed out in the preface of the first edition), and still is, to present a book that can serve as a basis for an introductory (graduate) course in this field. The choice of material covered in this book reects my personal point of view: students should learn the basic facts for linear ill-posed problems including some of the present classical concepts of regularization and also some important examples of more modern nonlinear inverse problems.

Although there has been considerable progress made on regularization concepts and convergence properties of iterative methods for abstract nonlinear inverse problems, I decided not to include these new developments in this monograph. One reason is that these theoretical results on nonlinear inverse problems are still not applicable to the inverse scattering problems that are my major field of interest. Instead, I refer the reader to the monographs [92, 149] where regularization methods for nonlinear problems are intensively treated.

Also, in my opinion, every nonlinear inverse problem has its own characteristic features that should be used for a successful solution. With respect to the inverse scattering problem to determine the shape of the support of the contrast, a whole class of methods has been developed during the last decade, sometimes subsumed under the name *Sampling Methods*. Because they are very popular not only in the field of inverse scattering theory but also in the field of electrical impedance tomography (EIT) I decided to include the *Factorization Method* as one of the prominent members in this monograph.

The Factorization Method is particularly simple for the problem of EIT and this field has attracted a lot of attention during the past decade, therefore a chapter on EIT has been added to this monograph as Chapter 5 and the chapter on inverse scattering theory now becomes Chapter 6.

The main changes of this second edition compared to the first edition concern only Chapters 5 and 6 and Appendix A. As just mentioned, in Chapter 5 we introduce the reader to the inverse problem of *electrical impedance tomography*. This area has become increasingly important because of its applications in medicine and engineering sciences. Also, the methods of EIT serve as tools and guidelines for the investigation of other areas of tomography such that optical and photoacoustic tomography.

The forward model of EIT is usually set up in the weak sense, that is, in appropriate Sobolev spaces. Although I expect that the reader is familiar with the basic facts on Sobolev spaces such as the trace theorem and Friedrich's inequality, a tutorial section on Sobolev spaces on the unit disk is added in Appendix A, Section A.5. The approach using Fourier techniques is not very common but fits well with the presentation of Sobolev spaces of fractional order on the boundary of the unit disk in Section A.4 of Appendix A. In Chapter 5 on electrical impedance tomography the Neumann–Dirichlet operator is introduced and its most important properties such as monotonicity, continuity, and differentiability are shown. Uniqueness of the inverse problem is proven for the linearized problem only because it was this example for which Calderón presented his famous proof of uniqueness. (The fairly recent uniqueness proof by Astala and Päivärinta in [10] is far too complicated to be treated in this introductory work.) As mentioned above, the Factorization Method was developed during the last decade. It is a completely new and mathematically elegant approach to characterize the shape of the domain where the conductivity differs from the background by the Neumann–Dirichlet operator. The Factorization Method is an example of an approach that uses special features of the nonlinear inverse problem under consideration and has no analogy for traditional linear inverse problems.

Major changes are also made in Chapter 6 on inverse scattering problems. A section on the Factorization Method has been added (Section 6.4) because inverse scattering problems are the type of problem for which it is perfectly applicable. The rigorous mathematical treatment of the Factorization Method makes it necessary to work with weak solutions of the scattering problem. Therefore, here we also have to use (local) Sobolev spaces rather than spaces of continuously differentiable functions. I took the opportunity to introduce the reader to a (in my opinion) very natural approach to prove existence of weak solutions by the Lippmann–Schwinger equation in $L^2(D)$ (where D contains the support of the contrast $n - 1$). The key is the fact that the volume potential with any L^2-density solves the corresponding inhomogeneous Helmholtz equation in the weak sense (just as in the case of smooth densities) and can easily be proved by using the classical result and a density argument. The notion of weak solutions has the advantage of allowing arbitrary L^∞-functions as indices of refraction but makes it necessary to modify almost all of the arguments in this chapter slightly. In Section 6.7 we dropped the motivating example for the uniqueness of the inverse scattering problem (Lemma 6.8 in the first edition) because it has already been presented for the uniqueness of the linearized inverse problem of impedance tomography.

Finally, I want to thank all the readers of the first edition of the monograph for their extraordinarily positive response. I hope that with this second edition I added some course material suitable for being presented in a graduate course on inverse problems. In particular I have found that my students like the problem of impedance tomography and, in particular, the Factorization Method and I hope that this is true for others!

Karlsruhe, Germany   Andreas Kirsch
March 2011

Preface to the First Edition

Following Keller [152] we call two problems *inverse* to each other if the formulation of each of them requires full or partial knowledge of the other. By this definition, it is obviously arbitrary which of the two problems we call the direct and which we call the inverse problem. But usually, one of the problems has been studied earlier and, perhaps, in more detail. This one is usually called the *direct* problem, whereas the other is the *inverse* problem. However, there is often another more important difference between these two problems. Hadamard (see [115]) introduced the concept of a *wellposed problem*, originating from the philosophy that the mathematical model of a physical problem has to have the properties of uniqueness, existence, and stability of the solution. If one of the properties fails to hold, he called the problem *ill-posed*. It turns out that many interesting and important inverse problems in science lead to ill-posed problems, whereas the corresponding direct problems are well-posed. Often, existence and uniqueness can be forced by enlarging or reducing the solution space (the space of "models"). For restoring stability, however, one has to change the topology of the spaces, which is in many cases impossible because of the presence of measurement errors. At first glance, it seems to be impossible to compute the solution of a problem numerically if the solution of the problem does not depend continuously on the data, that is, for the case of ill-posed problems. Under additional a priori information about the solution, such as smoothness and bounds on the derivatives, however, it is possible to restore stability and construct efficient numerical algorithms.

We make no claim to cover all of the topics in the theory of inverse problems. Indeed, with the rapid growth of this field and its relationship to many fields of natural and technical sciences, such a task would certainly be impossible for a single author in a single volume. The aim of this book is twofold: first, we introduce the reader to the basic notions and difficulties encountered with ill-posed problems. We then study the basic properties of regularization methods for *linear* ill-posed problems. These methods can roughly be classified into two groups, namely, whether the regularization parameter is chosen a priori or a posteriori. We study some of the most important regularization schemes in detail.

The second aim of this book is to give a first insight into two special *nonlinear* inverse problems that are of vital importance in many areas of the applied sciences. In both inverse spectral theory and inverse scattering theory, one tries to determine a coefficient in a differential equation from measurements of either the eigenvalues of the problem or the field "far away" from the scatterer. We hope that these two examples clearly show that a successful treatment of nonlinear inverse problems requires a solid knowledge of characteristic features of the corresponding direct problem. The combination of classical analysis and modern areas of applied and numerical analysis is, in the author's opinion, one of the fascinating features of this relatively new area of applied mathematics.

This book arose from a number of graduate courses, lectures, and survey talks during my time at the universities of Göttingen and Erlangen/ Nürnberg. It was my intention to present a fairly elementary and complete introduction to the field of inverse problems, accessible not only to mathematicians but also to physicists and engineers. I tried to include as many proofs as possible as long as they required knowledge only of classical differential and integral calculus. The notions of functional analysis make it possible to treat different kinds of inverse problems in a common language and extract its basic features. For the convenience of the reader, I have collected the basic definitions and theorems from linear and nonlinear functional analysis at the end of the book in an appendix. Results on nonlinear mappings, in particular for the Fréchet derivative, are only needed in Chapters 4 and 5.

The book is organized as follows. In Chapter 1, we begin with a list of pairs of direct and inverse problems. Many of them are quite elementary and should be well known. We formulate them from the point of view of inverse theory to demonstrate that the study of particular inverse problems has a long history. Sections 1.3 and 1.4 introduce the notions of ill-posedness and the worstcase error. Although ill-posedness of a problem (roughly speaking) implies that the solution cannot be computed numerically — which is a very pessimistic point of view — the notion of the worst-case error leads to the possibility that stability can be recovered if additional information is available. We illustrate these notions with several elementary examples.

In Chapter 2, we study the general regularization theory for linear ill-posed equations in Hilbert spaces. The general concept in Section 2.1 is followed by the most important special examples: Tikhonov regularization in Section 2.2, Landweber iteration in Section 2.3, and spectral cutoff in Section 2.4. These regularization methods are applied to a test example in Section 2.5. While in Sections 2.1–2.5 the regularization parameter has been chosen a priori, that is before starting the actual computation, Sections 2.6–2.8 are devoted to regularization methods in which the regularization parameter is chosen implicitly by the stopping rule of the algorithm. In

Sections 2.6 and 2.7, we study Morozov's discrepancy principle and, again, Landweber's iteration method. In contrast to these *linear* regularization schemes, we will investigate the conjugate gradient method in Section 2.8. This algorithm can be interpreted as a *nonlinear* regularization method and is much more difficult to analyze.

Chapter 2 deals with ill-posed problems in infinite-dimensional spaces. However, in practical situations, these problems are first discretized. The discretization of linear ill-posed problems leads to badly conditioned finite linear systems. This subject is treated in Chapter 3. In Section 3.1, we recall basic facts about general projection methods. In Section 3.2, we study several Galerkin methods as special cases and apply the results to Symm's integral equation in Section 3.3. This equation serves as a popular model equation in many papers on the numerical treatment of integral equations of the first kind with weakly singular kernels. We present a complete and elementary existence and uniqueness theory of this equation in Sobolev spaces and apply the results about Galerkin methods to this equation. In Section 3.4, we study collocation methods. Here, we restrict ourselves to two examples: the moment collocation and the collocation of Symm's integral equation with trigono-metric polynomials or piecewise constant functions as basis functions. In Section 3.5, we compare the different regularization techniques for a concrete numerical example of Symm's integral equation. Chapter 3 is completed by an investigation of the Backus–Gilbert method. Although this method does not quite fit into the general regularization theory, it is nevertheless widely used in the applied sciences to solve moment problems.

In Chapter 4, we study an *inverse eigenvalue problem* for a linear ordi-nary differential equation of second order. In Sections 4.2 and 4.3, we develop a careful analysis of the direct problem, which includes the asymptotic behaviour of the eigenvalues and eigenfunctions. Section 4.4 is devoted to the question of uniqueness of the inverse problem, that is, the problem of recovering the coefficient in the differential equation from the knowledge of one or two spectra. In Section 4.5, we show that this inverse problem is closely related to a parameter identification problem for parabolic equations. Section 4.6 describes some numerical reconstruction techniques for the inverse spectral problem.

In Chapter 5, we introduce the reader to the field of *inverse scattering theory*. Inverse scattering problems occur in several areas of science and technology, such as medical imaging, nondestructive testing of material, and geological prospecting. In Section 5.2, we study the direct problem and prove uniqueness, existence, and continuous dependence on the data. In Section 5.3, we study the asymptotic form of the scattered field as $r \to \infty$ and introduce the *far eld pattern*. The corresponding inverse scattering problem is to recover the *index of refraction* from a knowledge of the far field pattern. We give a complete proof of uniqueness of this inverse problem in Section 5.4. Finally, Section 5.5 is devoted to the study of some recent reconstruction techniques for the inverse scattering problem.

Chapter 5 differs from previous ones in the unavoidable fact that we have to use some results from scattering theory without giving proofs. We only formulate these results, and for the proofs we refer to easily accessible standard literature.

There exists a tremendous amount of literature on several aspects of inverse theory ranging from abstract regularization concepts to very concrete applications. Instead of trying to give a complete list of all relevant contributions, I mention only the monographs [17, 105, 110, 136, 168, 173, 174, 175, 182, 197, 198, 215, 263, 264], the proceedings, [5, 41, 73, 91, 117, 212, 237, 259], and survey articles [88, 148, 152, 155, 214] and refer to the references therein.

This book would not have been possible without the direct or indirect contributions of numerous colleagues and students. But, first of all, I would like to thank my father for his ability to stimulate my interest and love of mathematics over the years. Also, I am deeply indebted to my friends and teachers, Professor Dr. Rainer Kress and Professor David Colton, who introduced me to the field of scattering theory and inuenced my mathematical life in an essential way. This book is dedicated to my long friendship with them!

Particular thanks are given to Dr. Frank Hettlich, Dr. Stefan Ritter, and Dipl.-Math. Markus Wartha for carefully reading the manuscript. Furthermore, I would like to thank Professor William Rundell and Dr. Martin Hanke for their manuscripts on inverse Sturm–Liouville problems and conjugate gradient methods, respectively, on which parts of Chapters 4 and 2 are based.

Karlsruhe, Germany   Andreas Kirsch
April 1996

Contents

Chapter 1

Introduction and Basic Concepts

1.1 Examples of Inverse Problems

In this section, we present some examples of pairs of problems that are inverse to each other. We start with some simple examples that are normally not even recognized as inverse problems. Most of them are taken from the survey article [152] and the monograph [111].

Example 1.1
Find a polynomial p of degree n with given zeros $x_1, \ldots, x_n$. This problem is inverse to the direct problem: Find the zeros $x_1, \ldots, x_n$ of a given polynomial p. In this example, the inverse problem is easier to solve. Its solution is $p(x) = c(x - x_1) \ldots (x - x_n)$ with an arbitrary constant c.

Example 1.2
Find a polynomial p of degree n that assumes given values $y_1, \ldots, y_n \in \mathbb{R}$ at given points $x_1, \ldots, x_n \in \mathbb{R}$. This problem is inverse to the direct problem of calculating the given polynomial at given $x_1, \ldots, x_n$. The inverse problem is the *Lagrange interpolation problem*.

Example 1.3
Given a real symmetric $n \times n$ matrix A and n real numbers $\lambda_1, \ldots, \lambda_n$, find a diagonal matrix D such that $A + D$ has the eigenvalues $\lambda_1, \ldots, \lambda_n$. This problem is inverse to the direct problem of computing the eigenvalues of the given matrix $A + D$.

Example 1.4
This inverse problem is used with intelligence tests: Given the first few terms $a_1, a_2, \ldots, a_k$ of a sequence, find the law of formation of the sequence; that is, find a_n for all n! Usually, only the next two or three terms are asked for to show

© Springer Nature Switzerland AG 2021
A. Kirsch, *An Introduction to the Mathematical Theory of Inverse Problems*,
Applied Mathematical Sciences 120,
https://doi.org/10.1007/978-3-030-63343-1_1

that the law of formation has been found. The corresponding direct problem is to evaluate the sequence (a_n) given the law of formation. It is clear that such inverse problems always have many solutions (from the mathematical point of view), and for this reason their use on intelligence tests has been criticized.

Example 1.5 *(Geological prospecting)*
In general, this is the problem of determining the location, shape, and/or some parameters (such as conductivity) of geological anomalies in the Earth's interior from measurements at its surface. We consider a simple one-dimensional example and describe the following inverse problem.

Determine changes $\rho = \rho(x)$, $0 \leq x \leq 1$, of the mass density of an anomalous region at depth h from measurements of the vertical component $f_v(x)$ of the change of force at x. $\rho(x')\Delta x'$ is the mass of a "volume element" at x' and $\sqrt{(x - x')^2 + h^2}$ is its distance from the instrument. The change of gravity is described by Newton's law of gravity $f = \gamma \frac{m}{r^2}$ with gravitational constant γ. For the vertical component, we have

$$\Delta f_v(x) = \gamma \, \frac{\rho(x')\Delta x'}{(x - x')^2 + h^2} \, \cos \theta = \gamma \, \frac{h \, \rho(x')\Delta x'}{\left[(x - x')^2 + h^2\right]^{3/2}} \, .$$

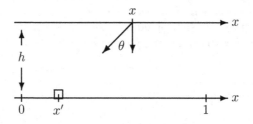

This yields the following integral equation for the determination of ρ:

$$f_v(x) = \gamma \, h \int\limits_0^1 \frac{\rho(x')}{\left[(x - x')^2 + h^2\right]^{3/2}} \, dx' \quad \text{for } 0 \leq x \leq 1. \tag{1.1}$$

We refer to [6, 105, 277] for further reading on this and related inverse problems in geological prospecting.

Example 1.6 *(Inverse scattering problem)*
Find the shape of a scattering object, given the intensity (and phase) of sound or electromagnetic waves scattered by this object. The corresponding direct problem is that of calculating the scattered wave for a given object.

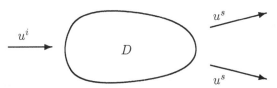

More precisely, the *direct problem* can be described as follows. Let a bounded region $D \subset \mathbb{R}^N$ ($N = 2$ or 3) be given with smooth boundary ∂D (the scattering object) and a plane *incident* wave $u^i(x) = e^{ik\hat{\theta}\cdot x}$, where $k > 0$ denotes the wave number and $\hat{\theta}$ is a unit vector that describes the direction of the incident wave. The direct problem is to find the *total field* $u = u^i + u^s$ as the sum of the incident field u^i and the *scattered field* u^s such that

$$\triangle u + k^2 u = 0 \quad \text{in } \mathbb{R}^N \setminus \overline{D}, \qquad u = 0 \quad \text{on } \partial D, \tag{1.2a}$$

$$\frac{\partial u^s}{\partial r} - iku^s = \mathcal{O}\left(r^{-(N+1)/2}\right) \quad \text{for } r = |x| \to \infty \text{ uniformly in } \frac{x}{|x|}. \tag{1.2b}$$

For *acoustic* scattering problems, $v(x,t) = u(x)e^{-i\omega t}$ describes the pressure and $k = \omega/c$ is the wave number with speed of sound c. For suitably polarized time harmonic *electromagnetic* scattering problems, Maxwell's equations reduce to the *two-dimensional Helmholtz equation* $\Delta u + k^2 u = 0$ for the components of the electric (or magnetic) field u. The wave number k is given in terms of the dielectric constant ε and permeability μ by $k = \sqrt{\varepsilon\mu}\,\omega$.

In both cases, the radiation condition (1.2b) yields the following asymptotic behavior:

$$u^s(x) = \frac{\exp(ik|x|)}{|x|^{(N-1)/2}}\, u_\infty(\hat{x}) + \mathcal{O}\left(|x|^{-(N+1)/2}\right) \quad \text{as } |x| \to \infty,$$

where $\hat{x} = x/|x|$. The *inverse problem* is to determine the shape of D when the *far field pattern* $u_\infty(\hat{x})$ is measured for all $\hat{x}$ on the unit sphere in $\mathbb{R}^N$.

These and related inverse scattering problems have various applications in computer tomography, seismic and electromagnetic exploration in geophysics, and nondestructive testing of materials, for example. An inverse scattering problem of this type is treated in detail in Chapter 7.

Standard literature on these direct and inverse scattering problems are the monographs [53, 55, 176] and the survey articles [50, 248].

Example 1.7 *(Computer tomography)*
The most spectacular application of the Radon transform is in medical imaging. For example, consider a fixed plane through a human body. Let $\rho(x,y)$ denote the change of density at the point (x,y), and let L be any line in the plane. Suppose that we direct a thin beam of X–rays into the body along L and measure how much of the intensity is attenuated by going through the body.

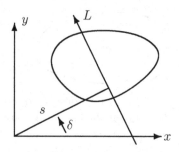

Let L be parametrized by (s,δ), where $s \in \mathbb{R}$ and $\delta \in [0,\pi)$. The ray $L_{s,\delta}$ has the coordinates

$$se^{i\delta} + iue^{i\delta} \in \mathbb{C}, \quad u \in \mathbb{R},$$

where we have identified $\mathbb{C}$ with $\mathbb{R}^2$. The attenuation of the intensity I is approximately described by $dI = -\gamma\rho I\, du$ with some constant γ. Integration along the ray yields

$$\ln I(u) = -\gamma \int_{u_0}^{u} \rho\left(se^{i\delta} + ite^{i\delta}\right) dt$$

or, assuming that ρ is of compact support, the relative intensity loss is given by

$$\ln I(\infty) = -\gamma \int_{-\infty}^{\infty} \rho\left(se^{i\delta} + ite^{i\delta}\right) dt.$$

In principle, from the attenuation factors we can compute all line integrals

$$(R\rho)(s,\delta) := \int_{-\infty}^{\infty} \rho\left(se^{i\delta} + iue^{i\delta}\right) du, \quad s \in \mathbb{R},\ \delta \in [0,\pi). \tag{1.3}$$

$R\rho$ is called the *Radon transform* of ρ. The *direct problem* is to compute the Radon transform $R\rho$ when ρ is given. The *inverse problem* is to determine the density ρ for a given Radon transform $R\rho$ (that is, measurements of all line integrals).

The problem simplifies in the following special case, where we assume that ρ is radially symmetric and we choose only vertical rays. Then $\rho = \rho(r)$, $r = \sqrt{x^2 + y^2}$, and the ray L_x passing through $(x,0)$ can be parametrized by (x,u), $u \in \mathbb{R}$. This leads to (the factor 2 is due to symmetry)

$$V(x) := \ln I(\infty) = -2\gamma \int_{0}^{\infty} \rho\left(\sqrt{x^2 + u^2}\right) du.$$

Again, we assume that ρ is of compact support in $\{x : |x| \le R\}$. The change of variables $u = \sqrt{r^2 - x^2}$ leads to

$$V(x) = -2\gamma \int_{x}^{\infty} \frac{r}{\sqrt{r^2 - x^2}}\, \rho(r)\, dr = -2\gamma \int_{x}^{R} \frac{r}{\sqrt{r^2 - x^2}}\, \rho(r)\, dr. \tag{1.4}$$

A further change of variables $z = R^2 - r^2$ and $y = R^2 - x^2$ transforms this equation into the following *Abel's integral equation* for the function $z \mapsto \rho(\sqrt{R^2 - z})$:

$$V(\sqrt{R^2 - y}) = -\gamma \int_0^y \frac{\rho(\sqrt{R^2 - z})}{\sqrt{y - z}} \, dz, \quad 0 \le y \le R. \tag{1.5}$$

The standard mathematical literature on the Radon transform and its applications are the monographs [128, 130, 206]. We refer also to the survey articles [131, 183, 185, 192].

The following example is due to Abel himself.

Example 1.8 *(Abel's integral equation)*
Let a mass element move along a curve Γ from a point p_1 on level $h > 0$ to a point p_0 on level $h = 0$. The only force acting on this mass element is the gravitational force mg.

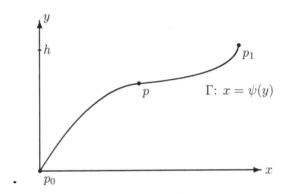

The *direct problem* is to determine the time T in which the element moves from p_1 to p_0 when the curve Γ is given. In the *inverse problem*, one measures the time $T = T(h)$ for several values of h and tries to determine the curve Γ. Let the curve be parametrized by $x = \psi(y)$. Let p have the coordinates $(\psi(y), y)$.

By conservation of energy, that is,

$$E + U = \frac{m}{2} v^2 + m g y = \text{const} = m g h,$$

we conclude for the velocity that

$$\frac{ds}{dt} = v = \sqrt{2g(h - y)}.$$

The total time T from p_1 to p_0 is

$$T = T(h) = \int_{p_0}^{p_1} \frac{ds}{v} = \int_0^h \sqrt{\frac{1 + \psi'(y)^2}{2g \, (h - y)}} \, dy \quad \text{for } h > 0.$$

Set $\phi(y) = \sqrt{1 + \psi'(y)^2}$ and let $f(h) := T(h)\sqrt{2g}$ be known (measured). Then we have to determine the unknown function ϕ from Abel's integral equation

$$\int_0^h \frac{\phi(y)}{\sqrt{h-y}}\, dy = f(h) \quad \text{for } h > 0. \tag{1.6}$$

A similar—but more important—problem occurs in seismology. One studies the problem to determine the velocity distribution c of the Earth from measurements of the travel times of seismic waves (see [29]).

For further examples of inverse problems leading to Abel's integral equations, we refer to the lecture notes by R. Gorenflo and S. Vessella [108], the monograph [198], and the papers [179, 270].

Example 1.9 *(Backwards heat equation)*
Consider the one-dimensional heat equation

$$\frac{\partial u(x,t)}{\partial t} = \frac{\partial^2 u(x,t)}{\partial x^2}, \quad (x,t) \in (0,\pi) \times \mathbb{R}_{>0}, \tag{1.7a}$$

with boundary conditions

$$u(0,t) = u(\pi,t) = 0, \quad t \geq 0, \tag{1.7b}$$

and initial condition

$$u(x,0) = u_0(x), \quad 0 \leq x \leq \pi. \tag{1.7c}$$

The separation of variables leads to the (formal) solution

$$u(x,t) = \sum_{n=1}^{\infty} a_n e^{-n^2 t} \sin(nx) \quad \text{with} \quad a_n = \frac{2}{\pi} \int_0^\pi u_0(y) \sin(n\mathring{y}) dy. \tag{1.8}$$

The *direct problem* is to solve the classical initial boundary value problem: Given the initial temperature distribution u_0 and the final time T, determine $u(\cdot, T)$. In the *inverse problem*, one measures the final temperature distribution $u(\cdot, T)$ and tries to determine the temperature at earlier times $t < T$, for example, the initial temperature $u(\cdot, 0)$.

From the solution formula (1.8), we see that we have to determine $u_0 := u(\cdot, 0)$ from the following integral equation:

$$u(x,T) = \frac{2}{\pi} \int_0^\pi k(x,y)\, u_0(y)\, dy, \quad 0 \leq x \leq \pi, \tag{1.9}$$

where

$$k(x,y) := \sum_{n=1}^{\infty} e^{-n^2 T} \sin(nx) \sin(ny). \tag{1.10}$$

We refer to the monographs [17, 175, 198] and papers [31, 43, 49, 80, 81, 94, 193, 247] for further reading on this subject.

Example 1.10 *(Diffusion in an inhomogeneous medium)*
The equation of diffusion in an inhomogeneous medium (now in two dimensions) is described by the equation

$$\frac{\partial u(x,t)}{\partial t} \;=\; \frac{1}{c} \operatorname{div}\left(\gamma(x)\nabla u(x,t)\right), \quad x \in D,\ t > 0, \tag{1.11}$$

where c is a constant and $\gamma = \gamma(x)$ is a parameter describing the medium. In the stationary case, this reduces to

$$\operatorname{div}\left(\gamma \nabla u\right) \;=\; 0 \quad \text{in } D. \tag{1.12}$$

The *direct problem* is to solve the boundary value problem for this equation for given boundary values $u|_{\partial D}$ and given function γ. In the *inverse problem*, one measures u and the flux $\gamma \partial u / \partial \nu$ on the boundary ∂D and tries to determine the unknown function γ in D. This is the problem of *impedance tomography* which we consider in more detail in Chapter 6.

The problem of impedance tomography is an example of a *parameter identification problem* for a partial differential equation. Among the extensive literature on parameter identification problems, we only mention the classical papers [166, 225, 224], the monographs [15, 17, 198], and the survey article [200].

Example 1.11 *(Sturm–Liouville eigenvalue problem)*
Let a string of length L and mass density $\rho = \rho(x) > 0$, $0 \leq x \leq L$, be fixed at the endpoints $x = 0$ and $x = L$. Plucking the string produces tones due to vibrations. Let $v(x,t)$, $0 \leq x \leq L$, $t > 0$, be the displacement at x and time t. It satisfies the *wave equation*

$$\rho(x)\,\frac{\partial^2 v(x,t)}{\partial t^2} \;=\; \frac{\partial^2 v(x,t)}{\partial x^2}, \quad 0 < x < L,\ t > 0, \tag{1.13}$$

subject to boundary conditions $v(0,t) = v(L,t) = 0$ for $t > 0$.
A periodic displacement of the form

$$v(x,t) \;=\; w(x)\left[a\cos\omega t + b\sin\omega t\right]$$

with frequency $\omega > 0$ is called a *pure tone*. This form of v solves the boundary value problem (1.13) if and only if w and ω satisfy the Sturm–Liouville eigenvalue problem

$$w''(x) \;+\; \omega^2 \rho(x)\, w(x) \;=\; 0,\ 0 < x < L, \quad w(0) = w(L) = 0. \tag{1.14}$$

The *direct problem* is to compute the eigenfrequencies ω and the corresponding eigenfunctions for known function ρ. In the *inverse problem*, one tries to determine the mass density ρ from a number of measured frequencies ω.

We see in Chapter 5 that parameter estimation problems for parabolic and hyperbolic initial boundary value problems are closely related to inverse spectral problems.

Example 1.12 *(Inverse Stefan problem)*
The physicist Stefan (see [253]) modeled the melting of arctic ice in the summer
by a simple one-dimensional model. In particular, consider a homogeneous block
of ice filling the region $x \geq \ell$ at time $t = 0$. The ice starts to melt by heating
the block at the left end. Thus, at time $t > 0$ the region between $x = 0$ and
$x = s(t)$ for some $s(t) > 0$ is filled with water, and the region $x \geq s(t)$ is filled
with ice.

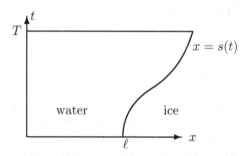

Let $u(x, t)$ be the temperature at $0 < x < s(t)$ and time t. Then u satisfies the
one-dimensional heat equation

$$\frac{\partial u(x,t)}{\partial t} = \frac{\partial^2 u(x,t)}{\partial x^2} \quad \text{in } D := \{(x,t) \in \mathbb{R}^2 : 0 < x < s(t), \ t > 0\} \qquad (1.15)$$

subject to *boundary conditions* $\frac{\partial}{\partial x} u(0, t) = f(t)$ and $u(s(t), t) = 0$ for $t \in [0, T]$
and *initial condition* $u(x, 0) = u_0(x)$ for $0 \leq x \leq \ell$.

Here, u_0 describes the initial temperature and $f(t)$ the heat flux at the left
boundary $x = 0$. The speed at which the interface between water and ice
moves is proportional to the heat flux. This is described by the following *Stefan
condition*:

$$\frac{ds(t)}{dt} = -\frac{\partial u(s(t), t)}{\partial x} \quad \text{for } t \in [0, T]. \qquad (1.16)$$

The *direct problem* is to compute the curve s when the boundary data f and u_0
are given. In the *inverse problem*, one has given a desired curve s and tries to
reconstruct u and f (or u_0).

We refer to the monographs [39, 198] and the classical papers [40, 95] for a
detailed introduction to Stefan problems.

In all of these examples, we can formulate the direct problem as the evaluation
of an operator K acting on a known "model" x in a model space X and the
inverse problem as the solution of the equation $K(x) = y$:

Direct problem: given x (and K), evaluate $K(x)$.
Inverse problem: given y (and K), solve $K(x) = y$ for x.

In order to formulate an inverse problem, the definition of the operator K,
including its domain and range, has to be given. The formulation as an operator equation allows us to distinguish among finite, semifinite, and infinite-
dimensional, linear and nonlinear problems.

In general, the evaluation of $K(x)$ means solving a boundary value problem for a differential equation or evaluating an integral.

For more general and "philosophical" aspects of inverse theory, we refer to [7, 214].

1.2   Ill-Posed Problems

For all of the pairs of problems presented in the last section, there is a fundamental difference between the direct and the inverse problems. In all cases, the inverse problem is *ill-posed* or *improperly posed* in the sense of Hadamard, while the direct problem is well-posed. In his lectures published in [115], Hadamard claims that a mathematical model for a physical problem (he was thinking in terms of a boundary value problem for a partial differential equation) has to be *properly posed* or *well-posed* in the sense that it has the following three properties:

1. There exists a solution of the problem (existence).

2. There is at most one solution of the problem (uniqueness).

3. The solution depends continuously on the data (stability).

Mathematically, the existence of a solution can be enforced by enlarging the solution space. The concept of distributional solutions of a differential equation is an example. If a problem has more than one solution, then information about the model is missing. In this case, additional properties, such as sign conditions, can be built into the model. The requirement of stability is the most important one. If a problem lacks the property of stability, then its solution is practically impossible to compute because any measurement or numerical computation is polluted by unavoidable errors: thus the data of a problem are always perturbed by noise! If the solution of a problem does not depend continuously on the data, then in general the computed solution has nothing to do with the true solution. Indeed, there is no way to overcome this difficulty unless additional information about the solution is available. Here, we remind the reader of the following statement (see Lanczos [171]):

A lack of information cannot be remedied by any mathematical trickery!

Mathematically, we formulate the notation of well-posedness in the following way.

Definition 1.13 *(Well-posedness)*
Let X and Y be normed spaces, and $K : X \rightarrow Y$ a linear operator. The equation $Kx = y$ is called properly posed *or* well-posed *if the following holds:*

1. Existence: *For every $y \in Y$, there is (at least one) $x \in X$ such that $Kx = y$.*

2. Uniqueness: *For every $y \in Y$, there is at most one $x \in X$ with $Kx = y$.*

3. Stability: *The solution x depends continuously on y; that is, for every sequence (x_n) in X with $Kx_n \to Kx$ $(n \to \infty)$, it follows that $x_n \to x$ $(n \to \infty)$.*

Equations for which (at least) one of these properties does not hold are called improperly posed or ill-posed.

In Chapter 4, we will extend this definition to local ill-posedness of nonlinear problems.

It is important to specify the full triple (X, Y, K) and their norms. Existence and uniqueness depend only on the algebraic nature of the spaces and the operator, that is, whether the operator is onto or one-to-one. Stability, however, depends also on the topologies of the spaces, that is, whether the inverse operator $K^{-1} : Y \to X$ is continuous.

These requirements are not independent of each other. For example, due to the open mapping theorem (see Theorem A.27 of Appendix A.3), the inverse operator K^{-1} is automatically continuous if K is linear and continuous and X and Y are Banach spaces.

As an example for an ill-posed problem, we study the classical example given by Hadamard in his famous paper [115].

Example 1.14 *(Cauchy's problem for the Laplace equation)*
Find a solution u of the Laplace equation

$$\Delta u(x,y) := \frac{\partial^2 u(x,y)}{\partial x^2} + \frac{\partial^2 u(x,y)}{\partial y^2} = 0 \quad \text{in } \mathbb{R} \times [0, \infty) \qquad (1.17a)$$

that satisfies the "initial conditions"

$$u(x,0) = f(x), \quad \frac{\partial}{\partial y}u(x,0) = g(x), \quad x \in \mathbb{R}, \qquad (1.17b)$$

where f and g are given functions. Obviously, the (unique) solution for $f(x) = 0$ and $g(x) = \frac{1}{n}\sin(nx)$ is given by

$$u(x,y) = \frac{1}{n^2}\sin(nx)\sinh(ny), \quad x \in \mathbb{R}, \ y \geq 0.$$

Therefore, we have

$$\sup_{x \in \mathbb{R}}\{|f(x)| + |g(x)|\} = \frac{1}{n} \longrightarrow 0, \quad n \to \infty,$$

but

$$\sup_{x \in \mathbb{R}}|u(x,y)| = \frac{1}{n^2}\sinh(ny) \longrightarrow \infty, \quad n \to \infty$$

for all $y > 0$. The error in the data tends to zero while the error in the solution u tends to infinity! Therefore, the solution does not depend continuously on the data, and the problem is improperly posed.

Many inverse problems and some of the examples of the last section (for further examples, we refer to [111]) lead to integral equations of the first kind with continuous or weakly singular kernels. Such integral operators are *compact* with respect to any reasonable topology. The following example will often serve as a model case in these lectures.

Example 1.15 *(Differentiation)*
The direct problem is to find the antiderivative y with $y(0) = 0$ of a given continuous function x on $[0, 1]$, that is, compute

$$y(t) = \int_0^t x(s)\,ds, \quad t \in [0, 1]. \tag{1.18}$$

In the inverse problem, we are given a continuously differentiable function y on $[0, 1]$ with $y(0) = 0$ and want to determine $x = y'$. This means we have to solve the integral equation $Kx = y$, where $K : C[0, 1] \to C[0, 1]$ is defined by

$$(Kx)(t) := \int_0^t x(s)\,ds, \quad t \in [0, 1], \quad \text{for } x \in C[0, 1]. \tag{1.19}$$

Here, we equip $C[0, 1]$ with the supremum norm $\|x\|_\infty := \max_{0 \le t \le 1} |x(t)|$. The solution of $Kx = y$ is just the derivative $x = y'$, provided $y(0) = 0$ and y is continuously differentiable! If x is the exact solution of $Kx = y$, and if we perturb y in the norm $\|\cdot\|_\infty$, then the perturbed right-hand side $\tilde{y}$ doesn't have to be differentiable, and even if it is the solution of the perturbed problem is not necessarily close to the exact solution. We can, for example, perturb y by $\delta \sin(t/\delta^2)$ for small δ. Then the error of the data (with respect to $\|\cdot\|_\infty$) is δ and the error in the solution is $1/\delta$. The problem $\big(K, C[0, 1], C[0, 1]\big)$ is therefore ill-posed.

Now we choose a different space $Y := \{y \in C^1[0, 1] : y(0) = 0\}$ for the right-hand side and equip Y with the stronger norm $\|x\|_{C^1} := \max_{0 \le t \le 1} |x'(t)|$. If the right-hand side is perturbed with respect to this norm $\|\cdot\|_{C^1}$, then the problem $\big(K, C[0, 1], Y\big)$ is well-posed because $K : C[0, 1] \to Y$ is boundedly invertible. This example again illustrates the fact that well-posedness depends on the topology.

In the numerical treatment of integral equations, a discretization error cannot be avoided. For integral equations of the first kind, a "naive" discretization usually leads to disastrous results as the following simple example shows (see also [267]).

Example 1.16
The integral equation

$$\int_0^1 e^{ts} x(s)\,ds = y(t), \quad 0 \le t \le 1, \tag{1.20}$$

with $y(t) = (\exp(t+1) - 1)/(t+1)$, is uniquely solvable by $x(t) = \exp(t)$. We approximate the integral by the trapezoidal rule

$$\int_0^1 e^{ts} x(s)\, ds \approx h\left(\frac{1}{2} x(0) + \frac{1}{2} e^t x(1) + \sum_{j=1}^{n-1} e^{jht} x(jh)\right)$$

with $h := 1/n$. For $t = ih$, we obtain the linear system

$$h\left(\frac{1}{2} x_0 + \frac{1}{2} e^{ih} x_n + \sum_{j=1}^{n-1} e^{jih^2} x_j\right) = y(ih), \quad i = 0, \ldots, n. \qquad (1.21)$$

Then x_i should be an approximation to $x(ih)$. The following table lists the error between the exact solution $x(t)$ and the approximate solution x_i for $t = 0$, 0.25, 0.5, 0.75, and 1. Here, i is chosen such that $ih = t$.

t	$n = 4$	$n = 8$	$n = 16$	$n = 32$
0	0.44	0.47	1.30	41.79
0.25	0.67	2.03	39.02	78.39
0.5	0.95	4.74	15.34	1.72
0.75	1.02	3.08	15.78	2.01
1	1.09	1.23	0.91	20.95

We see that the approximations have nothing to do with the true solution and become even worse for finer discretization schemes.

In the previous two examples, the problem was to solve integral equations of the first kind. Integral operators are *compact operators* in many natural topologies under very weak conditions on the kernels. The next theorem implies that linear equations of the form $Kx = y$ with compact operators K are *always* ill-posed.

Theorem 1.17 *Let X, Y be normed spaces and $K : X \to Y$ be a linear compact operator with nullspace $\mathcal{N}(K) := \{x \in X : Kx = 0\}$. Let the dimension of the factor space $X/\mathcal{N}(K)$ be infinite. Then there exists a sequence (x_n) in X such that $Kx_n \to 0$ but (x_n) does not converge. We can even choose (x_n) such that $\|x_n\|_X \to \infty$. In particular, if K is one-to-one, the inverse $K^{-1} : Y \supset \mathcal{R}(K) \to X$ is unbounded. Here, $\mathcal{R}(K) := \{Kx \in Y : x \in X\}$ denotes the range of K.*

Proof: We set $\mathcal{N} = \mathcal{N}(K)$ for abbreviation. The factor space $X/\mathcal{N}$ is a normed space with norm $\|[x]\| := \inf\{\|x + z\|_X : z \in \mathcal{N}\}$ since the nullspace is closed. The induced operator $\tilde{K} : X/\mathcal{N} \to Y$, defined by $\tilde{K}([x]) := Kx$, $[x] \in X/\mathcal{N}$, is well-defined, compact, and one-to-one. The inverse $\tilde{K}^{-1} : Y \supset \mathcal{R}(K) \to X/\mathcal{N}$ is unbounded since otherwise the identity $I = \tilde{K}^{-1}\tilde{K} : X/\mathcal{N} \to$

$X/\mathcal{N}$ would be compact as a composition of a bounded and a compact operator (see Theorem A.34). This would contradict the assumption that the dimension of $X/\mathcal{N}$ is infinite (see again Theorem A.34). Because $\tilde{K}^{-1}$ is unbounded, there exists a sequence $([z_n])$ in $X/\mathcal{N}$ with $Kz_n \to 0$ and $\|[z_n]\| = 1$. We choose $v_n \in \mathcal{N}$ such that $\|z_n + v_n\|_X \geq \frac{1}{2}$ and set $x_n := (z_n + v_n)/\sqrt{\|Kz_n\|}$. Then $Kx_n \to 0$ and $\|x_n\|_X \to \infty$. $\square$

1.3   The Worst-Case Error

We come back to Example 1.15 of the previous section: Determine $x \in C[0,1]$ such that $\int_0^t x(s)\,ds = y(t)$ for all $t \in [0,1]$. An obvious question is: How large could the error be in the worst case if the error in the right side y is at most δ? The answer is already given by Theorem 1.17: If the errors are measured in norms such that the integral operator is compact, then the solution error could be arbitrarily large. For the special Example 1.15, we have constructed explicit perturbations with this property.

However, the situation is different if additional information is available. Before we study the general case, we illustrate this observation for a model example.

Let y and $\tilde{y}$ be twice continuously differentiable and let a number $E > 0$ be available with

$$\|y''\|_\infty \leq E \quad \text{and} \quad \|\tilde{y}''\|_\infty \leq E. \qquad (1.22)$$

Set $z := \tilde{y} - y$ and assume that $z'(0) = z(0) = 0$ and $z'(t) \geq 0$ for $t \in [0,1]$. Then we estimate the error $\tilde{x} - x$ in the solution of Example 1.15 by

$$|\tilde{x}(t) - x(t)|^2 = z'(t)^2 = \int_0^t \frac{d}{ds}\left[z'(s)^2\right] ds = 2\int_0^t z'(s)\,z''(s)\,ds$$

$$\leq 4E\int_0^t z'(s)\,ds = 4E\,z(t).$$

Therefore, under the above assumptions on $z = \tilde{y} - y$ we have shown that $\|\tilde{x} - x\|_\infty \leq 2\sqrt{E\,\delta}$ if $\|\tilde{y} - y\|_\infty \leq \delta$ and E is a bound as in (1.22). In this example, $2\sqrt{E\,\delta}$ is a bound on the worst-case error for an error δ in the data and the additional information $\|x'\|_\infty = \|y''\|_\infty \leq E$ on the solution.

We define the following quite generally.

Definition 1.18 *Let $K : X \to Y$ be a linear bounded operator between Banach spaces, $\hat{X} \subset X$ a subspace, and $\|\cdot\|_{\hat{X}}$ a "stronger" norm on $\hat{X}$; that is, there exists $c > 0$ such that $\|x\|_X \leq c\|x\|_{\hat{X}}$ for all $x \in \hat{X}$. Then we define*

$$\mathcal{F}\big(\delta, E, \|\cdot\|_{\hat{X}}\big) := \sup\left\{\|x\|_X : x \in \hat{X},\ \|Kx\|_Y \leq \delta,\ \|x\|_{\hat{X}} \leq E\right\}, \qquad (1.23)$$

and call $\mathcal{F}\big(\delta, E, \|\cdot\|_{\hat{X}}\big)$ the worst-case error for the error δ in the data and a priori information $\|x\|_{\hat{X}} \leq E$.

$\mathcal{F}(\delta, E, \| \cdot \|_{\hat{X}})$ depends on the operator K and the norms in X, Y, and $\hat{X}$. It is desirable that this worst-case error not only converges to zero as δ tends to zero but that it is of order δ. This is certainly true (even without a priori information) for boundedly invertible operators, as is readily seen from the inequality $\|x\|_X \leq \|K^{-1}\|_{\mathcal{L}(Y,X)} \|Kx\|_Y$. For compact operators K, however, and norm $\| \cdot \|_{\hat{X}} = \| \cdot \|_X$, this worst-case error does not converge (see the following lemma), and one is forced to take a stronger norm $\| \cdot \|_{\hat{X}}$.

Lemma 1.19 *Let $K : X \to Y$ be linear and compact and assume that $X/\mathcal{N}(K)$ is infinite-dimensional. Then for every $E > 0$, there exists $c > 0$ and $\delta_0 > 0$ such that $\mathcal{F}(\delta, E, \| \cdot \|_X) \geq c$ for all $\delta \in (0, \delta_0)$.*

Proof: Assume that there exists a sequence $\delta_n \to 0$ such that $\mathcal{F}(\delta_n, E, \| \cdot \|_X) \to 0$ as $n \to \infty$. Let $\tilde{K} : X/\mathcal{N}(K) \to Y$ be again the induced operator in the factor space. We show that $\tilde{K}^{-1}$ is bounded: Let $\tilde{K}([x_m]) = Kx_m \to 0$. Then there exists a subsequence (x_{m_n}) with $\|Kx_{m_n}\|_Y \leq \delta_n$ for all n. We set

$$z_n := \begin{cases} x_{m_n}, & \text{if } \|x_{m_n}\|_X \leq E, \\ E \, \|x_{m_n}\|_X^{-1} \, x_{m_n}, & \text{if } \|x_{m_n}\|_X > E. \end{cases}$$

Then $\|z_n\|_X \leq E$ and $\|Kz_n\|_Y \leq \delta_n$ for all n. Because the worst-case error tends to zero, we also conclude that $\|z_n\|_X \to 0$. From this, we see that $z_n = x_{m_n}$ for sufficiently large n; that is, $x_{m_n} \to 0$ as $n \to \infty$. This argument, applied to every subsequence of the original sequence (x_m), yields that x_m tends to zero for $m \to \infty$; that is, $\tilde{K}^{-1}$ is bounded on the range $\mathcal{R}(K)$ of K. This, however, contradicts the assertion of Theorem 1.17. $\square$

In the following analysis, we make use of the singular value decomposition of the operator K (see Appendix A.6, Definition A.56). Therefore, we assume from now on that X and Y are Hilbert spaces. In many applications X and Y are *Sobolev spaces*; that is, spaces of measurable functions such that their (generalized) derivatives are square integrable. Sobolev spaces of functions of one variable can be characterized as follows:

$$H^p(a,b) := \left\{ x \in C^{p-1}[a,b] : x^{(p-1)}(t) = \alpha + \int_a^t \psi \, ds, \ \alpha \in \mathbb{R}, \ \psi \in L^2 \right\}$$

$$(1.24)$$

for $p \in \mathbb{N}$.

Example 1.20 *(Differentiation)*
As an example, we study differentiation and set $X = Y = L^2(0,1)$,

$$(Kx)(t) = \int_0^t x(s) \, ds, \quad t \in (0,1), \ x \in L^2(0,1),$$

and

$$\hat{X}_1 := \{x \in H^1(0,1) : x(1) = 0\}, \tag{1.25a}$$

$$\hat{X}_2 := \{x \in H^2(0,1) : x(1) = 0,\ x'(0) = 0\}. \tag{1.25b}$$

We define $\|x\|_1 := \|x'\|_{L^2}$ for $x \in \hat{X}_1$, and $\|x\|_2 := \|x''\|_{L^2}$ for $x \in \hat{X}_1$. Then the norms $\| \cdot \|_j$, $j = 1, 2$, are stronger than $\| \cdot \|_{L^2}$ (see Problem 1.2), and we can prove for every $E > 0$ and $\delta > 0$:

$$\mathcal{F}(\delta, E, \| \cdot \|_1) \leq \sqrt{\delta E} \quad \text{and} \quad \mathcal{F}(\delta, E, \| \cdot \|_2) \leq \delta^{2/3} E^{1/3}. \tag{1.26}$$

From this result, we observe that the possibility to reconstruct x is dependent on the smoothness of the solution. We come back to this remark in a more general setting (Theorem 1.21). We will also see that these estimates are asymptotically sharp; that is, the exponent of δ cannot be increased.

Proof of (1.26):   First, assume that $x \in H^1(0,1)$ with $x(1) = 0$. Partial integration, which is easily seen to be allowed for H^1-functions and the Cauchy–Schwarz inequality, yields

$$
\begin{aligned}
\|x\|_{L^2}^2 &= \int_0^1 x(t)\, x(t)\, dt \\
&= -\int_0^1 x'(t) \left[\int_0^t x(s)\, ds \right] dt + \left[x(t) \int_0^t x(s)\, ds \right]_{t=0}^{t=1} \\
&= -\int_0^1 x'(t)\, (Kx)(t)\, dt \ \leq\ \|Kx\|_{L^2} \|x'\|_{L^2}. \tag{1.27}
\end{aligned}
$$

This yields the first estimate. Now let $x \in H^2(0,1)$ such that $x(1) = 0$ and $x'(0) = 0$. Using partial integration again, we estimate

$$
\begin{aligned}
\|x'\|_{L^2}^2 &= \int_0^1 x'(t)\, x'(t)\, dt \\
&= -\int_0^1 x(t)\, x''(t)\, dt + \left[x(t)\, x'(t) \right]_{t=0}^{t=1} \\
&= -\int_0^1 x(t)\, x''(t)\, dt \ \leq\ \|x\|_{L^2} \|x''\|_{L^2}.
\end{aligned}
$$

Now we substitute this into the right-hand side of (1.27):

$$\|x\|_{L^2}^2 \leq \|Kx\|_{L^2} \|x'\|_{L^2} \leq \|Kx\|_{L^2} \sqrt{\|x\|_{L^2}} \sqrt{\|x''\|_{L^2}}.$$

From this, the second estimate of (1.26) follows.   □

This example is typical in the sense that integral operators are often smoothing. We can define an abstract "smoothness" of an element $x \in X$ with respect to a compact operator $K : X \to Y$ by requiring that $x \in \mathcal{R}(K^*)$ or $x \in \mathcal{R}(K^*K)$ or, more generally, $x \in \mathcal{R}\left((K^*K)^{\sigma/2}\right)$ for some real $\sigma > 0$. The operator $(K^*K)^{\sigma/2}$ from X into itself is defined as

$$(K^*K)^{\sigma/2}x \;=\; \sum_{j \in J} \mu_j^\sigma (x, x_j)_X x_j \,, \quad x \in X \,,$$

where $\{\mu_j, x_j, y_j \,:\, j \in J\}$ is a singular system for K (see Appendix A.6, Theorem A.57 and formula (A.47)). We note that $\mathcal{R}(K^*) = \mathcal{R}\left((K^*K)^{1/2}\right)$. Picard's Theorem (Theorem A.58) yields that $x \in \mathcal{R}\left((K^*K)^{\sigma/2}\right)$ is equivalent to $\sum_j \frac{|(x,x_j)_X|^2}{\mu_j^{2\sigma}} < \infty$ which is indeed a smoothness assumption in concrete applications. We refer to Example A.59 and the definition of Sobolev spaces of periodic functions (Section A.4) where smoothness is expressed by a decay of the Fourier coefficients.

Theorem 1.21 *Let X and Y be Hilbert spaces, and $K : X \to Y$ linear, compact, and one-to-one with dense range $\mathcal{R}(K)$. Let $K^* : Y \to X$ be the adjoint operator.*

(a) *Set $\hat{X}_1 := \mathcal{R}(K^*)$ and $\|x\|_1 := \left\|(K^*)^{-1}x\right\|_Y$ for $x \in \hat{X}_1$. Then*

$$\mathcal{F}\big(\delta, E, \|\cdot\|_1\big) \;\leq\; \sqrt{\delta E}\,.$$

Furthermore, for every $E > 0$ there exists a sequence $\delta_j \to 0$ such that $\mathcal{F}\big(\delta_j, E, \|\cdot\|_1\big) = \sqrt{\delta_j E}$; that is, this estimate is asymptotically sharp.

(b) *Set $\hat{X}_2 := \mathcal{R}(K^*K)$ and $\|x\|_2 := \left\|(K^*K)^{-1}x\right\|_X$ for $x \in \hat{X}_2$. Then*

$$\mathcal{F}\big(\delta, E, \|\cdot\|_2\big) \;\leq\; \delta^{2/3} E^{1/3}\,,$$

and for every $E > 0$ there exists a sequence $\delta_j \to 0$ such that $\mathcal{F}\big(\delta_j, E, \|\cdot\|_2\big) = \delta_j^{2/3} E^{1/3}$.

(c) *More generally, for some $\sigma > 0$ define $\hat{X}_\sigma := \mathcal{R}\left((K^*K)^{\sigma/2}\right)$ and $\|x\|_\sigma := \left\|(K^*K)^{-\sigma/2}x\right\|_X$ for $x \in \hat{X}_\sigma$. Then*

$$\mathcal{F}\big(\delta, E, \|\cdot\|_\sigma\big) \;\leq\; \delta^{\sigma/(\sigma+1)} E^{1/(\sigma+1)}\,,$$

and for every $E > 0$ there exists a sequence $\delta_j \to 0$ such that $\mathcal{F}\big(\delta_j, E, \|\cdot\|_\sigma\big) = \delta_j^{\sigma/(\sigma+1)} E^{1/(\sigma+1)}$.

The norms $\|\cdot\|_1$, $\|\cdot\|_2$, and $\|\cdot\|_\sigma$ are well-defined because K^* and $(K^*K)^{\sigma/2}$ are one-to-one. In concrete examples, the assumptions $x \in \mathcal{R}(K^*)$ and $x \in \mathcal{R}((K^*K)^{\sigma/2}$ are smoothness assumptions on the exact solution x (together with boundary conditions) as we have mentioned already before. In the preceding example, where $(Kx)(t) = \int_0^t x(s)\,ds$, the spaces $\mathcal{R}(K^*)$ and $\mathcal{R}(K^*K)$ coincide with the Sobolev spaces $\hat{X}_1$ and $\hat{X}_2$ defined in (1.25a) and (1.25b) (see Problem 1.3).

Proof of *Theorem* 1.21: (a) Let $x = K^*z \in \hat{X}_1$ with $\|Kx\|_Y \leq \delta$ and $\|x\|_1 \leq E$; that is, $\|z\|_Y \leq E$. Then

$$\|x\|_X^2 = (K^*z, x)_X = (z, Kx)_Y \leq \|z\|_Y \|Kx\|_Y \leq E\delta.$$

This proves the first estimate. Now let $\{\mu_j, x_j, y_j : j \in J\}$ be a singular system for K (see Appendix A.6, Theorem A.57). Set $\hat{x}_j = EK^*y_j = \mu_j E x_j$ and $\delta_j := \mu_j^2 E \to 0$. Then $\|\hat{x}_j\|_1 = E$, $\|K\hat{x}_j\|_Y = \delta_j$, and $\|\hat{x}_j\|_X = \mu_j E = \sqrt{\delta_j E}$. This proves part (a). Part (b) is proven similarly or as a special case ($\sigma = 2$) of part (c).

(c) With a singular system $\{\mu_j, x_j, y_j : j \in J\}$, we have $\|x\|_X^2 = \sum_{j \in J} |\rho_j|^2$ where $\rho_j = (x, x_j)_X$ are the expansion coefficients of x. In the following estimate, we use Hölder's inequality with $p = (\sigma+1)/\sigma$ and $q = \sigma+1$ (note that $1/p + 1/q = 1$):

$$\|x\|_X^2 = \sum_{j \in J} |\rho_j|^2 = \sum_{j \in J} (|\rho_j|\,\mu_j)^{2\sigma/(\sigma+1)} \left(|\rho_j|/\mu_j^\sigma\right)^{2/(\sigma+1)}$$

$$\leq \left(\sum_{j \in J} (|\rho_j|\,\mu_j)^{2p\sigma/(\sigma+1)}\right)^{1/p} \left(\sum_{j \in J} (|\rho_j|/\mu_j^\sigma)^{2q/(\sigma+1)}\right)^{1/q}$$

$$= \left(\sum_{j \in J} |\rho_j|^2 \mu_j^2\right)^{\sigma/(\sigma+1)} \left(\sum_{j \in J} |\rho_j|^2 \mu_j^{-2\sigma}\right)^{1/(\sigma+1)}$$

$$= \|Kx\|_Y^{2\sigma/(\sigma+1)} \|(K^*K)^{-\sigma/2}\|_X^{2/(\sigma+1)}.$$

This ends the proof. $\quad\square$

Next, we consider Example 1.9 again. We are given the parabolic initial boundary value problem

$$\frac{\partial u(x,t)}{\partial t} = \frac{\partial^2 u(x,t)}{\partial x^2}, \quad 0 < x < \pi,\ t > 0,$$

$$u(0,t) = u(\pi,t) = 0,\ t > 0, \quad u(x,0) = u_0(x),\ 0 < x < \pi.$$

In the inverse problem, we know the final temperature distribution $u(x,T)$, $0 \leq x \leq \pi$, and we want to determine the temperature $u(x,\tau)$ at time $\tau \in (0,T)$. As additional information, we also assume the knowledge of $E > 0$ with $\|u(\cdot,0)\|_{L^2} \leq E$.

The solution of the initial boundary value problem is given by the series

$$u(x,t) = \frac{2}{\pi} \sum_{n=1}^\infty e^{-n^2 t} \sin(nx) \int_0^\pi u_0(y)\sin(ny)\,dy, \quad 0 \leq x \leq \pi,\ t > 0.$$

We denote the unknown function by $v := u(\cdot, \tau)$, set $X = Y = L^2(0, \pi)$, and

$$\hat{X} := \left\{ v \in L^2(0, \pi) : v = \sum_{n=1}^{\infty} a_n \, e^{-n^2\tau} \sin(n\cdot) \text{ with } \sum_{n=1}^{\infty} a_n^2 < \infty \right\}$$

and $\|v\|_{\hat{X}} := \sqrt{\frac{\pi}{2} \sum_{n=1}^{\infty} a_n^2}$ for $v \in \hat{X}$. In this case, the operator $K : X \to Y$ is an integral operator with kernel

$$k(x, y) = \frac{2}{\pi} \sum_{n=1}^{\infty} e^{-n^2(T-\tau)} \sin(nx) \sin(ny), \quad x, y \in [0, \pi],$$

(see Example 1.9). Then we have for any $\tau \in (0, T)$:

$$\mathcal{F}(\delta, E, \|\cdot\|_{\hat{X}}) \leq E^{1-\tau/T} \, \delta^{\tau/T}. \tag{1.28}$$

This means that under the information $\|u(\cdot, 0)\|_{L^2} \leq E$, the solution $u(\cdot, \tau)$ can be determined from the final temperature distribution $u(\cdot, T)$, the determination being better the closer τ is to T.

Proof of (1.28): Let $v \in \hat{X}$. From the definition of $\hat{X}$ and

$$(Kv)(x) = \sum_{n=1}^{\infty} e^{-n^2 T} a_n \sin(nx),$$

we conclude that the Fourier coefficients of v are given by $\exp(-n^2\tau) \, a_n$ and those of Kv by $\exp(-n^2 T) \, a_n$. Therefore, we have to maximize

$$\|v\|_{L^2(0,\pi)}^2 = \frac{\pi}{2} \sum_{n=1}^{\infty} |a_n|^2 e^{-2n^2\tau}$$

subject to the constraints

$$\|v\|_{\hat{X}}^2 = \frac{\pi}{2} \sum_{n=1}^{\infty} |a_n|^2 \leq E^2 \quad \text{and} \quad \|Kv\|_{L^2(0,\pi)}^2 = \frac{\pi}{2} \sum_{n=1}^{\infty} |a_n|^2 e^{-2n^2 T} \leq \delta^2.$$

From the Hölder inequality, we have (for $p, q > 1$ with $1/p + 1/q = 1$ to be specified in a moment):

$$\frac{\pi}{2} \sum_{n=1}^{\infty} |a_n|^2 e^{-2n^2\tau} = \frac{\pi}{2} \sum_{n=1}^{\infty} |a_n|^{2/q} \left(|a_n|^{2/p} e^{-2n^2\tau} \right)$$

$$\leq \left(\frac{\pi}{2} \sum_{n=1}^{\infty} |a_n|^2 \right)^{1/q} \left(\frac{\pi}{2} \sum_{n=1}^{\infty} |a_n|^2 e^{-2pn^2\tau} \right)^{1/p}.$$

We now choose $p = T/\tau$. Then $1/p = \tau/T$ and $1/q = 1 - \tau/T$. This yields the assertion. □

   The next chapter is devoted to the construction of regularization schemes that are *asymptotically optimal* in the sense that, under the information $x \in \hat{X}$, $\|x\|_{\hat{X}} \leq E$, and $\|\tilde{y} - y\|_Y \leq \delta$, an approximation $\tilde{x}$ and a constant $c > 0$ are constructed such that $\|\tilde{x} - x\|_X \leq c \, \mathcal{F}(\delta, E, \|\cdot\|_{\hat{X}})$.

   As the first tutorial example, we consider the problem of numerical differentiation; see Examples 1.15 and 1.20.

Example 1.22

Let again, as in Examples 1.15 and 1.20, $(Kx)(t) = \int_0^t x(s)ds$, $t \in (0,1)$. Solving $Kx = y$ is equivalent to differentiating y. We fix $h \in (0,1/2)$ and define the one-sided difference quotient by

$$
v(t) \;=\; \begin{cases} \frac{1}{h}\left[y(t+h) - y(t)\right], & 0 < t < 1/2, \\[2mm] \frac{1}{h}\left[y(t) - y(t-h)\right], & 1/2 < t < 1, \end{cases}
$$

for any $y \in L^2(0,1)$. First, we estimate $\|v - y'\|_{L^2}$ for smooth functions y; that is, $y \in H^2(0,1)$. From Taylor's formula (see Problem 1.4), we have

$$
y(t \pm h) \;=\; y(t) \pm y'(t)\,h \;+\; \int_t^{t \pm h} (t \pm h - s)\,y''(s)\,ds \,;
$$

that is,

$$
\begin{aligned}
v(t) - y'(t) &= \frac{1}{h} \int_t^{t+h} (t + h - s)\,y''(s)\,ds \\[2mm]
&= \frac{1}{h} \int_0^h \tau\, y''(t + h - \tau)\,d\tau
\end{aligned}
$$

for $t \in (0,1/2)$ and analogously for $t \in (1/2,1)$. Hence, we estimate

$$
\begin{aligned}
& h^2 \int_0^{1/2} |v(t) - y'(t)|^2 dt \\[2mm]
&= \int_0^h \int_0^h \tau\, s \left[\int_0^{1/2} y''(t + h - \tau)\, y''(t + h - s)\, dt \right] d\tau\, ds \\[2mm]
&\leq \int_0^h \int_0^h \tau\, s \sqrt{\int_0^{1/2} |y''(t + h - \tau)|^2 dt} \; \sqrt{\int_0^{1/2} |y''(t + h - s)|^2 dt}\; d\tau\, ds \\[2mm]
&\leq \|y''\|_{L^2}^2 \left[\int_0^h \tau\, d\tau \right]^2 \;=\; \frac{1}{4}\, h^4 \, \|y''\|_{L^2}^2,
\end{aligned}
$$

and analogously for $h^2 \int_{1/2}^1 |v(t) - y'(t)|^2 dt$. Summing these estimates yields

$$\|v - y'\|_{L^2} \leq \frac{1}{\sqrt{2}} E h,$$

where E is some bound on $\|y''\|_{L_2}$.

Now we treat the situation with errors. Instead of $y(t)$ and $y(t \pm h)$, we measure $\tilde{y}(t)$ and $\tilde{y}(t \pm h)$, respectively. We assume that $\|\tilde{y} - y\|_{L^2} \leq \delta$. Instead of $v(t)$, we compute $\tilde{v}(t) = \pm[\tilde{y}(t \pm h) - \tilde{y}(t)]/h$ for $t \in (0, 1/2)$ or $t \in (1/2, 1)$, respectively. Because

$$|\tilde{v}(t) - v(t)| \leq \frac{|\tilde{y}(t \pm h) - y(t \pm h)|}{h} + \frac{|\tilde{y}(t) - y(t)|}{h},$$

we conclude that $\|\tilde{v} - v\|_{L^2} \leq 2\delta/h$. Therefore, the total error due to the error on the right-hand side and the discretization error is

$$\|\tilde{v} - y'\|_{L^2} \leq \|\tilde{v} - v\|_{L^2} + \|v - y'\|_{L^2} \leq \frac{2\delta}{h} + \frac{1}{\sqrt{2}} E h. \qquad (1.29)$$

By this estimate, it is desirable to choose the discretization parameter h as the minimum of the right-hand side of (1.29). Its minimum is obtained at $h = \sqrt{2\sqrt{2}\,\delta/E}$. This results in the optimal error $\|\tilde{v} - y'\|_{L^2} \leq 2\sqrt[4]{2}\sqrt{E\delta}$.

Summarizing, we note that the discretization parameter h should be of order $\sqrt{\delta/E}$ if the derivative of a function is computed by the one-sided difference quotient. With this choice, the method is asymptotically optimal under the information $\|x'\|_{L^2} \leq E$.

The two-sided difference quotient is optimal under the a priori information $\|x''\|_{L^2} \leq E$ and results in an algorithm of order $\delta^{2/3}$ (see Example 2.4 in the following chapter).

We have carried out the preceding analysis with respect to the L^2-norm rather that the maximum norm, mainly because we present the general theory in Hilbert spaces. For this example, however, estimates with respect to $\|\cdot\|_\infty$ are simpler to derive (see the estimates preceding Definition 1.18 of the worst-case error).

The result of this example is of practical importance: For many algorithms using numerical derivatives (for example, quasi-Newton methods in optimization), it is recommended that you choose the discretization parameter ε to be the square root of the floating-point precision of the computer because a one-sided difference quotient is used.

1.4   Problems

1.1 Show that equations (1.1) and (1.20) have at most one solution.
     Hints: Extend ρ in (1.1) by zero into $\mathbb{R}$, and apply the Fourier transform and the convolution theorem. For (1.20) use results of the Laplace transform.

1.2 Let the Sobolev spaces $\hat{X}_1$ and $\hat{X}_2$ be defined by (1.25a) and (1.25b), respectively. Define the bilinear forms by

$$(x,y)_1 := \int_0^1 x'(t)\,y'(t)\,dt \quad \text{and} \quad (x,y)_2 := \int_0^1 x''(t)\,y''(t)\,dt$$

on $\hat{X}_1$ and $\hat{X}_2$, respectively. Prove that $\hat{X}_j$ are Hilbert spaces with respect to the inner products $(\cdot,\cdot)_j$, $j = 1,2$, and that $\|x\|_{L^2} \leq \|x\|_j$ for all $x \in \hat{X}_j$, $j = 1,2$.

1.3 Let $K : L^2(0,1) \to L^2(0,1)$ be defined by (1.19). Show that the ranges $\mathcal{R}(K^*)$ and $\mathcal{R}(K^*K)$ coincide with the spaces $\hat{X}_1$ and $\hat{X}_2$ defined by (1.25a) and (1.25b), respectively.

1.4 Prove the following version of Taylor's formula by induction with respect to n and partial integration:

Let $y \in H^{n+1}(a,b)$ and $t, t+h \in [a,b]$. Then

$$y(t+h) = \sum_{k=0}^n \frac{y^{(k)}(t)}{k!}\,h^k + R_n(t;h),$$

where the error term is given by

$$R_n(t;h) = \frac{1}{n!} \int_t^{t+h} (t+h-s)^n\, y^{(n+1)}(s)\,ds\,.$$

1.5 Verify the assertions of Example A.59 of Appendix A.6.

Chapter 2

Regularization Theory for Equations of the First Kind

We saw in the previous chapter that many inverse problems can be formulated as operator equations of the form

$$K x = y \,,$$

where K is a linear compact operator between Hilbert spaces X and Y over the field $\mathbb{K} = \mathbb{R}$ or $\mathbb{C}$. We also saw that a successful reconstruction strategy requires additional a priori information about the solution.

This chapter is devoted to a systematic study of regularization strategies for solving $K x = y$. In particular, we wish to investigate under which conditions they are *asymptotically optimal*, that is, of the same asymptotic order as the worst-case error. In Section 2.1, we introduce the general concept of regularization. In Sections 2.2 and 2.3, we study Tikhonov's method and the Landweber iteration as two of the most important regularization strategies. In these three sections, the regularization parameter $\alpha = \alpha(\delta)$ is chosen a priori, that is, before we start to compute the regularized solution. We see that the optimal regularization parameter α depends on bounds of the exact solution; they are not known in advance. Therefore, it is advantageous to study strategies for the choice of α that depend on the numerical algorithm and are made during the algorithm (a posteriori). Different a posteriori choices are studied in Sections 2.5–2.7.

All of them are motivated by the idea that it is certainly sufficient to compute an approximation $x^{\alpha,\delta}$ of the solution x such that the norm of the defect $K x^{\alpha,\delta} - y^\delta$ is of the same order as the perturbation error δ of the right-hand side. The classical strategy, due to Morozov [194], determines α by solving a nonlinear scalar equation. To solve this equation, we still need a numerical algorithm such as the "regula falsi" or the Newton method. In Sections 2.6 and 2.7, we investigate two well-known iterative algorithms for solving linear (or nonlinear) equations: Landweber's method (see [172]), which is the steepest

© Springer Nature Switzerland AG 2021   23
A. Kirsch, *An Introduction to the Mathematical Theory of Inverse Problems*,
Applied Mathematical Sciences 120,
https://doi.org/10.1007/978-3-030-63343-1_2

descent method, and the conjugate gradient method. The choices of α are made implicitly by stopping the algorithm as soon as the defect $\|Kx^m - y^\delta\|_Y$ is less than $r\delta$. Here, $r > 1$ is a given parameter.

Landweber's method and Morozov's discrepancy principle are easy to investigate theoretically because they can be formulated as *linear* regularization methods. The study of the conjugate gradient method is more difficult because the choice of α depends *nonlinearly* on the right-hand side y. Because the proofs in Section 2.7 are very technical, we postpone them to an appendix (Appendix B).

2.1   A General Regularization Theory

For simplicity, we assume throughout this chapter that the compact operator K is one-to-one. This is not a serious restriction because we can always replace the domain X by the orthogonal complement of the kernel of K. We make the assumption that there exists a solution $x^* \in X$ of the unperturbed equation $Kx^* = y^*$. In other words, we assume that $y^* \in \mathcal{R}(K)$. The injectivity of K implies that this solution is unique.

In practice, the right-hand side $y^* \in Y$ is never known exactly but only up to an error of, say, $\delta > 0$. Therefore, we assume that we know $\delta > 0$ and $y^\delta \in Y$ with

$$\|y^* - y^\delta\|_Y \leq \delta. \tag{2.1}$$

It is our aim to "solve" the perturbed equation

$$Kx^\delta = y^\delta. \tag{2.2}$$

In general, (2.2) is not solvable because we cannot assume that the measured data y^δ are in the range $\mathcal{R}(K)$ of K. Therefore, the best we can hope is to determine an approximation $x^\delta \in X$ to the exact solution x^* that is "not much worse" than the worst-case error $\mathcal{F}(\delta, E, \|\cdot\|_{\hat{X}})$ of Definition 1.18.

An additional requirement is that the approximate solution x^δ should depend continuously on the data y^δ. In other words, it is our aim to construct a suitable bounded approximation $R : Y \to X$ of the (unbounded) inverse operator $K^{-1} : \mathcal{R}(K) \to X$.

Definition 2.1 *A regularization strategy is a family of linear and bounded operators*

$$R_\alpha : Y \longrightarrow X, \quad \alpha > 0,$$

such that

$$\lim_{\alpha \to 0} R_\alpha Kx = x \quad \text{for all } x \in X;$$

that is, the operators $R_\alpha K$ converge pointwise to the identity.

From this definition and the compactness of K, we conclude the following.

Theorem 2.2 *Let R_α be a regularization strategy for a compact and injective operator $K : X \to Y$ where $\dim X = \infty$. Then we have*

(1) The operators R_α are not uniformly bounded; that is, there exists a sequence (α_j) with $\|R_{\alpha_j}\|_{\mathcal{L}(Y,X)} \to \infty$ for $j \to \infty$.

(2) The sequence $(R_\alpha Kx)$ does not converge uniformly on bounded subsets of X; that is, there is no convergence $R_\alpha K$ to the identity I in the operator norm.

Proof: (1) Assume, on the contrary, that there exists $c > 0$ such that $\|R_\alpha\|_{\mathcal{L}(Y,X)} \leq c$ for all $\alpha > 0$. From $R_\alpha y \to K^{-1}y$ ($\alpha \to 0$) for all $y \in \mathcal{R}(K)$ and $\|R_\alpha y\|_X \leq c\|y\|_Y$ for $\alpha > 0$, we conclude that $\|K^{-1}y\|_X \leq c\|y\|_Y$ for every $y \in \mathcal{R}(K)$; that is, K^{-1} is bounded. This implies that $I = K^{-1}K : X \to X$ is compact, a contradiction to $\dim X = \infty$.

(2) Assume that $R_\alpha K \to I$ in $\mathcal{L}(X,X)$. From the compactness of $R_\alpha K$ and Theorem A.34, we conclude that I is also compact, which again would imply that $\dim X < \infty$. □

The notation of a regularization strategy is based on unperturbed data; that is, the regularizer $R_\alpha y^*$ converges to x^* for the exact right-hand side $y^* = Kx^*$.

Now let $y^* \in \mathcal{R}(K)$ be the exact right-hand side and $y^\delta \in Y$ be the measured data with $\|y^* - y^\delta\|_Y \leq \delta$. We define

$$x^{\alpha,\delta} := R_\alpha y^\delta \tag{2.3}$$

as an approximation of the solution x^* of $Kx^* = y^*$. Then the error splits into two parts by the following obvious application of the triangle inequality:

$$\begin{aligned}\|x^{\alpha,\delta} - x^*\|_X &\leq \|R_\alpha y^\delta - R_\alpha y^*\|_X + \|R_\alpha y^* - x^*\|_X \\ &\leq \|R_\alpha\|_{\mathcal{L}(Y,X)}\|y^\delta - y^*\|_Y + \|R_\alpha Kx^* - x^*\|_X\end{aligned}$$

and thus

$$\|x^{\alpha,\delta} - x^*\|_X \leq \delta\|R_\alpha\|_{\mathcal{L}(Y,X)} + \|R_\alpha Kx^* - x^*\|_X. \tag{2.4a}$$

Analogously, for the defect in the equation we have

$$\|Kx^{\alpha,\delta} - y^*\|_Y \leq \delta\|KR_\alpha\|_{\mathcal{L}(Y)} + \|KR_\alpha y^* - y^*\|_Y. \tag{2.4b}$$

These are our fundamental estimates, which we use often in the following.

We observe that the error between the exact and computed solutions consists of two parts: The first term on the right-hand side of (2.4a) describes the error in the data multiplied by the "condition number" $\|R_\alpha\|_{\mathcal{L}(Y,X)}$ of the regularized problem. By Theorem 2.2, this term tends to infinity as α tends to zero. The second term denotes the approximation error $\|(R_\alpha - K^{-1})y^*\|_X$ at the exact right-hand side $y^* = Kx^*$. By the definition of a regularization strategy, this term tends to zero with α. The following figure illustrates the situation (Figure 2.1).

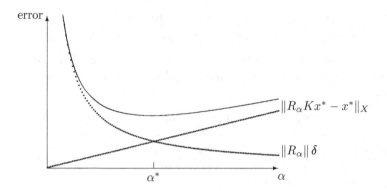

Figure 2.1: Behavior of the total error

We need a strategy to choose $\alpha = \alpha(\delta)$ dependent on δ in order to keep the total error as small as possible. This means that we would like to minimize

$$\delta \, \|R_\alpha\|_{\mathcal{L}(Y,X)} \; + \; \|R_\alpha K x^* - x^*\|_X \, .$$

The procedure is the same in every concrete situation: One has to estimate the quantities $\|R_\alpha\|_{\mathcal{L}(Y,X)}$ and $\|R_\alpha K x^* - x^*\|_X$ in terms of α and then minimize this upper bound with respect to α. Before we carry out these steps for two model examples, we introduce the following notation.

Definition 2.3 *A parameter choice* $\alpha = \alpha(\delta)$ *for the regularization strategy* R_α *is called* admissible *if* $\lim_{\delta \to 0} \alpha(\delta) = 0$ *and*

$$\sup\big\{\|R_{\alpha(\delta)} y^\delta - x\|_X : y^\delta \in Y, \; \|K x - y^\delta\|_Y \le \delta\big\} \to 0, \quad \delta \to 0,$$

for every $x \in X$.

From the fundamental estimate (2.4a), we note that a parameter choice $\alpha = \alpha(\delta)$ is admissible if $\alpha(\delta) \to 0$ and $\delta \|R_{\alpha(\delta)}\|_{\mathcal{L}(Y,X)} \to 0$ as $\delta \to 0$ and $\|R_\alpha K x - x\|_X \to 0$ as $\alpha \to 0$.

Example 2.4 *(Numerical differentiation by two-sided difference quotient)*
Let again $(Kx)(t) = \int_0^t x(s)\,ds$, $t \in (0,1)$; that is, solving $Kx = y$ is equivalent to differentiating y. It is our aim to compute the derivative of y by the two-sided difference quotient (see Example 1.22 for the one-sided difference quotient). Here $\alpha = h$ is the stepsize, and we define

$$(R_h y)(t) = \begin{cases} \frac{1}{h}\big[4\,y\big(t + \frac{h}{2}\big) - y(t + h) - 3\,y(t)\big], & 0 < t < \frac{h}{2}, \\[2mm] \frac{1}{h}\big[y\big(t + \frac{h}{2}\big) - y\big(t - \frac{h}{2}\big)\big], & \frac{h}{2} < t < 1 - \frac{h}{2}, \\[2mm] \frac{1}{h}\big[3\,y(t) + y(t - h) - 4\,y\big(t - \frac{h}{2}\big)\big], & 1 - \frac{h}{2} < t < 1, \end{cases}$$

for $y \in L^2(0,1)$. In order to prove that R_h defines a regularization strategy, it suffices to show that $R_h K$ are uniformly bounded with respect to h in the operator norm of $L^2(0,1)$ and that $\|R_h Kx - x\|_{L^2}$ tends to zero for smooth x (see Theorem A.29 of Appendix A.3). Later, we show convergence for $x \in H^2(0,1)$.

The fundamental theorem of calculus (or Taylor's formula from Problem 1.4 for $n = 0$) yields

$$(R_h y)(t) = \frac{1}{h} \int_{t-h/2}^{t+h/2} y'(s)\,ds = \frac{1}{h} \int_{-h/2}^{h/2} y'(s+t)\,ds, \quad \frac{h}{2} < t < 1 - \frac{h}{2},$$

and thus

$$\int_{h/2}^{1-h/2} |(R_h y)(t)|^2 dt = \frac{1}{h^2} \int_{-h/2}^{h/2} \int_{-h/2}^{h/2} \int_{h/2}^{1-h/2} y'(s+t)\,y'(\tau+t)\,dt\,d\tau\,ds.$$

The Cauchy–Schwarz inequality yields

$$\int_{h/2}^{1-h/2} |(R_h y)(t)|^2 dt \leq \|y'\|_{L^2}^2.$$

From $(R_h y)(t) = 4\big[y(t+h/2) - y(t)\big]/h - \big[y(t+h) - y(t)\big]/h$ for $0 < t < h/2$ and an analogous representation for $1 - h/2 < t < 1$, similar estimates yield the existence of $c > 0$ with

$$\|R_h\, y\|_{L^2} \leq c\,\|y'\|_{L^2}$$

for all $y \in H^1(0,1)$. For $y = Kx$, $x \in L^2(0,1)$, the uniform boundedness of $(R_h K)$ follows.

Now let $x \in H^2(0,1)$ and thus $y = Kx \in H^3(0,1)$. We apply Taylor's formula (see Problem 1.4) in the form (first again for $h/2 < t < 1 - h/2$)

$$y(t \pm h/2) - y(t) \mp \frac{h}{2} y'(t) - \frac{h^2}{8} y''(t) = \frac{1}{2} \int_0^{\pm h/2} s^2\, y'''(t \pm h/2 - s)\,ds.$$

Subtracting the formulas for $+$ and $-$ yields

$$(R_h y)(t) - y'(t) = \frac{1}{2h} \int_0^{h/2} s^2 \big[y'''(t + h/2 - s) + y'''(t - h/2 + s)\big]\,ds,$$

and thus by changing the orders of integration and using the Cauchy–Schwarz inequality

$$\int_{h/2}^{1-h/2} |(R_h y)(t) - y'(t)|^2 dt \leq \frac{1}{h^2} \|y'''\|_{L^2}^2 \left(\int_0^{h/2} s^2 ds\right)^2 = \frac{1}{24^2} \|y'''\|_{L^2}^2\, h^4.$$

Similar applications of Taylor's formula in the intervals $(0, h/2)$ and $(1 - h/2, 1)$ yield an estimate of the form

$$\|R_h Kx - x\|_{L^2} = \|R_h y - y'\|_{L^2} \leq c_1 E h^2$$

for all $x \in H^2(0, 1)$ with $\|x''\|_{L^2} \leq E$. Together with the uniform boundedness of $R_h K$, this implies that $R_h Kx \to x$ for all $x \in L^2(0, 1)$.

In order to apply the fundamental estimate (2.4a), we must estimate the first term, that is, the L^2-norm of R_h. It is easily checked that there exists $c_2 > 0$ with $\|R_h y\|_{L^2} \leq c_2 \|y\|_{L^2}/h$ for all $y \in L^2(0, 1)$. Estimate (2.4a) yields

$$\|R_h y^\delta - x^*\|_{L^2} \leq c_2 \frac{\delta}{h} + c_1 E h^2,$$

where E is a bound on $\|(x^*)''\|_{L^2} = \|(y^*)'''\|_{L^2}$. Minimization with respect to h of the expression on the right-hand side leads to

$$h(\delta) = c \sqrt[3]{\delta/E} \quad \text{and} \quad \|R_{h(\delta)} y^\delta - x^*\|_{L^2} \leq \tilde{c} E^{1/3} \delta^{2/3}$$

for some $c > 0$ and $\tilde{c} = c_2/c + c_1 c^2$.

We observe that this strategy is asymptotically optimal for the information $\|x''\|_{L^2} \leq E$ because it provides an approximation x^δ that is asymptotically not worse than the worst-case error (see Example 1.20).

The (one- or two-sided) difference quotient uses only local portions of the function y. An alternative approach is to first smooth the function y by mollification and then to differentiate the mollified function.

Example 2.5 *(Numerical differentiation by mollification)*
Again, we define the operator $(Kx)(t) = \int_0^t x(s)\,ds$, $t \in [0, 1]$, but now as an operator from the (closed) subspace

$$L_0^2(0, 1) := \left\{ z \in L^2(0, 1) : \int_0^1 z(s)\,ds = 0 \right\}$$

of $L^2(0, 1)$ into $L^2(0, 1)$.

We define the Gaussian kernel ψ_α by

$$\psi_\alpha(t) = \frac{1}{\alpha\sqrt{\pi}} \exp(-t^2/\alpha^2), \quad t \in \mathbb{R},$$

where $\alpha > 0$ denotes a parameter. Then $\int_{-\infty}^\infty \psi_\alpha(t)\,dt = 1$, and the convolution

$$(\psi_\alpha * y)(t) := \int_{-\infty}^{\infty} \psi_\alpha(t - s)\,y(s)\,ds = \int_{-\infty}^{\infty} \psi_\alpha(s)\,y(t - s)\,ds, \quad t \in \mathbb{R},$$

exists and is an L^2-function for every $y \in L^2(\mathbb{R})$. Furthermore, by Young's inequality (see [44], p. 102), we have

$$\|\psi_\alpha * y\|_{L^2(\mathbb{R})} \leq \|\psi_\alpha\|_{L^1(\mathbb{R})} \|y\|_{L^2(\mathbb{R})} = \|y\|_{L^2(\mathbb{R})} \quad \text{for all } y \in L^2(\mathbb{R}).$$

Therefore, the operators $y \mapsto \psi_\alpha * y$ are uniformly bounded in $L^2(\mathbb{R})$ with respect to α. We note that $\psi_\alpha * y$ is infinitely often differentiable on $\mathbb{R}$ for every $y \in L^2(\mathbb{R})$.

We need the two convergence properties

$$\|\psi_\alpha * z - z\|_{L^2(\mathbb{R})} \to 0 \quad \text{as } \alpha \to 0 \quad \text{for every } z \in L^2(0,1) \tag{2.5a}$$

and

$$\|\psi_\alpha * z - z\|_{L^2(\mathbb{R})} \leq \sqrt{2}\,\alpha\,\|z'\|_{L^2(0,1)} \tag{2.5b}$$

for every $z \in H^1(0,1)$ with $z(0) = z(1) = 0$. Here and in the following, we identify functions $z \in L^2(0,1)$ with functions $z \in L^2(\mathbb{R})$ where we think of them being extended by zero outside of $[0,1]$.

Proof of (2.5a), (2.5b): It is sufficient to prove (2.5b) because the space $\{z \in H^1(0,1) : z(0) = z(1) = 0\}$ is dense in $L^2(0,1)$, and the operators $z \mapsto \psi_\alpha * z$ are uniformly bounded from $L^2(0,1)$ into $L^2(\mathbb{R})$.

Let the Fourier transform be defined by

$$(\mathcal{F}z)(t) := \frac{1}{\sqrt{2\pi}} \int\limits_{-\infty}^{\infty} z(s)\,e^{ist}ds, \quad t \in \mathbb{R},$$

for $z \in \mathcal{S}$, where the Schwarz space $\mathcal{S}$ is defined by

$$\mathcal{S} := \left\{ z \in C^\infty(\mathbb{R}) : \sup_{t \in \mathbb{R}} \left| t^p z^{(q)}(t) \right| < \infty \text{ for all } p, q \in \mathbb{N}_0 \right\}.$$

With this normalization, Plancherel's theorem and the convolution theorem take the form (see [44])

$$\|\mathcal{F}z\|_{L^2(\mathbb{R})} = \|z\|_{L^2(\mathbb{R})}, \quad \mathcal{F}(u * z)(t) = \sqrt{2\pi}\,(\mathcal{F}u)(t)\,(\mathcal{F}z)(t), \ t \in \mathbb{R},$$

for all $z, u \in \mathcal{S}$. Because $\mathcal{S}$ is dense in $L^2(\mathbb{R})$ the first formula allows it to extend $\mathcal{F}$ to a bounded operator from $L^2(\mathbb{R})$ onto itself (see Theorem A.30), and both formulas hold also for $z \in L^2(\mathbb{R})$. Now we combine these properties and conclude that

$$\|\psi_\alpha * z - z\|_{L^2(\mathbb{R})} = \|\mathcal{F}(\psi_\alpha * z) - \mathcal{F}z\|_{L^2(\mathbb{R})} = \left\| \left[\sqrt{2\pi}\,\mathcal{F}(\psi_\alpha) - 1 \right] \mathcal{F}z \right\|_{L^2(\mathbb{R})}$$

for every $z \in L^2(0,1)$. Partial integration yields that

$$\mathcal{F}(z')(t) = \frac{1}{\sqrt{2\pi}} \int\limits_0^1 z'(s)\,e^{ist}ds = -\frac{it}{\sqrt{2\pi}} \int\limits_0^1 z(s)\,e^{ist}ds = (-it)\,(\mathcal{F}z)(t)$$

for all $z \in H^1(0,1)$ with $z(0) = z(1) = 0$. We define the function ϕ_α by

$$\phi_\alpha(t) := \frac{1}{it}\left[1 - \sqrt{2\pi}\,\mathcal{F}(\psi_\alpha)\right] = \frac{1}{it}\left[1 - e^{-\alpha^2 t^2/4}\right], \quad t \in \mathbb{R}.$$

Then we conclude that

$$\begin{aligned}
\|\psi_\alpha * z - z\|_{L^2(\mathbb{R})} &= \|\phi_\alpha \mathcal{F}(z')\|_{L^2(\mathbb{R})} \leq \|\phi_\alpha\|_\infty \|\mathcal{F}(z')\|_{L^2(\mathbb{R})} \\
&= \|\phi_\alpha\|_\infty \|z'\|_{L^2(0,1)}.
\end{aligned}$$

From

$$|\phi_\alpha(t)| = \frac{\alpha}{2}\frac{1}{(\alpha t)/2}\left[1 - e^{-(\alpha t/2)^2}\right]$$

and the elementary estimate $[1 - \exp(-\tau^2)]/\tau \leq 2\sqrt{2}$ for all $\tau > 0$, the desired estimate (2.5b) follows.

After these preparations, we define the regularization operators $R_\alpha : L^2(0,1) \to L_0^2(0,1)$ by

$$\begin{aligned}
(R_\alpha y)(t) &:= \frac{d}{dt}(\psi_\alpha * y)(t) - \int_0^1 \frac{d}{ds}(\psi_\alpha * y)(s)\,ds \\
&= (\psi_\alpha' * y)(t) - \int_0^1 (\psi_\alpha' * y)(s)\,ds
\end{aligned}$$

for $t \in (0,1)$ and $y \in L^2(0,1)$. First, we note that R_α is well-defined, that is, maps $L^2(0,1)$ into $L_0^2(0,1)$ and is bounded. To prove that R_α is a regularization strategy, we proceed as in the previous example and show that

(i) $\|R_\alpha y\|_{L^2} \leq \frac{4}{\alpha\sqrt{\pi}}\|y\|_{L^2}$ for all $\alpha > 0$ and $y \in L^2(0,1)$,

(ii) $\|R_\alpha K x\|_{L^2} \leq 2\|x\|_{L^2}$ for all $\alpha > 0$ and $x \in L_0^2(0,1)$, that is, the operators $R_\alpha K$ are uniformly bounded in $L_0^2(0,1)$, and

(iii) $\|R_\alpha K x - x\|_{L^2} \leq 2\sqrt{2}\,\alpha\,\|x'\|_{L^2}$ for all $\alpha > 0$ and $x \in H_{00}^1(0,1)$, where we have set

$$H_{00}^1(0,1) := \left\{ x \in H^1(0,1) : x(0) = x(1) = 0, \int_0^1 x(s)\,ds = 0 \right\}.$$

To prove part (i), we estimate with the Young's inequality

$$\begin{aligned}
\|R_\alpha y\|_{L^2(0,1)} &\leq 2\|\psi_\alpha' * y\|_{L^2(0,1)} \leq 2\|\psi_\alpha' * y\|_{L^2(\mathbb{R})} \\
&\leq 2\|\psi_\alpha'\|_{L^1(\mathbb{R})}\|y\|_{L^2(0,1)} \leq \frac{4}{\alpha\sqrt{\pi}}\|y\|_{L^2(0,1)}
\end{aligned}$$

for all $y \in L^2(0,1)$ because

$$\|\psi'_\alpha\|_{L^1(\mathbb{R})} = -2 \int_0^\infty \psi'_\alpha(s)\, ds = 2\,\psi_\alpha(0) = \frac{2}{\alpha\sqrt{\pi}}\,.$$

This proves part (i).

Now let $y \in H^1(0,1)$ with $y(0) = y(1) = 0$. Then, by partial integration,

$$(\psi'_\alpha * y)(t) = \int_0^1 \psi'_\alpha(t-s)\, y(s)\, ds = \int_0^1 \psi_\alpha(t-s)\, y'(s)\, ds = (\psi_\alpha * y')(t)\,.$$

Taking $y = Kx$, $x \in L_0^2(0,1)$ yields

$$(R_\alpha K x)(t) = (\psi_\alpha * x)(t) - \int_0^1 (\psi_\alpha * x)(s)\, ds\,.$$

Part (ii) now follows from Young's inequality.

Finally, we write

$$(R_\alpha K x)(t) - x(t) = (\psi_\alpha * x)(t) - x(t) - \int_0^1 \big[(\psi_\alpha * x)(s) - x(s)\big]\, ds$$

because $\int_0^1 x(s)\, ds = 0$. Therefore, by (2.5b),

$$\|R_\alpha K x - x\|_{L^2(0,1)} \leq 2\,\|\psi_\alpha * x - x\|_{L^2(0,1)} \leq 2\sqrt{2}\,\alpha\,\|x'\|_{L^2}$$

for all $x \in H_{00}^1(0,1)$. This proves part (iii).

Now we conclude that $R_\alpha K x$ converges to x for any $x \in L_0^2(0,1)$ by (ii), (iii), and the denseness of $H_{00}^1(0,1)$ in $L_0^2(0,1)$. Therefore, R_α defines a regularization strategy. From (i) and (iii), we rewrite the fundamental estimate (2.4a) as

$$\|R_\alpha y^\delta - x^*\|_{L^2} \leq \frac{4\,\delta}{\alpha\sqrt{\pi}} + 2\sqrt{2}\,\alpha\, E$$

if $x^* \in H_{00}^1(0,1)$ with $\|(x^*)'\|_{L^2} \leq E$, $y = Kx$, and $y^\delta \in L^2(0,1)$ such that $\|y^\delta - y^*\|_{L^2} \leq \delta$. The choice $\alpha = c\sqrt{\delta/E}$ again leads to the optimal order $\mathcal{O}(\sqrt{\delta E})$.

For further applications of the mollification method, we refer to the monograph by Murio [198]. There exists an enormous number of publications on numerical differentiation. We mention only the papers [3, 69, 74, 165] and, for more general Volterra equations of the first kind, [25, 26, 76, 77, 178].

A convenient method to construct classes of admissible regularization strategies is given by filtering singular systems. Let $K : X \rightarrow Y$ be a linear compact operator, and let $\{\mu_j, x_j, y_j : j \in J\}$ be a singular system for K (see Appendix A.6, Definition A.56, and Theorem A.57). As readily seen, the solution x of $Kx = y$ is given by Picard's theorem (see Theorem A.58 of Appendix A.6) as

$$x = \sum_{j=1}^{\infty} \frac{1}{\mu_j} (y, y_j)_Y \, x_j \tag{2.6}$$

provided the series converges, that is, $y \in \mathcal{R}(K)$. This result illustrates again the influence of errors in y because the large factors $1/\mu_j$ (note that $\mu_j \to 0$ as $j \to \infty$) amplify the errors in the expansion coefficients $(y, y_j)_Y$. We construct regularization strategies by damping the factors $1/\mu_j$.

Theorem 2.6 *Let $K : X \to Y$ be compact and one-to-one with singular system $\{\mu_j, x_j, y_j : j \in \mathbb{N}\}$ and*

$$q : (0, \infty) \times \big(0, \|K\|_{\mathcal{L}(X,Y)}\big] \longrightarrow \mathbb{R}$$

be a function with the following properties:

(1) $|q(\alpha, \mu)| \leq 1$ for all $\alpha > 0$ and $0 < \mu \leq \|K\|_{\mathcal{L}(X,Y)}$.

(2) For every $\alpha > 0$, there exists $c(\alpha)$ such that

$$|q(\alpha, \mu)| \leq c(\alpha)\, \mu \quad \text{for all } 0 < \mu \leq \|K\|_{\mathcal{L}(X,Y)}.$$

(3a) $\lim_{\alpha \to 0} q(\alpha, \mu) = 1$ for every $0 < \mu \leq \|K\|_{\mathcal{L}(X,Y)}$.

Then the operator $R_\alpha : Y \to X$, $\alpha > 0$, defined by

$$R_\alpha y := \sum_{j=1}^{\infty} \frac{q(\alpha, \mu_j)}{\mu_j} (y, y_j)_Y \, x_j, \quad y \in Y, \tag{2.7}$$

is a regularization strategy with $\|R_\alpha\|_{\mathcal{L}(Y,X)} \leq c(\alpha)$ and $\|KR_\alpha\|_{\mathcal{L}(Y)} \leq 1$. A choice $\alpha = \alpha(\delta)$ is admissible if $\alpha(\delta) \to 0$ and $\delta\, c(\alpha(\delta)) \to 0$ as $\delta \to 0$. The function q is called a regularizing filter *for K.*

Proof: The operators R_α are bounded because we have by assumption (2) that

$$\|R_\alpha y\|_X^2 = \sum_{j=1}^{\infty} [q(\alpha, \mu_j)]^2 \frac{1}{\mu_j^2} |(y, y_j)_Y|^2$$

$$\leq c(\alpha)^2 \sum_{j=1}^{\infty} |(y, y_j)_Y|^2 \leq c(\alpha)^2 \|y\|_Y^2;$$

that is, $\|R_\alpha\|_{\mathcal{L}(Y,X)} \leq c(\alpha)$. From

$$KR_\alpha y = \sum_{j=1}^{\infty} \frac{q(\alpha, \mu_j)}{\mu_j} (y, y_j)_Y \, Kx_j = \sum_{j=1}^{\infty} q(\alpha, \mu_j)\, (y, y_j)_Y \, y_j,$$

we conclude that $\|KR_\alpha y\|_Y^2 = \sum_{j=1}^\infty |q(\alpha, \mu_j)|^2 |(y, y_j)_Y|^2 \le \|y\|_Y^2$ and thus $\|KR_\alpha\|_{\mathcal{L}(Y)} \le 1$. Furthermore, from

$$R_\alpha K x = \sum_{j=1}^\infty \frac{q(\alpha, \mu_j)}{\mu_j} (Kx, y_j)_Y \, x_j \quad \text{and} \quad x = \sum_{j=1}^\infty (x, x_j)_X \, x_j \, ,$$

and $(Kx, y_j)_Y = (x, K^* y_j)_X = \mu_j (x, x_j)_X$, we conclude that

$$\|R_\alpha K x - x\|_X^2 = \sum_{j=1}^\infty \left[q(\alpha, \mu_j) - 1\right]^2 |(x, x_j)_X|^2 . \tag{2.8}$$

Here, K^* denotes the adjoint of K (see Theorem A.24). This fundamental representation will be used quite often in the following. Now let $x \in X$ be arbitrary but fixed. For $\epsilon > 0$ there exists $N \in \mathbb{N}$ such that

$$\sum_{j=N+1}^\infty |(x, x_j)_X|^2 < \frac{\epsilon^2}{8} .$$

By (3a) there exists $\alpha_0 > 0$ such that

$$[q(\alpha, \mu_j) - 1]^2 < \frac{\epsilon^2}{2\|x\|_X^2} \quad \text{for all } j = 1, \ldots, N \text{ and } 0 < \alpha \le \alpha_0 .$$

With (1) we conclude that

$$\begin{aligned}
\|R_\alpha K x - x\|_X^2 &= \sum_{j=1}^N \left[q(\alpha, \mu_j) - 1\right]^2 |(x, x_j)_X|^2 \\
&\quad + \sum_{j=N+1}^\infty \left[q(\alpha, \mu_j) - 1\right]^2 |(x, x_j)_X|^2 \\
&< \frac{\epsilon^2}{2\|x\|_X^2} \sum_{j=1}^N |(x, x_j)_X|^2 + \frac{\epsilon^2}{2} \le \epsilon^2
\end{aligned}$$

for all $0 < \alpha \le \alpha_0$. Thus, we have shown that

$$\lim_{\alpha \to 0} R_\alpha K x = x \quad \text{for every } x \in X.$$

Using this and $\delta \|R_{\alpha(\delta)}\|_{\mathcal{L}(Y,X)} \le \delta c(\alpha(\delta)) \to 0$ in the fundamental estimate, (2.4a) ends the proof. □

In this theorem, we showed convergence of $R_\alpha y$ to the solution x. As Examples 2.4 and 2.5 indicate, we are particularly interested in optimal strategies, that is, those that converge of the same order as the worst-case error. We see in the next theorem that a proper replacement of assumption (3a) together with the abstract smoothness assumption that x belongs to the range of K^* or $(K^* K)^{\sigma/2}$, respectively, leads to such optimal strategies.

Theorem 2.7 *Let the assumptions (1) and (2) of the previous theorem hold. Let (3a) be replaced by the stronger assumption:*

(3b) For every $\sigma \geq 0$, there exists a continuous function $\omega_\sigma : (0, \infty) \to (0, \infty)$ with $\lim_{\alpha \to 0} \omega_\sigma(\alpha) = 0$ and

$$\mu^\sigma |q(\alpha, \mu) - 1| \leq \omega_\sigma(\alpha) \quad \text{for all } \alpha > 0 \text{ and } 0 < \mu \leq \|K\|_{\mathcal{L}(X,Y)}. \quad (2.9)$$

Then

$$\|R_\alpha Kx - x\|_X \leq \omega_0(\alpha) \|x\|_X, \qquad (2.10a)$$
$$\|KR_\alpha Kx - Kx\|_Y \leq \omega_1(\alpha) \|x\|_X. \qquad (2.10b)$$

If, furthermore, $x \in \mathcal{R}\big((K^*K)^{\sigma/2}\big)$, then

$$\|R_\alpha Kx - x\|_X \leq \omega_\sigma(\alpha) \|z\|_X, \qquad (2.11a)$$
$$\|KR_\alpha Kx - Kx\|_Y \leq \omega_{\sigma+1}(\alpha) \|z\|_X, \qquad (2.11b)$$

where $x = (K^*K)^{\sigma/2} z$. We note that the powers $(K^*K)^{\sigma/2}$ are defined by the singular system; see (A.47). In the case $\sigma = 1$ we can replace the assumption $x = (K^*K)^{1/2} z$ by $x = Kz$ for some $z \in Y$, and (2.11a), (2.11b) hold with $\|z\|_Y$ replacing $\|z\|_X$.

Proof: Estimate (2.10a) follows immediately from (2.8) and the definition of $\omega_0(\alpha)$. Furthermore, from

$$KR_\alpha Kx - Kx = \sum_{j=1}^{\infty} [q(\alpha, \mu_j) - 1] (x, x_j)_X Kx_j$$
$$= \sum_{j=1}^{\infty} [q(\alpha, \mu_j) - 1] \mu_j (x, x_j)_X y_j$$

we conclude that $\|KR_\alpha Kx - Kx\|_Y^2 \leq \omega_1(\alpha)^2 \|x\|_X^2$ which proves (2.10b). Let now $x = (K^*K)^{\sigma/2} z$ for some $z \in X$. Then $(x, x_j)_X = \mu_j^\sigma (z, x_j)_X$ and, again from formula (2.8),

$$\|R_\alpha Kx - x\|_X^2 = \sum_{j=1}^{\infty} [q(\alpha, \mu_j) - 1]^2 \mu_j^{2\sigma} |(z, x_j)_X|^2 \leq \omega_\sigma(\alpha)^2 \|z\|_X^2.$$

Estimate (2.11b) is proven analogously.   □

We substitute these estimates into the fundamental estimates (2.4a) and (2.4b). Therefore, under the assumptions of the previous theorem,

$$\|x^{\alpha,\delta} - x^*\|_X \leq \delta \|R_\alpha\|_{\mathcal{L}(Y,X)} + \omega_\sigma(\alpha) \|z\|_X, \qquad (2.12a)$$
$$\|Kx^{\alpha,\delta} - y^*\|_Y \leq \delta + \omega_{\sigma+1}(\alpha) \|z\|_X, \qquad (2.12b)$$

for $\sigma \geq 0$ where $x^* = (K^*K)^{\sigma/2}z$. Note that we used $\|KR_\alpha\|_{\mathcal{L}(Y)} \leq 1$ (see Theorem 2.6).

There are many examples of functions $q : (0, \infty) \times (0, \|K\|_{\mathcal{L}(X,Y)}] \to \mathbb{R}$ that satisfy assumptions (1), (2), and (3a–b) of the preceding theorems. We study two of the following three filter functions in the next sections in more detail.

Theorem 2.8 *The following three functions q satisfy the assumptions (1), (2), (3a), and (3b) of Theorems 2.6 or 2.7, respectively:*

(a) $q(\alpha, \mu) = \mu^2/(\alpha + \mu^2)$. *This choice satisfies (2) with $c(\alpha) = 1/(2\sqrt{\alpha})$. Assumption (3b) holds with $w_\sigma(\alpha) = c_\sigma \alpha^{\sigma/2}$ if $\sigma \leq 2$ and $w_\sigma(\alpha) \leq c_\sigma \alpha$ if $\sigma > 2$. Here c_σ is independent of α. It is $c_1 = 1/2$ and $c_2 = 1$.*

(b) $q(\alpha, \mu) = 1 - (1 - a\,\mu^2)^{1/\alpha}$ *for some $0 < a < 1/\|K\|^2_{\mathcal{L}(X,Y)}$. In this case (2) holds with $c(\alpha) = \sqrt{a/\alpha}$, and (3b) is satisfied with $w_\sigma(\alpha) = \left(\frac{\sigma}{2a}\right)^{\sigma/2} \alpha^{\sigma/2}$ for all $\sigma, \alpha > 0$.*

(c) *Let q be defined by*

$$q(\alpha, \mu) = \begin{cases} 1, & \mu^2 \geq \alpha, \\ 0, & \mu^2 < \alpha. \end{cases}$$

In this case (2) holds with $c(\alpha) = 1/\sqrt{\alpha}$, and (3b) is satisfied with $w_\sigma(\alpha) = \alpha^{\sigma/2}$ for all $\sigma, \alpha > 0$.

Therefore, all of the functions q defined in (a), (b), and (c) are regularizing filters.

Proof: For all three cases, properties (1) and (3a) are obvious.

(a) Property (2) follows from the elementary estimate $\frac{\mu}{\alpha + \mu^2} \leq \frac{1}{2\sqrt{\alpha}}$ for all $\alpha, \mu > 0$ (which is equivalent to $(\mu - \sqrt{\alpha})^2 \geq 0$).

We observe that $1 - q(\alpha, \mu) = \alpha/(\alpha + \mu^2)$. For fixed $\alpha > 0$ and $\sigma > 0$, we define the function

$$f(\mu) = \mu^\sigma (1 - q(\alpha, \mu)) = \frac{\alpha\,\mu^\sigma}{\alpha + \mu^2}, \quad 0 \leq \mu \leq \mu_0,$$

and compute its derivative as

$$f'(\mu) = \alpha\,\mu^{\sigma-1} \frac{(\sigma - 2)\,\mu^2 + \alpha\sigma}{(\alpha + \mu^2)^2}.$$

Then f is monotonically increasing for $\sigma \geq 2$ and thus $f(\mu) \leq f(\mu_0) \leq \mu_0^{\sigma-2}\alpha$. If $\sigma < 2$, then we compute its maximum as $\mu^2_{max} = \frac{\alpha\sigma}{2-\sigma}$ with value $f(\mu_{max}) \leq c_\sigma \alpha^{\sigma/2}$. We leave the details to the reader.

(b) Property (2) follows immediately from Bernoulli's inequality:

$$1 - \left(1 - a\,\mu^2\right)^{1/\alpha} \leq 1 - \left(1 - \frac{a\,\mu^2}{\alpha}\right) = \frac{a\,\mu^2}{\alpha},$$

thus $|q(\alpha,\mu)| \leq \sqrt{|q(\alpha,\mu)|} \leq \sqrt{a/\alpha}\,\mu$.

(3b) is shown by the same method as in (a) when we define

$$f(\mu) = \mu^\sigma\left(1 - q(\alpha,\mu)\right) = \mu^\sigma(1 - a\,\mu^2)^{1/\alpha}, \ 0 \leq \mu \leq a,$$

with derivative $f'(\mu) = \mu^{\sigma-1}(1 - a\,\mu^2)^{1/\alpha-1}\left[\sigma(1 - a\mu^2) - 2a\mu^2\right]$. Then $a\mu_{max}^2 = \frac{\alpha\sigma}{2+\alpha\sigma}$ with value $f(\mu_{max}) \leq c_\sigma\,\alpha^{\sigma/2}$. Again, we leave the details to the reader.

(c) For property (2), it is sufficient to consider the case $\mu^2 \geq \alpha$. In this case, $q(\alpha,\mu) = 1 \leq \mu/\sqrt{\alpha}$. For (3b), we consider only the case $\mu^2 < \alpha$ and have $\mu^\sigma\left(1 - q(\alpha,\mu)\right) = \mu^\sigma \leq \alpha^{\sigma/2}$. $\square$

We will see later that the regularization methods for the first two choices of q admit a characterization that avoids knowledge of the singular system. The choice (c) of q is called the *spectral cutoff*. The spectral cutoff solution $x^{\alpha,\delta} \in X$ is therefore defined by

$$x^{\alpha,\delta} = \sum_{\mu_j^2 \geq \alpha} \frac{1}{\mu_j}\,(y^\delta, y_j)_Y\,x_j.$$

For this spectral cutoff solution, we combine the fundamental estimates (2.12a), (2.12b) with the previous theorem and show the following result.

Theorem 2.9 *(a) Let $K : X \to Y$ be a compact and injective operator with singular system $\{\mu_j, x_j, y_j : j \in \mathbb{N}\}$. The operators*

$$R_\alpha y := \sum_{\mu_j^2 \geq \alpha} \frac{1}{\mu_j}(y, y_j)_Y\,x_j, \quad y \in Y, \tag{2.13}$$

define a regularization strategy with $\|R_\alpha\|_{\mathcal{L}(Y,X)} \leq 1/\sqrt{\alpha}$. Thus the parameter choice $\alpha = \alpha(\delta)$ is admissible if $\alpha(\delta) \to 0$ $(\delta \to 0)$ and $\delta^2/\alpha(\delta) \to 0$ $(\delta \to 0)$.

(b) Let $Kx^* = y^*$ and $y^\delta \in Y$ be such that $\|y^\delta - y^*\|_Y \leq \delta$. Furthermore, let $x^* = K^*z \in \mathcal{R}(K^*)$ with $\|z\|_Y \leq E$ and $c > 0$. For the choice $\alpha(\delta) = c\,\delta/E$, we have the following error estimates for the spectral cutoff regularization.

$$\left\|x^{\alpha(\delta),\delta} - x^*\right\|_X \leq \left(\frac{1}{\sqrt{c}} + \sqrt{c}\right)\sqrt{\delta\,E}, \tag{2.14a}$$

$$\left\|Kx^{\alpha(\delta),\delta} - y^*\right\|_Y \leq (1 + c)\,\delta. \tag{2.14b}$$

(c) Let $x^* = (K^*K)^{\sigma/2}z \in \mathcal{R}\big((K^*K)^{\sigma/2}\big)$ for some $\sigma > 0$ with $\|z\|_X \leq E$. The choice $\alpha(\delta) = c\,\delta^{2/(\sigma+1)}$ leads to the estimates

$$\left\|x^{\alpha(\delta),\delta} - x^*\right\|_X \leq \left(\frac{1}{\sqrt{c}} + c^{\sigma/2}\right) \delta^{\sigma/(\sigma+1)} E^{1/(\sigma+1)}, \qquad (2.14c)$$

$$\left\|Kx^{\alpha(\delta),\delta} - y^*\right\|_Y \leq \left(1 + c^{(\sigma+1)/2}\right) \delta. \qquad (2.14d)$$

Therefore, the spectral cutoff regularization is optimal under the information $\|(K^*)^{-1}x^*\|_Y \leq E$ or $\|(K^*K)^{-\sigma/2}x^*\|_X \leq E$, respectively (if K^* is one-to-one).

Proof: Combining the fundamental estimates (2.12a), (2.12b) with Theorem 2.8 (part (c)) yields the error estimates

$$\|x^{\alpha,\delta} - x^*\|_X \leq \frac{\delta}{\sqrt{\alpha}} + \sqrt{\alpha}\,\|z\|_Y,$$

$$\|Kx^{\alpha,\delta} - y^*\|_Y \leq \delta + \alpha\|z\|_Y,$$

for part (b) and

$$\|x^{\alpha,\delta} - x^*\|_X \leq \frac{\delta}{\sqrt{\alpha}} + \alpha^{\sigma/2}\,\|z\|_X,$$

$$\|Kx^{\alpha,\delta} - y^*\|_Y \leq \delta + \alpha^{(\sigma+1)/2}\,\|z\|_X,$$

for part (c). The choices $\alpha(\delta) = c\,\delta/E$ and $\alpha(\delta) = c(\delta/E)^{2/(\sigma+1)}$, respectively, lead to the estimates (2.14a), (2.14b) and (2.14c), (2.14d), respectively. $\qquad\square$

The general regularization concept discussed in this section can be found in many books on inverse theory [17, 110, 182]. It was not the aim of this section to study the most general theory. This concept has been extended in several directions. For example, in [84] (see also [88]) the notations of strong and weak convergence and divergence are defined, and in [182] different notations of optimality of regularization schemes are discussed .

The idea of using filters has a long history [109, 265] and is very convenient for theoretical purposes. For given concrete integral operators, however, one often wants to avoid the computation of a singular system. In the next sections, we give equivalent characterizations for the first two examples without using singular systems.

2.2 Tikhonov Regularization

A common method to deal with overdetermined finite linear systems of the form $Kx = y$ is to determine the best fit in the sense that one tries to minimize the defect $\|Kx - y\|_Y$ with respect to $x \in X$ for some norm in Y. If X is infinite-dimensional and K is compact, this minimization problem is also ill-posed by the following lemma.

Lemma 2.10 *Let X and Y be Hilbert spaces, $K : X \to Y$ be linear and bounded, and $y^* \in Y$. There exists $\hat{x} \in X$ with $\|K\hat{x} - y^*\|_Y \leq \|Kx - y^*\|_Y$ for all $x \in X$ if and only if $\hat{x} \in X$ solves the* normal equation $K^*K\hat{x} = K^*y^*$. *Here, $K^* : Y \to X$ denotes the adjoint of K.*

Proof: A simple application of the binomial theorem yields

$$
\begin{aligned}
\|Kx - y^*\|_Y^2 - \|K\hat{x} - y^*\|_Y^2 &= 2\,\mathrm{Re}\big(K\hat{x} - y^*, K(x - \hat{x})\big)_Y + \|K(x - \hat{x})\|_Y^2 \\
&= 2\,\mathrm{Re}\big(K^*(K\hat{x} - y^*), x - \hat{x}\big)_X + \|K(x - \hat{x})\|_Y^2
\end{aligned}
$$

for all $x, \hat{x} \in X$. If $\hat{x}$ satisfies $K^*K\hat{x} = K^*y^*$, then $\|Kx-y^*\|_Y^2-\|K\hat{x}-y^*\|_Y^2 \geq 0$, that is, $\hat{x}$ minimizes $\|Kx-y^*\|_Y$. If, on the other hand, $\hat{x}$ minimizes $\|Kx-y^*\|_Y$, then we substitute $x = \hat{x} + tz$ for any $t > 0$ and $z \in X$ and arrive at

$$
0 \leq 2t\,\mathrm{Re}\big(K^*(K\hat{x} - y^*), z\big)_X + t^2\|Kz\|_Y^2 \,.
$$

Division by $t > 0$ and $t \to 0$ yields $\mathrm{Re}\big(K^*(K\hat{x} - y^*), z\big)_X \geq 0$ for all $z \in X$; that is, $K^*(K\hat{x} - y^*) = 0$, and $\hat{x}$ solves the normal equation. $\qquad\square$

As a consequence of this lemma, we should penalize the defect (in the language of optimization theory) or replace the equation of the first kind $K^*K\hat{x} = K^*y^*$ by an equation of the second kind (in the language of integral equation theory). Both viewpoints lead to the following minimization problem.

Given the linear, bounded operator $K : X \to Y$ and $y \in Y$, determine $x^\alpha \in X$ that minimizes the *Tikhonov functional*

$$
J_\alpha(x) := \|Kx - y\|_Y^2 + \alpha\|x\|_X^2 \quad \text{for } x \in X. \tag{2.15}
$$

We prove the following theorem.

Theorem 2.11 *Let $K : X \to Y$ be a linear and bounded operator between Hilbert spaces and $\alpha > 0$. Then the Tikhonov functional J_α has a unique minimum $x^\alpha \in X$. This minimum x^α is the unique solution of the normal equation*

$$
\alpha x^\alpha + K^*Kx^\alpha = K^*y. \tag{2.16}
$$

The operator $\alpha I + K^*K$ is an isomorphism from X onto itself for every $\alpha > 0$.

Proof: We use the following formula as in the proof of the previous lemma:

$$
\begin{aligned}
J_\alpha(x) - J_\alpha(x^\alpha) &= 2\,\mathrm{Re}\big(Kx^\alpha - y, K(x - x^\alpha)\big)_Y + 2\alpha\,\mathrm{Re}(x^\alpha, x - x^\alpha)_X \\
&\quad + \|K(x - x^\alpha)\|_Y^2 + \alpha\|x - x^\alpha\|_X^2 \\
&= 2\,\mathrm{Re}\big(K^*(Kx^\alpha - y) + \alpha x^\alpha, x - x^\alpha\big)_X \\
&\quad + \|K(x - x^\alpha)\|_Y^2 + \alpha\|x - x^\alpha\|_X^2
\end{aligned} \tag{2.17}
$$

for all $x \in X$. From this, the equivalence of the normal equation with the minimization problem for J_α is shown exactly as in the proof of Lemma 2.10. Next, we show that $\alpha I + K^* K$ is one-to-one for every $\alpha > 0$. Let $\alpha x + K^* K x = 0$. Multiplication by x yields $\alpha(x, x)_X + (Kx, Kx)_Y = 0$, that is, $x = 0$. Finally, we show that $\alpha I + K^* K$ is onto. Since $\alpha I + K^* K$ is one-to-one and self-adjoint, we conclude that its range is dense in X. It remains to show that the range is closed. Let $z_n = \alpha x_n + K^* K x_n$ converge to some $z \in X$. Then $z_n - z_m = \alpha(x_n - x_m) + K^* K(x_n - x_m)$. Multiplication of this equation by $x_n - x_m$ yields

$$
\begin{aligned}
\alpha \|x_n - x_m\|_X^2 + \|K(x_n - x_m)\|_Y^2 &= (z_n - z_m, x_n - x_m)_X \\
&\leq \|z_n - z_m\|_X \|x_n - x_m\|_X .
\end{aligned}
$$

From this, we conclude that $\alpha \|x_n - x_m\|_X \leq \|z_n - z_m\|_X$. Therefore, (x_n) is a Cauchy sequence and thus convergent to some $x \in X$ which obviously satisfies $\alpha x + K^* K x = z$. $\qquad \square$

The solution x^α of equation (2.16) can be written in the form $x^\alpha = R_\alpha y$ with

$$
R_\alpha := (\alpha I + K^* K)^{-1} K^* : Y \longrightarrow X . \tag{2.18}
$$

Choosing a singular system $\{\mu_j, x_j, y_j : j \in \mathbb{N}\}$ for the compact and injective operator K, we see that $R_\alpha y$ has the representation

$$
R_\alpha y = \sum_{n=0}^{\infty} \frac{\mu_j}{\alpha + \mu_j^2} (y, y_j)_Y x_j = \sum_{n=0}^{\infty} \frac{q(\alpha, \mu_j)}{\mu_j} (y, y_j)_Y x_j , \quad y \in Y , \tag{2.19}
$$

with $q(\alpha, \mu) = \mu^2/(\alpha + \mu^2)$. This function q is exactly the filter function that was studied in Theorem 2.8, part (a). Therefore, applications of Theorems 2.6 and 2.7 yield the following.

Theorem 2.12 *Let $K : X \to Y$ be a linear, compact, and injective operator and $\alpha > 0$ and $x^* \in X$ be the exact solution of $Kx^* = y^*$. Furthermore, let $y^\delta \in Y$ with $\|y^\delta - y^*\|_Y \leq \delta$.*

(a) The operators $R_\alpha : Y \to X$ from (2.18) form a regularization strategy with $\|R_\alpha\|_{\mathcal{L}(Y,X)} \leq 1/(2\sqrt{\alpha})$. It is called the Tikhonov regularization method. *$R_\alpha y^\delta$ is determined as the unique solution $x^{\alpha,\delta} \in X$ of the equation of the second kind*

$$
\alpha x^{\alpha,\delta} + K^* K x^{\alpha,\delta} = K^* y^\delta . \tag{2.20}
$$

Every choice $\alpha(\delta) \to 0$ ($\delta \to 0$) with $\delta^2/\alpha(\delta) \to 0$ ($\delta \to 0$) is admissible.

(b) Let $x^* = K^* z \in \mathcal{R}(K^*)$ with $\|z\|_Y \leq E$. We choose $\alpha(\delta) = c \delta/E$ for some $c > 0$. Then the following estimates hold:

$$
\|x^{\alpha(\delta),\delta} - x^*\|_X \leq \frac{1}{2}(1/\sqrt{c} + \sqrt{c}) \sqrt{\delta E} , \tag{2.21a}
$$

$$
\|K x^{\alpha(\delta),\delta} - y^*\|_Y \leq (1 + c) \delta . \tag{2.21b}
$$

*(c) For some $\sigma \in (0,2]$, let $x^* = (K^*K)^{\sigma/2}z \in \mathcal{R}((K^*K)^{\sigma/2})$ with $\|z\|_X \leq E$.
The choice $\alpha(\delta) = c\,(\delta/E)^{2/(\sigma+1)}$ for $c > 0$ leads to the error estimates*

$$\|x^{\alpha(\delta),\delta} - x^*\|_X \;\leq\; \left(\frac{1}{2\sqrt{c}} + c_\sigma\,c^{\sigma/2}\right)\delta^{\sigma/(\sigma+1)}E^{1/(\sigma+1)}, \quad (2.21c)$$

$$\|Kx^{\alpha(\delta),\delta} - y^*\|_Y \;\leq\; \left(1 + c_{\sigma+1}\,c^{(\sigma+1)/2}\right)\delta. \quad\quad\quad\quad (2.21d)$$

Here, c_σ are the constants for the choice of q of part (a) of Theorem 2.8. Therefore, for $\sigma \leq 2$ Tikhonov's regularization method is optimal for the information $\|(K^*)^{-1}x^*\|_Y \leq E$ or $\|(K^*K)^{-\sigma/2}x^*\|_X \leq E$, respectively (provided K^* is one-to-one).

Proof: Combining the fundamental estimates (2.12a), (2.12b) with Theorem 2.8 (part (a)) yields the error estimates

$$\|x^{\alpha,\delta} - x^*\|_X \;\leq\; \frac{\delta}{2\sqrt{\alpha}} + \frac{\sqrt{\alpha}}{2}\|z\|_Y$$

$$\|Kx^{\alpha,\delta} - y^*\|_X \;\leq\; \delta + \alpha\|z\|_Y$$

for part (b) and

$$\|x^{\alpha,\delta} - x^*\|_X \;\leq\; \frac{\delta}{2\sqrt{\alpha}} + c_\sigma\,\alpha^{\sigma/2}\|z\|_X$$

$$\|Kx^{\alpha,\delta} - y^*\|_X \;\leq\; \delta + c_{\sigma+1}\alpha^{(\sigma+1)/2}\|z\|_X$$

for part (c). The choices $\alpha(\delta) = c\,\delta/E$ and $\alpha(\delta) = c(\delta/E)^{2/(\sigma+1)}$, respectively, lead to the estimates (2.21a), (2.21b) and (2.21c), (2.21d), respectively. $\square$

From Theorem 2.12, we observe that α has to be chosen to depend on δ in such a way that it converges to zero as δ tends to zero but not as fast as δ^2. From parts (b) and (c), we conclude that the smoother the solution x^* is, the slower α has to tend to zero. On the other hand, the convergence can be arbitrarily slow if no a priori assumption about the solution x^* (such as (b) or (c)) is available (see [243]).

The case $\sigma = 2$ leads to the order $\mathcal{O}(\delta^{2/3})$ for $\|x^{\alpha(\delta),\delta} - x^*\|_X$. It is surprising to note that this order of convergence of Tikhonov's regularization method cannot be improved even if $x^* \in \mathcal{R}((K^*K)^{\sigma/2})$ for $\sigma > 2$. Indeed, we prove the following result.

Theorem 2.13 *Let $K : X \to Y$ be linear, compact, and one-to-one such that the range $\mathcal{R}(K)$ is infinite-dimensional. Furthermore, let $x \in X$, and assume that there exists a continuous function $\alpha : [0,\infty) \to [0,\infty)$ with $\alpha(0) = 0$ such that*

$$\lim_{\delta \to 0}\|x^{\alpha(\delta),\delta} - x\|_X\,\delta^{-2/3} \;=\; 0$$

for every $y^\delta \in Y$ with $\|y^\delta - Kx\|_Y \leq \delta$, where $x^{\alpha(\delta),\delta} \in X$ solves (2.20) for $\alpha = \alpha(\delta)$. Then $x = 0$.

Proof: Assume, on the contrary, that $x \neq 0$.

First, we show that $\alpha(\delta)\, \delta^{-2/3} \to 0$. Set $y = Kx$. From

$$\left(\alpha(\delta)\, I + K^*K\right)\left(x^{\alpha(\delta),\delta} - x\right) \;=\; K^*(y^\delta - y) \;-\; \alpha(\delta)\, x\,,$$

we estimate

$$|\alpha(\delta)|\, \|x\|_X \;\leq\; \|K\|_{\mathcal{L}(X,Y)}\delta \;+\; \left(\alpha(\delta) + \|K\|^2_{\mathcal{L}(X,Y)}\right) \|x^{\alpha(\delta),\delta} - x\|_X\,.$$

We multiply this equation by $\delta^{-2/3}$ and use the assumption that $x^{\alpha(\delta),\delta}$ tends to x faster than $\delta^{2/3}$ to zero, that is,

$$\|x^{\alpha(\delta),\delta} - x\|_X\, \delta^{-2/3} \;\to\; 0\,.$$

This yields $\alpha(\delta)\, \delta^{-2/3} \to 0$.

In the second part we construct a contradiction. Let $\{\mu_j, x_j, y_j : j \in \mathbb{N}\}$ be a singular system for K. Define

$$\delta_j \;:=\; \mu_j^3 \quad \text{and} \quad y^{\delta_j} \;:=\; y + \delta_j\, y_j\,, \quad j \in \mathbb{N}\,.$$

Then $\delta_j \to 0$ as $j \to \infty$ and, with $\alpha_j := \alpha(\delta_j)$ and $x^{\alpha_j} := (\alpha_j I + K^*K)^{-1}y$,

$$\begin{aligned}
x^{\alpha_j,\delta_j} - x &= \left(x^{\alpha_j,\delta_j} - x^{\alpha_j}\right) + \left(x^{\alpha_j} - x\right) \\
&= \left(\alpha_j I + K^*K\right)^{-1}K^*(\delta_j y_j) + \left(x^{\alpha_j} - x\right) \\
&= \frac{\delta_j\, \mu_j}{\alpha_j + \mu_j^2}\, x_j + \left(x^{\alpha_j} - x\right).
\end{aligned}$$

Because also $\|x^{\alpha_j} - x\|_X\, \delta_j^{-2/3} \to 0$, we conclude that

$$\frac{\delta_j^{1/3}\, \mu_j}{\alpha_j + \mu_j^2} \;\longrightarrow\; 0\,, \quad j \to \infty\,.$$

But, on the other hand,

$$\frac{\delta_j^{1/3}\, \mu_j}{\alpha_j + \mu_j^2} \;=\; \frac{\mu_j^2}{\alpha_j + \mu_j^2} \;=\; \left(1 + \alpha_j\, \delta_j^{-2/3}\right)^{-1} \;\longrightarrow\; 1\,, \quad j \to \infty\,.$$

This is a contradiction.   □

This result shows that Tikhonov's regularization method is not optimal for stronger "smoothness" assumptions $x^* \in \mathcal{R}\left((K^*K)^{\sigma/2}\right)$ for $\sigma > 2$. This is in contrast to, for example, the spectral cutoff regularization (see Theorem 2.9 above) or Landweber's method or the conjugate gradient method, which are discussed later.

The choice of α in Theorem 2.12 is made *a priori*, that is, before starting the computation of $x^{\alpha,\delta}$ by solving the least squares problem. In Sections 2.5

to 2.7 we study *a posteriori* choices of α, that is, choices of α made during the process of computing $x^{\alpha,\delta}$.

It is possible to choose stronger norms in the penalty term of the Tikhonov functional. Instead of (2.15), one can minimize the functional

$$\|Kx - y^\delta\|_y^2 \; + \; \alpha \|x\|_1^2 \quad \text{on } X_1,$$

where $\| \cdot \|_1$ is a stronger norm (or only seminorm) on a subspace $X_1 \subset X$. This was originally done by Phillips [217] and Tikhonov [261, 262] (see also [97]) for linear integral equations of the first kind. They chose the seminorm $\|x\|_1 := \|x'\|_{L^2}$ or the H^1-norm $\|x\|_1 := \left(\|x\|_{L^2}^2 + \|x'\|_{L^2}^2\right)^{1/2}$. By characterizing $\| \cdot \|_1$ through a singular system for K, one obtains similar convergence results as above in the stronger norm $\| \cdot \|_1$. For further aspects of regularization with differential operators or stronger norms, we refer to [70, 119, 180, 205] and the monographs [110, 111, 182]. The interpretation of regularization by smoothing norms in terms of reproducing kernel Hilbert spaces has been observed in [133].

2.3   Landweber Iteration

Landweber [172], Fridman [98], and Bialy [18] suggested to rewrite the equation $Kx = y$ in the form $x = (I - a\,K^*K)\,x + a\,K^*y$ for some $a > 0$ and iterate this equation, that is, compute

$$x^0 \; := \; 0 \quad \text{and} \quad x^m \; = \; (I - a\,K^*K)\,x^{m-1} \; + \; a\,K^*y \qquad (2.22)$$

for $m = 1, 2, \dots$. This iteration scheme can be interpreted as the steepest descent algorithm applied to the quadratic functional $x \mapsto \|Kx - y\|_Y^2$ as the following lemma shows.

Lemma 2.14 *Let the sequence (x^m) be defined by (2.22) and define the functional $\psi : X \to \mathbb{R}$ by $\psi(x) = \frac{1}{2}\|Kx - y\|_Y^2$, $x \in X$. Then ψ is Fréchet differentiable in every $z \in X$ and*

$$\psi'(z)x \; = \; \text{Re}(Kz - y, Kx)_Y \; = \; \text{Re}\big(K^*(Kz - y), x\big)_X, \qquad x \in X. \qquad (2.23)$$

The linear functional $\psi'(z)$ can be identified with $K^*(Kz - y) \in X$ in the Hilbert space X over the field $\mathbb{R}$. Therefore, $x^m = x^{m-1} - a\,K^*(Kx^{m-1} - y)$ is the steepest descent step with stepsize a.

Proof: The binomial formula yields

$$\psi(z + x) \; - \; \psi(z) \; - \; \text{Re}(Kz - y, Kx)_Y \; = \; \frac{1}{2} \, \|Kx\|_Y^2$$

and thus

$$\big|\psi(z + x) \; - \; \psi(z) \; - \; \text{Re}(Kz - y, Kx)_Y\big| \; \leq \; \frac{1}{2} \, \|K\|_{\mathcal{L}(X,Y)}^2 \, \|x\|_X^2,$$

which proves that the mapping $x \mapsto \operatorname{Re}(Kz - y, Kx)_Y$ is the Fréchet derivative of ψ at z. $\quad\square$

Equation (2.22) is a linear recursion formula for x^m. By induction with respect to m, it is easily seen that x^m has the explicit form $x^m = R_m y$, where the operator $R_m : Y \to X$ is defined by

$$R_m := a \sum_{k=0}^{m-1} (I - aK^*K)^k K^* \quad \text{for } m = 1, 2, \dots . \tag{2.24}$$

Choosing a singular system $\{\mu_j, x_j, y_j : j \in \mathbb{N}\}$ for the compact and injective operator K, we see that $R_m y$ has the representation

$$
\begin{aligned}
R_m y &= a \sum_{j=1}^{\infty} \mu_j \sum_{k=0}^{m-1} (1 - a\mu_j^2)^k \, (y, y_j)_Y \, x_j \\
&= \sum_{j=1}^{\infty} \frac{1}{\mu_j} \left[1 - (1 - a\mu_j^2)^m \right] (y, y_j)_Y \, x_j \\
&= \sum_{n=0}^{\infty} \frac{q(m, \mu_j)}{\mu_j} \, (y, y_j)_Y \, x_j, \quad y \in Y,
\end{aligned}
\tag{2.25}
$$

with $q(m, \mu) = 1 - (1 - a\mu^2)^m$. We studied this filter function q in Theorem 2.8, part (b), when we define $\alpha = 1/m$. Therefore, applications of Theorems 2.6 and 2.7 yield the following result.

Theorem 2.15 *Again let $K : X \to Y$ be a compact and injective operator and let $0 < a < 1/\|K\|_{\mathcal{L}(X,Y)}^2$. Let $x^* \in X$ be the exact solution of $Kx^* = y^*$. Furthermore, let $y^\delta \in Y$ with $\|y^\delta - y^*\|_Y \leq \delta$.*

(a) Define the linear and bounded operators $R_m : Y \to X$ by (2.24). These operators R_m define a regularization strategy with discrete regularization parameter $\alpha = 1/m$, $m \in \mathbb{N}$, and $\|R_m\|_{\mathcal{L}(Y,X)} \leq \sqrt{a\,m}$. The sequence $x^{m,\delta} = R_m y^\delta$ is computed by the iteration (2.22), that is,

$$x^{0,\delta} = 0 \quad \text{and} \quad x^{m,\delta} = (I - a\,K^*K)x^{m-1,\delta} + a\,K^* y^\delta \tag{2.26}$$

for $m = 1, 2, \dots$. Every strategy $m(\delta) \to \infty$ ($\delta \to 0$) with $\delta^2\, m(\delta) \to 0$ ($\delta \to 0$) is admissible.

(b) Again let $x^* = K^* z \in \mathcal{R}(K^*)$ with $\|z\|_Y \leq E$ and $0 < c_1 < c_2$. For every choice $m(\delta)$ with $c_1 \frac{E}{\delta} \leq m(\delta) \leq c_2 \frac{E}{\delta}$, the following estimates hold:

$$
\begin{aligned}
\|x^{m(\delta),\delta} - x^*\|_X &\leq c_3 \sqrt{\delta\, E}, & (2.27a) \\
\|Kx^{m(\delta),\delta} - y^*\|_Y &\leq (1 + 1/(ac_1))\, \delta, & (2.27b)
\end{aligned}
$$

for some c_3 depending on c_1, c_2, and a. Therefore, the Landweber iteration is optimal under the information $\|(K^*)^{-1} x^*\|_Y \leq E$.

(c) For some $\sigma > 0$, let $x^* = (K^*K)^{\sigma/2}z \in \mathcal{R}((K^*K)^{\sigma/2})$ with $\|z\|_X \leq E$ and let $0 < c_1 < c_2$. For every choice $m(\delta)$ with $c_1(E/\delta)^{2/(\sigma+1)} \leq m(\delta) \leq c_2(E/\delta)^{2/(\sigma+1)}$, we have

$$\|x^{m(\delta),\delta} - x^*\|_X \leq c_3 \, \delta^{\sigma/(\sigma+1)} \, E^{1/(\sigma+1)}, \qquad (2.27c)$$

$$\|Kx^{m(\delta),\delta} - y^*\|_Y \leq c_3 \, \delta, \qquad (2.27d)$$

for some c_3 depending on c_1, c_2, σ, and a. Therefore, the Landweber iteration is also optimal for the information $\|(K^*K)^{-\sigma/2}x^*\|_X \leq E$ for every $\sigma > 0$.

Proof: Combining the fundamental estimates (2.4a), (2.12a), and (2.12b) with Theorem 2.8 (part (b)) yields the error estimates

$$\|x^{m,\delta} - x^*\|_X \leq \delta \sqrt{a\,m} + \frac{1}{\sqrt{2am}} \|z\|_Y,$$

$$\|Kx^{m,\delta} - y^*\|_Y \leq \delta + \frac{1}{am} \|z\|_Y,$$

for part (b) and

$$\|x^{m,\delta} - x^*\|_X \leq \delta \sqrt{a\,m} + \left(\frac{\sigma}{2am}\right)^{\sigma/2} \|z\|_X,$$

$$\|Kx^{m,\delta} - y^*\|_Y \leq \delta + \left(\frac{\sigma+1}{2am}\right)^{(\sigma+1)/2} \|z\|_X, \qquad (2.28)$$

for part (c). Replacing m in the first term by the upper bound and in the second by the lower bound yields estimates (2.27a), (2.27b) and (2.27c), (2.27d), respectively. $\square$

The choice $x^0 = 0$ is made to simplify the analysis. In general, the explicit iteration x^m is given by

$$x^m = a \sum_{k=0}^{m-1} (I - aK^*K)^k K^*y + (I - aK^*K)^m x^0, \quad m = 1, 2, \ldots .$$

In this case, R_m is *affine linear*, that is, of the form $R_m y = z^m + S_m y$, $y \in Y$, for some $z^m \in X$ and some linear operator $S_m : Y \to X$.

For this method, we observe again that high precision (ignoring the presence of errors) requires a large number m of iterations but stability forces us to keep m small enough.

We come back to the Landweber iteration in the next chapter, where we show that an optimal choice of $m(\delta)$ can be made a posteriori through a proper stopping rule.

Other possibilities for regularizing first kind equations $Kx = y$ with compact operators K, which we have not discussed, are methods using positivity or more general convexity constraints (see [27, 30, 235, 236]).

2.4   A Numerical Example

In this section, we demonstrate the regularization methods by Tikhonov and Landweber for the following integral equation of the first kind:

$$\int_0^1 (1+ts)\, e^{ts}\, x(s)\, ds \; = \; e^t, \quad 0 \le t \le 1, \tag{2.29}$$

with unique solution $x^*(t) = 1$ (see Problem 2.1). The operator $K : L^2(0,1) \to L^2(0,1)$ is given by

$$(Kx)(t) \; = \; \int_0^1 (1+ts)\, e^{ts}\, x(s)\, ds$$

and is self-adjoint, that is, $K^* = K$. We note that x^* does not belong to the range of K (see Problem 2.1). For the numerical evaluation of Kx, we use Simpson's rule. With $t_i = i/n$, $i = 0, \ldots, n$, n even, we replace $(Kx)(t_i)$ by

$$\sum_{j=0}^n w_j\, (1+t_it_j)\, e^{t_it_j}\, x(t_j) \quad \text{where} \quad w_j \; = \; \begin{cases} \frac{1}{3n}, & j = 0 \text{ or } n, \\ \frac{4}{3n}, & j = 1,3,\ldots,n-1, \\ \frac{2}{3n}, & j = 2,4,\ldots,n-2. \end{cases}$$

We note that the corresponding matrix A is not symmetric. This leads to the discretized Tikhonov equation $\alpha\, x^{\alpha,\delta} + A^2 x^{\alpha,\delta} = A\, y^\delta$. Here, $y^\delta = \left(y_i^\delta\right) \in \mathbb{R}^{n+1}$ is a *perturbation* (uniformly distributed random vector) of the discrete right-hand $y_i^* = \exp(i/n)$ such that

$$|y^* - y^\delta|_2 \; := \; \sqrt{\frac{1}{n+1} \sum_{i=0}^n (y_i^* - y_i^\delta)^2} \; \le \; \delta.$$

The average results of ten computations are given in the following tables, where we have listed the discrete norms $|1 - x^{\alpha,\delta}|_2$ of the errors between the exact solution $x^*(t) = 1$ and Tikhonov's approximation $x^{\alpha,\delta}$ (Table 2.1).

Table 2.1: Tikhonov regularization for $\delta = 0$

α	$n = 8$	$n = 16$
10^{-1}	$2.4 * 10^{-1}$	$2.3 * 10^{-1}$
10^{-2}	$7.2 * 10^{-2}$	$6.8 * 10^{-2}$
10^{-3}	$2.6 * 10^{-2}$	$2.4 * 10^{-2}$
10^{-4}	$1.3 * 10^{-2}$	$1.2 * 10^{-2}$
10^{-5}	$2.6 * 10^{-3}$	$2.3 * 10^{-3}$
10^{-6}	$9.3 * 10^{-4}$	$8.7 * 10^{-4}$
10^{-7}	$3.5 * 10^{-4}$	$4.4 * 10^{-4}$
10^{-8}	$1.3 * 10^{-3}$	$3.2 * 10^{-5}$
10^{-9}	$1.6 * 10^{-3}$	$9.3 * 10^{-5}$
10^{-10}	$3.9 * 10^{-3}$	$2.1 * 10^{-4}$

Table 2.2: Tikhonov regularization for $\delta > 0$

α	$\delta = 0.0001$	$\delta = 0.001$	$\delta = 0.01$	$\delta = 0.1$
10^{-1}	0.2317	0.2317	0.2310	0.2255
10^{-2}	0.0681	0.0677	0.0692	0.1194
10^{-3}	0.0238	0.0240	0.0268	0.1651
10^{-4}	0.0119	0.0127	0.1172	1.0218
10^{-5}	0.0031	0.0168	0.2553	3.0065
10^{-6}	0.0065	0.0909	0.6513	5.9854
10^{-7}	0.0470	0.2129	2.4573	30.595
10^{-8}	0.1018	0.8119	5.9775	
10^{-9}	0.1730	1.8985	16.587	
10^{-10}	1.0723	14.642		

In the first table, we have chosen $\delta = 0$; that is, only the discretization error for Simpson's rule is responsible for the increase of the error for small α. This difference between discretization parameters $n = 8$ and $n = 16$ is noticeable for $\alpha \leq 10^{-8}$. We refer to [267] for further examples (Table 2.2).

In the second table, we always took $n = 16$ and observed that the total error first decreases with decreasing α up to an optimal value and then increases again. This is predicted by the theory, in particular by estimates (2.21a) and (2.21b).

In the following table, we list results corresponding to the iteration steps for Landweber's method with parameter $a = 0.5$ and again $n = 16$ (Table 2.3).

Table 2.3: Landweber iteration

m	$\delta = 0.0001$	$\delta = 0.001$	$\delta = 0.01$	$\delta = 0.1$
1	0.8097	0.8097	0.8088	0.8135
2	0.6274	0.6275	0.6278	0.6327
3	0.5331	0.5331	0.5333	0.5331
4	0.4312	0.4311	0.4322	0.4287
5	0.3898	0.3898	0.3912	0.3798
6	0.3354	0.3353	0.3360	0.3339
7	0.3193	0.3192	0.3202	0.3248
8	0.2905	0.2904	0.2912	0.2902
9	0.2838	0.2838	0.2845	0.2817
10	0.2675	0.2675	0.2677	0.2681
100	0.0473	0.0474	0.0476	0.0534
200	0.0248	0.0248	0.0253	0.0409
300	0.0242	0.0242	0.0249	0.0347
400	0.0241	0.0241	0.0246	0.0385
500	0.0239	0.0240	0.0243	0.0424

We observe that the error decreases quickly in the first few steps and then slows down. To compare Tikhonov's method and Landweber's iteration, we note that the error corresponding to iteration number m has to be compared with the error corresponding to $\alpha = 1/(2m)$ (see the estimates in the proofs of Theorems 2.15 and 2.12). Taking this into account, we observe that both methods are comparable where precision is concerned. We note, however, that the computation time of Landweber's method is considerably higher than for Tikhonov's method, in particular if the error δ is small. On the other hand, Landweber's method is stable with respect to perturbations of the right-hand side and gives very good results even for large errors δ.

We refer also to Section 3.5, where these regularization methods are compared with those to be discussed in the subsequent sections for Symm's integral equation.

2.5 The Discrepancy Principle of Morozov

The following three sections are devoted to a posteriori choices of the regularization parameter α. In this section, we study a discrepancy principle based on the Tikhonov regularization method. Throughout this section, we assume again that $K : X \longrightarrow Y$ is a compact and injective operator between Hilbert spaces X and Y with dense range $\mathcal{R}(K) \subset Y$. Again, we study the equation

$$Kx = y^{\delta}$$

for given perturbations $y^{\delta} \in Y$ of the exact right-hand side $y^* = Kx^*$. The Tikhonov regularization of this equation was investigated in Section 2.2. It corresponds to the regularization operators

$$R_{\alpha} = (\alpha I + K^*K)^{-1}K^* \quad \text{for } \alpha > 0$$

that approximate the unbounded inverse of K on $\mathcal{R}(K)$. We have seen that $x^{\alpha} = R_{\alpha}y$ exists and is the unique minimum of the Tikhonov functional

$$J_{\alpha}(x) := \|Kx - y\|_Y^2 + \alpha \|x\|_X^2, \quad x \in X, \quad \alpha > 0. \tag{2.30}$$

More facts about the dependence on α and y are proven in the following theorem.

Theorem 2.16 *Let* $y \in Y$, $\alpha > 0$, *and* x^{α} *be the unique solution of the equation*

$$\alpha x^{\alpha} + K^*Kx^{\alpha} = K^*y. \tag{2.31}$$

Then x^{α} *depends continuously on* y *and* α. *The mapping* $\alpha \mapsto \|x^{\alpha}\|_X$ *is monotonously nonincreasing and*

$$\lim_{\alpha \to \infty} x^{\alpha} = 0.$$

The mapping $\alpha \mapsto \|Kx^{\alpha} - y\|_Y$ *is monotonically nondecreasing and*

$$\lim_{\alpha \to 0} Kx^{\alpha} = y.$$

If $y \neq 0$, *then strict monotonicity holds in both cases.*

Proof: We proceed in five steps.

(i) Using the definition of J_α and the optimality of x^α, we conclude that

$$\alpha \|x^\alpha\|_X^2 \;\leq\; J_\alpha(x^\alpha) \;\leq\; J_\alpha(0) \;=\; \|y\|_Y^2\,,$$

that is, $\|x^\alpha\|_X \leq \|y\|_Y/\sqrt{\alpha}$. This proves that $x^\alpha \to 0$ as $\alpha \to \infty$.

(ii) We choose $\alpha > 0$ and $\beta > 0$ and subtract the equations for x^α and x^β:

$$\alpha\,(x^\alpha - x^\beta) \;+\; K^*K(x^\alpha - x^\beta) \;+\; (\alpha - \beta)x^\beta \;=\; 0. \qquad (2.32)$$

Multiplication by $x^\alpha - x^\beta$ yields

$$\alpha\|x^\alpha - x^\beta\|_X^2 \;+\; \|K(x^\alpha - x^\beta)\|_Y^2 \;=\; (\beta - \alpha)\,(x^\beta, x^\alpha - x^\beta)_X\,. \qquad (2.33)$$

From this equation, we first conclude that

$$\alpha\|x^\alpha - x^\beta\|_X^2 \;\leq\; |\beta - \alpha|\,|(x^\beta, x^\alpha - x^\beta)_X| \;\leq\; |\beta - \alpha|\,\|x^\beta\|_X\,\|x^\alpha - x^\beta\|_X\,,$$

that is,

$$\alpha\|x^\alpha - x^\beta\|_X \;\leq\; |\beta - \alpha|\,\|x^\beta\|_X \;\leq\; |\beta - \alpha|\,\frac{\|y\|_Y}{\sqrt{\beta}}\,.$$

This proves continuity of the mapping $\alpha \mapsto x^\alpha$.

(iii) Now let $\beta > \alpha > 0$. From (2.33) we conclude that $(x^\beta, x^\alpha - x^\beta)_X$ is (real and) positive (if zero then $x^\alpha = x^\beta$ and thus $x^\alpha = x^\beta = 0$ from (2.32). This would contradict the assumption $y \neq 0$). Therefore, $\|x^\beta\|_X^2 < (x^\beta, x^\alpha)_X \leq \|x^\beta\|_X\,\|x^\alpha\|_X$, that is, $\|x^\beta\|_X < \|x^\alpha\|_X$ which proves strict monotonicity of $\alpha \mapsto \|x^\alpha\|_X$.

(iv) We multiply the normal equation for x^β by $x^\alpha - x^\beta$. This yields

$$\beta\,(x^\beta, x^\alpha - x^\beta)_X \;+\; \big(Kx^\beta - y, K(x^\alpha - x^\beta)\big)_Y \;=\; 0\,.$$

Now let $\alpha > \beta$. From (2.33), we see that $(x^\beta, x^\alpha - x^\beta)_X < 0$; that is,

$$0 \;<\; \big(Kx^\beta - y, K(x^\alpha - x^\beta)\big)_Y \;=\; (Kx^\beta - y, Kx^\alpha - y)_Y - \|Kx^\beta - y\|_Y^2\,.$$

The Cauchy–Schwarz inequality yields $\|Kx^\beta - y\|_Y < \|Kx^\alpha - y\|_Y$.

(v) Finally, let $\varepsilon > 0$. Because the range of K is dense in Y, there exists $x \in X$ with $\|Kx - y\|_Y^2 \leq \varepsilon^2/2$. Choose α_0 such that $\alpha_0\|x\|_X^2 \leq \varepsilon^2/2$. Then

$$\|Kx^\alpha - y\|_Y^2 \;\leq\; J_\alpha(x^\alpha) \;\leq\; J_\alpha(x) \;\leq\; \varepsilon^2\,;$$

that is, $\|Kx^\alpha - y\|_Y \leq \varepsilon$ for all $\alpha \leq \alpha_0$.   □

Now we consider the determination of $\alpha(\delta)$ from the discrepancy principle; see [194–196]. We compute $\alpha = \alpha(\delta) > 0$ such that the corresponding Tikhonov solution $x^{\alpha,\delta}$, that is, the solution of the equation

$$\alpha\,x^{\alpha,\delta} \;+\; K^*Kx^{\alpha,\delta} \;=\; K^*y^\delta\,,$$

that is, the minimum of

$$J_{\alpha,\delta}(x) := \|Kx - y^\delta\|_Y^2 + \alpha\|x\|_X^2,$$

satisfies the equation

$$\|Kx^{\alpha,\delta} - y^\delta\|_Y = \delta. \tag{2.34}$$

Note that this choice of α by the discrepancy principle guarantees that, on the one side, the error of the defect is δ and, on the other side, α is not too small.

Equation (2.34) is uniquely solvable, provided $\delta < \|y^\delta\|_Y$ because by the previous theorem

$$\lim_{\alpha\to\infty} \|Kx^{\alpha,\delta} - y^\delta\|_Y = \|y^\delta\|_Y > \delta$$

and

$$\lim_{\alpha\to 0} \|Kx^{\alpha,\delta} - y^\delta\|_Y = 0 < \delta.$$

Furthermore, $\alpha \mapsto \|Kx^{\alpha,\delta} - y^\delta\|_Y$ is continuous and strictly increasing.

Theorem 2.17 *Let $K : X \to Y$ be linear, compact, and one-to-one with dense range in Y. Let $Kx^* = y^*$ with $x^* \in X$, $y^* \in Y$, and $y^\delta \in Y$ such that $\|y^\delta - y^*\|_Y \leq \delta < \|y^\delta\|_Y$. Let the Tikhonov solution $x^{\alpha(\delta)}$ satisfy $\|Kx^{\alpha(\delta),\delta} - y^\delta\|_Y = \delta$ for all $\delta \in (0,\delta_0)$. Then*

(a) $x^{\alpha(\delta),\delta} \to x^*$ for $\delta \to 0$; that is, the discrepancy principle is admissible.

(b) Let $x^* = K^*z \in K^*(Y)$ with $\|z\|_Y \leq E$. Then

$$\|x^{\alpha(\delta),\delta} - x^*\|_X \leq 2\sqrt{\delta}\, E.$$

Therefore, the discrepancy principle is an optimal regularization strategy under the information $\|(K^*)^{-1}x^*\|_Y \leq E$.

Proof: $x^\delta := x^{\alpha(\delta),\delta}$ minimizes the Tikhonov functional

$$J^{(\delta)}(x) := J_{\alpha(\delta),\delta}(x) = \alpha(\delta)\|x\|_X^2 + \|Kx - y^\delta\|_Y^2.$$

Therefore, we conclude that

$$\begin{aligned}
\alpha(\delta)\|x^\delta\|_X^2 + \delta^2 &= J^{(\delta)}(x^\delta) \leq J^{(\delta)}(x^*) \\
&= \alpha(\delta)\|x^*\|_X^2 + \|y^* - y^\delta\|_Y^2 \\
&\leq \alpha(\delta)\|x^*\|_X^2 + \delta^2,
\end{aligned}$$

and hence $\|x^\delta\|_X \leq \|x^*\|_X$ for all $\delta > 0$. This yields the following important estimate:

$$\begin{aligned}
\|x^\delta - x^*\|_X^2 &= \|x^\delta\|_X^2 - 2\,\mathrm{Re}(x^\delta, x^*)_X + \|x^*\|_X^2 \\
&\leq 2\left[\|x^*\|_X^2 - \mathrm{Re}(x^\delta, x^*)_X\right] = 2\,\mathrm{Re}(x^* - x^\delta, x^*)_X.
\end{aligned}$$

First, we prove part (b): Let $x^* = K^*z$, $z \in Y$. Then

$$
\begin{aligned}
\|x^\delta - x^*\|_X^2 &\leq 2 \operatorname{Re}(x^* - x^\delta, K^*z)_X = 2 \operatorname{Re}(y^* - Kx^\delta, z)_Y \\
&\leq 2 \operatorname{Re}(y^* - y^\delta, z)_Y + 2 \operatorname{Re}(y^\delta - Kx^\delta, z)_Y \\
&\leq 2\delta\|z\|_Y + 2\delta\|z\|_Y = 4\delta\|z\|_Y \leq 4\delta E.
\end{aligned}
$$

(a) Now let $x^* \in X$ and $\varepsilon > 0$ be arbitrary. The range $\mathcal{R}(K^*)$ is dense in X because K is one-to-one. Therefore, there exists $\hat{x} = K^*z \in \mathcal{R}(K^*)$ such that $\|\hat{x} - x^*\|_X \leq \varepsilon/3$. Then we conclude by similar arguments as above that

$$
\begin{aligned}
\|x^\delta - x^*\|_X^2 &\leq 2 \operatorname{Re}(x^* - x^\delta, x^* - \hat{x})_X + 2 \operatorname{Re}(x^* - x^\delta, K^*z)_X \\
&\leq 2\|x^* - x^\delta\|_X \frac{\varepsilon}{3} + 2 \operatorname{Re}(y^* - Kx^\delta, z)_Y \\
&\leq 2\|x^* - x^\delta\|_X \frac{\varepsilon}{3} + 4\delta\|z\|_Y.
\end{aligned}
$$

This can be rewritten as $\left(\|x^* - x^\delta\|_X - \varepsilon/3\right)^2 \leq \varepsilon^2/9 + 4\delta\|z\|_Y$.

Now we choose $\delta > 0$ such that the right-hand side is less than $4\varepsilon^2/9$. Taking the square root, we conclude that $\|x^* - x^\delta\|_X \leq \varepsilon$ for this δ. $\qquad\square$

The condition $\|y^\delta\|_Y > \delta$ certainly makes sense because otherwise the right-hand side would be less than the error level δ, and $x^\delta = 0$ would be an acceptable approximation to x^*.

The determination of $\alpha(\delta)$ is thus equivalent to the problem of finding the zero of the monotonic function $\phi(\alpha) := \|Kx^{\alpha,\delta} - y^\delta\|_Y^2 - \delta^2$ (for fixed $\delta > 0$). It is not necessary to satisfy the equation $\|Kx^{\alpha,\delta} - y^\delta\|_Y = \delta$ exactly. An inclusion of the form

$$
c_1\delta \leq \|Kx^{\alpha,\delta} - y^\delta\|_Y \leq c_2\delta
$$

for fixed $1 \leq c_1 < c_2$ is sufficient to prove the assertions of the previous theorem.

The computation of $\alpha(\delta)$ can be carried out with Newton's method. The derivative of the mapping $\alpha \mapsto x^{\alpha,\delta}$ is given by the solution of the equation $(\alpha I + K^*K)\frac{d}{d\alpha}x^{\alpha,\delta} = -x^{\alpha,\delta}$, as is easily seen by differentiating (2.31) with respect to α.

In the following theorem, we prove that the order of convergence $\mathcal{O}(\sqrt{\delta})$ is best possible for the discrepancy principle. Therefore, by the results of Example 1.20, it cannot be optimal under the information $\|(K^*K)^{-\sigma/2}x\|_X \leq E$ for $\sigma > 1$.

Theorem 2.18 *Let K be one-to-one and compact and assume that there exists $\sigma > 0$ such that for every $x \in \mathcal{R}\left((K^*K)^{\sigma/2}\right)$ with $y = Kx \neq 0$, and all sequences $\delta_n \to 0$ and $y^{\delta_n} \in Y$ with $\|y - y^{\delta_n}\|_Y \leq \delta_n$ and $\|y^{\delta_n}\|_Y > \delta_n$ for all n, the Tikhonov solutions $x^n = x^{\alpha(\delta_n),\delta_n}$ (where $\alpha(\delta_n)$ is chosen by the discrepancy principle) converge to x faster than $\sqrt{\delta_n}$ to zero, that is,*

$$
\frac{1}{\sqrt{\delta_n}}\|x^n - x\|_X \longrightarrow 0 \quad \text{as } n \to \infty. \tag{2.35}
$$

Then the range $\mathcal{R}(K)$ has to be finite-dimensional.

Proof: We show first that the choice of $\alpha(\delta)$ by the discrepancy principle implies the boundedness of $\alpha(\delta)/\delta$. Abbreviating $x^\delta := x^{\alpha(\delta),\delta}$, we write for $\delta \leq \frac{1}{3}\|y\|_Y$

$$
\begin{aligned}
\frac{1}{3}\|y\|_Y &= \left(1 - \frac{2}{3}\right)\|y\|_Y \leq \|y\|_Y - 2\delta \\
&\leq \|y - y^\delta\|_Y + \|y^\delta\|_Y - 2\delta \leq \|y^\delta\|_Y - \delta \\
&= \|y^\delta\|_Y - \|y^\delta - Kx^\delta\|_Y \leq \|Kx^\delta\|_Y \\
&= \frac{1}{\alpha(\delta)}\|KK^*(y^\delta - Kx^\delta)\|_Y \leq \frac{\delta}{\alpha(\delta)}\|K\|^2_{\mathcal{L}(X,Y)},
\end{aligned}
$$

where we applied K to (2.31). Thus we have shown that there exists $c > 0$ with $\alpha(\delta) \leq c\delta$ for all sufficiently small δ.

Now we assume that $\dim \mathcal{R}(K) = \infty$ and construct a contradiction. Let $\{\mu_j, x_j, y_j : j \in \mathbb{N}\}$ be a singular system of K and define

$$
x := \frac{1}{\mu_1} x_1 \quad \text{and} \quad y^{\delta_n} := y_1 + \delta_n y_n \quad \text{with} \quad \delta_n := \mu_n^2 .
$$

Then $y = Kx = y_1$ and $\delta_n \to 0$ as $n \to \infty$ and $x \in \mathcal{R}\big((K^*K)^{\sigma/2}\big)$ for every $\sigma > 0$ and $\|y^{\delta_n} - y\|_Y = \delta_n < \sqrt{1 + \delta_n^2} = \|y^{\delta_n}\|_Y$. Therefore, the assumptions for the discrepancy principle are satisfied and thus (2.35) holds.

The solution of $\alpha(\delta_n)x^n + K^*Kx^n = K^*y^{\delta_n}$ is given by

$$
x^n = \frac{\mu_1}{\alpha(\delta_n) + \mu_1^2} x_1 + \frac{\mu_n \delta_n}{\alpha(\delta_n) + \mu_n^2} x_n .
$$

We compute

$$
x^n - x = -\frac{\alpha(\delta_n)}{\mu_1\big(\alpha(\delta_n) + \mu_1^2\big)} x_1 + \frac{\mu_n \delta_n}{\alpha(\delta_n) + \mu_n^2} x_n
$$

and hence

$$
\frac{1}{\sqrt{\delta_n}}\|x^n - x\|_X \geq \frac{\mu_n \sqrt{\delta_n}}{\alpha(\delta_n) + \mu_n^2} = \frac{\delta_n}{\alpha(\delta_n) + \delta_n} = \frac{1}{1 + \alpha(\delta_n)/\delta_n} \geq \frac{1}{1 + c} .
$$

This contradicts (2.35). $\square$

We remark that the estimate $\alpha(\delta) \leq \delta\|K\|^2_{\mathcal{L}(X,Y)}/\big(\|y^\delta\|_Y - \delta\big)$ derived in the previous proof suggests to use $\delta\|K\|^2_{\mathcal{L}(X,Y)}/\big(\|y^\delta\|_Y - \delta\big)$ as a starting value for Newton's method to determine $\alpha(\delta)$!

There has been an enormous effort to modify the original discrepancy principle while still retaining optimal orders of convergence. We refer to [86, 93, 102, 209, 238].

2.6 Landweber's Iteration Method with Stopping Rule

It is very natural to use the following stopping criteria, which can be implemented in every iterative algorithm for the solution of $Kx = y$.

Let $r > 1$ be a fixed number. Stop the algorithm at the first occurrence of $m \in \mathbb{N}_0$ with $\|Kx^{m,\delta} - y^\delta\|_Y \leq r\delta$. The following theorem shows that this choice of m is possible for Landweber's method and leads to an admissible and even optimal regularization strategy.

Theorem 2.19 *Let $K : X \to Y$ be linear, compact, and one-to-one with dense range. Let $Kx^* = y^*$ and $y^\delta \in Y$ be perturbations with $\|y^* - y^\delta\|_Y \leq \delta$ and $\|y^\delta\|_Y \geq r\delta$ for all $\delta \in (0, \delta_0)$ where $r > 1$ is some fixed parameter (independent of δ). Let the sequence $x^{m,\delta}$, $m = 0, 1, 2, \ldots$, be determined by Landweber's method; that is, $x^{0,\delta} = 0$ and*

$$x^{m+1,\delta} = x^{m,\delta} + a\,K^* \left(y^\delta - Kx^{m,\delta} \right), \quad m = 0, 1, 2, \ldots, \tag{2.36}$$

for some $0 < a < 1/\|K\|^2_{\mathcal{L}(X,Y)}$. Then the following assertions hold:

(1) $\lim_{m\to\infty} \|Kx^{m,\delta} - y^\delta\|_Y = 0$ *for every $\delta > 0$; that is, the following stopping rule is well-defined: Let $m = m(\delta) \in \mathbb{N}_0$ be the smallest integer with $\|Kx^{m,\delta} - y^\delta\|_Y \leq r\delta$.*

(2) $\delta^2 m(\delta) \to 0$ *for $\delta \to 0$, that is, this choice of $m(\delta)$ is admissible. Therefore, by the assertions of Theorem 2.15, the sequence $x^{m(\delta),\delta}$ converges to x^* as δ tends to zero.*

(3) If $x^* = K^*z \in \mathcal{R}(K^*)$ or $x^* = (K^*K)^{\sigma/2}z \in \mathcal{R}\big((K^*K)^{\sigma/2}\big)$ for some $\sigma > 0$ and some z, then we have the following orders of convergence:

$$\|x^{m(\delta),\delta} - x^*\|_X \leq c\sqrt{E\,\delta} \quad or \tag{2.37a}$$

$$\|x^{m(\delta),\delta} - x^*\|_X \leq c\,E^{1/(\sigma+1)}\,\delta^{\sigma/(\sigma+1)}, \tag{2.37b}$$

respectively, for some $c > 0$ where again $E = \|z\|$. This means that this choice of $m(\delta)$ is optimal for all $\sigma > 0$.

Proof: In (2.25), we showed the representation

$$R_m y = \sum_{j=1}^{\infty} \frac{1 - (1 - a\mu_j^2)^m}{\mu_j} (y, y_j)_Y \, x_j$$

for every $y \in Y$ and thus

$$\|KR_m y - y\|_Y^2 = \sum_{j=1}^{\infty} (1 - a\mu_j^2)^{2m} \left| (y, y_j)_Y \right|^2.$$

From $|1 - a\mu_j^2| < 1$, we conclude that $\|KR_m - I\|_{\mathcal{L}(Y)} \leq 1$. Application to y^δ instead of y yields

$$\|Kx^{m,\delta} - y^\delta\|_Y^2 = \sum_{j=1}^{\infty} (1 - a\mu_j^2)^{2m} |(y^\delta, y_j)_Y|^2.$$

(1) Let $\varepsilon > 0$ be given. Choose $j_1 \in \mathbb{N}$ with

$$\sum_{j=j_1+1}^{\infty} |(y^\delta, y_j)_Y|^2 < \frac{\varepsilon^2}{2}.$$

Because $|1 - a\mu_j^2|^{2m} \to 0$ as $m \to \infty$ uniformly for $j = 1, \ldots, j_1$, we can find $m_0 \in \mathbb{N}$ with

$$\sum_{j=1}^{j_1} (1 - a\mu_j^2)^{2m} |(y^\delta, y_j)_Y|^2$$
$$\leq \max_{j=1,\ldots,j_1} (1 - a\mu_j^2)^{2m} \sum_{j=1}^{j_1} |(y^\delta, y_j)_Y|^2 \leq \frac{\varepsilon^2}{2} \quad \text{for } m \geq m_0.$$

This implies that $\|Kx^{m,\delta} - y^\delta\|_Y^2 \leq \varepsilon^2$ for $m \geq m_0$; that is, the method is admissible.

It is sufficient to prove assertion (2) only for the case $m(\delta) \to \infty$. We set $m := m(\delta)$ for abbreviation and derive an upper bound of m. By the choice of $m(\delta)$, we have with $y^* = Kx^*$

$$\|KR_{m-1}y^* - y^*\|_Y \geq \|KR_{m-1}y^\delta - y^\delta\|_Y - \|(KR_{m-1} - I)(y^* - y^\delta)\|_Y$$
$$\geq r\delta - \|KR_{m-1} - I\|_{\mathcal{L}(Y)}\, \delta \geq (r-1)\,\delta, \qquad (2.38)$$

and hence

$$m\,(r-1)^2\,\delta^2 \leq m \sum_{j=1}^{\infty} (1 - a\mu_j^2)^{2m-2} |(y^*, y_j)_Y|^2$$
$$= \sum_{j=1}^{\infty} m\,(1 - a\mu_j^2)^{2m-2}\, \mu_j^2\, |(x^*, x_j)_X|^2. \qquad (2.39)$$

We show that the series converges to zero as $\delta \to 0$. (The dependence on δ is hidden in m.) First we note that $m\mu^2(1 - a\mu^2)^{2m-2} \leq 1/a$ for all $m \geq 1$ and all $\mu \geq 0$ (see Problem 2.6). Now we again split the series into a finite sum and a remaining series and estimate in the "long tail", the expression $m(1 - a\mu_j^2)^{2m-2}\mu_j^2$ by $1/a$ and note that $m(1 - a\mu_j^2)^{2m-2}$ tends to zero as $m \to \infty$ uniformly in $j \in \{1, \ldots, j_1\}$. This proves convergence and thus part (2).

For part (3) we remind the reader of the fundamental estimate (2.4a), which we need in the following form:

$$\|x^{m,\delta} - x^*\|_X \;\le\; \delta \sqrt{a\,m} \;+\; \|R_m K x^* - x^*\|_X\,. \qquad (2.40)$$

We restrict ourselves to the case that $x^* = (K^*K)^{\sigma/2}z$ for some $\sigma > 0$. We set again $E = \|z\|_X$ and estimate $m = m(\delta)$ from above by using (2.38) and (2.28) for y^* instead of y^δ (that is, $\delta = 0$) and $m - 1$ instead of m.

$$(r-1)\delta \;\le\; \|K R_{m-1} y^* - y^*\|_Y \;\le\; \left(\frac{\sigma+1}{2a(m-1)}\right)^{(\sigma+1)/2} E\,.$$

Solving for $m - 1$ yields an estimate of the form

$$m(\delta) \;\le\; 2(m(\delta) - 1) \;\le\; c\left(\frac{E}{\delta}\right)^{2/(\sigma+1)}$$

for some c which depends solely on σ, r, and a. Substituting this into the first term of (2.40) for $m = m(\delta)$ yields $\delta\sqrt{a\,m(\delta)} \le \sqrt{ac}\,E^{1/(\sigma+1)}\delta^{\sigma/(\sigma+1)}$. Next, we estimate the second term of (2.40). In the following estimates, we use again that $x^* = (K^*K)^{\sigma/2}z$ and also Hölder's inequality with $p = \frac{\sigma+1}{\sigma}$ and $q = \sigma+1$.

$$\|R_m K x^* - x^*\|_X^2$$
$$= \sum_{j=1}^{\infty}(1 - a\mu_j^2)^{2m}|(x^*, x_j)_X|^2 \;=\; \sum_{j=1}^{\infty}(1 - a\mu_j^2)^{2m}\mu_j^{2\sigma}|(z, x_j)_X|^2$$
$$= \sum_{j=1}^{\infty}\left[(1 - a\mu_j^2)^{2m}\mu_j^{2\sigma+2}|(z, x_j)_X|^2\right]^{1/p}\left[(1 - a\mu_j^2)^{2m}|(z, x_j)_X|^2\right]^{1/q}$$
$$\le \left[\sum_{j=1}^{\infty}(1 - a\mu_j^2)^{2m}\mu_j^{2\sigma+2}|(z, x_j)_X|^2\right]^{1/p}\left[\sum_{j=1}^{\infty}(1 - a\mu_j^2)^{2m}|(z, x_j)_X|^2\right]^{1/q}$$
$$\le \|K R_m y^* - y^*\|_Y^{2\sigma/(\sigma+1)}\,E^{2/(\sigma+1)}$$
$$\le E^{2/(\sigma+1)}\left[\|K R_m y^\delta - y^\delta\|_Y + \|(K R_m - I)(y^\delta - y^*)\|_Y\right]^{2\sigma/(\sigma+1)}.$$

Now we use the stopping rule for $m = m(\delta)$ in the first term and the estimate $\|K R_m - I\|_{\mathcal{L}(Y)} \le 1$ in the second one. This yields

$$\|R_{m(\delta)} K x^* - x^*\|_X^2 \;\le\; (r+1)^{2\sigma/(\sigma+1)}\,E^{2/(\sigma+1)}\delta^{2\sigma/(\sigma+1)}\,.$$

Substituting this into (2.40) for $m = m(\delta)$ yields the assertion. $\qquad\square$

It is also possible to formulate a similar stopping criterion for Morozov's discrepancy principle. Choose an arbitrary monotonically decreasing sequence (α_m) in $\mathbb{R}$ with $\lim_{m\to\infty}\alpha_m = 0$. Determine $m = m(\delta)$ as the smallest integer m with $\|K x^{\alpha_m,\delta} - y^\delta\|_Y \le r\delta$. For details, we refer the reader to [89] or [182].

One can construct more general classes of methods through the spectral representation of the solution x^*.

Comparing the regularizer x^δ of Landweber's method with the true solution x^*, we observe that the function $\phi(\mu) = 1/\mu$, $\mu > 0$ is approximated by the polynomial $\mathbb{P}_m(\mu) = \left[1 - (1 - a\mu^2)^m\right]/\mu$. It is certainly possible to choose better polynomial approximations of the function $\mu \mapsto 1/\mu$. Orthogonal polynomials are particularly useful. This leads to the ν-methods; see [21, 118], or [120].

A common feature of these methods that is very crucial in the analysis is the fact that all of the polynomials $\mathbb{P}_m$ are independent of y and y^δ. For the important conjugate gradient algorithm discussed in the next section, this is not the case, and that makes an error analysis much more difficult to obtain.

2.7 The Conjugate Gradient Method

In this section, we study the regularizing properties of the conjugate gradient method. Because the proofs of the theorems are rather technical, we only state the results and transfer the proofs to the Appendix B.

First, we recall the conjugate gradient method for least squares problems for overdetermined systems of linear equations of the form $Kx = y$. Here, $K \in \mathbb{R}^{m \times n}$ and $y \in \mathbb{R}^m$ with $m \geq n$ are given. Because it is hopeless to satisfy all equations simultaneously, one minimizes the defect $f(x) := \|Kx - y\|^2$, $x \in \mathbb{R}^n$, where $\| \cdot \|$ denotes the Euclidean norm in $\mathbb{R}^m$. Standard algorithms for solving least squares problems are the QR-algorithm or the conjugate gradient method; see [71, 106, 132]. Because we assume that the latter is known for systems of equations, we formulate it now (in Figure 2.2) for the operator equation $Kx = y$, where $K : X \to Y$ is a bounded, linear, and injective operator between Hilbert spaces X and Y with adjoint $K^* : Y \to X$.

Theorem 2.20 *(Fletcher–Reeves)*
Let $K : X \to Y$ be a compact, linear, and injective operator between Hilbert spaces X and Y. Then the conjugate gradient method is well-defined and either stops or produces sequences (x^m), (p^m) in X with the properties

$$\left(\nabla f(x^m), \nabla f(x^j) \right)_X = 0 \quad \text{for all } j \neq m \tag{2.41a}$$

and

$$(Kp^m, Kp^j)_Y = 0 \quad \text{for all } j \neq m; \tag{2.41b}$$

that is, the gradients are orthogonal and the directions p^m are K-conjugate. Furthermore,

$$\left(\nabla f(x^j), K^* Kp^m \right)_X = 0 \quad \text{for all } j < m. \tag{2.41c}$$

Define again the function

$$f(x) := \|Kx - y\|_Y^2 = (Kx - y, Kx - y)_Y, \quad x \in X.$$

We abbreviate $\nabla f(x) := 2 K^*(Kx - y) \in X$ and note that $\nabla f(x)$ is indeed the Riesz representation (see Theorem A.23) of the Fréchet derivative $f'(x)$ of f at x (see Lemma 2.14). We call two elements $p, q \in X$ K-conjugate if $(Kp, Kq)_Y = 0$. If K is one-to-one, this bilinear form has the properties of an inner product on X.

The following theorem gives an interesting and different interpretation of the elements x^m.

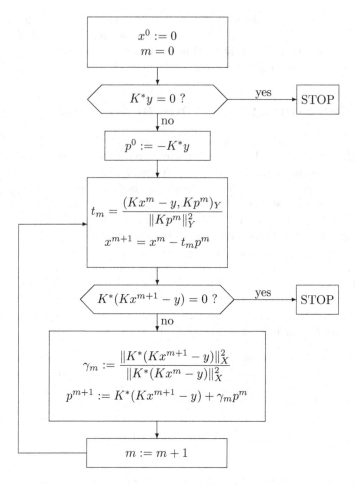

Figure 2.2: The conjugate gradient method

Theorem 2.21 *Let (x^m) and (p^m) be the sequences of the conjugate gradient method. Define the space $V_m := \text{span}\{p^0, \ldots, p^m\}$. Then we have the following equivalent characterizations of V_m:*

$$
\begin{aligned}
V_m &= \text{span}\left\{\nabla f(x^0), \ldots, \nabla f(x^m)\right\} & \text{(2.42a)} \\
&= \text{span}\left\{p^0, K^*Kp^0, \ldots, (K^*K)^m p^0\right\} & \text{(2.42b)}
\end{aligned}
$$

for $m = 0, 1, \ldots$. The spaces V_m are called **Krylov spaces**. Furthermore, x^m is the minimum of f on V_{m-1} for every $m \geq 1$.

By this result, we can write x^m in the form

$$x^m = -\mathbb{P}_{m-1}(K^*K)p^0 = \mathbb{P}_{m-1}(K^*K)K^*y \qquad (2.43)$$

with a well-defined polynomial $\mathbb{P}_{m-1} \in \mathcal{P}_{m-1}$ of degree at most $m-1$ (which depends itself on y). Analogously, we write the defect in the form

$$y - Kx^m = y - K\mathbb{P}_{m-1}(K^*K)K^*y = y - KK^*\mathbb{P}_{m-1}(KK^*)y$$
$$= \mathbb{Q}_m(KK^*)y$$

with the polynomial $\mathbb{Q}_m(t) := 1 - t\,\mathbb{P}_{m-1}(t)$ of degree m.

Let $\{\mu_j, x_j, y_j : j \in \mathbb{N}\}$ be a singular system for K. If it happens that

$$y = \sum_{j=1}^{N} \alpha_j y_j \in W_N := \mathrm{span}\,\{y_1, \ldots, y_N\}$$

for some $N \in \mathbb{N}$, then all iterates $x^m \in A_N := \mathrm{span}\,\{x_1, \ldots, x_N\}$ because

$$x^m = \mathbb{P}_{m-1}(K^*K)K^*y = \sum_{j=1}^{N} \alpha_j\, \mathbb{P}_{m-1}(\mu_j^2)\, \mu_j\, x_j\,.$$

In this exceptional case, the algorithm terminates after at most N iterations because the dimension of A_N is at most N and the gradients $\nabla f(x^i)$ are orthogonal to each other. This is the reason why the conjugate gradient method applied to matrix equations stops after finitely many iterations. For operator equations in infinite-dimensional Hilbert spaces, this method produces sequences of, in general, infinitely many elements.

The following characterizations of $\mathbb{Q}_m$ are useful.

Lemma 2.22 (a) The polynomial $\mathbb{Q}_m$ minimizes the functional

$$H(\mathbb{Q}) := \|\mathbb{Q}(KK^*)y\|_Y^2 \quad on \quad \{\mathbb{Q} \in \mathcal{P}_m : \mathbb{Q}(0) = 1\}$$

and satisfies

$$H(\mathbb{Q}_m) = \|Kx^m - y\|_Y^2\,.$$

(b) For $k \neq \ell$, the following orthogonality relation holds:

$$\langle \mathbb{Q}_k, \mathbb{Q}_\ell \rangle := \sum_{j=1}^{\infty} \mu_j^2\, \mathbb{Q}_k(\mu_j^2)\, \mathbb{Q}_\ell(\mu_j^2)\, \big|(y, y_j)_Y\big|^2 = 0\,. \qquad (2.44)$$

If $y \notin \mathrm{span}\,\{y_1, \ldots, y_N\}$ for any $N \in \mathbb{N}$, then $\langle \cdot, \cdot \rangle$ defines an inner product on the space $\mathcal{P}$ of all polynomials.

Without a priori information, the sequence (x^m) does not converge to the solution x of $Kx = y$. The images, however, do converge to y. This is the subject of the next theorem.

Theorem 2.23 *Let K and K^* be one-to-one, and assume that the conjugate gradient method does not stop after finitely many steps. Then*

$$Kx^m \longrightarrow y \quad as \quad m \to \infty$$

for every $y \in Y$.

We give a proof of this theorem because it is a simple conclusion of the previous lemma.

Proof: Let $\mathbb{Q} \in \mathcal{P}_m$ be an arbitrary polynomial with $\mathbb{Q}(0) = 1$. Then, by the previous lemma,

$$\|Kx^m - y\|_Y^2 = H(\mathbb{Q}_m) \leq H(\mathbb{Q}) = \sum_{j=1}^{\infty} \mathbb{Q}(\mu_j^2)^2 |(y, y_j)_Y|^2. \qquad (2.45)$$

Now let $\varepsilon > 0$ be arbitrary. Choose $j_1 \in \mathbb{N}$ such that

$$\sum_{j=j_1+1}^{\infty} |(y, y_j)_Y|^2 < \frac{\varepsilon^2}{2}$$

and choose a function $R \in C[0, \mu_1^2]$ with $R(0) = 1$, $\|R\|_\infty \leq 1$ and $R(\mu_j^2) = 0$ for $j = 1, 2, \ldots, j_1$. By the theorem of Weierstrass, there exist polynomials $\tilde{\mathbb{Q}}_m \in \mathcal{P}_m$ with $\|R - \tilde{\mathbb{Q}}_m\|_\infty \to 0$ as $m \to \infty$. We set $\hat{\mathbb{Q}}_m = \tilde{\mathbb{Q}}_m / \tilde{\mathbb{Q}}_m(0)$, which is possible for sufficiently large m because $R(0) = 1$. Then $\hat{\mathbb{Q}}_m$ converges to R as $m \to \infty$ and $\hat{\mathbb{Q}}_m(0) = 1$. Substituting this into (2.45) yields

$$\|Kx^m - y\|_Y^2 \leq H(\hat{\mathbb{Q}}_m)$$

$$\leq \sum_{j=1}^{j_1} |\hat{\mathbb{Q}}_m(\mu_j^2) - \underbrace{R(\mu_j^2)}_{=0}|^2 |(y, y_j)_Y|^2 + \|\hat{\mathbb{Q}}_m\|_\infty^2 \sum_{j > j_1} |(y, y_j)_Y|^2$$

$$\leq \|\hat{\mathbb{Q}}_m - R\|_\infty^2 \|y\|_Y^2 + \frac{\varepsilon^2}{2} \|\hat{\mathbb{Q}}_m\|_\infty^2.$$

This expression is less than ε^2 for sufficiently large m. $\qquad \square$

Now we return to the regularization of the operator equation $Kx^* = y^*$. The operator $\mathbb{P}_{m-1}(K^*K)K^* : Y \to X$ corresponds to the regularization operator R_α of the general theory. But this operator certainly depends on the right-hand side y. The mapping $y \mapsto \mathbb{P}_{m-1}(K^*K)K^*y$ is therefore nonlinear.

So far, we have formulated and studied the conjugate gradient method for unperturbed right-hand sides. Now we consider the situation where we know only an approximation y^δ of y^* such that $\|y^\delta - y^*\|_Y \leq \delta$. We apply the algorithm to y^δ instead of y. This yields a sequence $x^{m,\delta}$ and polynomials $\mathbb{P}_m^\delta$ and $\mathbb{Q}_m^\delta$. There is no a priori strategy $m = m(\delta)$ such that $x^{m(\delta),\delta}$ converges to x^* as δ tends to zero; see [78]. An a posteriori choice as in the previous section, however, again leads to an optimal strategy. We stop the algorithm with the smallest $m = m(\delta)$ such that the defect $\|Kx^{m,\delta} - y^\delta\|_Y \leq r\delta$, where $r > 1$ is some given parameter. From now on, we make the assumption that y^δ is never a finite linear combination of the y_j. Then, by Theorem 2.23, the defect tends to zero, and this stopping rule is well-defined. We want to show that the choice $m = m(\delta)$ leads to an optimal algorithm. The following analysis, which we learned from [121] (see also [226]), is more elementary than, for example, in [21, 181], or [182]. We carry out the complete analysis but, again, postpone the proofs to Appendix B because they are rather technical.

We recall that by our stopping rule

$$\|Kx^{m(\delta),\delta} - y^\delta\|_Y \leq r\delta < \|Kx^{m(\delta)-1,\delta} - y^\delta\|_Y. \qquad (2.46)$$

The following theorem establishes the optimality of the conjugate gradient method with this stopping rule.

Theorem 2.24 *Assume that $y^* = Kx^*$ and y^δ do not belong to the linear span of finitely many y_j. Let the sequence $x^{m(\delta),\delta}$ be constructed by the conjugate gradient method with stopping rule (2.46) for fixed parameter $r > 1$. Let $x^* = (K^*K)^{\sigma/2}z \in \mathcal{R}\big((K^*K)^{\sigma/2}\big)$ for some $\sigma > 0$ and $z \in X$. Then there exists $c > 0$ with*

$$\|x^{m(\delta),\delta} - x^*\|_X \leq c\,\delta^{\sigma/(\sigma+1)}\,E^{1/(\sigma+1)}. \qquad (2.47)$$

where $E = \|z\|_X$.

As Landweber's method and the regularization method by the spectral cut-off (but in contrast to Tikhonov's method), the conjugate gradient method is optimal for all $\sigma > 0$ under the a priori information $\|(K^*K)^{-\sigma/2}x^*\|_X \leq E$.

There is a much simpler implementation of the conjugate gradient method for self-adjoint positive definite operators $K : X \rightarrow X$. For such K, there exists a unique self-adjoint positive operator $A : X \rightarrow X$ with $A^2 = K$. Let $Kx = y$ and set $z := Ax$, that is, $Az = y$. We apply the conjugate gradient method to the equation $Ax = z$ without knowing z. In the process of the algorithm, only the elements $A^*z = y$, $\|Ap^m\|^2 = (Kp^m, p^m)$, and $A^*(Ax^m - z) = Kx^m - y$ have to be computed. The square root A and the quantity z do not have to be known explicitly, and the method is much simpler to implement.

Actually, the conjugate gradient method presented here is only one member of a large class of conjugate gradient methods. For a detailed study of these methods in connection with ill-posed problems, we refer to [104, 124, 207, 208, 210] and, in particular, the work [122].

2.8   Problems

2.1 Let $K : L^2(0,1) \to L^2(0,1)$ be the integral operator

$$(Kx)(t) := \int_0^1 (1+ts)\, e^{ts}\, x(s)\, ds, \quad 0 < t < 1.$$

Show by induction that

$$\frac{d^n}{dt^n}(Kx)(t) = \int_0^1 (n+1+ts)\, s^n\, e^{ts}\, x(s)\, ds, \quad 0 < t < 1, \ n = 0, 1, \ldots .$$

Prove that K is one-to-one and that the constant functions do not belong to the range of K.

2.2 Apply Tikhonov's method of Section 2.2 to the integral equation

$$\int_0^t x(s)\, ds = y(t), \quad 0 \le t \le 1.$$

Prove that for $y \in H^1(0,1)$ with $y(0) = 0$, Tikhonov's solution x^α is given by the solution of the boundary value problem

$$-\alpha\, x''(t) + x(t) = y'(t), \ 0 < t < 1, \quad x(1) = 0, \ x'(0) = 0.$$

2.3 Let $K : X \to Y$ be compact and one-to-one. For any $\sigma \ge 0$ let the spaces X_σ be defined by $X_\sigma = \mathcal{R}\big((K^*K)^{\sigma/2}\big)$ equipped with the inner product

$$\big((K^*K)^{\sigma/2} z_1, (K^*K)^{\sigma/2} z_2\big)_{X_\sigma} = (z_1, z_2)_X, \ z_1, z_2 \in X.$$

Prove that X_σ are Hilbert spaces and that X_{σ_2} is compactly embedded in X_{σ_1} for $\sigma_2 > \sigma_1$.

2.4 The *iterated Tikhonov regularization* $x^{m,\alpha,\delta}$ of order $m \in \mathbb{N}$ (see [87, 153]) is iteratively defined by

$$x^{0,\alpha,\delta} = 0, \quad (\alpha I + K^*K) x^{m+1,\alpha,\delta} = K^* y^\delta + \alpha\, x^{m,\alpha,\delta}$$

for $m = 0, 1, 2, \ldots$. (Note that $x^{1,\alpha,\delta}$ is the ordinary Tikhonov regularization.)

(a) Show that $q^{(m)}(\alpha, \mu) := 1 - \left(\frac{\alpha}{\alpha+\mu^2}\right)^m$ is the corresponding filter function.

(b) Prove that this filter function leads to a regularizing operator $R_\alpha^{(m)}$ with $\|R_\alpha^{(m)}\|_{\mathcal{L}(Y,X)} \le m/(2\sqrt{\alpha})$, and $q^{(m)}$ satisfies (2.9) from Theorem 2.7 with

$$\omega_\sigma^{(m)}(\alpha) = c\, \alpha^{\min\{\sigma/2, m\}}$$

where c depends only on m and σ.

(c) Show that the iterated Tikhonov regularization of order m is asymptotically optimal under the information $\|(K^*K)^{-\sigma/2}x^*\|_X \leq E$ for every $\sigma \leq 2m$.

2.5 Fix y^δ with $\|y^\delta - y^*\|_Y \leq \delta$ and let $x^{\alpha,\delta}$ be the Tikhonov solution corresponding to $\alpha > 0$. The curve

$$\alpha \mapsto \begin{pmatrix} f(\alpha) \\ g(\alpha) \end{pmatrix} := \begin{pmatrix} \|Kx^{\alpha,\delta} - y^\delta\|_Y^2 \\ \|x^{\alpha,\delta}\|_X^2 \end{pmatrix}, \quad \alpha > 0,$$

in $\mathbb{R}^2$ is called an L-curve because it has often the shape of the letter L; see [90, 125, 127].

Show by using a singular system that $f'(\alpha) = -\alpha\, g'(\alpha)$. Furthermore, compute the curvature

$$C(\alpha) := \frac{|f'(\alpha)\, g''(\alpha) - g'(\alpha)\, f''(\alpha)|}{\left(f'(\alpha)^2 + g'(\alpha)^2\right)^{3/2}}$$

and show that the curvature increases monotonically for $0 < \alpha \leq 1/\|K\|_{\mathcal{L}(X,Y)}^2$.

2.6 Show that

$$m\mu^2(1 - a\mu^2)^{2m-2} \leq \frac{1}{a} \quad \text{for all } m \geq 1 \text{ and } 0 \leq \mu \leq \frac{1}{\sqrt{a}}.$$

Chapter 3

Regularization by Discretization

In this chapter, we study a different approach to regularizing operator equations of the form $Kx = y$, where x and y are elements of certain function spaces. This approach is motivated by the fact that for the numerical treatment of such equations, one has to discretize the continuous problem and reduce it to a finite system of (linear or nonlinear) equations. We see in this chapter that the discretization schemes themselves are regularization strategies in the sense of Chapter 2.

In Section 3.1, we study the general concept of projection methods and give a necessary and sufficient condition for convergence. Although we have in mind the treatment of integral equations of the first kind, we treat the general case where K is a linear, bounded, not necessarily compact operator between (real or complex) Banach or Hilbert spaces. Section 3.2 is devoted to Galerkin methods. As special cases, we study least squares and dual least squares methods in Subsections 3.2.1 and 3.2.2. In Subsection 3.2.3, we investigate the Bubnov–Galerkin method for the case where the operator satisfies Gårding's inequality. In Section 3.3, we illustrate the Galerkin methods for Symm's integral equation of the first kind. This equation arises in potential theory and serves as a model equation for more complicated situations. Section 3.4 is devoted to collocation methods. We restrict ourselves to the moment method in Subsection 3.4.1 and to collocation by piecewise constant functions in Subsection 3.4.2, where the analysis is carried out only for Symm's integral equation. In Section 3.5, we present numerical results for various regularization techniques (Tikhonov, Landweber, conjugate gradient, projection, and collocation methods) tested for Dirichlet boundary value problems for the Laplacian in an ellipse. Finally, we study the Backus–Gilbert method in Section 3.6. Although not very popular among mathematicians, this method is extensively used by scientists in geophysics and other applied sciences. The general ideas of Sections 3.1 and 3.2 can also be found in, for example, [17, 168, 182, 226].

© Springer Nature Switzerland AG 2021

A. Kirsch, *An Introduction to the Mathematical Theory of Inverse Problems*,
Applied Mathematical Sciences 120,
https://doi.org/10.1007/978-3-030-63343-1_3

3.1   Projection Methods

First, we recall the definition of a projection operator.

Definition 3.1 *Let X be a normed space over the field $\mathbb{K}$ where $\mathbb{K} = \mathbb{R}$ or $\mathbb{K} = \mathbb{C}$. Let $U \subset X$ be a closed subspace. A linear bounded operator $P : X \to X$ is called a* projection operator *on U if*

- *$Px \in U$ for all $x \in X$ and*

- *$Px = x$ for all $x \in U$.*

We now summarize some obvious properties of projection operators.

Theorem 3.2 *Every nontrivial projection operator satisfies $P^2 = P$ and $\|P\|_{\mathcal{L}(X)} \geq 1$.*

Proof: $P^2 x = P(Px) = Px$ follows from $Px \in U$. Furthermore, $\|P\|_{\mathcal{L}(X)} = \|P^2\|_{\mathcal{L}(X)} \leq \|P\|^2_{\mathcal{L}(X)}$ and $P \neq 0$. This implies $\|P\|_{\mathcal{L}(X)} \geq 1$.   □

In the following two examples, we introduce the most important projection operators.

Example 3.3
(a) (*Orthogonal projection*) Let X be a pre-Hilbert space over $\mathbb{K} = \mathbb{R}$ or $\mathbb{K} = \mathbb{C}$ and $U \subset X$ be a complete subspace. Let $Px \in U$ be the best approximation to x in U; that is, Px satisfies

$$\|Px - x\|_X \;\leq\; \|u - x\|_X \quad \text{for all } u \in U. \tag{3.1}$$

By the projection theorem (Theorem A.13 of Appendix A.1), $P : X \to U$ is linear and $Px \in U$ is characterized by the abstract "normal equation" $(x - Px, u)_X = 0$ for all $u \in U$, that is, $x - Px \in U^{\perp}$. In this example, by the binomial theorem we have

$$
\begin{aligned}
\|x\|^2_X &= \|Px + (x - Px)\|^2_X \\
&= \|Px\|^2_X + \|x - Px\|^2_X + 2 \underbrace{\operatorname{Re}(x - Px, Px)_X}_{=0} \geq \|Px\|^2_X,
\end{aligned}
$$

that is, $\|P\|_{\mathcal{L}(X)} = 1$. Important examples of subspaces U are spaces of splines or finite elements.

(b) (*Interpolation operator*) Let $X = C[a, b]$ be the space of real-valued continuous functions on $[a, b]$ supplied with the supremum norm. Then X is a normed space over $\mathbb{R}$. Let $U = \operatorname{span}\{u_1, \ldots, u_n\}$ be an n-dimensional subspace and $t_1, \ldots, t_n \in [a, b]$ such that the interpolation problem in U is uniquely solvable; that is, $\det[u_j(t_k)] \neq 0$. We define $Px \in U$ by the interpolant of $x \in C[a, b]$ in U, i.e., $u = Px \in U$ satisfies $u(t_j) = x(t_j)$ for all $j = 1, \ldots, n$. Then $P : X \to U$ is a projection operator.

Examples for U are spaces of algebraic or trigonometric polynomials. As a drawback of these choices, we note that from the results of Faber (see [203]) the interpolating polynomials of continuous functions x do not, in general, converge to x as the degree of the polynomials tends to infinity. For smooth periodic functions, however, trigonometric interpolation at equidistant points converges with an optimal order of convergence. We use this fact in Subsection 3.4.2. Here, as an example, we recall the interpolation by linear splines. For simplicity, we formulate only the case where the endpoints are included in the set of interpolation points.

Let $a = t_1 < \cdots < t_n = b$ be given points, and let $U \subset C[a,b]$ be defined by

$$U = \mathcal{S}_1(t_1, \ldots, t_n)$$

$$:= \left\{ x \in C[a,b] : x|_{[t_j, t_{j+1}]} \in \mathcal{P}_1, \; j = 1, \ldots, n-1 \right\}, \qquad (3.2)$$

where $\mathcal{P}_1$ denotes the space of polynomials of degree at most one. Then the interpolation operator $Q_n : C[a,b] \to \mathcal{S}_1(t_1, \ldots, t_n)$ is given by

$$Q_n x = \sum_{j=1}^{n} x(t_j)\, \hat{y}_j \quad \text{for } x \in C[a,b],$$

where the basis functions $\hat{y}_j \in \mathcal{S}_1(t_1, \ldots, t_n)$ are defined by

$$\hat{y}_j(t) = \begin{cases} \dfrac{t - t_{j-1}}{t_j - t_{j-1}}, & t \in [t_{j-1}, t_j] \quad (\text{if } j \geq 2), \\[2mm] \dfrac{t_{j+1} - t}{t_{j+1} - t_j}, & t \in [t_j, t_{j+1}] \quad (\text{if } j \leq n-1), \\[2mm] 0, & t \notin [t_{j-1}, t_{j+1}], \end{cases} \qquad (3.3)$$

for $j = 1, \ldots, n$. In this example $\|Q_n\|_{\mathcal{L}(C[a,b])} = 1$ (see Problem 3.1).

For general interpolation operators, $\|Q_n\|_{\mathcal{L}(X)}$ exceeds one and $\|Q_n\|_{\mathcal{L}(X)}$ does not have to be bounded with respect to n.

Now we define the class of projection methods.

Definition 3.4 *Let X and Y be Banach spaces and $K : X \to Y$ be bounded and one-to-one. Furthermore, let $X_n \subset X$ and $Y_n \subset Y$ be finite-dimensional subspaces of dimension n and $Q_n : Y \to Y_n$ be a projection operator. For given $y \in Y$, the projection method for solving the equation $Kx = y$ is to solve the equation*

$$Q_n K x_n = Q_n y \quad \text{for } x_n \in X_n. \qquad (3.4)$$

Equation (3.4) reduces to a finite linear system by choosing bases $\{\hat{x}_1, \ldots, \hat{x}_n\}$ and $\{\hat{y}_1, \ldots, \hat{y}_n\}$ of X_n and Y_n, respectively. One possibility is to represent $Q_n y$ and every $Q_n K \hat{x}_j$, $j = 1, \ldots, n$, in the forms

$$Q_n y = \sum_{i=1}^{n} \beta_i\, \hat{y}_i \quad \text{and} \quad Q_n K \hat{x}_j = \sum_{i=1}^{n} A_{ij}\, \hat{y}_i, \quad j = 1, \ldots, n, \qquad (3.5)$$

with β_i, $A_{ij} \in \mathbb{K}$. The linear combination $x_n = \sum_{j=1}^{n} \alpha_j \hat{x}_j$ solves (3.4) if and only if $\alpha = (\alpha_1, \ldots, \alpha_n)^\top \in \mathbb{K}^n$ solves the finite system of linear equations

$$\sum_{j=1}^{n} A_{ij} \alpha_j = \beta_i, \quad i = 1, \ldots, n; \quad \text{that is,} \quad A\alpha = \beta. \tag{3.6}$$

There is a second possibility to reduce (3.4) to a finite system which is used for Galerkin methods. Let Y^* be the dual space of Y with dual pairing $\langle \cdot, \cdot \rangle_{Y^*,Y}$. We choose elements $\hat{y}_i^* \in Y^*$ for $i = 1, \ldots, n$, such that the matrix $Y \in \mathbb{K}^{n \times n}$ given by $Y_{ij} = \langle \hat{y}_i^*, \hat{y}_j \rangle_{Y^*,Y}$ is regular. Then, representing x_n again as $x_n = \sum_{j=1}^{n} \alpha_j \hat{x}_j$ equation (3.4) is equivalent to

$$\sum_{j=1}^{n} A_{ij} \alpha_j = \beta_i, \quad i = 1, \ldots, n; \quad \text{that is,} \quad A\alpha = \beta, \tag{3.7}$$

where now

$$A_{ij} = \langle \hat{y}_i^*, Q_n K \hat{x}_j \rangle_{Y^*,Y} \quad \text{and} \quad \beta_i = \langle \hat{y}_i^*, Q_n y \rangle_{Y^*,Y}. \tag{3.8}$$

The orthogonal projection and the interpolation operator from Example 3.3 lead to the following important classes of projection methods, which are studied in more detail in the next sections.

Example 3.5

Let $K : X \to Y$ be bounded and one-to-one.

(a) (*Galerkin method*) Let X and Y be pre-Hilbert spaces and $X_n \subset X$ and $Y_n \subset Y$ be finite-dimensional subspaces with $\dim X_n = \dim Y_n = n$. Let $Q_n : Y \to Y_n$ be the *orthogonal* projection. Then the projected equation $Q_n K x_n = Q_n y$ is equivalent to

$$(K x_n, z_n)_Y = (y, z_n)_Y \quad \text{for all } z_n \in Y_n. \tag{3.9a}$$

Again let $X_n = \text{span}\{\hat{x}_1, \ldots, \hat{x}_n\}$ and $Y_n = \text{span}\{\hat{y}_1, \ldots, \hat{y}_n\}$. Looking for a solution of (3.9a) in the form $x_n = \sum_{j=1}^{n} \alpha_j \hat{x}_j$ leads to the system

$$\sum_{j=1}^{n} \alpha_j (K \hat{x}_j, \hat{y}_i)_Y = (y, \hat{y}_i)_Y \quad \text{for } i = 1, \ldots, n, \tag{3.9b}$$

or $A\alpha = \beta$, where $A_{ij} := (K \hat{x}_j, \hat{y}_i)_Y$ and $\beta_i = (y, \hat{y}_i)_Y$. This corresponds to (3.7) with $\hat{y}_i^* = \hat{y}_i$ after the identification of Y^* with Y (Theorem A.23 of Riesz).

(b) (*Collocation method*) Let X be a Banach space, $Y = C[a, b]$, and $K : X \to C[a, b]$ be a bounded operator. Let $a = t_1 < \cdots < t_n = b$ be given points (*collocation points*) and $Y_n = \mathcal{S}_1(t_1, \ldots, t_n)$ be the corresponding space (3.2) of linear splines with interpolation operator $Q_n y = \sum_{j=1}^{n} y(t_j) \hat{y}_j$. Let $y \in C[a, b]$ and some n-dimensional subspace $X_n \subset X$ be given. Then $Q_n K x_n = Q_n y$ is equivalent to

$$(K x_n)(t_i) = y(t_i) \quad \text{for all } i = 1, \ldots, n. \tag{3.10a}$$

If we denote by $\{\hat{x}_1, \ldots, \hat{x}_n\}$ a basis of X_n, then looking for a solution of (3.10a) in the form $x_n = \sum_{j=1}^n \alpha_j \hat{x}_j$ leads to the finite linear system

$$\sum_{j=1}^n \alpha_j (K\hat{x}_j)(t_i) = y(t_i), \quad i = 1, \ldots, n, \tag{3.10b}$$

or $A\alpha = \beta$, where $A_{ij} = (K\hat{x}_j)(t_i)$ and $\beta_i = y(t_i)$.

We are particularly interested in the study of integral equations of the form

$$\int_a^b k(t, s)\, x(s)\, ds = y(t), \quad t \in [a, b], \tag{3.11}$$

in $L^2(a, b)$ or $C[a, b]$ for some continuous or weakly singular function k. (3.9b) and (3.10b) now take the form

$$A\alpha = \beta, \tag{3.12}$$

where $x = \sum_{j=1}^n \alpha_j \hat{x}_j$ and

$$A_{ij} = \int_a^b \int_a^b k(t, s)\, \hat{x}_j(s)\, \hat{y}_i(t)\, ds\, dt \tag{3.13a}$$

$$\beta_i = \int_a^b y(t)\, \hat{y}_i(t)\, dt \tag{3.13b}$$

for the Galerkin method, and

$$A_{ij} = \int_a^b k(t_i, s)\, \hat{x}_j(s)\, ds \tag{3.13c}$$

$$\beta_i = y(t_i) \tag{3.13d}$$

for the collocation method.

Comparing the systems of equations in (3.12), we observe that the computation of the matrix elements (3.13c) is less expensive than for those of (3.13a) due to the double integration for every matrix element in (3.13a). For this reason, collocation methods are generally easier to implement than Galerkin methods. On the other hand, Galerkin methods have convergence properties of high order in weak norms (superconvergence) which are of practical importance in many cases, such as boundary element methods for the solution of boundary value problems.

For the remaining part of this section, we make the following assumption.

Assumption 3.6 *Let $K : X \to Y$ be a linear, bounded, and injective operator between Banach spaces, $X_n \subset X$ and $Y_n \subset Y$ be finite-dimensional subspaces of dimension n, and $Q_n : Y \to Y_n$ be a projection operator. We assume that $\bigcup_{n \in \mathbb{N}} X_n$ is dense in X and that $Q_n K|_{X_n} : X_n \to Y_n$ is one-to-one and thus invertible. Let $x \in X$ be the solution of*

$$Kx = y. \tag{3.14}$$

By $x_n \in X_n$, we denote the unique solutions of the equations

$$Q_n K x_n = Q_n y \tag{3.15}$$

for $n \in \mathbb{N}$.

We can represent the solutions $x_n \in X_n$ of (3.15) in the form $x_n = R_n y$, where $R_n : Y \to X_n \subset X$ is defined by

$$R_n := \left(Q_n K|_{X_n} \right)^{-1} Q_n : Y \longrightarrow X_n \subset X. \tag{3.16}$$

The projection method is called *convergent* if the approximate solutions $x_n \in X_n$ of (3.15) converge to the exact solution $x \in X$ of (3.14) for every $y \in \mathcal{R}(K)$, that is, if

$$R_n K x = \left(Q_n K|_{X_n} \right)^{-1} Q_n K x \longrightarrow x, \quad n \to \infty, \tag{3.17}$$

for every $x \in X$.

We observe that this definition of convergence coincides with Definition 2.1 of a regularization strategy for the equation $Kx = y$ with regularization parameter $\alpha = 1/n$. Therefore, the projection method converges if and only if R_n is a regularization strategy for the equation $Kx = y$.

Obviously, we can only expect convergence if we require that $\bigcup_{n \in \mathbb{N}} X_n$ is dense in X and $Q_n y \to y$ for all $y \in \mathcal{R}(K)$. But, in general, this is not sufficient for convergence if K is compact. We have to assume the following boundedness condition.

Theorem 3.7 *Let Assumption 3.6 be satisfied. The solution $x_n = R_n y \in X_n$ of (3.15) converges to x for every $y = Kx$ if and only if there exists $c > 0$ such that*

$$\|R_n K\|_{\mathcal{L}(X)} \leq c \quad \text{for all } n \in \mathbb{N}. \tag{3.18}$$

If (3.18) is satisfied the following error estimate holds:

$$\|x_n - x\|_X \leq (1 + c) \min_{z_n \in X_n} \|z_n - x\|_X \tag{3.19}$$

with the same constant c as in (3.18).

Proof: Let the projection method be convergent. Then $R_n K x \to x$ for every $x \in X$. The assertion follows directly from the principle of uniform boundedness (Theorem A.28 of Appendix A.3).

Now let $\|R_n K\|_{\mathcal{L}(X)}$ be bounded. The operator $R_n K$ is a projection operator onto X_n because for $z_n \in X_n$, we have $R_n K z_n = (Q_n K|_{X_n})^{-1} Q_n K z_n = z_n$. Thus we conclude that

$$x_n - x = (R_n K - I)x = (R_n K - I)(x - z_n) \quad \text{for all } z_n \in X_n .$$

This yields

$$\|x_n - x\|_X \leq (c+1)\|x - z_n\|_X \quad \text{for all } z_n \in X_n$$

and proves (3.19). Convergence $x_n \to x$ follows because $\bigcup_{n \in \mathbb{N}} X_n$ is dense in X. $\square$

We show now a perturbation result: It is sufficient to study the question of convergence for the "principal part" of an operator K. In particular, if the projection method converges for an operator K, then convergence and the error estimates hold also for $K + C$, where C is compact relative to K (that is, $K^{-1}C$ is compact).

Theorem 3.8 *Let Assumption 3.6 hold. Let $C : X \to Y$ be a linear operator with $\mathcal{R}(C) \subset \mathcal{R}(K)$ such that $K + C$ is one-to-one and $K^{-1}C$ is compact in X. Assume, furthermore, that the projection method converges for K, that is, that $R_n K x \to x$ for every $x \in X$, where again*

$$R_n = [Q_n K|_{X_n}]^{-1} Q_n .$$

Then it converges also for $K + C$; that is, $\tilde{R}_n(K+C)x \to x$ for every $x \in X$, where now

$$\tilde{R}_n = [Q_n(K+C)|_{X_n}]^{-1} Q_n .$$

Let $x^* \in X$ be the solution of $(K+C)x^* = y^*$ and $\tilde{x}_n = \tilde{R}_n y^* \in X_n$ be the solution of the corresponding projected equation $Q_n(K+C)\tilde{x}_n = Q_n y^*$. Then there exists $c > 0$ with

$$\|\tilde{x}_n - x^*\|_X \leq c\left[\|R_n y^* - K^{-1}y^*\|_X + \|R_n C x^* - K^{-1}C x^*\|_X\right] \quad (3.20)$$

for all sufficiently large n.

Proof: First we note that $I + K^{-1}C = K^{-1}(K+C)$ is one-to-one and thus, because of the compactness of $K^{-1}C$, an isomorphism from X onto itself (see Theorem A.36). We write the equation $Q_n(K+C)\tilde{x}_n = Q_n y^*$ as $(I+R_n C)\tilde{x}_n = R_n y^*$ and have

$$(I+R_n C)(\tilde{x}_n - x^*) = R_n y^* - R_n C x^* - x^* = [R_n y^* - K^{-1}y^*] - [R_n C x^* - K^{-1}C x^*]$$

where we have used that $x^* + K^{-1}C x^* = K^{-1}y^*$. Now we note that the operators $R_n C = [R_n K]K^{-1}C$ converge to $K^{-1}C$ in the operator norm because $R_n K x \to x$ for every $x \in X$ and $K^{-1}C$ is compact in X (see Theorem A.34, part

(d), of Appendix A.3). Therefore, by the general Theorem A.37 of Appendix A.3, we conclude that $(I + R_n C)^{-1}$ exist for sufficiently large n and their operator norms are uniformly bounded by some $c > 0$. This yields the estimate (3.20).
□

The first term on the right-hand side of (3.20) is just the error of the projection method for the equation $Kx = y^*$, the second for the equation $Kx = Cx^*$. By the previous Theorem 3.7, these terms are estimated by

$$\|R_n y^* - K^{-1} y^*\|_X \;\leq\; (1+c) \min_{z_n \in X_n} \|K^{-1} y^* - z_n\|_X \,,$$

$$\|R_n Cx^* - K^{-1} Cx^*\|_X \;\leq\; (1+c) \min_{z_n \in X_n} \|K^{-1} Cx^* - z_n\|_X$$

where c is a bound of $\|R_n K\|_{\mathcal{L}(X)}$.

So far, we have considered the case where the right-hand side $y^* = Kx^*$ is known exactly. Now we study the case where the right-hand side is known only approximately. We understand the operator R_n from (3.16) as a regularization operator in the sense of the previous chapter. We have to distinguish between two kinds of errors on the right-hand side. The first kind corresponds to the kind of perturbation discussed in Chapter 2. Instead of the exact right-hand side y^*, only $y^\delta \in Y$ is given with $\|y^\delta - y^*\|_Y \leq \delta$. We call this the *continuous perturbation* of the right-hand side. A simple application of the triangle inequality yields the following result.

Theorem 3.9 *Let Assumption 3.6 be satisfied and let again $R_n = (Q_n K|_{X_n})^{-1}$ $Q_n : Y \to X_n \subset X$ as in (3.16). Let $x^* \in X$ the solution of the unperturbed equation $Kx^* = y^*$. Furthermore, we assume that the projection method converges; that is by Theorem 3.7, $\|R_n K\|_{\mathcal{L}(X)}$ are uniformly bounded with respect to n. Furthermore, let $y^\delta \in Y$ with $\|y^\delta - y^*\|_Y \leq \delta$ and $x_n^\delta = R_n y^\delta$ the solution of the projected equation $Q_n K x_n^\delta = Q_n y^\delta$. Then*

$$\|x_n^\delta - x^*\|_X \;\leq\; \|x_n^\delta - R_n y^*\|_X \;+\; \|R_n y^* - x^*\|_X$$
$$\leq\; \|R_n\|_{\mathcal{L}(Y,X)} \|y^\delta - y^*\|_Y \;+\; \|R_n Kx^* - x^*\|_X \,. \quad (3.21)$$

This estimate corresponds to the fundamental estimate from (2.4a). The first term reflects the ill-posedness of the equation: The (continuous) error δ of the right-hand side is multiplied by the norm of R_n. The second term describes the discretization error for exact data.

In practice, one solves the discrete systems (3.6) or (3.7) where the coefficients β_i are replaced by perturbed coefficients $\beta_i^\delta \in \mathbb{K}$; that is, one solves

$$\sum_{j=1}^n A_{ij} \alpha_j^\delta \;=\; \beta_i^\delta \,, \quad i = 1, \ldots, n; \quad \text{that is,} \quad A\alpha^\delta = \beta^\delta \,, \qquad (3.22)$$

instead of $A\alpha = \beta$ where now

$$|\beta^\delta - \beta|^2 = \sum_{i=1}^{n} |\beta_i^\delta - \beta_i|^2 \leq \delta^2.$$

Recall, that the elements A_{ij} of the matrix $A \in \mathbb{K}^{n \times n}$ and the exact coefficients β_i of $\beta \in \mathbb{K}^n$ are given by (3.5) or (3.8). We call this the *discrete perturbation* of the right-hand side. Then $x_n^\delta \in X_n$ is defined by

$$x_n^\delta = \sum_{j=1}^{n} \alpha_j^\delta \hat{x}_j.$$

In this case, the choices of basis functions $\hat{x}_j \in X_n$ and $\hat{y}_j \in Y_n$ (and the dual basis functions $\hat{y}_i^*$) are essential. We will also see that the condition number of A reflects the ill-conditioning of the equation $Kx = y$. We do not carry out the analysis for these two forms of discrete perturbations in the general case but do it only for Galerkin methods in the next section.

3.2   Galerkin Methods

In this section, we assume that X and Y are (real or complex) Hilbert spaces; $K : X \to Y$ is linear, bounded, and one-to-one; $X_n \subset X$ and $Y_n \subset Y$ are finite-dimensional subspaces with $\dim X_n = \dim Y_n = n$; and $Q_n : Y \to Y_n$ is the *orthogonal projection operator* onto Y_n. Then equation $Q_n K x_n = Q_n y$ reduces to the Galerkin equations (see Example 3.5)

$$(K x_n, z_n)_Y = (y, z_n)_Y \quad \text{for all } z_n \in Y_n. \tag{3.23}$$

If we choose bases $\{\hat{x}_1, \ldots, \hat{x}_n\}$ and $\{\hat{y}_1, \ldots, \hat{y}_n\}$ of X_n and Y_n, respectively, then this leads to a finite system for the coefficients of $x_n = \sum_{j=1}^{n} \alpha_j \hat{x}_j$ (compare with (3.9b)):

$$\sum_{i=1}^{n} A_{ij} \alpha_j = \beta_i, \quad i = 1, \ldots, n, \tag{3.24}$$

where

$$A_{ij} = (K\hat{x}_j, \hat{y}_i)_Y \quad \text{and} \quad \beta_i = (y, \hat{y}_i)_Y. \tag{3.25}$$

The Galerkin method is also known as the *Petrov–Galerkin method* (see [216]) because Petrov was the first to consider the general situation of (3.23). The special case $X = Y$ and $X_n = Y_n$ was studied by Bubnov in 1913 and later by Galerkin in 1915 (see [99]). For this reason, this special case is also known as the *Bubnov–Galerkin method*. In the case when the operator K is self-adjoint and positive definite, we will see that the Bubnov–Galerkin method coincides with the *Rayleigh–Ritz method*; see [221] and [228].

Theorem 3.10 *Let Assumption 3.6 be satisfied and let again $R_n = (Q_n K|_{X_n})^{-1} Q_n : Y \to X_n \subset X$ as in (3.16). Let $x^* \in X$ the solution of the unperturbed equation $Kx^* = y^*$. Furthermore, we assume that the Galerkin method converges; that is by Theorem 3.7, $\|R_n K\|_{\mathcal{L}(X)}$ are uniformly bounded with respect to n.*

(a) *Let $y^\delta \in Y$ with $\|y^\delta - y^*\|_Y \leq \delta$ and $x_n^\delta = R_n y^\delta$ the solution of the projected equation $Q_n K x_n^\delta = Q_n y^\delta$. Then*

$$\|x_n^\delta - x^*\|_X \leq \|R_n\|_{\mathcal{L}(Y,X)} \delta + \|R_n K x^* - x^*\|_X . \qquad (3.26)$$

(b) *Let $Q_n y^* = \sum_{i=1}^n \beta_i \hat{y}_i$ and $\beta_i^\delta \in \mathbb{K}$ with $|\beta^\delta - \beta| = \sqrt{\sum_{i=1}^n |\beta_i^\delta - \beta_i|^2} \leq \delta$ and let $\alpha^\delta \in \mathbb{K}^n$ be the solution of $A\alpha^\delta = \beta^\delta$. Then, with $x_n^\delta = \sum_{j=1}^n \alpha_j^\delta \hat{x}_j$,*

$$\|x_n^\delta - x^*\|_X \leq \frac{a_n}{\lambda_n} \delta + \|R_n K x^* - x^*\|_X , \qquad (3.27a)$$

$$\|x_n^\delta - x^*\|_X \leq \|R_n\|_{\mathcal{L}(Y,X)} b_n \delta + \|R_n K x^* - x^*\|_X , \qquad (3.27b)$$

where

$$a_n = \max\left\{ \left\| \sum_{j=1}^n \rho_j \hat{x}_j \right\|_X : \sum_{j=1}^n |\rho_j|^2 = 1 \right\}, \qquad (3.28a)$$

$$b_n = \max\left\{ \sqrt{\sum_{i=1}^n |\rho_i|^2} : \left\| \sum_{i=1}^n \rho_i \hat{y}_i \right\|_Y = 1 \right\}, \qquad (3.28b)$$

and $\lambda_n > 0$ denotes the smallest singular value of the matrix A. We note that if X or Y are Hilbert spaces and $\{\hat{x}_j : j = 1, \ldots, n\}$ or $\{\hat{y}_i : i = 1, \ldots, n\}$, respectively, are orthonormal systems then $a_n = 1$ or $b_n = 1$, respectively.

Proof: (a) has been shown before.

(b) Again we use $\|x_n^\delta - x^*\|_X \leq \|x_n^\delta - R_n y^*\|_X + \|R_n y^* - x^*\|_X$ and estimate the first term. We note that $R_n y^* = \sum_{j=1}^n \alpha_j \hat{x}_j$ with $A\alpha = \beta$ and thus

$$\|x_n^\delta - R_n y^*\|_X = \left\| \sum_{j=1}^n (\alpha_j^\delta - \alpha_j) \hat{x}_j \right\|_X \leq a_n \sqrt{\sum_{j=1}^n (\alpha^\delta - \alpha)^2}$$

$$= a_n |A^{-1}(\beta^\delta - \beta)| \leq a_n |A^{-1}|_2 |\beta^\delta - \beta| \leq \frac{a_n}{\lambda_n} \delta ,$$

where $|A^{-1}|_2$ denotes the spectral norm of A^{-1}, that is, the inverse of the smallest singular value of A. This yields (3.27a).

To prove (3.27b), we choose $y_n^\delta \in Y_n$ such that $(y_n^\delta, \hat{y}_i)_Y = \beta_i^\delta$ for $i = 1, \ldots, n$. Then $R_n y_n^\delta = x_n^\delta$ and thus

$$
\begin{aligned}
\|x_n^\delta - R_n y^*\|_X &\leq \|R_n\|_{\mathcal{L}(Y,X)} \|y_n^\delta - Q_n y^*\|_Y \\
&= \|R_n\|_{\mathcal{L}(Y,X)} \sup_{z_n \in Y_n} \frac{|(y_n^\delta - Q_n y^*, z_n)_Y|}{\|z_n\|_Y} \\
&= \|R_n\|_{\mathcal{L}(Y,X)} \sup_{\rho_j} \frac{\left|\sum_{j=1}^n \overline{\rho_j}(y_n^\delta - Q_n y^*, \hat{y}_j)_Y\right|}{\|\sum_{j=1}^n \rho_j \hat{y}_j\|_Y} \\
&= \|R_n\|_{\mathcal{L}(Y,X)} \sup_{\rho_j} \frac{\left|\sum_{j=1}^n \overline{\rho_j}(\beta_j^\delta - \beta_j)\right|}{\|\sum_{j=1}^n \rho_j \hat{y}_j\|_Y} \\
&\leq \|R_n\|_{\mathcal{L}(Y,X)} |\beta^\delta - \beta| \sup_{\rho_j} \frac{\sqrt{\sum_{j=1}^n |\rho_j|^2}}{\|\sum_{j=1}^n \rho_j \hat{y}_j\|_Y} \\
&\leq \|R_n\|_{\mathcal{L}(Y,X)} \, b_n \, \delta \,. \qquad\qquad \square
\end{aligned}
$$

This ends the proof. $\square$

In the following three subsections, we derive error estimates for three special choices for the finite-dimensional subspaces X_n and Y_n. The cases where X_n and Y_n are coupled by $Y_n = K(X_n)$ or $X_n = K^*(Y_n)$ lead to the *least squares method* or the *dual least squares method*, respectively. Here, $K^* : Y \to X$ denotes the adjoint of K. In Subsection 3.2.3, we study the Bubnov–Galerkin method for the case where K satisfies Gårding's inequality. In all of the subsections, we formulate the Galerkin equations for the perturbed cases first without using particular bases and then with respect to given bases in X_n and Y_n.

3.2.1   The Least Squares Method

An obvious method to solve an equation of the kind $Kx = y$ is the following: Given a finite-dimensional subspace $X_n \subset X$, determine $x_n \in X_n$ such that

$$\|Kx_n - y\|_Y \leq \|Kz_n - y\|_Y \quad \text{for all } z_n \in X_n \,. \tag{3.29}$$

Existence and uniqueness of $x_n \in X_n$ follow easily because X_n is finite-dimensional and K is one-to-one. The solution $x_n \in X_n$ of this least squares problem is characterized by

$$(Kx_n, Kz_n)_Y = (y, Kz_n)_Y \quad \text{for all } z_n \in X_n \,. \tag{3.30a}$$

We observe that this method is a special case of the Galerkin method when we set $Y_n := K(X_n)$.

Choosing a basis $\{\hat{x}_j : j = 1, \ldots, n\}$ of X_n leads to the finite system

$$\sum_{j=1}^n \alpha_j (K\hat{x}_j, K\hat{x}_i)_Y = \beta_i = (y, K\hat{x}_i)_Y \quad \text{for } i = 1, \ldots, n \,, \tag{3.30b}$$

or $A\alpha = \beta$. This has the form of (3.25) for $\hat{y}_i = K\hat{x}_i$. The corresponding matrix $A \in \mathbb{K}^{n \times n}$ with $A_{ij} = (K\hat{x}_j, K\hat{x}_i)_Y$ is symmetric (if $\mathbb{K} = \mathbb{R}$) or Hermitian (if $\mathbb{K} = \mathbb{C}$) and positive definite because K is also one-to-one.

Again, we study the case where the exact right-hand side y^* is perturbed by an error. For continuous perturbations, let $x_n^\delta \in X_n$ be the solution of

$$(Kx_n^\delta, Kz_n)_Y = (y^\delta, Kz_n)_Y \quad \text{for all } z_n \in X_n, \tag{3.31a}$$

where $y^\delta \in Y$ is the perturbed right-hand side with $\|y^\delta - y^*\|_Y \leq \delta$.

For the discrete perturbation, we assume that $\beta_i = (y^*, K\hat{x}_i)_Y$, $i = 1, \ldots, n$, is replaced by a vector $\beta^\delta \in \mathbb{K}^n$ with $|\beta^\delta - \beta| \leq \delta$, where $|\cdot|$ denotes the Euclidean norm in $\mathbb{K}^n$. This leads to the following finite system of equations for the coefficients of $x_n^\delta = \sum_{j=1}^n \alpha_j^\delta \hat{x}_j$:

$$\sum_{j=1}^n \alpha_j^\delta (K\hat{x}_j, K\hat{x}_i)_Y = \beta_i^\delta \quad \text{for } i = 1, \ldots, n. \tag{3.31b}$$

This system is uniquely solvable because the matrix A is positive definite. For least squares methods, the boundedness condition (3.18) is not satisfied without additional assumptions. We refer to [246] or [168], Problem 17.2, for an example. However, we can prove the following theorem.

Theorem 3.11 *Let $K : X \to Y$ be a linear, bounded, and injective operator between Hilbert spaces and $X_n \subset X$ be finite-dimensional subspaces such that $\bigcup_{n \in \mathbb{N}} X_n$ is dense in X. Let $x^* \in X$ be the solution of $Kx^* = y^*$ and $x_n^\delta \in X_n$ be the least squares solution from (3.31a) or (3.31b). Define*

$$\sigma_n := \max\{\|z_n\|_X : z_n \in X_n, \|Kz_n\|_Y = 1\} \tag{3.32}$$

and let there exist $c > 0$, independent of n, such that

$$\min_{z_n \in X_n} \{\|x - z_n\|_X + \sigma_n\|K(x - z_n)\|_Y\} \leq c\|x\|_X \quad \text{for all } x \in X. \tag{3.33}$$

Then the least squares method is convergent and $\|R_n\|_{\mathcal{L}(Y,X)} \leq \sigma_n$. In this case, we have the error estimate

$$\|x^* - x_n^\delta\|_X \leq b_n \sigma_n \delta + \tilde{c} \min\{\|x^* - z_n\|_X : z_n \in X_n\} \tag{3.34}$$

for some $\tilde{c} > 0$. Here, $b_n = 1$ if $x_n^\delta \in X_n$ solves (3.31a); that is, δ measures the continuous perturbation $\|y^\delta - y^*\|_Y$. If δ measures the discrete error $|\beta^\delta - \beta|$ in the Euclidean norm and $x_n^\delta = \sum_{j=1}^n \alpha_j^\delta \hat{x}_j \in X_n$, where α^δ solves (3.31b), then b_n is given by

$$b_n = \max\left\{\sqrt{\sum_{j=1}^n |\rho_j|^2} : \left\|K\left(\sum_{j=1}^n \rho_j \hat{x}_j\right)\right\|_Y = 1\right\}. \tag{3.35}$$

If $\{\hat{x}_j : j = 1, \ldots, n\}$ is an orthonormal basis of X_n then $b_n = \sigma_n$.

Proof: It is the aim to apply Theorem 3.10. First we prove that $\|R_n K\|_{\mathcal{L}(X)}$ are bounded uniformly in n. Let $x \in X$ and $x_n := R_n K x$. Then x_n satisfies $(Kx_n, Kz_n)_Y = (Kx, Kz_n)_Y$ for all $z_n \in X_n$. This yields

$$\begin{aligned}
\|K(x_n - z_n)\|_Y^2 &= \left(K(x_n - z_n), K(x_n - z_n)\right)_Y \\
&= \left(K(x - z_n), K(x_n - z_n)\right)_Y \\
&\leq \|K(x - z_n)\|_Y \, \|K(x_n - z_n)\|_Y
\end{aligned}$$

and thus $\|K(x_n - z_n)\|_Y \leq \|K(x - z_n)\|_Y$ for all $z_n \in X_n$. Using this and the definition of σ_n, we conclude that

$$\|x_n - z_n\|_X \leq \sigma_n \|K(x_n - z_n)\|_Y \leq \sigma_n \|K(x - z_n)\|_Y,$$

and thus

$$\begin{aligned}
\|x_n\|_X &\leq \|x_n - z_n\|_X + \|z_n - x\|_X + \|x\|_X \\
&\leq \|x\|_X + \left[\|z_n - x\|_X + \sigma_n \|K(x - z_n)\|_Y\right].
\end{aligned}$$

This holds for all $z_n \in X_n$. Taking the minimum, we have by Assumption (3.33) that $\|x_n\|_X \leq (1 + c)\|x\|_X$. Thus the boundedness condition (3.18) is satisfied. The application of Theorem 3.7 proves convergence.

Analogously we prove the estimate for $\|R_n\|_{\mathcal{L}(Y,X)}$. Let $y \in Y$ and set $x_n := R_n y$. Then from (3.30a), we have

$$\|Kx_n\|_Y^2 = (Kx_n, Kx_n)_Y = (y, Kx_n)_Y \leq \|y\|_Y \|Kx_n\|_Y$$

and thus

$$\|x_n\|_X \leq \sigma_n \|Kx_n\|_Y \leq \sigma_n \|y\|_Y.$$

This proves the estimate $\|R_n\|_{\mathcal{L}(Y,X)} \leq \sigma_n$.

The error estimates (3.34) follow directly from Theorem 3.10 and the estimates (3.26) and (3.27b) for $\hat{y}_j = K\hat{x}_j$. $\qquad\square$

For further numerical aspects of least squares methods, we refer to [79, 82, 107, 150, 188, 189, 201, 202].

3.2.2 The Dual Least Squares Method

As the next example for the Galerkin method, we study the dual least squares method. We will see that the boundedness condition (3.18) is always satisfied. We assume in addition to the general assumptions of this section that the range $\mathcal{R}(K)$ of K is dense in Y.

Given any finite-dimensional subspace $Y_n \subset Y$, determine $u_n \in Y_n$ such that

$$(KK^*u_n, z_n)_Y = (y, z_n)_Y \quad \text{for all } z_n \in Y_n, \tag{3.36}$$

where $K^* : Y \to X$ denotes the adjoint of K. Then $x_n := K^*u_n$ is called the *dual least squares solution.* It is a special case of the Galerkin method when we set $X_n := K^*(Y_n)$. Writing equation (3.36) for $y = Kx$ in the form

$$(K^*u_n, K^*z_n)_X = (x, K^*z_n)_X \quad \text{for all } z_n \in Y_n,$$

we observe that the dual least squares method is just the least squares method for the equation $K^*u = x$. This explains the name.

We assume again that the exact right-hand side is perturbed. Let $y^\delta \in Y$ with $\|y^\delta - y^*\|_Y \leq \delta$. Instead of equation (3.36), one determines $x_n^\delta = K^*u_n^\delta \in X_n$ with

$$(K^*u_n^\delta, K^*z_n)_X = (y^\delta, z_n)_Y \quad \text{for all } z_n \in Y_n . \tag{3.37}$$

For discrete perturbations, we choose a basis $\{\hat{y}_j : j = 1, \ldots, n\}$ of Y_n and assume that the right-hand sides $\beta_i = (y^*, \hat{y}_i)_Y$, $i = 1, \ldots, n$, of the Galerkin equations are perturbed by a vector $\beta^\delta \in \mathbb{K}^n$ with $|\beta^\delta - \beta| \leq \delta$ where $|\cdot|$ denotes the Euclidean norm in $\mathbb{K}^n$. Instead of (3.36), we determine

$$x_n^\delta = K^*u_n^\delta = \sum_{j=1}^n \alpha_j^\delta K^*\hat{y}_j ,$$

where $\alpha^\delta \in \mathbb{K}^n$ solves

$$\sum_{j=1}^n \alpha_j^\delta (K^*\hat{y}_j, K^*\hat{y}_i)_X = \beta_i^\delta \quad \text{for } i = 1, \ldots, n . \tag{3.38}$$

First we show that equations (3.37) and (3.38) are uniquely solvable. $K^* : Y \to X$ is one-to-one because the range $\mathcal{R}(K)$ is dense in Y. Thus the dimensions of Y_n and X_n coincide and K^* is an isomorphism from Y_n onto X_n. It is sufficient to prove the uniqueness of a solution to (3.37). Let $u_n \in Y_n$ with $(K^*u_n, K^*z_n)_X = 0$ for all $z_n \in Y_n$. For $z_n = u_n$ we conclude that $0 = (K^*u_n, K^*u_n)_X = \|K^*u_n\|_X^2$, that is, $K^*u_n = 0$ or $u_n = 0$.

Convergence and error estimates are proven in the following theorem.

Theorem 3.12 *Let X and Y be Hilbert spaces and $K : X \to Y$ be linear, bounded, and one-to-one such that the range $\mathcal{R}(K)$ is dense in Y. Let $Y_n \subset Y$ be finite-dimensional subspaces such that $\bigcup_{n\in\mathbb{N}} Y_n$ is dense in Y. Let $x^* \in X$ be the solution of $Kx^* = y^*$. Then the Galerkin equations (3.37) and (3.38) are uniquely solvable for every right-hand side and every $n \in \mathbb{N}$. The dual least squares method is convergent and*

$$\|R_n\|_{\mathcal{L}(Y,X)} \leq \sigma_n := \max\{\|z_n\|_Y : z_n \in Y_n, \|K^*z_n\|_X = 1\} . \tag{3.39}$$

Furthermore, we have the error estimates

$$\|x^* - x_n^\delta\|_X \leq b_n\sigma_n\delta + c \min\{\|x^* - z_n\|_X : z_n \in K^*(Y_n)\} \tag{3.40}$$

for some $c > 0$. Here, $b_n = 1$ if $x_n^\delta \in X_n$ solves (3.37); that is, δ measures the norm $\|y^\delta - y^*\|_Y$ in Y. If δ measures the discrete error $|\beta^\delta - \beta|$ and $x_n^\delta = \sum_{j=1}^n \alpha_j^\delta K^*\hat{y}_j \in X_n$, where α^δ solves (3.38), then b_n is given by

$$b_n = \max\left\{\sqrt{\sum_{j=1}^n |\rho_j|^2} : \left\|\sum_{j=1}^n \rho_j\hat{y}_j\right\|_Y = 1\right\} . \tag{3.41}$$

We note that $b_n = 1$ if $\{\hat{y}_j : j = 1, \ldots, n\}$ forms an orthonormal system in Y.

Proof: We have seen already that (3.37) and (3.38) are uniquely solvable for every right-hand side and every $n \in \mathbb{N}$.

Now we prove the estimate $\|R_n K\|_{\mathcal{L}(X)} \leq 1$, that is condition (3.18) with $c = 1$. Let $x \in X$ and set $x_n := R_n K x \in X_n$. Then $x_n = K^* u_n$, and $u_n \in Y_n$ satisfies

$$(K^* u_n, K^* z_n)_X = (Kx, z_n)_Y \quad \text{for all } z_n \in Y_n.$$

For $z_n = u_n$ this implies

$$\|x_n\|_X^2 = \|K^* u_n\|_X^2 = (Kx, u_n)_Y = (x, K^* u_n)_X \leq \|x\|_X \|x_n\|_X,$$

which proves the desired estimate. If we replace Kx by y in the preceding arguments, we have

$$\|x_n\|_X^2 \leq \|y\|_Y \|u_n\|_Y \leq \sigma_n \|y\|_Y \|K^* u_n\|_X = \sigma_n \|y\|_Y \|x_n\|_X,$$

which proves (3.39).

Finally, we show that $\bigcup_{n \in \mathbb{N}} X_n$ is dense in X. Let $x \in X$ and $\varepsilon > 0$. Because $K^*(Y)$ is dense in X, there exists $y \in Y$ with $\|x - K^* y\|_X < \varepsilon/2$. Because $\bigcup_{n \in \mathbb{N}} Y_n$ is dense in Y, there exists $y_n \in Y_n$ with $\|y - y_n\|_Y < \varepsilon/(2\|K\|_{\mathcal{L}(X,Y)})$. The triangle inequality yields that for $x_n := K^* y_n \in X_n$,

$$\|x - x_n\|_X \leq \|x - K^* y\|_X + \|K^*(y - y_n)\|_X \leq \varepsilon.$$

The application of Theorem 3.10 and the estimates (3.26) and (3.27b) proves (3.40).   □

3.2.3   The Bubnov–Galerkin Method for Coercive Operators

In this subsection, we assume that $Y = X$ coincides, and $K : X \to X$ is a linear and bounded operator and X_n, $n \in \mathbb{N}$, are finite-dimensional subspaces. The Galerkin method reduces to the problem of determining $x_n \in X_n$ such that

$$(Kx_n, z_n)_X = (y, z_n)_X \quad \text{for all } z_n \in X_n. \tag{3.42}$$

This special case is called the *Bubnov–Galerkin method*. Again, we consider two kinds of perturbations of the right-hand side. If $y^\delta \in X$ with $\|y^\delta - y^*\|_X \leq \delta$ is a perturbed right-hand side, then instead of (3.42) we study the equation

$$(Kx_n^\delta, z_n)_X = (y^\delta, z_n)_X \quad \text{for all } z_n \in X_n. \tag{3.43}$$

The other possibility is to choose a basis $\{\hat{x}_j : j = 1, \ldots, n\}$ of X_n and assume that the right-hand sides $\beta_i = (y^*, \hat{x}_i)_X$, $i = 1, \ldots, n$, of the Galerkin equations are perturbed by a vector $\beta^\delta \in \mathbb{K}^n$ with $|\beta^\delta - \beta| \leq \delta$, where $|\cdot|$ denotes again the Euclidean norm in $\mathbb{K}^n$. In this case, instead of (3.42), we have to solve

$$\sum_{j=1}^n \alpha_j^\delta (K\hat{x}_j, \hat{x}_i)_X = \beta_i^\delta \quad \text{for } i = 1, \ldots, n, \tag{3.44}$$

for $\alpha^\delta \in \mathbb{K}^n$ and set $x_n^\delta = \sum_{j=1}^n \alpha_j^\delta \hat{x}_j$.

Before we prove a convergence result for this method, we briefly describe the Rayleigh–Ritz method and show that it is a special case of the Bubnov–Galerkin method.

Let $K : X \to X$ also be self-adjoint and positive definite, that is, $(Kx, y)_X = (x, Ky)_X$ and $(Kx, x)_X > 0$ for all $x, y \in X$ with $x \neq 0$. We define the functional

$$\psi(z) := (Kz, z)_X - 2 \operatorname{Re}(y, z)_X \quad \text{for } z \in X. \tag{3.45}$$

From the equation

$$\psi(z) - \psi(x) = 2 \operatorname{Re}(Kx - y, z - x)_X + (K(z - x), z - x)_X \tag{3.46}$$

and the positivity of K, we easily conclude (see Problem 3.2) that $x \in X$ is the unique minimum of ψ if and only if x solves $Kx = y$. The Rayleigh–Ritz method is to minimize ψ over the finite-dimensional subspace X_n. From (3.46), we see that if $x_n \in X_n$ minimizes ψ on X_n, then, for $z_n = x_n \pm \varepsilon u_n$ with $u_n \in X_n$ and $\varepsilon > 0$, we have that

$$0 \leq \psi(z_n) - \psi(x_n) = \pm \varepsilon \, 2 \operatorname{Re}(Kx_n - y, u_n)_X + \varepsilon^2 (Ku_n, u_n)_X$$

for all $u_n \in X_n$. Dividing by $\varepsilon > 0$ and letting $\varepsilon \to 0$ yields that $x_n \in X_n$ satisfies the Galerkin equation (3.42). If, on the other hand, $x_n \in X_n$ solves (3.42), then from (3.46)

$$\psi(z_n) - \psi(x_n) = (K(z_n - x_n), z_n - x_n)_X \geq 0$$

for all $z_n \in X_n$. Therefore, the Rayleigh–Ritz method is identical to the Bubnov–Galerkin method.

Now we generalize the Rayleigh–Ritz method and study the Bubnov–Galerkin method for the important class of coercive operators in Gelfand triples $V \subset X \subset V'$. For the definition of a Gelfand triple and coercive operators $K : V' \to V$, we refer to Definition A.26 of Appendix A.3. We just recall that $V' = \{\ell : V \to \mathbb{K} : \bar{\ell} \in V^*\}$ is the space of anti-linear functionals on V with corresponding bounded sesquilinear form $\langle \cdot, \cdot \rangle : V' \times V \to \mathbb{K}$ which extends the inner product in X, that is, $\langle x, v \rangle = (x, v)_X$ for all $v \in V$ and $x \in X$. Coercivity of an operator K from V' into V means that there exists a constant $\gamma > 0$ with

$$|\langle x, Kx \rangle| \geq \gamma \|x\|_{V'}^2 \quad \text{for all } x \in V'.$$

This condition implies that K is an isomorphism from V' onto V; see the remark following Definition A.26. If V is compactly imbedded in X, that is, $j : V \hookrightarrow X$ is compact, then $K|_X$ is a compact operator from X into itself. Therefore, in this subsection we measure the compactness by the "smoothing" of K from V' onto V rather than by a decay of the singular values.

Now we can prove the main theorem about convergence of the Bubnov–Galerkin method for coercive operators in Gelfand triples.

Theorem 3.13 *Let $V \subset X \subset V'$ be a Gelfand triple, and $X_n \subset V$ be finite-dimensional subspaces such that $\bigcup_{n\in\mathbb{N}} X_n$ is dense in X. Let $K : V' \to V$ be coercive in the sense of Definition A.26 with constant $\gamma > 0$. Let $x^* \in X$ be the solution of $Kx^* = y^*$ for some $y^* \in V$. Then we have the following:*

(a) There exist unique solutions of the Galerkin equations (3.42)–(3.44). The Bubnov–Galerkin solutions $x_n \in X_n$ of (3.42) converge in V' with

$$\|x^* - x_n\|_{V'} \leq c \min\{\|x^* - z_n\|_{V'} : z_n \in X_n\} \tag{3.47}$$

for some $c > 0$.

(b) Define $\rho_n > 0$ by

$$\rho_n = \max\{\|z_n\|_X : z_n \in X_n, \ \|z_n\|_{V'} = 1\} \tag{3.48}$$

and the orthogonal projection operator P_n from X onto X_n. The Bubnov–Galerkin solutions converge in X (rather than in the weaker norm of V') if there exists $c > 0$ with

$$\|x - P_n x\|_{V'} \leq \frac{c}{\rho_n} \|x\|_X \quad \text{for all } x \in X. \tag{3.49}$$

In this case, we have the estimates

$$\|R_n\|_{\mathcal{L}(X)} \leq \frac{\rho_n^2}{\gamma} \tag{3.50}$$

and

$$\|x^* - x_n^\delta\|_X \leq c \left[b_n \frac{\rho_n^2}{\gamma} \delta + \min\{\|x^* - z_n\|_X : z_n \in X_n\} \right] \tag{3.51}$$

for some $c > 0$. Here, $b_n = 1$ if $x_n^\delta \in X_n$ solves (3.43); that is, δ measures the norm $\|y^\delta - y^*\|_X$ in X. If δ measures the discrete error $|\beta^\delta - \beta|$ in the Euclidean norm and $x_n^\delta = \sum_{j=1}^n \alpha_j^\delta \hat{x}_j \in X_n$, where α^δ solves (3.44), then b_n is given by

$$b_n = \max\left\{ \sqrt{\sum_{j=1}^n |a_j|^2} : \left\| \sum_{j=1}^n a_j \hat{x}_j \right\|_X = 1 : \right\}. \tag{3.52}$$

Again, we note that $b_n = 1$ if $\{\hat{x}_j : j = 1, \ldots, n\}$ forms an orthonormal system in X.

Proof: (a) We apply Theorem 3.7 to the equations $Kx = y$, $x \in V'$, and $P_n K x_n = P_n y$, $x_n \in X_n$, where we consider K as an operator from V' into V. We observe that the projection operator P_n is also bounded from V into X_n where we consider X_n as a subspace of V. This follows from the observation that on the finite-dimensional space X_n the norms $\|\cdot\|_X$ and $\|\cdot\|_V$ are equivalent and thus

$$\|P_n u\|_V \leq c \|P_n u\|_X \leq c \|u\|_X \leq \tilde{c} \|u\|_V \quad \text{for } u \in V.$$

The constants c, and thus $\tilde{c}$, depend on n. Because V is dense in X and X is dense in V', we conclude that also $\bigcup_{n \in \mathbb{N}} X_n$ is dense in V'. To apply Theorem 3.7, we have to show that (3.42) is uniquely solvable in X_n and that $R_n K : V' \to X_n \subset V'$ is uniformly bounded with respect to n.

Because (3.42) is a finite-dimensional quadratic system, it is sufficient to prove uniqueness. Let $x_n \in X_n$ satisfy (3.42) for $y = 0$. Because K is coercive, we have

$$\gamma \|x_n\|_{V'}^2 \leq |\langle x_n, K x_n \rangle| = |(x_n, K x_n)_X| = 0;$$

thus $x_n = 0$.

Now let $x \in V'$ and set $x_n = R_n K x$. Then $x_n \in X_n$ satisfies

$$(K x_n, z_n)_X = (K x, z_n)_X \quad \text{for all } z_n \in X_n. \tag{3.53}$$

Again, we conclude that

$$
\begin{aligned}
\gamma \|x_n\|_{V'}^2 &\leq |\langle x_n, K x_n \rangle| = |(x_n, K x_n)_X| \\
&= |(x_n, K x)_X| = |\langle x_n, K x \rangle| \leq \|K x\|_V \|x_n\|_{V'}
\end{aligned}
$$

and thus

$$\|x_n\|_{V'} \leq \frac{1}{\gamma} \|K x\|_V \leq \frac{1}{\gamma} \|K\|_{\mathcal{L}(V',V)} \|x\|_{V'}.$$

Because this holds for all $x \in V'$, we conclude that

$$\|R_n K\|_{\mathcal{L}(V')} \leq \frac{1}{\gamma} \|K\|_{\mathcal{L}(V',V)}.$$

Then the assumptions of Theorem 3.7 are satisfied for $K : V' \to V$.

(b) In this part we wish to apply Theorem 3.10. Let $x \in X$ and $x_n = R_n K x$. Using the estimates (3.47) and (3.49), we conclude that

$$
\begin{aligned}
\|x - x_n\|_X &\leq \|x - P_n x\|_X + \|P_n x - x_n\|_X \\
&\leq \|x - P_n x\|_X + \rho_n \|P_n x - x_n\|_{V'} \\
&\leq \|x - P_n x\|_X + \rho_n \|P_n x - x\|_{V'} + \rho_n \|x - x_n\|_{V'} \\
&\leq \|x - P_n x\|_X + \rho_n \|P_n x - x\|_{V'} + c \rho_n \min_{z_n \in X_n} \|x - z_n\|_{V'} \\
&\leq \|x - P_n x\|_X + (c+1) \rho_n \|P_n x - x\|_{V'} \\
&\leq 2 \|x\|_X + c_1 \|x\|_X = (2 + c_1) \|x\|_X,
\end{aligned}
$$

and thus $\|x_n\|_X \leq \|x_n - x\|_X + \|x\|_X \leq (3 + c_1) \|x\|_X$. Therefore, $\|R_n K\|_{\mathcal{L}(X)}$ are uniformly bounded.

Next, we prove the estimate of R_n in $\mathcal{L}(X)$. Let $y \in X$ and $x_n = R_n y$. We estimate

$$
\begin{aligned}
\gamma \|x_n\|_{V'}^2 &\leq |\langle x_n, K x_n \rangle| = |(x_n, K x_n)_X| = |(x_n, y)_X| \\
&\leq \|y\|_X \|x_n\|_X \leq \rho_n \|y\|_X \|x_n\|_{V'}
\end{aligned}
$$

and thus

$$\|x_n\|_X \le \rho_n \|x_n\|_{V'} \le \frac{1}{\gamma} \rho_n^2 \|y\|_X,$$

which proves the estimate (3.50). The application of Theorem 3.10 yields the estimate (3.51). □

From our general perturbation theorem (Theorem 3.8), we observe that the assumption of K being coercive can be weakened. It is sufficient to assume that K is one-to-one and satisfies Gårding's inequality. We formulate the result in the next theorem.

Theorem 3.14 *The assertions of Theorem 3.13 also hold if $K : V' \to V$ is one-to-one and satisfies Gårding's inequality with some compact operator $C : V' \to V$.*

For further reading, we refer to [204] and the monographs [17, 168, 182].

3.3 Application to Symm's Integral Equation of the First Kind

In this section, we apply the Galerkin methods to an integral equation of the first kind that occurs in potential theory. We study the Dirichlet problem for harmonic functions, that is, solutions of the *Laplace equation* satisfying a boundary condition; that is,

$$\Delta u = 0 \quad \text{in } \Omega, \qquad u = f \quad \text{on } \partial\Omega, \tag{3.54}$$

where $\Omega \subset \mathbb{R}^2$ is some bounded, simply connected region with analytic boundary $\partial\Omega$ and $f \in C(\partial\Omega)$ is some given function. The single layer potential

$$u(x) = -\frac{1}{\pi} \int_{\partial\Omega} \phi(y) \ln|x - y| \, ds(y), \quad x \in \Omega, \tag{3.55}$$

solves the boundary value problem (3.54) if and only if the density $\phi \in C(\partial\Omega)$ solves *Symm's equation*

$$-\frac{1}{\pi} \int_{\partial\Omega} \phi(y) \ln|x - y| \, ds(y) = f(x) \quad \text{for } x \in \partial\Omega; \tag{3.56}$$

see [53]. It is well-known (see [141]) that in general the corresponding integral operator is not one-to-one. One has to make assumptions on the transfinite diameter of Ω; see [274]. We give a more elementary assumption in the following theorem.

Theorem 3.15 *Suppose there exists $z_0 \in \Omega$ with $|x - z_0| \ne 1$ for all $x \in \partial\Omega$. Then the only solution $\phi \in C(\partial\Omega)$ of Symm's equation (3.56) for $f = 0$ is $\phi = 0$; that is, the integral operator is one-to-one.*

Proof: We give a more elementary proof than in [142], but we still need a few results from potential theory.

From the continuity of $x \mapsto |x - z_0|$, we conclude that either $|x - z_0| < 1$ for all $x \in \partial\Omega$ or $|x - z_0| > 1$ for all $x \in \partial\Omega$. Assume first that $|x - z_0| < 1$ for all $x \in \partial\Omega$ and choose a small disk $A \subset \Omega$ with center z_0 such that $|x - z| < 1$ for all $x \in \partial\Omega$ and $z \in A$. Let $\phi \in C(\partial\Omega)$ satisfy (3.56) for $f = 0$ and define u by

$$u(x) \;=\; -\frac{1}{\pi} \int\limits_{\partial\Omega} \phi(y) \ln|x - y| \, ds(y) \quad \text{for } x \in \mathbb{R}^2 \,.$$

From potential theory (see [53]), we conclude that u is continuous in $\mathbb{R}^2$, harmonic in $\mathbb{R}^2 \setminus \partial\Omega$, and vanishes on $\partial\Omega$. The maximum principle for harmonic functions implies that u vanishes in Ω. We show that u also vanishes in the exterior Ω^e of Ω. The main part is to prove that

$$\hat{\phi} \;:=\; \int\limits_{\partial\Omega} \phi(y) \, ds(y) \;=\; 0 \,.$$

Without loss of generality, we can assume that $\hat{\phi} \geq 0$. We study the harmonic function v defined by

$$v(x) \;:=\; u(x) + \frac{\hat{\phi}}{\pi} \ln|x - z| \;=\; \frac{1}{\pi} \int\limits_{\partial\Omega} \phi(y) \ln\frac{|x - z|}{|x - y|} \, ds(y), \quad x \in \Omega^e \,,$$

for some $z \in A$. From the choice of A, we have

$$v(x) \;=\; \frac{\hat{\phi}}{\pi} \ln|x - z| \;\leq\; 0 \quad \text{for } x \in \partial\Omega \,.$$

We study the asymptotic behavior of $v(x)$ as $|x|$ tends to infinity. Elementary calculations show that

$$\frac{|x - z|}{|x - y|} \;=\; 1 \;+\; \frac{1}{|x|} \hat{x} \cdot (y - z) \;+\; \mathcal{O}\big(1/|x|^2\big)$$

for $|x| \to \infty$ uniformly in $y \in \partial\Omega$, $z \in A$, and $\hat{x} := x/|x|$. In particular, $v(x)$ tends to zero as $|x|$ tends to infinity. The maximum principle applied to v in Ω^e yields that $v(x) \leq 0$ for all $x \in \Omega^e$. From the asymptotic formula and $\ln(1 + \varepsilon) = \varepsilon + \mathcal{O}(\varepsilon^2)$, we conclude furthermore that

$$v(x) \;=\; \frac{1}{\pi|x|} \hat{x} \cdot \int\limits_{\partial\Omega} \phi(y) \, (y - z) \, ds(y) \;+\; \mathcal{O}\big(1/|x|^2\big)$$

and thus

$$\hat{x} \cdot \int\limits_{\partial\Omega} \phi(y) \, (y - z) \, ds(y) \;\leq\; 0 \quad \text{for all } |\hat{x}| = 1 \,.$$

This implies that

$$\int_{\partial\Omega} \phi(y)\, y\, ds(y) = z \int_{\partial\Omega} \phi(y)\, ds(y)\,.$$

Because this holds for all $z \in A$, we conclude that $\int_{\partial\Omega} \phi(y)\, ds(y) = 0$.
Now we see from the definition of v (for any fixed $z \in A$) that

$$u(x) = v(x) \rightarrow 0 \quad \text{as } |x| \rightarrow \infty\,.$$

The maximum principle again yields $u = 0$ in Ω^e.
Finally, the jump conditions of the normal derivative of the single layer potential operator (see [53]) yield

$$2\,\phi(x) = \lim_{\varepsilon \to 0+} \left[\nabla u\big(x - \varepsilon\nu(x)\big) - \nabla u\big(x + \varepsilon\nu(x)\big)\right] \cdot \nu(x) = 0$$

for $x \in \partial\Omega$, where $\nu(x)$ denotes the unit normal vector at $x \in \partial\Omega$ directed into the exterior of Ω.
This ends the proof for the case that $\max_{x \in \partial\Omega} |x - z_0| < 1$. The case $\min_{x \in \partial\Omega} |x - z_0| > 1$ is settled by the same arguments. □

Now we assume that the boundary $\partial\Omega$ has a parametrization of the form

$$x = \gamma(s), \quad s \in [0, 2\pi]\,,$$

for some 2π-periodic analytic function $\gamma : [0, 2\pi] \rightarrow \mathbb{R}^2$ that satisfies $|\dot{\gamma}(s)| > 0$ for all $s \in [0, 2\pi]$. Then Symm's equation (3.56) takes the form

$$-\frac{1}{\pi} \int_0^{2\pi} \psi(s) \ln|\gamma(t) - \gamma(s)|\, ds = g(t) := f(\gamma(t)) \quad \text{for } t \in [0, 2\pi] \quad (3.57)$$

for the transformed density $\psi(s) := \phi(\gamma(s))|\dot{\gamma}(s)|$, $s \in [0, 2\pi]$.
For the special case where Ω is the disk with center 0 and radius $a > 0$, we have $\gamma_a(s) = a\,(\cos s, \sin s)$ and thus

$$\ln|\gamma_a(t) - \gamma_a(s)| = \ln a + \frac{1}{2} \ln\left(4 \sin^2 \frac{t - s}{2}\right). \quad (3.58)$$

For general boundaries, we can split the kernel in the form

$$-\frac{1}{\pi} \ln|\gamma(t) - \gamma(s)| = -\frac{1}{2\pi} \ln\left(4 \sin^2 \frac{t - s}{2}\right) + k(t, s), \quad t \neq s, \quad (3.59)$$

for some function k that is analytic for $t \neq s$. From the mean value theorem, we conclude that

$$\lim_{s \to t} k(t, s) = -\frac{1}{\pi} \ln|\dot{\gamma}(t)|\,.$$

This implies that k has an analytic continuation onto $[0, 2\pi] \times [0, 2\pi]$. With this, splitting the integral equation (3.57) takes the form

$$-\frac{1}{2\pi} \int_0^{2\pi} \psi(s) \ln\left(4 \sin^2 \frac{t-s}{2}\right) ds + \int_0^{2\pi} \psi(s) k(t, s) ds = g(t) \qquad (3.60)$$

for $t \in [0, 2\pi]$. We want to apply the results of the previous section on Galerkin methods to this integral equation.

As the Hilbert space X, we choose $X = L^2(0, 2\pi)$. The operators K, K_0, and C are defined by

$$(K\psi)(t) = -\frac{1}{\pi} \int_0^{2\pi} \psi(s) \ln|\gamma(t) - \gamma(s)| ds, \qquad (3.61a)$$

$$(K_0\psi)(t) = -\frac{1}{2\pi} \int_0^{2\pi} \psi(s) \left[\ln\left(4 \sin^2 \frac{t-s}{2}\right) - 1\right] ds, \qquad (3.61b)$$

$$C\psi = K\psi - K_0\psi \qquad (3.61c)$$

for $t \in [0, 2\pi]$ and $\psi \in L^2(0, 2\pi)$. First, we observe that K, K_0, and C are well-defined and compact operators in $L^2(0, 2\pi)$ because the kernels are weakly singular (see Theorem A.35 of Appendix A.3). They are also self-adjoint in $L^2(0, 2\pi)$. Then Symm's equation (3.57) takes the form $K\psi = g$ in the Hilbert space $L^2(0, 2\pi)$.

As finite-dimensional subspaces X_n and Y_n, we choose the spaces of truncated Fourier series, that is,

$$X_n = Y_n = \left\{ \sum_{j=-n}^{n} \alpha_j \hat{\psi}_j : \alpha_j \in \mathbb{C} \right\}, \qquad (3.62)$$

where $\hat{\psi}_j(t) = e^{ijt}$ for $t \in [0, 2\pi]$ and $j \in \mathbb{Z}$. The corresponding orthogonal projection operators P_n from $L^2(0, 2\pi)$ into $X_n = Y_n$ are given as

$$P_n\psi = \sum_{j=-n}^{n} \alpha_j \hat{\psi}_j \qquad (3.63)$$

where $\psi = \sum_{j=-\infty}^{\infty} \alpha_j \hat{\psi}_j$.

To investigate the mapping properties of K, we need the following technical result (see [168], Lemma 8.21).

Lemma 3.16

$$\frac{1}{2\pi} \int_0^{2\pi} e^{ins} \ln\left(4 \sin^2 \frac{s}{2}\right) ds = \begin{cases} -1/|n|, & n \in \mathbb{Z}, \ n \neq 0, \\ 0, & n = 0. \end{cases} \qquad (3.64)$$

Proof: It suffices to study the case $n \in \mathbb{N}_0$. First let $n \in \mathbb{N}$. Integrating the geometric sum

$$1 + 2 \sum_{j=1}^{n-1} e^{ijs} + e^{ins} = i\left(1 - e^{ins}\right) \cot \frac{s}{2}, \quad 0 < s < 2\pi,$$

yields

$$\int_0^{2\pi} \left(e^{ins} - 1\right) \cot \frac{s}{2} \, ds = 2\pi i.$$

Integration of the identity

$$\frac{d}{ds}\left[\left(e^{ins} - 1\right) \ln\left(4\sin^2 \frac{s}{2}\right)\right] = in\, e^{ins} \ln\left(4\sin^2 \frac{s}{2}\right) + \left(e^{ins} - 1\right) \cot \frac{s}{2}$$

yields

$$\int_0^{2\pi} e^{ins} \ln\left(4\sin^2 \frac{s}{2}\right) ds = -\frac{1}{in} \int_0^{2\pi} \left(e^{ins} - 1\right) \cot \frac{s}{2} \, ds = -\frac{2\pi}{n},$$

which proves the assertion for $n \in \mathbb{N}$.

It remains to study the case where $n = 0$. Define

$$I := \int_0^{2\pi} \ln\left(4\sin^2 \frac{s}{2}\right) ds.$$

Then we conclude that

$$
\begin{aligned}
2I &= \int_0^{2\pi} \ln\left(4\sin^2 \frac{s}{2}\right) ds + \int_0^{2\pi} \ln\left(4\cos^2 \frac{s}{2}\right) ds \\
&= \int_0^{2\pi} \ln\left(16\sin^2 \frac{s}{2} \cos^2 \frac{s}{2}\right) ds \\
&= \int_0^{2\pi} \ln\left(4\sin^2 s\right) ds = \frac{1}{2}\int_0^{4\pi} \ln\left(4\sin^2 \frac{s}{2}\right) ds = I
\end{aligned}
$$

and thus $I = 0$. $\square$

This lemma shows that the functions

$$\hat{\psi}_n(t) := e^{int}, \quad t \in [0, 2\pi], \quad n \in \mathbb{Z}, \tag{3.65}$$

are eigenfunctions of K_0:

$$K_0 \hat{\psi}_n = \frac{1}{|n|} \hat{\psi}_n \quad \text{for } n \neq 0 \quad \text{and} \tag{3.66a}$$

$$K_0 \hat{\psi}_0 = \hat{\psi}_0. \tag{3.66b}$$

Now can prove the mapping properties of the operators.

Theorem 3.17 *Suppose there exists $z_0 \in \Omega$ with $|x - z_0| \neq 1$ for all $x \in \partial\Omega$. Let the operators K and K_0 be given by (3.61a) and (3.61b), respectively. By $H^s_{per}(0, 2\pi)$, we denote the Sobolev spaces of order s (see Section A.4 of Appendix A).*

(a) *The operators K and K_0 can be extended to isomorphisms from $H^{s-1}_{per}(0, 2\pi)$ onto $H^s_{per}(0, 2\pi)$ for every $s \in \mathbb{R}$.*

(b) *The operator K_0 is coercive from $H^{-1/2}_{per}(0, 2\pi)$ onto $H^{1/2}_{per}(0, 2\pi)$.*

(c) *The operator $C = K - K_0$ is compact from $H^{s-1}_{per}(0, 2\pi)$ into $H^s_{per}(0, 2\pi)$ for every $s \in \mathbb{R}$.*

Proof: Let $\psi \in L^2(0, 2\pi)$. Then ψ has the representation

$$\psi(t) = \sum_{n \in \mathbb{Z}} \alpha_n e^{int} \quad \text{with} \quad \sum_{n \in \mathbb{Z}} |\alpha_n|^2 < \infty.$$

From (3.66a) and (3.66b), we have

$$(K_0\psi)(t) = \alpha_0 + \sum_{n \neq 0} \frac{1}{|n|} \alpha_n e^{int}$$

and thus for any $s \in \mathbb{R}$:

$$\|K_0\psi\|^2_{H^s_{per}} = |\alpha_0|^2 + \sum_{n \neq 0} (1 + n^2)^s \frac{1}{n^2} |\alpha_n|^2$$

$$(\psi, K_0\psi)_{L^2} = 2\pi \left[|\alpha_0|^2 + \sum_{n \neq 0} \frac{1}{|n|} |\alpha_n|^2 \right]$$

$$\geq 2\pi \sum_{n \in \mathbb{Z}} (1 + n^2)^{-1/2} |\alpha_n|^2 = 2\pi \|\psi\|^2_{H^{-1/2}_{per}}.$$

From the elementary estimate

$$(1 + n^2)^{s-1} \leq \frac{(1 + n^2)^s}{n^2} \leq \frac{(1 + n^2)^s}{\frac{1}{2}(1 + n^2)} = 2(1 + n^2)^{s-1}, \quad n \neq 0,$$

we see that K_0 can be extended to an isomorphism from $H^{s-1}_{per}(0, 2\pi)$ onto $H^s_{per}(0, 2\pi)$ and is coercive for $s = 1/2$. The operator C is bounded from

$H^r_{per}(0, 2\pi)$ into $H^s_{per}(0, 2\pi)$ for all $r, s \in \mathbb{R}$ by Theorem A.47 of Appendix A.4. This proves part (c) and that $K = K_0 + C$ is bounded from $H^{s-1}_{per}(0, 2\pi)$ into $H^s_{per}(0, 2\pi)$. It remains to show that K is also an isomorphism from $H^{s-1}_{per}(0, 2\pi)$ onto $H^s_{per}(0, 2\pi)$. From the Riesz theory (Theorem A.36), it is sufficient to prove injectivity. Let $\psi \in H^{s-1}_{per}(0, 2\pi)$ with $K\psi = 0$. From $K_0\psi = -C\psi$ and the mapping properties of C, we conclude that $K_0\psi \in H^r_{per}(0, 2\pi)$ for all $r \in \mathbb{R}$, that is, $\psi \in H^r_{per}(0, 2\pi)$ for all $r \in \mathbb{R}$. In particular, this implies that ψ is continuous and the transformed function $\phi(\gamma(t)) = \psi(t)/|\dot{\gamma}(t)|$ satisfies Symm's equation (3.56) for $f = 0$. The application of Theorem 3.15 yields $\phi = 0$.   □

We are now in a position to apply all of the Galerkin methods of the previous section to Symm's equation (3.56), that is, $K\psi = g$ in $L^2(0, 2\pi)$.

We have seen that the convergence results require estimates of the condition numbers of K on the finite-dimensional spaces X_n and also approximation properties of $P_n\psi$. In Lemma A.45 of Appendix A, we show the following estimates for any $r \geq s$.

$$\|\psi_n\|_{H^r_{per}} \;\leq\; c\, n^{r-s}\, \|\psi_n\|_{H^s_{per}} \quad \text{for all } \psi_n \in X_n, \tag{3.67a}$$

$$\|P_n\psi - \psi\|_{H^s_{per}} \;\leq\; \frac{c}{n^{r-s}}\, \|\psi\|_{H^r_{per}} \quad \text{for all } \psi_n \in H^r_{per}(0, 2\pi), \tag{3.67b}$$

and all $n \in \mathbb{N}$. Estimate (3.67a) is sometimes called the *stability property* (see [142]). From (3.67a) and the continuity of K^{-1} from $L^2(0, 2\pi)$ into $H^{-1}_{per}(0, 2\pi)$, we conclude that

$$\|\psi_n\|_{L^2} \;\leq\; c\, n\, \|K\psi_n\|_{L^2} \quad \text{for all } \psi_n \in X_n. \tag{3.67c}$$

Indeed, this follows from

$$\begin{aligned}
\|\psi_n\|_{L^2} &\;\leq\; c\, n\, \|\psi_n\|_{H^{-1}_{per}} \;=\; c\, n\, \|K^{-1}K\psi_n\|_{H^{-1}_{per}} \\
&\;\leq\; c\, n\, \|K^{-1}\|_{\mathcal{L}(L^2, H^{-1}_{per})}\, \|K\psi_n\|_{L^2}.
\end{aligned}$$

Combining these estimates with the convergence results of the previous section, we have (almost) shown the following[1].

Theorem 3.18 *Let* $\psi^* \in H^r_{per}(0, 2\pi)$ *be the unique solution of (3.57), that is,*

$$(K\psi^*)(t) := -\frac{1}{\pi} \int_0^{2\pi} \psi^*(s) \ln|\gamma(t) - \gamma(s)|\, ds \;=\; g^*(t) := f(\gamma(t)),$$

for $t \in [0, 2\pi]$ *and some* $g^* \in H^{r+1}_{per}(0, 2\pi)$ *for some* $r \geq 0$*. Let* $g^\delta \in L^2(0, 2\pi)$ *with* $\|g^\delta - g^*\|_{L^2} \leq \delta$ *and let* X_n *be defined by (3.62).*

---

[1]Note that in this theorem, ψ^* denotes the exact solution in accordance with the notation of the general theory. It must not be mixed up with the complex conjugate.

(a) *Let $\psi_n^\delta \in X_n$ be the least squares solution, that is, the solution of*

$$(K\psi_n^\delta, K\phi_n)_{L^2} = (g^\delta, K\phi_n)_{L^2} \quad \text{for all } \phi_n \in X_n \qquad (3.68a)$$

  or

(b) *Let $\psi_n^\delta = K\tilde{\psi}_n^\delta$ with $\tilde{\psi}_n^\delta \in X_n$ be the dual least squares solution, that is, $\tilde{\psi}_n^\delta$ solves*

$$(K\tilde{\psi}_n^\delta, K\phi_n)_{L^2} = (g^\delta, \phi_n)_{L^2} \quad \text{for all } \phi_n \in X_n \qquad (3.68b)$$

  or

(c) *Let $\psi_n^\delta \in X_n$ be the Bubnov–Galerkin solution, that is, the solution of*

$$(K\psi_n^\delta, \phi_n)_{L^2} = (g^\delta, \phi_n)_{L^2} \quad \text{for all } \phi_n \in X_n. \qquad (3.68c)$$

Then there exists $c > 0$ with

$$\|\psi_n^\delta - \psi^*\|_{L^2} \le c \left(n\,\delta + \frac{1}{n^r} \|\psi^*\|_{H_{per}^r} \right) \qquad (3.69)$$

for all $n \in \mathbb{N}$.

Proof: We apply Theorems 3.11, 3.12, and 3.14 (the latter with $V = H_{per}^{1/2}(0, 2\pi)$ and $V' = H_{per}^{-1/2}(0, 2\pi)$).

For the least squares method (Theorem 3.11), we have to show Assumption (3.33). By (3.67c) we have

$$\sigma_n = \max\{\|\phi_n\|_{L^2} : \phi_n \in X_n, \|K\phi_n\|_{L^2} = 1\} \le cn, \qquad (3.70)$$

and thus, using (3.67b) for $r = 0$ and $s = -1$,

$$\min_{\phi_n \in X_n} \left\{ \|\psi - \phi_n\|_{L^2} + \sigma_n \|K(\psi - \phi_n)\|_{L^2} \right\}$$
$$\le \|\psi - P_n\psi\|_{L^2} + \sigma_n \|K(\psi - P_n\psi)\|_{L^2}$$
$$\le \|\psi\|_{L^2} + cn\,\|K\|_{\mathcal{L}(H_{per}^{-1}, L^2)} \|\psi - P_n\psi\|_{H_{per}^{-1}} \le c'\|\psi\|_{L^2}.$$

Furthermore,

$$\min\{\|\psi^* - \phi_n\|_{L^2} : \phi_n \in X_n\} \le \|\psi^* - P_n\psi^*\|_{L^2} \le \frac{c}{n^r} \|\psi^*\|_{H_{per}^r}$$

which proves the result for the least squares method.

The estimate for the dual least squares method follows immediately from (3.40) and $\sigma_n \le cn$.

For the Bubnov–Galerkin method, we have to estimate ρ_n from (3.48).

$$\rho_n = \max\{\|\phi_n\|_{L^2} : \phi_n \in X_n, \|\phi_n\|_{H_{per}^{-1/2}} = 1\} \le c\sqrt{n}$$

by (3.67a) for $r = 0$ and $s = -1/2$. This ends the proof.   □

It is interesting to note that different error estimates hold for discrete perturbations of the right-hand side. Let us denote by β_k the right-hand sides of (3.68a), (3.68b), or (3.68c), respectively, for $\phi_k(t) = \exp(ikt)$. Assume that β is perturbed by a vector $\beta^\delta \in \mathbb{C}^{2n+1}$ with $|\beta - \beta^\delta| \leq \delta$. We have to compute b_n of (3.35), (3.41), and (3.52), respectively. Because the functions $\hat{\psi}_k(t) = \exp(ikt)$, $k = -n, \ldots, n$, are orthogonal, we compute b_n for (3.41) and (3.52) by

$$\max \left\{ \sqrt{\sum_{j=-n}^{n} |\rho_j|^2} : \left\| \sum_{j=-n}^{n} \rho_j \hat{\psi}_j \right\|_{L^2} = 1 \right\} = \frac{1}{\sqrt{2\pi}}.$$

For the least squares method, however, we have to compute

$$\begin{aligned}
b_n^2 &= \max \left\{ \sum_{j=-n}^{n} |\rho_j|^2 : \left\| \sum_{j=-n}^{n} \rho_j K_0 \hat{\psi}_j \right\|_{L^2} = 1 \right\} \\
&= \max \left\{ \sum_{j=-n}^{n} |\rho_j|^2 : 2\pi \left(|\rho_0|^2 + \sum_{j \neq 0} \frac{1}{j^2} |\rho_j|^2 \right) = 1 \right\} \\
&= \frac{n^2}{2\pi};
\end{aligned}$$

that is, for discrete perturbations of the right-hand side, the estimate (3.69) is asymptotically the same for the dual least squares method and the Bubnov–Galerkin method, while for the least squares method it has to be replaced by

$$\|\psi_n^\delta - \psi^*\|_{L^2} \leq c \left(n^2 \delta + \frac{1}{n^r} \|\psi^*\|_{H_{per}^r} \right). \tag{3.71}$$

The error estimates (3.69) are optimal under the a priori information $\psi^* \in H_{per}^r(0, 2\pi)$ and $\|\psi^*\|_{H_{per}^r} \leq 1$. This is seen by choosing $n \sim (1/\delta)^{1/(r+1)}$, which gives the asymptotic estimate

$$\|\psi_{n(\delta)}^\delta - \psi^*\|_{L^2} \leq c \, \delta^{r/(r+1)}.$$

This is optimal by Problem 3.4.

From the preceding analysis, it is clear that the convergence property

$$\min \left\{ \|\psi^* - \phi_n\|_{H_{per}^s} : \phi_n \in X_n \right\} \leq c \left(\frac{1}{n} \right)^{r-s} \|\psi^*\|_{H_{per}^r}, \quad \psi^* \in H_{per}^r(0, 2\pi),$$

and the stability property

$$\|\phi_n\|_{H_{per}^r} \leq c \, n^{r-s} \|\phi_n\|_{H_{per}^s}, \quad \phi_n \in X_n,$$

for $r \geq s$ and $n \in \mathbb{N}$, are the essential tools in the proofs. For regions Ω with nonsmooth boundaries, finite element spaces for X_n are more suitable. They satisfy these conditions for a certain range of values of r and s (depending on the smoothness of the solution and the order of the finite elements). We refer to [65, 68, 141–143, 268] for more details and boundary value problems for more complicated partial differential equations.

We refer to Problem 3.5 and Section 3.5, where the Galerkin methods are explicitly compared for special cases of Symm's equation.

For further literature on Symm's and related integral equations, we refer to [11, 12, 22, 83, 223, 249, 275].

3.4   Collocation Methods

We have seen that collocation methods are subsumed under the general theory of projection methods through the use of interpolation operators. This requires the space Y to be a reproducing kernel Hilbert space, that is, a Hilbert space in which all the evaluation functionals $y \mapsto y(t)$ for $y \in Y$ and $t \in [a, b]$ are bounded.

Instead of presenting a general theory as in [202], we avoid the explicit introduction of reproducing kernel Hilbert spaces and investigate only two special, but important, cases in detail. First, we study the minimum norm collocation method. It turns out that this is a special case of a least squares method and can be treated by the methods of the previous section. In Subsection 3.4.2, we investigate a second collocation method for the important example of Symm's equation. We derive a complete and satisfactory error analysis for two choices of ansatz functions.

First, we formulate the general collocation method again and derive an error estimate in the presence of discrete perturbations of the right-hand side.

Let X be a Hilbert space over the field $\mathbb{K}$, $X_n \subset X$ be finite-dimensional subspaces with $\dim X_n = n$, and $a \leq t_1 < \cdots < t_n \leq b$ be the collocation points. Let $K : X \to C[a, b]$ be bounded and one-to-one. Let $Kx^* = y^*$, and assume that the collocation equations

$$(Kx_n)(t_i) \;=\; y(t_i), \quad i = 1, \ldots, n, \tag{3.72}$$

are uniquely solvable in X_n for every right-hand side. Choosing a basis $\{\hat{x}_j : j = 1, \ldots, n\}$ of X_n, we rewrite this as a system $A\alpha = \beta$, where $x_n = \sum_{j=1}^{n} \alpha_j \hat{x}_j$ and

$$A_{ij} \;=\; (K\hat{x}_j)(t_i), \qquad \beta_i = y(t_i). \tag{3.73}$$

The following main theorem is the analogue of Theorem 3.10 for collocation methods. We restrict ourselves to the important case of discrete perturbations of the right-hand side. Continuous perturbations could also be handled but are not of particular interest because point evaluation is no longer possible when the right-hand side is perturbed in the L^2-sense. This would require stronger norms in the range space and leads to the concept of reproducing kernel Hilbert spaces (see [168]).

Theorem 3.19 *Let $Kx^* = y^*$ and let $\{t_1^{(n)}, \ldots, t_n^{(n)}\} \subset [a, b]$, $n \in \mathbb{N}$, be a sequence of collocation points. Assume that $\bigcup_{n \in \mathbb{N}} X_n$ is dense in X and that the collocation method converges. Let $x_n^\delta = \sum_{j=1}^n \alpha_j^\delta \hat{x}_j \in X_n$, where α^δ solves $A\alpha^\delta = \beta^\delta$. Here, $\beta^\delta \in \mathbb{K}^n$ satisfies $|\beta - \beta^\delta| \leq \delta$ where $\beta_i = y^*(t_i)$ and $|\cdot|$ denotes again the Euclidean norm in $\mathbb{K}^n$. Then the following error estimate holds:*

$$\|x_n^\delta - x^*\|_X \leq c\left(\frac{a_n}{\lambda_n}\delta + \inf\{\|x^* - z_n\|_X : z_n \in X_n\}\right), \tag{3.74}$$

where

$$a_n = \max\left\{\left\|\sum_{j=1}^n \rho_j \hat{x}_j\right\|_X : \sum_{j=1}^n |\rho_j|^2 = 1\right\} \tag{3.75}$$

and λ_n denotes the smallest singular value of A.

Proof: Again we write $\|x_n^\delta - x^*\|_X \leq \|x_n^\delta - x_n\|_X + \|x_n - x^*\|_X$, where $x_n = R_n y^*$ solves the collocation equation for β instead of β^δ. The second term is estimated by Theorem 3.7. We estimate the first term by

$$\begin{aligned}\|x_n^\delta - x_n\|_X &\leq a_n|\alpha^\delta - \alpha| = a_n|A^{-1}(\beta^\delta - \beta)| \\ &\leq a_n|A^{-1}|_2|\beta^\delta - \beta| \leq \frac{a_n}{\lambda_n}\delta.\end{aligned} \qquad \square$$

Again we remark that $a_n = 1$ if $\{\hat{x}_j : j = 1, \ldots, n\}$ forms an orthonormal system in X.

3.4.1   Minimum Norm Collocation

Again, let $K : X \to C[a, b]$ be a linear, bounded, and injective operator from the Hilbert space X into the space $C[a, b]$ of continuous functions on $[a, b]$. We assume that there exists a unique solution $x^* \in X$ of $Kx^* = y^*$. Let $a \leq t_1 < \cdots < t_n \leq b$ be the set of collocation points. Solving the equations (3.72) in X is certainly not enough to specify the solution x_n uniquely. An obvious choice is to determine $x_n \in X$ from the set of solutions of (3.72) that has a minimal L^2-norm among all solutions.

Definition 3.20 $x_n \in X$ *is called the* moment solution *of (3.72) with respect to the collocation points $a \leq t_1 < \cdots < t_n \leq b$ if x_n satisfies (3.72) and*

$$\|x_n\|_X = \min\{\|z_n\|_X : z_n \in X \text{ satisfies (3.72)}\}.$$

We can interpret this moment solution as a least squares solution. Because $z \mapsto (Kz)(t_i)$ is bounded from X into $\mathbb{K}$, Theorem A.23 by Riesz yields the existence of $k_i \in X$ with $(Kz)(t_i) = (z, k_i)_X$ for all $z \in X$ and $i = 1, \ldots, n$.

If, for example, $X = L^2(a, b)$ and K is the integral operator

$$(Kz)(t) = \int_a^b k(t, s)\, z(s)\, ds, \quad t \in [a, b],\ z \in L^2(a, b),$$

with real-valued kernel k then $k_i \in L^2(a, b)$ is explicitly given by $k_i(s) = k(t_i, s)$.

Going back to the general case, we rewrite the moment equation (3.72) in the form

$$(x_n, k_i)_X = y(t_i) = (x, k_i)_X, \quad i = 1, \ldots, n.$$

The minimum norm solution x_n of the set of equations is characterized by the projection theorem (see Theorem A.13 of Appendix A.1) and is given by the solution of (3.72) in the space $X_n := \text{span}\{k_j : j = 1, \ldots, n\}$.

Now we define the Hilbert space Y by $Y := K(X) = \mathcal{R}(K)$ with inner product

$$(y, z)_Y := (K^{-1}y, K^{-1}z)_X \quad \text{for } y, z \in K(X) = \mathcal{R}(K).$$

We omit the simple proof of the following lemma.

Lemma 3.21 *Y is a Hilbert space that is continuously embedded in $C[a, b]$. Furthermore, K is an isomorphism from X onto Y.*

Now we can rewrite (3.72) in the form.

$$\left(Kx_n, Kk_i\right)_Y = \left(y, Kk_i\right)_Y, \quad i = 1, \ldots, n.$$

Comparing this equation with (3.31a), we observe that (3.72) is the Galerkin equation for the least squares method with respect to X_n. Thus we have shown that the moment solution can be interpreted as the least squares solution for the operator $K : X \to Y$. The application of Theorem 3.11 yields the following theorem.

Theorem 3.22 *Let K be one-to-one and $\{k_j : j = 1, \ldots, n\}$ be linearly independent where $k_j \in X$ are such that $(Kz)(t_j) = (z, k_j)_X$ for all $z \in X$, $j = 1, \ldots, n$. Then there exists one and only one moment solution x_n of (3.72). x_n is given by*

$$x_n = \sum_{j=1}^{n} \alpha_j k_j, \tag{3.76}$$

where $\alpha \in \mathbb{K}^n$ solves the linear system $A\alpha = \beta$ with

$$A_{ij} = (Kk_j)(t_i) = (k_j, k_i)_X \quad \text{and} \quad \beta_i = y(t_i). \tag{3.77}$$

Let $\{t_1^{(n)}, \ldots, t_n^{(n)}\} \subset [a, b]$, $n \in \mathbb{N}$, be a sequence of collocation points such that $\bigcup_{n \in \mathbb{N}} X_n$ is dense in X where

$$X_n := \text{span}\{k_j^{(n)} : j = 1, \ldots, n\}.$$

Then the moment method converges; that is, the moment solution $x_n \in X_n$ of (3.72) converges in X to the solution $x^* \in X$ of $Kx^* = y^*$. If $x_n^\delta =$

$\sum_{j=1}^{n} \alpha_j^\delta k_j^{(n)}$, where $\alpha^\delta \in \mathbb{K}^n$ solves $A\alpha^\delta = \beta^\delta$ with $|\beta - \beta^\delta| \leq \delta$, then the following error estimate holds:

$$\|x^* - x_n^\delta\|_X \leq \frac{a_n}{\lambda_n}\,\delta \;+\; c \min\{\|x^* - z_n\|_X : z_n \in X_n\}\,, \qquad (3.78)$$

where

$$a_n \;=\; \max\left\{\left\|\sum_{j=1}^{n} \rho_j k_j^{(n)}\right\|_X : \sum_{j=1}^{n} |\rho_j|^2 = 1\right\} \qquad (3.79)$$

and where λ_n denotes the smallest singular value of A.

Proof: The definition of $\|\cdot\|_Y$ implies that $\sigma_n = 1$, where σ_n is given by (3.32). Assumption (3.33) for the convergence of the least squares method is obviously satisfied because

$$\min_{z_n \in X_n}\left\{\|x^* - z_n\|_X + \sigma_n\|K(x^* - z_n)\|_Y\right\} \;\leq\; \|x^*\|_X + \sigma_n\|x^*\|_X \;=\; 2\,\|x^*\|_X\,.$$

The application of Theorem 3.11 yields the assertion.   □

As an example, we again consider numerical differentiation.

Example 3.23
Let $X = L^2(0,1)$ and K be defined by

$$(Kx)(t) \;=\; \int_0^t x(s)\,ds \;=\; \int_0^1 k(t,s)\,x(s)\,ds, \quad t \in [0,1]\,,$$

with $k(t,s) = \begin{cases} 1, & s \leq t, \\ 0, & s > t. \end{cases}$

We choose equidistant nodes, that is, $t_j = \frac{j}{n}$ for $j = 0, \ldots, n$. The moment method is to minimize $\|x\|_{L^2}^2$ under the restrictions that

$$\int_0^{t_j} x(s)\,ds \;=\; y(t_j), \quad j = 1, \ldots, n\,. \qquad (3.80)$$

The solution x_n is piecewise constant because it is a linear combination of the piecewise constant functions $k(t_j, \cdot)$. Therefore, the finite-dimensional space X_n is given by

$$X_n \;=\; \{z_n \in L^2(0,1) : z_n|_{(t_{j-1},t_j)} \text{ constant}, \; j = 1, \ldots, n\}\,. \qquad (3.81)$$

As basis functions $\hat{x}_j$ of X_n we choose $\hat{x}_j(s) = k(t_j, s)$.
    Then $x_n = \sum_{j=1}^{n} \alpha_j\, k(t_j, \cdot)$ is the moment solution, where α solves $A\alpha = \beta$ with $\beta_i = y(t_i)$ and

$$A_{ij} \;=\; \int_0^1 k(t_i, s)\,k(t_j, s)\,ds \;=\; \frac{1}{n}\,\min\{i, j\}\,.$$

It is not difficult to see that the moment solution is just the one-sided difference quotient

$$x_n(t_1) = \frac{1}{h}\, y(t_1), \quad x_n(t_j) = \frac{1}{h}\left[y(t_j) - y(t_{j-1}) \right], \quad j = 2, \ldots, n \,,$$

for $h = 1/n$.

We have to check the assumptions of Theorem 3.22. First, K is one-to-one and $\{k(t_j, \cdot) : j = 1 \ldots, n\}$ are linearly independent. The union $\bigcup_{n \in \mathbb{N}} X_n$ is dense in $L^2(0,1)$ (see Problem 3.6). We have to estimate a_n from (3.79), the smallest eigenvalue λ_n of A, and $\min\{\|x - z_n\|_{L^2} : z_n \in X_n\}$.

Let $\rho \in \mathbb{R}^n$ with $\sum_{j=1}^n \rho_j^2 = 1$. Using the Cauchy–Schwarz inequality, we estimate

$$\int_0^1 \left| \sum_{j=1}^n \rho_j k(t_j, s) \right|^2 ds \;\le\; \int_0^1 \sum_{j=1}^n k(t_j, s)^2 ds \;=\; \sum_{j=1}^n t_j \;=\; \frac{n+1}{2}\,.$$

Thus $a_n \le \sqrt{(n+1)/2}$.

It is straightforward to check that the inverse of A is given by the tridiagonal matrix

$$A^{-1} = n \begin{bmatrix} 2 & -1 & & & \\ -1 & 2 & -1 & & \\ & \ddots & \ddots & \ddots & \\ & & -1 & 2 & -1 \\ & & & -1 & 1 \end{bmatrix}.$$

We estimate the largest eigenvalue $\mu_{\max}$ of A^{-1} by the maximum absolute row sum $\mu_{\max} \le 4n$. This is asymptotically sharp because we can give a lower estimate of $\mu_{\max}$ by the trace formula

$$n\,\mu_{\max} \;\ge\; \operatorname{trace}(A^{-1}) \;=\; \sum_{j=1}^n (A^{-1})_{jj} \;=\; (2n-1)\,n\,;$$

that is, we have an estimate of λ_n of the form

$$\frac{1}{4n} \;\le\; \lambda_n \;\le\; \frac{1}{2n-1}\,.$$

In Problem 3.6, it is shown that

$$\min\{\|x - z_n\|_{L^2} : z_n \in X_n\} \;\le\; \frac{1}{n}\,\|x'\|_{L^2}\,.$$

Thus we have proven the following theorem.

Theorem 3.24 *The moment method for (3.80) converges. The following error estimate holds:*

$$\|x^* - x_n^\delta\|_{L^2} \;\le\; \sqrt{\frac{n+1}{2}}\,\delta \;+\; \frac{c}{n}\,\|(x^*)'\|_{L^2}$$

if $x^* \in H^1(0,1)$. Here, δ is the discrete error on the right-hand side, that is,
$\sum_{j=1}^{n} |\beta_j^\delta - y^*(t_j)|^2 \leq \delta^2$ *and* $x_n^\delta = \sum_{j=1}^{n} \alpha_j^\delta \hat{x}_j$, *where* $\alpha^\delta \in \mathbb{R}^n$ *solves* $A\alpha^\delta = \beta^\delta$.

The choice $X_n = S_1(t_1, \ldots, t_n)$ of linear splines leads to the two-sided difference quotient (see Problem 3.8). We refer to [85, 201, 202] for further reading on moment collocation.

3.4.2 Collocation of Symm's Equation

We now study the numerical treatment of Symm's equation (3.57), that is,

$$(K\psi^*)(t) := -\frac{1}{\pi} \int_0^{2\pi} \psi^*(s) \ln|\gamma(t) - \gamma(s)| \, ds = g^*(t) \qquad (3.82)$$

for $0 \leq t \leq 2\pi$ by collocation methods. The integral operator K from (3.82) is well-defined and bounded from $L^2(0, 2\pi)$ into $H_{per}^1(0, 2\pi)$. We assume throughout this subsection that K is one-to-one (see Theorem 3.15). Then we have seen in Theorem 3.17 that equation (3.82) is uniquely solvable in $L^2(0, 2\pi)$ for every $g \in H_{per}^1(0, 2\pi)$; that is, K is an isomorphism. We define equidistant collocation points by

$$t_k := k\frac{\pi}{n} \quad \text{for } k = 0, \ldots, 2n - 1.$$

There are several choices for the space $X_n \subset L^2(0, 2\pi)$ of basis functions. Before we study particular cases, let $X_n = \text{span}\{\hat{x}_j : j \in J_n\} \subset L^2(0, 2\pi)$ be arbitrary. $J_n \subset \mathbb{Z}$ denotes a set of indices with $2n$ elements. We assume that $\{\hat{x}_j : j \in J_n\}$ forms an orthonormal system in $L^2(0, 2\pi)$.

The collocation equations (3.72) take the form

$$-\frac{1}{\pi} \int_0^{2\pi} \psi_n(s) \ln|\gamma(t_k) - \gamma(s)| \, ds = g^*(t_k), \quad k = 0, \ldots, 2n - 1, \qquad (3.83)$$

with $\psi_n \in X_n$. Let $Q_n : H_{per}^1(0, 2\pi) \to Y_n$ be the trigonometric interpolation operator into the $2n$-dimensional space

$$Y_n := \left\{ \sum_{m=-n}^{n-1} a_m e^{imt} : a_m \in \mathbb{C} \right\}. \qquad (3.84)$$

We recall some approximation properties of the interpolation operator $Q_n : H_{per}^1(0, 2\pi) \to Y_n$. First, it is easily checked (see also Theorem A.46 of Appendix A.4) that Q_n is given by

$$Q_n \psi = \sum_{k=0}^{2n-1} \psi(t_k) \, \hat{y}_k$$

with Lagrange interpolation basis functions

$$\hat{y}_k(t) \;=\; \frac{1}{2n} \sum_{m=-n}^{n-1} e^{im(t-t_k)}, \quad k = 0, \ldots, 2n-1. \qquad (3.85)$$

From Theorem A.46 of Appendix A.4, we have also the estimates

$$\|\psi - Q_n\psi\|_{L^2} \;\leq\; \frac{c}{n} \|\psi\|_{H^1_{per}} \quad \text{for all } \psi \in H^1_{per}(0, 2\pi), \qquad (3.86a)$$

$$\|Q_n\psi\|_{H^1_{per}} \;\leq\; c\,\|\psi\|_{H^1_{per}} \quad \text{for all } \psi \in H^1_{per}(0, 2\pi). \qquad (3.86b)$$

Now we can reformulate the collocation equations (3.83) as

$$Q_n K \psi_n \;=\; Q_n g^* \quad \text{with } \psi_n \in X_n. \qquad (3.87)$$

We use the perturbation result of Theorem 3.8 again and split K into the form $K = K_0 + C$ with

$$(K_0\psi)(t) \;:=\; -\frac{1}{2\pi} \int_0^{2\pi} \psi(s) \left[\ln\left(4 \sin^2 \frac{t-s}{2} \right) - 1 \right] ds. \qquad (3.88)$$

Now we specify the spaces X_n. As a first example, we choose the orthonormal functions

$$\hat{x}_j(t) \;=\; \frac{1}{\sqrt{2\pi}} e^{ijt} \quad \text{for } j = -n, \ldots, n-1. \qquad (3.89)$$

We prove the following convergence result.

Theorem 3.25 *Let $\hat{x}_j$, $j = -n, \ldots, n-1$, be given by (3.89). The collocation method is convergent; that is, the solution $\psi_n \in X_n$ of (3.83) converges to the solution $\psi^* \in L^2(0, 2\pi)$ of (3.82) in $L^2(0, 2\pi)$.*

Let the right-hand side of (3.83) be replaced by $\beta^\delta \in \mathbb{C}^{2n}$ with

$$\sum_{k=0}^{2n-1} |\beta_k^\delta - g^*(t_k)|^2 \;\leq\; \delta^2.$$

Let $\alpha^\delta \in \mathbb{C}^{2n}$ be the solution of $A\alpha^\delta = \beta^\delta$, where $A_{kj} = (K\hat{x}_j)(t_k)$. Then the following error estimate holds:

$$\|\psi_n^\delta - \psi^*\|_{L^2} \;\leq\; c \left[\sqrt{n}\,\delta \;+\; \min\{\|\psi^* - \phi_n\|_{L^2} : \phi_n \in X_n\} \right], \qquad (3.90)$$

where

$$\psi_n^\delta(t) \;=\; \frac{1}{\sqrt{2\pi}} \sum_{j=-n}^{n-1} \alpha_j^\delta\, e^{ijt}.$$

If $\psi^* \in H^r_{per}(0, 2\pi)$ for some $r > 0$, then

$$\|\psi_n^\delta - \psi^*\|_{L^2} \;\leq\; c \left[\sqrt{n}\,\delta \;+\; \frac{1}{n^r} \|\psi^*\|_{H^r_{per}} \right]. \qquad (3.91)$$

Proof: By the perturbation Theorem 3.8, it is sufficient to prove the result for K_0 instead of K. By (3.66a) and (3.66b), the operator K_0 maps X_n into $Y_n = X_n$. Therefore, the collocation equation (3.87) for K_0 reduces to

$$K_0\psi_n = Q_n g^*.$$

We want to apply Theorem 3.7 and have to estimate $R_n K_0$ where in this case $R_n = \left(K_0|_{X_n}\right)^{-1} Q_n$. Because $K_0 : L^2(0, 2\pi) \to H^1_{per}(0, 2\pi)$ is invertible, we conclude that

$$\|R_n g\|_{L^2} = \|\psi_n\|_{L^2} \leq c_1 \|K_0 \psi_n\|_{H^1_{per}} = c_1 \|Q_n g\|_{H^1_{per}} \leq c_2 \|g\|_{H^1_{per}}$$

for all $g \in H^1_{per}(0, 2\pi)$, and thus

$$\|R_n K\psi\|_{L^2} \leq c_2 \|K\psi\|_{H^1_{per}} \leq c_3 \|\psi\|_{L^2}$$

for all $\psi \in L^2(0, 2\pi)$. The application of Theorem 3.7 yields convergence.

To prove the error estimate (3.90), we want to apply Theorem 3.19 and hence have to estimate the singular values of the matrix B defined by

$$B_{kj} = (K_0 \hat{x}_j)(t_k), \quad k, j = -n, \ldots, n - 1,$$

with $\hat{x}_j$ from (3.89). From (3.66a) and (3.66b), we observe that

$$B_{kj} = \frac{1}{\sqrt{2\pi}} \frac{1}{|j|} e^{ijk\frac{\pi}{n}}, \quad k, j = -n, \ldots, n - 1,$$

where $1/|j|$ has to be replaced by 1 if $j = 0$. Because the singular values of B are the square roots of the eigenvalues of B^*B, we compute

$$\left(B^*B\right)_{\ell j} = \sum_{k=-n}^{n-1} \overline{B_{k\ell}} \, B_{kj} = \frac{1}{2\pi} \frac{1}{|\ell||j|} \sum_{k=-n}^{n-1} e^{ik(j-\ell)\frac{\pi}{n}} = \frac{n}{\pi} \frac{1}{\ell^2} \delta_{\ell j},$$

where again $1/\ell^2$ has to be replaced by 1 for $\ell = 0$. From this, we see that the singular values of B are given by $\sqrt{n}/(\pi \ell^2)$ for $\ell = 1, \ldots, n$. The smallest singular value is $1/\sqrt{n\pi}$. Estimate (3.74) of Theorem 3.19 yields the assertion. (3.91) follows from Theorem A.46.   □

Comparing the estimate (3.91) with the corresponding error estimate (3.69) for the Galerkin methods, it seems that the estimate for the collocation method is better because the error δ is only multiplied by $\sqrt{n}$ instead of n. Let us now compare the errors of the continuous perturbation $\|y^* - y^\delta\|_{L^2}$ with the discrete perturbation for both methods. To do this, we have to "extend" the discrete vector β^δ to a function $y_n^\delta \in X_n$. For the collocation method, we have to use the interpolation operator Q_n and define $y_n^\delta \in X_n$ by $y_n^\delta = \sum_{j=1}^{2n-1} \beta_j^\delta \hat{y}_j$, where $\hat{y}_j$ are the Lagrange basis functions (3.85). Then $y_n^\delta(t_k) = \beta_k^\delta$, and we estimate

$$\|y_n^\delta - y^*\|_{L^2} \leq \|y_n^\delta - Q_n y^*\|_{L^2} + \|Q_n y^* - y^*\|_{L^2}.$$

Writing

$$y_n^\delta(t) - Q_n y^*(t) = \sum_{j=-n}^{n-1} \rho_j e^{ijt},$$

a simple computation shows that

$$\sum_{k=0}^{n-1} |\beta_k^\delta - y^*(t_k)|^2 = \sum_{k=0}^{n-1} |\beta_k^\delta - Q_n y^*(t_k)|^2 = \sum_{k=0}^{n-1} \left| \sum_{j=-n}^{n-1} \rho_j e^{ikj\frac{\pi}{n}} \right|^2$$

$$= 2n \sum_{j=-n}^{n-1} |\rho_j|^2 = \frac{n}{\pi} \|y_n^\delta - Q_n y^*\|_{L^2}^2 . \qquad (3.92)$$

Therefore, for the collocation method we have to compare the continuous error δ with the discrete error $\delta \sqrt{n/\pi}$. This gives an extra factor of $\sqrt{n}$ in the first terms of (3.90) and (3.91).

For Galerkin methods, however, we define $y_n^\delta(t) = \frac{1}{2\pi} \sum_{j=-n}^{n} \beta_j^\delta \exp(ijt)$. Then $\left(y_n^\delta, e^{ij\cdot}\right)_{L^2} = \beta_j^\delta$. Let P_n be the orthogonal projection onto X_n. In

$$\|y_n^\delta - y^*\|_{L^2} \leq \|y_n^\delta - P_n y^*\|_{L^2} + \|P_n y^* - y^*\|_{L^2},$$

we estimate the first term as

$$\|y_n^\delta - P_n y^*\|_{L^2}^2 = \frac{1}{2\pi} \sum_{j=-n}^{n} \left| \beta_j^\delta - (y^*, e^{ij\cdot})_{L^2} \right|^2 .$$

In this case, the continuous and discrete errors are of the same order.

Choosing trigonometric polynomials as basis functions is particularly suitable for smooth boundary data. If $\partial\Omega$ or the right-hand side f of the boundary value problem (3.54) is not smooth, then spaces of piecewise constant functions are more appropriate. We now study the case where the basis functions $\hat{x}_j \in L^2(0, 2\pi)$ are defined by

$$\hat{x}_0(t) = \begin{cases} \sqrt{\frac{n}{\pi}}, & \text{if } t < \frac{\pi}{2n} \text{ or } t > 2\pi - \frac{\pi}{2n}, \\ 0, & \text{if } \frac{\pi}{2n} < t < 2\pi - \frac{\pi}{2n}, \end{cases} \qquad (3.93a)$$

$$\hat{x}_j(t) = \begin{cases} \sqrt{\frac{n}{\pi}}, & \text{if } |t - t_j| < \frac{\pi}{2n}, \\ 0, & \text{if } |t - t_j| > \frac{\pi}{2n}, \end{cases} \qquad (3.93b)$$

for $j = 1, \ldots, 2n - 1$. Then $\hat{x}_j$, $j = 0, \ldots, 2n - 1$, are also orthonormal in $L^2(0, 2\pi)$. In the following lemma, we collect some approximation properties of the corresponding spaces X_n.

Lemma 3.26 *Let $X_n = \mathrm{span}\{\hat{x}_j : j = 0, \ldots, 2n-1\}$, where $\hat{x}_j$ are defined by (3.93a) and (3.93b). Let $P_n : L^2(0, 2\pi) \to X_n$ be the orthogonal projection operator. Then $\bigcup_{n \in \mathbb{N}} X_n$ is dense in $L^2(0, 2\pi)$ and there exists $c > 0$ with*

$$\|\psi - P_n\psi\|_{L^2} \;\leq\; \frac{c}{n}\|\psi\|_{H^1_{per}} \quad \text{for all } \psi \in H^1_{per}(0, 2\pi), \tag{3.94a}$$

$$\|K(\psi - P_n\psi)\|_{L^2} \;\leq\; \frac{c}{n}\|\psi\|_{L^2} \quad \text{for all } \psi \in L^2(0, 2\pi). \tag{3.94b}$$

Proof: Estimate (3.94a) is left as an exercise. To prove estimate (3.94b), we use (implicitly) a duality argument:

$$\|K(\psi - P_n\psi)\|_{L^2} \;=\; \sup_{\phi \neq 0} \frac{\left(K(\psi - P_n\psi), \phi\right)_{L^2}}{\|\phi\|_{L^2}}$$

$$=\; \sup_{\phi \neq 0} \frac{\left(\psi - P_n\psi, K\phi\right)_{L^2}}{\|\phi\|_{L^2}}$$

$$=\; \sup_{\phi \neq 0} \frac{\left(\psi, (I - P_n)K\phi\right)_{L^2}}{\|\phi\|_{L^2}}$$

$$\leq\; \|\psi\|_{L^2} \sup_{\phi \neq 0} \frac{\|(I - P_n)K\phi\|_{L^2}}{\|\phi\|_{L^2}}$$

$$\leq\; \frac{\tilde{c}}{n}\|\psi\|_{L^2} \sup_{\phi \neq 0} \frac{\|K\phi\|_{H^1_{per}}}{\|\phi\|_{L^2}} \;\leq\; \frac{c}{n}\|\psi\|_{L^2}. \qquad \square$$

Before we prove a convergence theorem, we compute the singular values of the matrix B defined by

$$B_{kj} \;=\; (K_0\hat{x}_j)(t_k) \;=\; -\frac{1}{2\pi} \int\limits_0^{2\pi} \hat{x}_j(s) \left[\ln\left(4 \sin^2 \frac{t_k - s}{2}\right) - 1 \right] ds. \tag{3.95}$$

Lemma 3.27 *B is symmetric and positive definite. The singular values of B coincide with the eigenvalues and are given by*

$$\mu_0 \;=\; \sqrt{\frac{n}{\pi}} \quad \text{and} \quad \mu_m \;=\; \sqrt{\frac{n}{\pi}} \, \frac{\sin\frac{m\pi}{2n}}{2n\pi} \sum_{j \in \mathbb{Z}} \frac{1}{\left(\frac{m}{2n} + j\right)^2} \tag{3.96a}$$

for $m = 1, \ldots, 2n-1$. Furthermore, there exists $c > 0$ with

$$\frac{1}{\sqrt{\pi n}} \;\leq\; \mu_m \;\leq\; c\sqrt{n} \quad \text{for all } m = 0, \ldots, 2n-1. \tag{3.96b}$$

We observe that the condition number of B, that is, the ratio between the largest and smallest singular values, is again bounded by n.

Proof: We write

$$B_{kj} = -\frac{1}{2\pi}\sqrt{\frac{n}{\pi}}\int_{t_j-\frac{\pi}{2n}}^{t_j+\frac{\pi}{2n}}\left[\ln\left(4\sin^2\frac{s-t_k}{2}\right)-1\right]ds = b_{j-k}$$

with

$$b_\ell = -\frac{1}{2\pi}\sqrt{\frac{n}{\pi}}\int_{t_\ell-\frac{\pi}{2n}}^{t_\ell+\frac{\pi}{2n}}\left[\ln\left(4\sin^2\frac{s}{2}\right)-1\right]ds,$$

where we extended the definition of t_ℓ to all $\ell \in \mathbb{Z}$. Therefore, B is circulant and symmetric. The eigenvectors $x^{(m)}$ and eigenvalues μ_m of B are given by

$$x^{(m)} = \left(e^{imk\frac{\pi}{n}}\right)_{k=0}^{2n-1} \quad\text{and}\quad \mu_m = \sum_{k=0}^{2n-1} b_k\, e^{imk\frac{\pi}{n}},$$

respectively, for $m = 0,\ldots,2n-1$, as is easily checked. We write μ_m in the form

$$\mu_m = -\frac{1}{2\pi}\int_0^{2\pi}\psi_m(s)\left[\ln\left(4\sin^2\frac{s}{2}\right)-1\right]ds = K_0\psi_m(0)$$

with

$$\psi_m(s) = \sqrt{\frac{n}{\pi}}\,e^{imk\frac{\pi}{n}} \quad\text{for } |s-t_k|\le\frac{\pi}{2n},\ k\in\mathbb{Z}.$$

Let $\psi_m(t)=\sum_{k\in\mathbb{Z}}\rho_{m,k}\exp(ikt)$. Then by (3.66a) and (3.66b), we have

$$\mu_m = \rho_{m,0} + \sum_{k\ne0}\frac{\rho_{m,k}}{|k|}.$$

Therefore, we have to compute the Fourier coefficients $\rho_{m,k}$ of ψ_m. They are given by

$$\rho_{m,k} = \frac{1}{2\pi}\int_0^{2\pi}\psi_m(s)\,e^{-iks}ds = \sqrt{\frac{n}{\pi}}\frac{1}{2\pi}\sum_{j=0}^{2n-1}e^{imj\frac{\pi}{n}}\int_{t_j-\frac{\pi}{2n}}^{t_j+\frac{\pi}{2n}}e^{-iks}ds.$$

For $k = 0$, this reduces to

$$\rho_{m,0} = \sqrt{\frac{n}{\pi}}\frac{1}{2n}\sum_{j=0}^{2n-1}e^{imj\frac{\pi}{n}} = \begin{cases}\sqrt{n/\pi} & \text{if } m=0,\\ 0 & \text{if } m=1,\ldots,2n-1,\end{cases}$$

and for $k\ne0$ to

$$\rho_{m,k} = \sqrt{\frac{n}{\pi}}\frac{\sin\frac{\pi k}{2n}}{\pi k}\sum_{j=0}^{2n-1}e^{i(m-k)j\frac{\pi}{n}} = \begin{cases}\sqrt{\frac{n}{\pi}}\frac{2n}{\pi k}\sin\frac{\pi k}{2n}, & \text{if } m-k\in2n\mathbb{Z},\\ 0, & \text{if } m-k\notin2n\mathbb{Z}.\end{cases}$$

Thus we have $\mu_0 = \sqrt{n/\pi}$ and

$$\mu_m = \sqrt{\frac{n}{\pi}}\frac{2n}{\pi}\sum_{k-m\in 2n\mathbb{Z}}\frac{|\sin\frac{\pi k}{2n}|}{k^2} = \sqrt{\frac{n}{\pi}}\frac{\sin\frac{\pi m}{2n}}{2n\pi}\sum_{j\in\mathbb{Z}}\frac{1}{\left(\frac{m}{2n}+j\right)^2}$$

for $m = 1,\ldots,2n-1$. This proves (3.96a). Because all eigenvalues are positive, the matrix B is positive definite and the eigenvalues coincide with the singular values. We set $x = m/(2n) \in (0,1)$ and separate the first two terms in the series. This yields

$$\mu_m = \sqrt{\frac{n}{\pi}}\frac{\sin\pi x}{2n\pi}\left(\frac{1}{x^2} + \frac{1}{(x-1)^2}\right) \tag{3.97}$$

$$+ \sqrt{\frac{n}{\pi}}\frac{\sin\pi x}{2n\pi}\sum_{j=1}^{\infty}\frac{1}{(x+j)^2} + \sqrt{\frac{n}{\pi}}\frac{\sin\pi x}{2n\pi}\sum_{j=2}^{\infty}\frac{1}{(x-j)^2}$$

$$\geq \sqrt{\frac{n}{\pi}}\frac{1}{2n\pi}\left(\frac{\sin\pi x}{x^2} + \frac{\sin\pi(1-x)}{(1-x)^2}\right).$$

From the elementary estimate

$$\frac{\sin\pi x}{x^2} + \frac{\sin\pi(1-x)}{(1-x)^2} \geq 8, \quad x\in(0,1),$$

we conclude that

$$\mu_m \geq \frac{4}{\pi}\frac{1}{\sqrt{\pi n}} \geq \frac{1}{\sqrt{\pi n}}$$

for $m = 1,\ldots,2n-1$. The upper estimate of (3.96b) is proven analogously. Indeed, from (3.97) we estimate, using also $|\sin t/t| \leq 1$,

$$\mu_m \leq \sqrt{\frac{n}{\pi}}\frac{1}{2n\pi}\left(\frac{\sin\pi x}{x^2} + \frac{\sin\pi(1-x)}{(1-x)^2} + 2\sum_{j=1}^{\infty}\frac{1}{j^2}\right)$$

$$\leq \sqrt{\frac{n}{\pi}}\frac{1}{2n}\left(\frac{1}{x} + \frac{1}{1-x} + \frac{2}{\pi}\sum_{j=1}^{\infty}\frac{1}{j^2}\right) \leq c\sqrt{n}$$

for some $c > 0$.   □

Now we can prove the following convergence result.

Theorem 3.28 *Let $\hat{x}_j$, $j = 0,\ldots,2n-1$, be defined by (3.93a) and (3.93b). The collocation method is convergent; that is, the solution $\psi_n \in X_n$ of (3.83) converges to the solution $\psi^* \in L^2(0,2\pi)$ of (3.82) in $L^2(0,2\pi)$.*

Let the right-hand side be replaced by $\beta^\delta \in \mathbb{C}^{2n}$ with

$$\sum_{j=0}^{2n-1}|\beta_j^\delta - g^*(t_j)|^2 \leq \delta^2.$$

Let $\alpha^\delta \in \mathbb{C}^{2n}$ be the solution of $A\alpha^\delta = \beta^\delta$, where $A_{kj} = K\hat{x}_j(t_k)$. Then the following error estimate holds:

$$\|\psi_n^\delta - \psi^*\|_{L^2} \leq c\left[\sqrt{n}\,\delta + \min\{\|\psi^* - \phi_n\|_{L^2} : \phi_n \in X_n\}\right], \qquad (3.98)$$

where $\psi_n^\delta = \sum_{j=0}^{2n-1} \alpha_j^\delta \hat{x}_j$. If $\psi^* \in H^1_{per}(0, 2\pi)$, then

$$\|\psi_n^\delta - \psi^*\|_{L^2} \leq c\left[\sqrt{n}\,\delta + \frac{1}{n}\|\psi^*\|_{H^1_{per}}\right]. \qquad (3.99)$$

Proof: By the perturbation theorem (Theorem 3.8), it is sufficient to prove the result for K_0 instead of K. Again set

$$R_n = \left[Q_n K_0|_{X_n}\right]^{-1} Q_n : H^1_{per}(0, 2\pi) \longrightarrow X_n \subset L^2(0, 2\pi),$$

let $\psi \in H^1_{per}(0, 2\pi)$, and set $\psi_n = R_n\psi = \sum_{j=0}^{2n-1} \alpha_j \hat{x}_j$. Then $\alpha \in \mathbb{C}^{2n}$ solves $B\alpha = \beta$ with $\beta_k = \psi(t_k)$, and thus by (3.96b)

$$\|\psi_n\|_{L^2} = |\alpha| \leq |B^{-1}|_2|\beta| \leq \sqrt{\pi n}\left[\sum_{k=0}^{2n-1} |\psi(t_k)|^2\right]^{1/2}$$

where $|\cdot|$ again denotes the Euclidean norm in $\mathbb{C}^n$. Using this estimate and (3.92) for $\beta_k^\delta = 0$, we conclude that

$$\|R_n\psi\|_{L^2} = \|\psi_n\|_{L^2} \leq n\|Q_n\psi\|_{L^2} \qquad (3.100)$$

for all $\psi \in H^1_{per}(0, 2\pi)$. Thus

$$\|R_n K_0\psi\|_{L^2} \leq n\|Q_n K_0\psi\|_{L^2}$$

for all $\psi \in L^2(0, 2\pi)$. Now we estimate $\|R_n K_0\psi\|_{L^2}$ by the L^2-norm of ψ itself.

Let $\tilde{\psi}_n = P_n\psi \in X_n$ be the orthogonal projection of $\psi \in L^2(0, 2\pi)$ in X_n. Then $R_n K_0\tilde{\psi}_n = \tilde{\psi}_n$ and $\|\tilde{\psi}_n\|_{L^2} \leq \|\psi\|_{L^2}$, and thus

$$\begin{aligned}
\|R_n K_0\psi - \tilde{\psi}_n\|_{L^2} &= \|R_n K_0(\psi - \tilde{\psi}_n)\|_{L^2} \leq n\|Q_n K_0(\psi - \tilde{\psi}_n)\|_{L^2} \\
&\leq n\|Q_n K_0\psi - K_0\psi\|_{L^2} + n\|K_0\psi - K_0\tilde{\psi}_n\|_{L^2} \\
&\quad + n\|K_0\tilde{\psi}_n - Q_n K_0\tilde{\psi}_n\|_{L^2}.
\end{aligned}$$

Now we use the error estimates (3.86a), (3.94a), and (3.94b) of Lemma 3.26. This yields

$$\begin{aligned}
\|R_n K_0\psi - \tilde{\psi}_n\|_{L^2} &\leq c_1\left[\|K_0\psi\|_{H^1_{per}} + \|\psi\|_{L^2} + \|K_0\tilde{\psi}_n\|_{H^1_{per}}\right] \\
&\leq c_2\left[\|\psi\|_{L^2} + \|\tilde{\psi}_n\|_{L^2}\right] \leq c_3\|\psi\|_{L^2},
\end{aligned}$$

that is, $\|R_n K_0\psi\|_{L^2} \leq c_4\|\psi\|_{L^2}$ for all $\psi \in L^2(0, 2\pi)$. Therefore, the assumptions of Theorem 3.7 are satisfied. The application of Theorem 3.19 yields the error estimate (3.99). $\qquad\square$

Among the extensive literature on collocation methods for Symm's integral equation and related equations, we mention only the work of [8, 66, 67, 140, 147, 241, 242]. Symm's equation has also been numerically treated by quadrature methods; see [90, 169, 239, 240, 250, 251]. For more general problems, we refer to [9, 68].

3.5   Numerical Experiments for Symm's Equation

In this section, we apply all of the previously investigated regularization strategies to Symm's integral equation

$$(K\psi)(t) := -\frac{1}{\pi} \int_0^{2\pi} \psi(s) \ln |\gamma(t) - \gamma(s)| \, ds = g(t), \quad 0 \le t \le 2\pi,$$

where in this example $\gamma(s) = (\cos s, 2\sin s), 0 \le s \le 2\pi$, denotes the parametrization of the ellipse with semiaxes 1 and 2. First, we discuss the numerical computation of $K\psi$. We write $K\psi$ in the form (see (3.60))

$$(K\psi)(t) = -\frac{1}{2\pi} \int_0^{2\pi} \psi(s) \ln\left(4\sin^2 \frac{t-s}{2}\right) ds + \int_0^{2\pi} \psi(s) \, k(t,s) \, ds,$$

for $0 \le t \le 2\pi$, with the analytic function

$$k(t,s) = -\frac{1}{2\pi} \ln \frac{|\gamma(t) - \gamma(s)|^2}{4\sin^2 \frac{t-s}{2}}, \quad t \ne s,$$

$$k(t,t) = -\frac{1}{\pi} \ln |\dot{\gamma}(t)|, \quad 0 \le t \le 2\pi.$$

We use the trapezoidal rule for periodic functions (see [168]). Let $t_j = j\frac{\pi}{n}$, $j = 0, \ldots, 2n - 1$. The smooth part is approximated by

$$\int_0^{2\pi} k(t,s)\,\psi(s)\,ds \approx \frac{\pi}{n} \sum_{j=0}^{2n-1} k(t,t_j)\,\psi(t_j), \quad 0 \le t \le 2\pi.$$

For the weakly singular part, we replace ψ by its trigonometric interpolation polynomial $Q_n\psi = \sum_{j=0}^{2n-1} \psi(t_j)\, L_j$ into the $2n$-dimensional space

$$\left\{ \sum_{j=0}^{n} a_j \cos(jt) + \sum_{j=1}^{n-1} b_j \sin(jt) : a_j, b_j \in \mathbb{R} \right\}$$

over $\mathbb{R}$ (see Section A.4 of Appendix A.4). From (A.37) and Lemma 3.16, we conclude that

$$-\frac{1}{2\pi} \int_0^{2\pi} \psi(s) \ln\left(4\sin^2\frac{t-s}{2}\right) ds \;\approx\; -\frac{1}{2\pi} \int_0^{2\pi} (Q_n\psi)(s) \ln\left(4\sin^2\frac{t-s}{2}\right) ds$$

$$= \sum_{j=0}^{2n-1} \psi(t_j)\, R_j(t), \quad 0 \le t \le 2\pi,$$

where

$$R_j(t) \;=\; -\frac{1}{2\pi} \int_0^{2\pi} L_j(s) \ln\left(4\sin^2\frac{t-s}{2}\right) ds$$

$$=\; \frac{1}{n}\left\{\frac{1}{2n}\cos n(t-t_j) \;+\; \sum_{m=1}^{n-1}\frac{1}{m}\cos m(t-t_j)\right\}$$

for $j = 0, \ldots, 2n - 1$. Therefore, the operator K is replaced by

$$(K_n\psi)(t) \;:=\; \sum_{j=0}^{2n-1} \psi(t_j)\left[R_j(t) \;+\; \frac{\pi}{n}k(t,t_j)\right], \quad 0 \le t \le 2\pi.$$

It is well-known (see [168]) that $K_n\psi$ converges uniformly to $K\psi$ for every 2π-periodic continuous function ψ. Furthermore, the error $\|K_n\psi - K\psi\|_\infty$ is exponentially decreasing for analytic functions ψ. For $t = t_k$, $k = 0, \ldots, 2n-1$, we have $(K_n\psi)(t_k) = \sum_{j=0}^{2n-1} A_{kj}\,\psi(t_j)$ with the symmetric matrix

$$A_{kj} \;:=\; R_{|k-j|} \;+\; \frac{\pi}{n}k(t_k,t_j), \quad k,j = 0\ldots, 2n - 1,$$

where
$$R_\ell \;=\; \frac{1}{n}\left\{\frac{(-1)^\ell}{2n} \;+\; \sum_{m=1}^{n-1}\frac{1}{m}\cos\frac{m\ell\pi}{n}\right\}, \quad \ell = 0, \ldots, 2n - 1.$$

For the numerical example, we take $\psi(s) = \exp(3\sin s)$, $0 \le s \le 2\pi$, and $g = K\psi$ or, discretized, $\tilde\psi_j = \exp(3\sin t_j)$, $j = 0, \ldots, 2n-1$, and $\tilde g = A\tilde\psi$. We take $n = 60$ and add uniformly distributed random noise on the data $\tilde g$. All the results show the average of 10 computations. The errors are measured in the discrete norm $|z|_2^2 := \frac{1}{2n}\sum_{j=0}^{2n-1}|z_j|^2$, $z \in \mathbb{C}^{2n}$.

First, we consider *Tikhonov's regularization method* for $\delta = 0.1$, $\delta = 0.01$, $\delta = 0.001$, and $\delta = 0$. In Figure 3.1, we plot the errors $\left|\tilde\psi^{\alpha,\delta} - \tilde\psi\right|_2$ and $\left|A\tilde\psi^{\alpha,\delta} - \tilde g\right|_2$ in the solution and the right-hand side, respectively, versus the regularization parameter α.

We clearly observe the expected behavior of the errors: For $\delta > 0$ the error in the solution has a well-defined minimum that depends on δ, while the defect always converges to zero as α tends to zero.

The minimal values err_δ of the errors in the solution are approximately 0.351, 0.0909, and 0.0206 for $\delta = 0.1$, 0.01, and 0.001, respectively. From this, we observe the order of convergence: increasing the error by factor 10 should increase the error by factor $10^{2/3} \approx 4.64$, which roughly agrees with the numerical results where $err_{\delta=0.1}/err_{\delta=0.01} \approx 3.86$ and $err_{\delta=0.01}/err_{\delta=0.001} \approx 4.41$.

In Figure 3.2, we show the results for the *Landweber iteration* with $a = 0.5$ for the same example where again $\delta = 0.1$, $\delta = 0.01$, $\delta = 0.001$, and $\delta = 0$. The errors in the solution and the defects are now plotted versus the iteration number m.

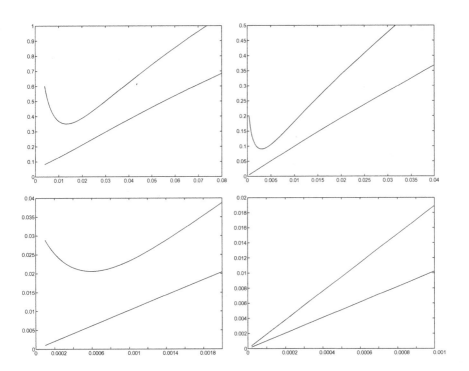

Figure 3.1: Error for Tikhonov's regularization method.

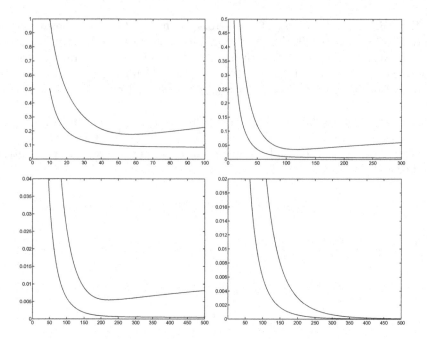

Figure 3.2: Error for Landweber's method ($a = 0.5$).

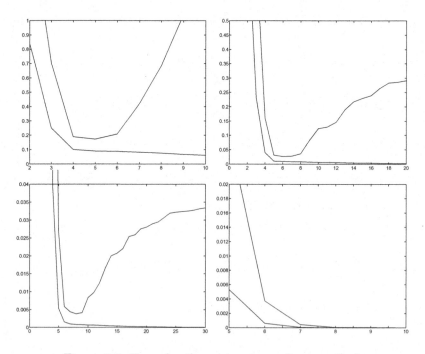

Figure 3.3: Error for the conjugate gradient method.

In Figure 3.3, we show the results for the *conjugate gradient method* for the same example where again $\delta = 0.1$, $\delta = 0.01$, $\delta = 0.001$, and $\delta = 0$. The errors in the solution and the defects are again plotted versus the iteration number m.

Here, we observe the same behavior as for Tikhonov's method. We note the difference in the results for the Landweber method and the conjugate gradient method. The latter decreases the errors very quickly but is very sensitive to the exact stopping rule, while the Landweber iteration is slow but very stable with respect to the stopping parameter τ. The minimal values are $err_{\delta=0.1} \approx 0.177$, $err_{\delta=0.01} \approx 0.0352$, and $err_{\delta=0.001} \approx 0.0054$ for the Landweber iteration and $err_{\delta=0.1} \approx 0.172$, $err_{\delta=0.01} \approx 0.0266$, and $err_{\delta=0.001} \approx 0.0038$ for the conjugate gradient method. The corresponding factors are considerably larger than $10^{2/3} \approx 4.64$ indicating the optimality of these methods also for smooth solutions (see the remarks following Theorem 2.15).

Table 3.1: Least squares method

n	$\delta = 0.1$	$\delta = 0.01$	$\delta = 0.001$	$\delta = 0$
1	38.190	38.190	38.190	38.190
2	15.772	15.769	15.768	15.768
3	5.2791	5.2514	5.2511	5.2511
4	1.6209	1.4562	1.4541	1.4541
5	1.0365	0.3551	$3.433 * 10^{-1}$	$3.432 * 10^{-1}$
6	1.1954	0.1571	$7.190 * 10^{-2}$	$7.045 * 10^{-2}$
10	2.7944	0.2358	$2.742 * 10^{-2}$	$4.075 * 10^{-5}$
12	3.7602	0.3561	$3.187 * 10^{-2}$	$5.713 * 10^{-7}$
15	4.9815	0.4871	$4.977 * 10^{-2}$	$5.570 * 10^{-10}$
20	7.4111	0.7270	$7.300 * 10^{-2}$	$3.530 * 10^{-12}$

Table 3.2: Bubnov–Galerkin method

n	$\delta = 0.1$	$\delta = 0.01$	$\delta = 0.001$	$\delta = 0$
1	38.190	38.190	38.190	38.190
2	15.771	15.769	15.768	15.768
3	5.2752	5.2514	5.2511	5.2511
4	1.6868	1.4565	1.4541	1.4541
5	1.1467	0.3580	$3.434 * 10^{-1}$	$3.432 * 10^{-1}$
6	1.2516	0.1493	$7.168 * 10^{-2}$	$7.045 * 10^{-2}$
10	2.6849	0.2481	$2.881 * 10^{-2}$	$4.075 * 10^{-5}$
12	3.3431	0.3642	$3.652 * 10^{-2}$	$5.713 * 10^{-7}$
15	4.9549	0.4333	$5.719 * 10^{-2}$	$5.570 * 10^{-10}$
20	7.8845	0.7512	$7.452 * 10^{-2}$	$3.519 * 10^{-12}$

Next, we compute the same example using some projection methods. First, we list the results for the least squares method and the Bubnov–Galerkin method of Subsections 3.2.1 and 3.2.3 in Tables 3.1 and 3.2. We observe that both methods produce almost the same results, which reflect the estimates of Theorem 3.18. Note that for $\delta = 0$ the error decreases exponentially with m. This reflects the fact that the best approximation $\min\{\|\psi - \phi_n\|_{L^2} : \phi_n \in X_n\}$ converges to zero exponentially due to the analyticity of the solution $\psi(s) = \exp(3 \sin s)$ (see [168], Theorem 11.7).

Now we turn to the collocation methods of Section 3.4. To implement the collocation method (3.83) for Symm's integral equation and the basis functions (3.89), (3.93a), and (3.93b), we have to compute the integrals

$$-\frac{1}{\pi} \int_0^{2\pi} e^{ijs} \ln|\gamma(t_k) - \gamma(s)|\, ds\,, \tag{3.101a}$$

$j = -m, \ldots, m - 1,\ k = 0, \ldots, 2m - 1$, and

$$-\frac{1}{\pi} \int_0^{2\pi} \hat{x}_j(s) \ln|\gamma(t_k) - \gamma(s)|\, ds\,, \quad j, k = 0, \ldots, 2m - 1\,, \tag{3.101b}$$

respectively. For the first integral (3.101a), we write using (3.64),

$$
\begin{aligned}
&-\frac{1}{\pi} \int_0^{2\pi} e^{ijs} \ln|\gamma(t_k) - \gamma(s)|\, ds \\
&= -\frac{1}{2\pi} \int_0^{2\pi} e^{ijs} \ln\left(4 \sin^2 \frac{t_k - s}{2}\right) ds - \frac{1}{2\pi} \int_0^{2\pi} e^{ijs} \ln \frac{|\gamma(t_k) - \gamma(s)|^2}{4 \sin^2(t_k - s)/2}\, ds \\
&= \varepsilon_j\, e^{ijt_k} - \frac{1}{2\pi} \int_0^{2\pi} e^{ijs} \ln \frac{|\gamma(t_k) - \gamma(s)|^2}{4 \sin^2(t_k - s)/2}\, ds\,,
\end{aligned}
$$

where $\varepsilon_j = 0$ for $j = 0$ and $\varepsilon_j = 1/|j|$ otherwise. The remaining integral is computed by the trapezoidal rule.

The computation of (3.101b) is more complicated. By Definition (3.93a), (3.93b) of $\hat{x}_j$, we have to calculate

$$\int_{t_j - \pi/(2m)}^{t_j + \pi/(2m)} \ln|\gamma(t_k) - \gamma(s)|^2\, ds = \int_{-\pi/(2m)}^{\pi/(2m)} \ln|\gamma(t_k) - \gamma(s + t_j)|^2\, ds\,.$$

For $j \neq k$, the integrand is analytic, and we use Simpson's rule

$$\int_{-\pi/(2m)}^{\pi/(2m)} g(s)\, ds \approx \sum_{\ell=0}^{n} w_\ell\, g(s_\ell),$$

where

$$s_\ell = \ell\,\frac{\pi}{mn} - \frac{\pi}{2m}, \qquad w_\ell = \frac{\pi}{3mn} \cdot \begin{cases} 1, & \ell = 0 \text{ or } n, \\ 4, & \ell = 1,3,\ldots,n-1, \\ 2, & \ell = 2,4,\ldots,n-2, \end{cases}$$

$\ell = 0,\ldots,n$. For $j = k$, the integral has a weak singularity at $s = 0$. We split the integrand into

$$\int_{-\pi/(2m)}^{\pi/(2m)} \ln\!\left(4\sin^2\frac{s}{2}\right) ds + \int_{-\pi/(2m)}^{\pi/(2m)} \ln \frac{|\gamma(t_k) - \gamma(s+t_k)|^2}{4\sin^2(s/2)}\, ds$$

$$= -2 \int_{\pi/(2m)}^{\pi} \ln\!\left(4\sin^2\frac{s}{2}\right) ds + \int_{-\pi/(2m)}^{\pi/(2m)} \ln \frac{|\gamma(t_k) - \gamma(s+t_k)|^2}{4\sin^2(s/2)}\, ds$$

because $\ln\!\left(4\sin^2(s/2)\right)$ is even and $\int_0^\pi \ln\!\left(4\sin^2(s/2)\right) ds = 0$ by (3.64). Both integrals are approximated by Simpson's rule. For the same example as earlier, with 100 integration points for Simpson's rule we obtain the following results for basis functions (3.89) (in Table 3.3) and basis functions (3.93a), (3.93b) (in Table 3.4).

Table 3.3: Collocation method for basis functions (3.89)

m	$\delta = 0.1$	$\delta = 0.01$	$\delta = 0.001$	$\delta = 0$
1	6.7451	6.7590	6.7573	6.7578
2	1.4133	1.3877	1.3880	1.3879
3	0.3556	$2.791 * 10^{-1}$	$2.770 * 10^{-1}$	$2.769 * 10^{-1}$
4	0.2525	$5.979 * 10^{-2}$	$5.752 * 10^{-2}$	$5.758 * 10^{-2}$
5	0.3096	$3.103 * 10^{-2}$	$1.110 * 10^{-2}$	$1.099 * 10^{-2}$
6	0.3404	$3.486 * 10^{-2}$	$3.753 * 10^{-3}$	$1.905 * 10^{-3}$
10	0.5600	$5.782 * 10^{-2}$	$5.783 * 10^{-3}$	$6.885 * 10^{-7}$
12	0.6974	$6.766 * 10^{-2}$	$6.752 * 10^{-3}$	$8.135 * 10^{-9}$
15	0.8017	$8.371 * 10^{-2}$	$8.586 * 10^{-3}$	$6.436 * 10^{-12}$
20	1.1539	$1.163 * 10^{-1}$	$1.182 * 10^{-2}$	$1.806 * 10^{-13}$

Table 3.4: Collocation method for basis functions (3.93a) and (3.93b)

m	$\delta = 0.1$	$\delta = 0.01$	$\delta = 0.001$	$\delta = 0$
1	6.7461	6.7679	6.7626	6.7625
2	1.3829	1.3562	1.3599	1.3600
3	0.4944	$4.874 * 10^{-1}$	$4.909 * 10^{-1}$	$4.906 * 10^{-1}$
4	0.3225	$1.971 * 10^{-1}$	$2.000 * 10^{-1}$	$2.004 * 10^{-1}$
5	0.3373	$1.649 * 10^{-1}$	$1.615 * 10^{-1}$	$1.617 * 10^{-1}$
6	0.3516	$1.341 * 10^{-1}$	$1.291 * 10^{-1}$	$1.291 * 10^{-1}$
10	0.5558	$8.386 * 10^{-2}$	$6.140 * 10^{-2}$	$6.107 * 10^{-2}$
12	0.6216	$7.716 * 10^{-2}$	$4.516 * 10^{-2}$	$4.498 * 10^{-2}$
15	0.8664	$9.091 * 10^{-2}$	$3.137 * 10^{-2}$	$3.044 * 10^{-2}$
20	1.0959	$1.168 * 10^{-1}$	$2.121 * 10^{-2}$	$1.809 * 10^{-2}$
30	1.7121	$1.688 * 10^{-1}$	$1.862 * 10^{-2}$	$8.669 * 10^{-3}$

The difference for $\delta = 0$ reflects the fact that the best approximation

$$\min\{\|\psi - \phi_n\|L^2 : \phi_n \in \text{span}\{\hat{x}_j : j \in J\}\}$$

converges to zero exponentially for $\hat{x}_j$ defined by (3.89), while it converges to zero only of order $1/n$ for $\hat{x}_j$ defined by (3.93a) and (3.93b) (see Theorem 3.28).

We have seen in this section that the theoretical investigations of the regularization strategies are confirmed by the numerical results for Symm's integral equation.

3.6   The Backus–Gilbert Method

In this section, we study a different numerical method for "solving" finite moment problems of the following type:

$$\int_a^b k_j(s)\, x(s)\, ds \; = \; y_j\,, \quad j = 1, \ldots, n\,. \tag{3.102}$$

Here, $y_j \in \mathbb{R}$ are any given numbers and $k_j \in L^2(a, b)$ arbitrary given functions. Certainly, we have in mind that $y_j = y(t_j)$ and $k_j = k(t_j, \cdot)$. In Section 3.4, we studied the moment solution of such problems; see [184, 229]. We saw that the moment solution x_n is a finite linear combination of the functions $\{k_1, \ldots, k_n\}$. Therefore, the moment solution x_n is as smooth as the functions k_j even if the true solution is smoother.

The concept originally proposed by Backus and Gilbert ([13, 14]) does not primarily wish to solve the moment problem but rather wants to determine how well all possible models x can be recovered pointwise.

Define the finite-dimensional operator $K : L^2(a,b) \to \mathbb{R}^n$ by

$$(Kx)_j = \int_a^b k_j(s)\, x(s)\, ds, \quad j = 1, \dots, n, \quad x \in L^2(a,b). \tag{3.103}$$

We try to find a left inverse S, that is, a linear operator $S : \mathbb{R}^n \to L^2(a,b)$ such that

$$SKx \approx x \quad \text{for all } x \in L^2(a,b). \tag{3.104}$$

Therefore, SKx should be a simultaneous approximation to all possible $x \in L^2(a,b)$. Of course, we have to make clear the meaning of the approximation.

The general form of a linear operator $S : \mathbb{R}^n \to L^2(a,b)$ has to be

$$(Sy)(t) = \sum_{j=1}^n y_j\, \varphi_j(t), \quad t \in (a,b), \quad y = (y_j) \in \mathbb{R}^n, \tag{3.105}$$

for some $\varphi_j \in L^2(a,b)$ that are to be determined from the requirement (3.104):

$$\begin{aligned}
(SKx)(t) &= \sum_{j=1}^n \varphi_j(t) \int_a^b k_j(s)\, x(s)\, ds \\
&= \int_a^b \left[\sum_{j=1}^n k_j(s)\, \varphi_j(t) \right] x(s)\, ds.
\end{aligned}$$

The requirement $SKx \approx x$ leads to the problem of approximating Dirac's delta distribution $\delta(s - t)$ by linear combinations of the form $\sum_{j=1}^n k_j(s)\, \varphi_j(t)$. For example, one can show that the minimum of

$$\int_a^b \int_a^b \left| \sum_{j=1}^n k_j(s)\, \varphi_j(t) - \delta(s - t) \right|^2 ds\, dt$$

(in the sense of distributions) is attained at $\varphi(s) = A^{-1} k(s)$, where $k(s) = \big(k_1(s), \dots, k_n(s)\big)^\top$ and $A_{ij} = \int_a^b k_i(s)\, k_j(s)\, ds$, $i, j = 1, \dots, n$. For this minimization criterion, $x = \sum_{j=1}^n y_j \varphi_j$ is again the moment solution of Subsection 3.4.1. In [184], it is shown that minimizing with respect to an H_{per}^{-s}-norm for $s > 1/2$ leads to projection methods in H_{per}^s-spaces. We refer also to [272] for a comparison of several minimization criteria.

The Backus–Gilbert method is based on a pointwise minimization criterion: Treat $t \in [a,b]$ as a fixed parameter and determine the numbers $\varphi_j = \varphi_j(t)$ for $j = 1, \dots, n$, as the solution of the following minimization problem:

$$\text{minimize} \quad \int_a^b |s - t|^2 \left| \sum_{j=1}^n k_j(s)\, \varphi_j \right|^2 ds \tag{3.106a}$$

subject to $\varphi \in \mathbb{R}^n$ and

$$\int_a^b \sum_{j=1}^n k_j(s)\,\varphi_j\,ds \;=\; 1\,. \tag{3.106b}$$

Using the matrix-vector notation, we rewrite this problem in short form:

$$\text{minimize} \quad \varphi^\top Q(t)\,\varphi \quad \text{subject to} \quad r \cdot \varphi = 1\,,$$

where

$$Q(t)_{ij} \;=\; \int_a^b |s - t|^2 k_i(s)\,k_j(s)\,ds, \quad i,j = 1,\ldots,n,$$

$$r_j \;=\; \int_a^b k_j(s)\,ds, \quad j = 1,\ldots,n\,.$$

This is a quadratic minimization problem with one linear equality constraint. We assume that $r \neq 0$ because otherwise the constraint (3.106b) cannot be satisfied. Uniqueness and existence are assured by the following theorem, which also gives a characterization by the Lagrange multiplier rule.

Theorem 3.29 *Assume that $\{k_1, \ldots, k_n\}$ are linearly independent. Then the symmetric matrix $Q(t) \in \mathbb{R}^{n \times n}$ is positive definite for every $t \in [a,b]$. The minimization problem (3.106a), (3.106b) is uniquely solvable. $\varphi \in \mathbb{R}^n$ is a solution of (3.106a) and (3.106b) if and only if there exists a number $\lambda \in \mathbb{R}$ (the Lagrange multiplier) such that $(\varphi, \lambda) \in \mathbb{R}^n \times \mathbb{R}$ solves the linear system*

$$Q(t)\varphi - \lambda r \;=\; 0 \quad and \quad r \cdot \varphi \;=\; 1\,. \tag{3.107}$$

$\lambda = \varphi^\top Q(t)\,\varphi$ *is the minimal value of this problem.*

Proof: From

$$\varphi^\top Q(t)\,\varphi \;=\; \int_a^b |s - t|^2 \left| \sum_{j=1}^n k_j(s)\,\varphi_j \right|^2 ds,$$

we conclude first that $\varphi^\top Q(t)\,\varphi \geq 0$ and second that $\varphi^\top Q(t)\,\varphi = 0$ implies that $\sum_{j=1}^n k_j(s)\,\varphi_j = 0$ for almost all $s \in (a,b)$. Because $\{k_j\}$ are linearly independent, $\varphi_j = 0$ for all j follows. Therefore, $Q(t)$ is positive definite. Existence, uniqueness, and equivalence to (3.107) are elementary results from optimization theory; see [269].   □

Definition 3.30 *We denote by $\left(\varphi_j(t)\right)_{j=1}^n \in \mathbb{R}^n$ the unique solution $\varphi \in \mathbb{R}^n$ of (3.106a) and (3.106b). The Backus–Gilbert solution x_n of*

$$\int_a^b k_j(s)\,x_n(s)\,ds \;=\; y_j, \quad j = 1,\ldots,n,$$

is defined as

$$x_n(t) = \sum_{j=1}^{n} y_j \varphi_j(t), \quad t \in [a,b]. \qquad (3.108)$$

The minimal value $\lambda = \lambda(t) = \varphi(t)^\top Q(t) \varphi(t)$ *is called the* spread.

We remark that, in general, the Backus–Gilbert solution $x_n = \sum_{j=1}^{n} y_j \varphi_j$ is not a solution of the moment problem, that is, $\int_a^b k_j(s)\, x_n(s)\, ds \neq y_j$! This is certainly a disadvantage. On the other hand, the solution x is analytic in $[a,b]$—even for nonsmooth data k_j. We can prove the following lemma.

Lemma 3.31 φ_j *and* λ *are rational functions. More precisely, there exist polynomials* $p_j, q \in \mathcal{P}_{2(n-1)}$ *and* $\rho \in \mathcal{P}_{2n}$ *such that* $\varphi_j = p_j/q,\ j = 1,\ldots,n$, *and* $\lambda = \rho/q$. *The polynomial* q *has no zeros in* $[a,b]$.

Proof: Obviously, $Q(t) = Q_0 - 2t\, Q_1 + t^2\, Q_2$ with symmetric matrices Q_0, Q_1, Q_2. We search for a polynomial solution $p \in \left[\mathcal{P}_m\right]^n$ and $\rho \in \mathcal{P}_{m+2}$ of $Q(t)p(t) - \rho(t)\, r = 0$ with $m = 2(n-1)$. Because the number of equations is $n(m+3) = 2n^2 + n$ and the number of unknowns is $n(m+1) + (m+3) = 2n^2 + n + 1$, there exists a nontrivial solution $p \in \left[\mathcal{P}_m\right]^n$ and $\rho \in \mathcal{P}_{m+2}$. If $p(\hat{t}) = 0$ for some $\hat{t} \in [a,b]$, then $\rho(\hat{t}) = 0$ because $r \neq 0$. In this case, we divide the equation by $(t - \hat{t})$. Therefore, we can assume that p has no zero in $[a,b]$.

Now we define $q(t) := r \cdot p(t)$ for $t \in [a,b]$. Then $q \in \mathcal{P}_m$ has no zero in $[a,b]$ because otherwise we would have

$$0 = \rho(\hat{t})\, r \cdot p(\hat{t}) = p(\hat{t})^\top Q(\hat{t})\, p(\hat{t});$$

thus $p(\hat{t}) = 0$, a contradiction. Therefore, $\varphi := p/q$ and $\lambda := \rho/q$ solves (3.107). By the uniqueness result, this is the only solution.   □

For the following error estimates, we assume two kinds of a priori information on x depending on the norm of the desired error estimate. Let

$$X_n = \operatorname{span}\{k_j : j = 1,\ldots,n\}.$$

Theorem 3.32 *Let* $x \in L^2(a,b)$ *be any solution of the finite moment problem* (3.102) *and* $x_n = \sum_{j=1}^{n} y_j \varphi_j$ *be the Backus–Gilbert solution. Then the following error estimates hold:*

(a) *Assume that* x *is Lipschitz continuous with constant* $\ell > 0$, *that is,*

$$|x(t) - x(s)| \leq \ell |s - t| \quad \text{for all } s, t \in [a,b].$$

Then

$$|x_n(t) - x(t)| \leq \ell \sqrt{b-a}\; \epsilon_n(t) \qquad (3.109)$$

for all $n \in \mathbb{N}$, $t \in [a,b]$, *where* $\epsilon_n(t)$ *is defined by*

$$\epsilon_n^2(t) := \min\left\{ \int_a^b |s-t|^2\, |z_n(s)|^2 ds : z_n \in X_n,\ \int_a^b z_n(s)\, ds = 1 \right\}. \quad (3.110)$$

(b) Let $x \in H^1(a,b)$. Then there exists $c > 0$, independent of x, such that

$$\|x_n - x\|_{L^2} \leq c \|x'\|_{L^2} \|\epsilon_n\|_\infty \quad \text{for all } n \in \mathbb{N}. \tag{3.111}$$

Proof: By the definition of the Backus–Gilbert solution and the constraint on φ, we have

$$x_n(t) - x(t) = \sum_{j=1}^{n} y_j \, \varphi_j(t) - x(t) \int_a^b \sum_{j=1}^{n} k_j(s) \, \varphi_j(t) \, ds$$

$$= \sum_{j=1}^{n} \int_a^b k_j(s) \, [x(s) - x(t)] \, \varphi_j(t) \, ds \, .$$

Thus

$$|x_n(t) - x(t)| \leq \int_a^b \left| \sum_{j=1}^{n} k_j(s) \, \varphi_j(t) \right| |x(s) - x(t)| \, ds \, .$$

Now we distinguish between parts (a) and (b):

(a) Let $|x(t) - x(s)| \leq \ell|t - s|$. Then, by the Cauchy–Schwarz inequality and the definition of φ_j,

$$|x_n(t) - x(t)| \leq \ell \int_a^b 1 \cdot \left| \sum_{j=1}^{n} k_j(s) \, \varphi_j(t) \right| |t - s| \, ds$$

$$\leq \ell \sqrt{b - a} \left[\int_a^b \left| \sum_{j=1}^{n} k_j(s) \, \varphi_j(t) \right|^2 |t - s|^2 \, ds \right]^{1/2}$$

$$= \ell \sqrt{b - a} \, \epsilon_n(t) \, .$$

(b) First, we define the cutoff function λ_δ on $[a,b] \times [a,b]$ by

$$\lambda_\delta(t, s) = \begin{cases} 1, & |t - s| \geq \delta, \\ 0, & |t - s| < \delta. \end{cases} \tag{3.112}$$

Then, by the Cauchy–Schwarz inequality again,

$$\left[\int_a^b \lambda_\delta(t, s) \left| \sum_{j=1}^{n} k_j(s) \, \varphi_j(t) \right| |x(s) - x(t)| \, ds \right]^2$$

$$= \left[\int_a^b \left| \sum_{j=1}^{n} k_j(s) \, \varphi_j(t) \, (t - s) \, \lambda_\delta(t, s) \right| \left| \frac{x(s) - x(t)}{t - s} \right| \, ds \right]^2$$

$$\leq \epsilon_n(t)^2 \int_a^b \lambda_\delta(t, s) \left| \frac{x(s) - x(t)}{s - t} \right|^2 \, ds \, .$$

Integration with respect to t yields

$$\int\limits_a^b \left[\int\limits_a^b \lambda_\delta(t,s) \left|\sum_{j=1}^n k_j(s)\,\varphi_j(t)\right| |x(s) - x(t)|\,ds\right]^2 dt$$

$$\leq \|\epsilon_n\|_\infty^2 \int\limits_a^b \int\limits_a^b \left|\frac{x(s) - x(t)}{s - t}\right|^2 \lambda_\delta(t,s)\,ds\,dt\,.$$

The following technical lemma from the theory of Sobolev spaces yields the assertion. $\quad\square$

Lemma 3.33 *There exists $c > 0$ such that*

$$\int\limits_a^b \int\limits_a^b \left|\frac{x(s) - x(t)}{s - t}\right|^2 \lambda_\delta(t,s)\,ds\,dt \;\leq\; c\|x'\|_{L^2}^2$$

for all $\delta > 0$ and $x \in H^1(a,b)$. Here, the cutoff function λ_δ is defined by (3.112).

Proof: First, we estimate

$$|x(s) - x(t)|^2 \;=\; \left|\int\limits_s^t 1 \cdot x'(\tau)\,d\tau\right|^2 \;\leq\; |t - s|\left|\int\limits_s^t |x'(\tau)|^2\,d\tau\right|$$

and thus, for $s \neq t$,

$$\left|\frac{x(s) - x(t)}{s - t}\right|^2 \;\leq\; \frac{1}{|s - t|}\left|\int\limits_s^t |x'(\tau)|^2\,d\tau\right|\,.$$

Now we fix $t \in (a,b)$ and write

$$\int\limits_a^b \left|\frac{x(s) - x(t)}{s - t}\right|^2 \lambda_\delta(t,s)\,ds$$

$$\leq \int\limits_a^t \frac{\lambda_\delta(t,s)}{t - s} \int\limits_s^t |x'(\tau)|^2\,d\tau\,ds \;+\; \int\limits_t^b \frac{\lambda_\delta(t,s)}{s - t} \int\limits_t^s |x'(\tau)|^2\,d\tau\,ds$$

$$= \int\limits_a^b |x'(s)|^2\, A_\delta(t,s)\,ds\,,$$

where

$$A_\delta(t,s) \;=\; \begin{cases} \int\limits_a^s \frac{\lambda_\delta(t,\tau)}{|t-\tau|}\,d\tau, & a \leq s < t, \\[2mm] \int\limits_s^b \frac{\lambda_\delta(t,\tau)}{|t-\tau|}\,d\tau, & t < s \leq b, \end{cases} \;=\; \begin{cases} \ln \frac{t-a}{\max(\delta,t-s)}, & s \leq t, \\[2mm] \ln \frac{b-t}{\max(\delta,s-t)}, & s \geq t. \end{cases}$$

Finally, we estimate

$$\int\limits_a^b \int\limits_a^b \frac{|x(s) - x(t)|^2}{|s-t|^2}\, \lambda_\delta(t,s)\, ds\, dt \;\leq\; \int\limits_a^b |x'(s)|^2 \left(\int\limits_a^b A_\delta(t,s)\, dt\right) ds$$

and

$$\int\limits_a^b A_\delta(t,s)\, dt \;\leq\; c \quad \text{for all } s \in (a,b) \text{ and } \delta > 0$$

which is seen by elementary integration.   □

From these error estimates, we observe that the rate of convergence depends on the magnitude of ϵ_n, that is, how well the kernels approximate the delta distribution. Finally, we study the question of convergence for $n \to \infty$.

Theorem 3.34 *Assume that $\{k_j : j \in \mathbb{N}\}$ is linearly independent and dense in $L^2(a,b)$. Then*

$$\|\epsilon_n\|_\infty \longrightarrow 0 \quad \text{for } n \to \infty.$$

Proof: For fixed $t \in [a,b]$ and arbitrary $\delta \in \big(0, (b-a)/2\big)$, we define

$$\tilde{v}(s) := \begin{cases} \frac{1}{|s-t|}, & |s-t| \geq \delta, \\ 0, & |s-t| < \delta, \end{cases} \quad \text{and} \quad v(s) := \left[\int\limits_a^b \tilde{v}(\tau)\, d\tau\right]^{-1} \tilde{v}(s).$$

Then $v \in L^2(a,b)$ and $\int_a^b v(s)\, ds = 1$. Because $\bigcup X_n$ is dense in $L^2(a,b)$, there exists a sequence $\tilde{v}_n \in X_n$ with $\tilde{v}_n \to v$ in $L^2(a,b)$. This implies also that $\int_a^b \tilde{v}_n(s)\, ds \to \int_a^b v(s)\, ds = 1$. Therefore, the functions

$$v_n := \left[\int_a^b \tilde{v}_n(s)\, ds\right]^{-1} \tilde{v}_n \;\in\; X_n$$

converge to v in $L^2(a,b)$ and are normalized by $\int_a^b v_n(s)\, ds = 1$. Thus v_n is admissible, and we conclude that

$$\epsilon_n(t)^2 \;\leq\; \int\limits_a^b |s-t|^2\, v_n(s)^2\, ds$$

$$= \int\limits_a^b |s-t|^2\, v(s)^2\, ds \;+\; 2\int\limits_a^b |s-t|^2\, v(s)\,[v_n(s) - v(s)]\, ds$$

$$+ \int\limits_a^b |s-t|^2\, [v_n(s) - v(s)]^2\, ds$$

$$\leq\; (b-a)\left[\int\limits_a^b \tilde{v}(s)\, ds\right]^{-2} + (b-a)^2 \big[2\,\|v\|_{L^2}\, \|v_n - v\|_{L^2} + \|v_n - v\|_{L^2}^2\big].$$

This shows that

$$\limsup_{n\to\infty} \epsilon_n(t) \leq \sqrt{b-a}\left[\int_a^b \tilde{v}(s)\,ds\right]^{-1} \quad \text{for all } t \in [a,b].$$

Direct computation yields

$$\int_a^b \tilde{v}(s)\,ds \geq c + |\ln\delta|$$

for some c independent of δ; thus

$$\limsup_{n\to\infty} \epsilon_n(t) \leq \frac{\sqrt{b-a}}{c+|\ln\delta|} \quad \text{for all } \delta \in \big(0,(b-a)/2\big).$$

This yields pointwise convergence, that is, $\epsilon_n(t) \to 0$ $(n \to \infty)$ for every $t \in [a,b]$. Because $\epsilon_n(t)$ is monotonic with respect to n, Dini's well-known theorem from classical analysis (see, for example, [231]) yields uniform convergence. □

For further aspects of the Backus–Gilbert method, we refer to [32, 114, 144, 162, 163, 244, 272, 273].

3.7   Problems

3.1 Let $Q_n : C[a,b] \to \mathcal{S}_1(t_1,\ldots,t_n)$ be the interpolation operator from Example 3.3. Prove that $\|Q_n\|_{\mathcal{L}(C[a,b])} = 1$ and derive an estimate of the form

$$\|Q_n x - x\|_\infty \leq c\,h\,\|x'\|_\infty$$

for $x \in C^1[a,b]$, where $h = \max\{t_j - t_{j-1} : t = 2,\ldots,n\}$.

3.2 Let $K : X \to X$ be self-adjoint and positive definite and let $y \in X$. Define $\psi(x) = (Kx,x)_X - 2\,\mathrm{Re}(y,x)_X$ for $x \in X$. Prove that $x^* \in X$ is a minimum of ψ if and only if x^* solves $Kx^* = y$.

3.3 Define the space X_n by

$$X_n = \left\{\sum_{|j|\leq n} a_j\,e^{ijt} : a_j \in \mathbb{C}\right\}$$

and let $P_n : L^2(0,2\pi) \to X_n$ be the orthogonal projection operator. Prove that for $r \geq s$ there exists $c > 0$ such that

$$\|\psi_n\|_{H_{per}^r} \leq c\,n^{r-s}\|\psi_n\|_{H_{per}^s} \quad \text{for all } \psi_n \in X_n,$$

$$\|P_n\psi - \psi\|_{H_{per}^s} \leq c\,\frac{1}{n^{r-s}}\|\psi\|_{H_{per}^r} \quad \text{for all } \psi \in H_{per}^r(0,2\pi).$$

3.4 Show that the worst-case error of Symm's equation under the information $\|\psi\|_{H^s_{per}} \leq E$ for some $s > 0$ is given by

$$\mathcal{F}\big(\delta, E, \|\cdot\|_{H^s_{per}}\big) \leq c\,\delta^{s/(s+1)}.$$

3.5 Let $\Omega \subset \mathbb{R}^2$ be the disk of radius $a = \exp(-1/2)$. Then $\psi = 1$ is the unique solution of Symm's integral equation (3.57) for $f = 1$. Compute explicitly the errors of the least squares solution, the dual least squares solution, and the Bubnov–Galerkin solution as in Section 3.3, and verify that the error estimates of Theorem 3.18 are asymptotically sharp.

3.6 Let $t_k = k/n$, $k = 1, \ldots, n$, be equidistant collocation points. Let X_n be the space of piecewise constant functions as in (3.81) and $P_n : L^2(0,1) \to X_n$ be the orthogonal projection operator. Prove that $\bigcup X_n$ is dense in $L^2(0,1)$ and

$$\|x - P_n x\|_{L^2} \leq \frac{1}{n}\|x'\|_{L^2}$$

for all $x \in H^1(0,1)$ (see Problem 3.1).

3.7 Show that the moment solution can also be interpreted as the solution of a dual least squares method.

3.8 Consider moment collocation of the equation

$$\int_0^t x(s)\,ds = y(t), \quad t \in [0,1],$$

in the space $X_n = \mathcal{S}_1(t_1, \ldots, t_n)$ of linear splines. Show that the moment solution x_n coincides with the two-sided difference quotient, that is,

$$x_n(t_j) = \frac{1}{2h}\big[y\big(t_{j+1} + h\big) - y\big(t_{j-1} - h\big)\big],$$

where $h = 1/n$. Derive an error estimate for $\|x_n^\delta - x\|_{L^2}$ as in Example 3.23.

Chapter 4

Nonlinear Inverse Problems

In the previous chapters, we considered linear problems which we wrote as $Kx = y$, where K was a linear and (often) compact operator between Hilbert spaces. Needless to say that most problems in applications are nonlinear. For example, even in the case of a linear differential equation of the form $-u'' + cu = f$ for the function u the dependence of u on the parameter function c is nonlinear; that is, the mapping $c \mapsto u$ is nonlinear.[1] In Chapters 5, 6, and 7 we will study particular nonlinear problems to determine parameters of an ordinary or partial differential equation from the knowledge of the solution. Although we believe that the best strategies for solving nonlinear problems are intrinsically linked to the particular nature of the underlying problem, there are general methods for solving these nonlinear problems if they can be written in the form $K(x) = y$, where K is now a nonlinear operator between Hilbert spaces or Banach spaces. Guided by the structure of Chapter 2, we will study the nonlinear form of the Tikhonov regularization in Section 4.2 and the extension of the Landweber method in Section 4.3. Since the investigation of the latter one is already rather complicated, we do not present the extension of the conjugate gradient method or methods of Newton type, but refer the interested reader to the monograph [149] of Kaltenbacher, Neubauer, and Scherzer.

We start this chapter with a clarification of the notion of ill-posedness in Section 4.1 and its relation to the ill-posedness of the linearized problem. In Section 4.2, we study Tikhonov's regularization method for nonlinear problems. In contrast to the linear case, the question of existence of a global minimum of the Tikhonov functional is not obvious and requires more advanced tools from functional analysis, in particular, on weak topologies. We include the arguments in Subsection 4.2.1 but note already here that this Subsection is not central for the understanding of the further theory. It can be omitted because we formulate the existence of a minimum also as Assumption 4.9 at the beginning of Subsection 4.2.2. After the application of the general theory to the abovementioned parameter identification problem for a boundary value

---

[1]The differential equation has to be complemented by initial or boundary conditions, of course.

© Springer Nature Switzerland AG 2021
A. Kirsch, *An Introduction to the Mathematical Theory of Inverse Problems*,
Applied Mathematical Sciences 120,
https://doi.org/10.1007/978-3-030-63343-1_4

problem for an ordinary differential equation, in Subsection 4.2.4, we present some of the basic ideas for Tikhonov's method in Banach spaces and more general metrics to penalize the discrepancy and to measure the error. As a particular example, we consider the determination of a sparse solution of a linear equation. Further, tools from convex analysis are needed, such as the subdifferential and the Bregman distance.

Finally, in Section 4.3, we return to the Hilbert space setting and extend the Landweber method from Sections 2.3 and 2.6 to the nonlinear case.

4.1   Local Illposedness

In this chapter, we assume that X and Y are normed spaces (in most cases Hilbert spaces), $K : X \supset \mathcal{D}(K) \to Y$ a nonlinear mapping with domain of definition $\mathcal{D}(K) \subset X$. Let $x^* \in \mathcal{D}(K)$ and $y^* \in Y$ such that $K(x^*) = y^*$. It is the aim—as in the linear case—to determine an approximate solution to x^* when the right-hand side y^* is perturbed; that is, replaced by $y^\delta \in Y$ with $\|y^\delta - y^*\|_Y \leq \delta$. In the following, let $B(x,r) = \{z : \|x - z\|_X < r\}$ denote the open ball centered at x with radius r. The following notion of local ill-posedness goes back to Hofmann and Scherzer (see, e.g., [137]).

Definition 4.1 *Let $x^* \in \mathcal{D}(K)$ and $y^* = K(x^*)$. The equation $K(x) = y$ is called* locally improperly-posed *or* locally ill-posed *at x^* if for any sufficiently small $\rho > 0$ there exists a sequence $x_n \in \mathcal{D}(K) \cap B(x^*, \rho)$ such that $K(x_n) \to K(x^*)$ but (x_n) does not converge to x^* as n tends to infinity.*

Example 4.2
Let $k \in C^1([0,1] \times [0,1] \times \mathbb{R})$, $k = k(t, s, r)$, and let there exist $c_1 > 0$ with $|\partial k(t, s, r)/\partial r| \leq c_1$ for all $(t, s, r) \in [0,1] \times [0,1] \times \mathbb{R}$. Define

$$K(x)(t) = \int_0^1 k(t, s, x(s)) \, ds \,, \quad t \in [0,1], \quad \text{for } x \in L^2(0,1) \,.$$

Then K is well-defined from $L^2(0,1)$ into itself, and the equation $K(x) = y$ is locally ill-posed at $x = 0$.

Proof: Let $x \in L^2(0,1)$. The application of the fundamental theorem of calculus in the form

$$k(t, s, r) = k(t, s, 0) + \int_0^r \frac{\partial k}{\partial r}(t, s, r') \, dr' \,, \quad r \in \mathbb{R}, \tag{4.1}$$

implies $|k(t, s, r)| \leq |k(t, s, 0)| + c_1|r|$, thus $|k(t, s, r)|^2 \leq 2|k(t, s, 0)|^2 + 2c_1^2 r^2$, thus, using the Cauchy-Schwarz inequality,

$$|K(x)(t)|^2 \leq \int_0^1 |k(t, s, x(s))|^2 \, ds \leq 2 \int_0^1 [|k(t, s, 0)|^2 + c_1^2 \, x(s)^2] \, ds$$

for all $t \in [0,1]$. Therefore, $K(x)$ is measurable and $|K(x)(t)|^2$ is bounded which implies $K(x) \in L^2(0,1)$.

Let now $\rho > 0$ be arbitrary and $x_n(t) = \rho\sqrt{2n+1}\,t^n$, $t \in [0,1]$. Then $\|x_n\|_{L^2(0,1)}^2 = \rho^2(2n+1)\int_0^1 t^{2n}dt = \rho^2$ and, with (4.1),

$$
\begin{aligned}
|K(x_n)(t) - K(0)(t)| &\leq c_1 \int_0^1 |x_n(s)|\,ds = c_1\rho\sqrt{2n+1}\int_0^1 s^n\,ds \\
&= \frac{c_1\rho\sqrt{2n+1}}{n+1} \to 0 \quad \text{as } n \to \infty .
\end{aligned}
$$

Therefore, the equation $K(x) = y$ is locally improperly-posed at $x = 0$.

A second, and more concrete, example is formulated as Problem 4.2.

If K is continuously Fréchet-differentiable at x^* with Lipschitz continuous derivative then local illposedness of the nonlinear problem implies the illposedness of the linearization.

Theorem 4.3 *Let K be Fréchet-differentiable in the ball $B(x^*, \hat\rho)$ and let there exists $\gamma > 0$ with $\|K'(x) - K'(x^*)\|_{\mathcal{L}(X,Y)} \leq \gamma\|x - x^*\|_X$ for all $x \in B(x^*, \hat\rho)$. Let the equation $K(x) = y$ be locally ill-posed at x^*. Then $K'(x^*)$ is not boundedly invertible; that is, the linear equation $K'(x^*)h = z$ is also ill-posed.*

Proof: We assume on the contrary that $K'(x^*)$ is boundedly invertible and choose $\rho \in (0, \hat\rho)$ such that $\frac{\rho\gamma}{2}\|K'(x^*)^{-1}\|_{\mathcal{L}(Y,X)} =: q < 1$. For this ρ we choose a sequence $x_n \in B(x^*, \rho)$ by Definition 4.1. Lemma A.63 of the Appendix A.7 implies the representation

$$
K(x_n) - K(x^*) = K'(x^*)(x_n - x^*) + r_n \quad \text{with} \quad \|r_n\|_Y \leq \frac{\gamma}{2}\|x_n - x^*\|_X^2
$$

for all $n \in \mathbb{N}$; that is,

$$
K'(x^*)^{-1}[K(x_n) - K(x^*)] = x_n - x^* + K'(x^*)^{-1}r_n \quad \text{for all } n \in \mathbb{N},
$$

thus

$$
\begin{aligned}
\|x_n - x^*\|_X &\leq \left\|K'(x^*)^{-1}[K(x_n) - K(x^*)]\right\|_X + \|K'(x^*)^{-1}r_n\|_X \\
&\leq \|K'(x^*)^{-1}\|_{\mathcal{L}(Y,X)}\|K(x_n) - K(x^*)\|_Y \\
&\quad + \frac{\gamma}{2}\|K'(x^*)^{-1}\|_{\mathcal{L}(Y,X)}\|x_n - x^*\|_X^2 \\
&\leq \|K'(x^*)^{-1}\|_{\mathcal{L}(Y,X)}\|K(x_n) - K(x^*)\|_Y + q\|x_n - x^*\|_X ,
\end{aligned}
$$

thus

$$
(1 - q)\|x_n - x^*\|_X \leq \|K'(x^*)^{-1}\|_{\mathcal{L}(Y,X)}\|K(x_n) - K(x^*)\|_Y ,
$$

and this expression converges to zero because of $K(x_n) \to K(x^*)$. This contradicts (x_n) does not converge to x^*. $\qquad\square$

We will see in Section 4.3 that also the reverse assertion is true provided an addition condition ("tangential cone condition") is satisfied, see Lemma 4.35.

4.2 The Nonlinear Tikhonov Regularization

In this section (except of Subsection 4.2.4), we assume that X and Y are Hilbert spaces with inner products $(\cdot,\cdot)_X$ and $(\cdot,\cdot)_Y$, respectively, and corresponding norms. Let again $x^* \in \mathcal{D}(K)$ and $y^* \in Y$ with $K(x^*) = y^*$ and $y^\delta \in Y$ with $\|y^\delta - y^*\|_Y \leq \delta$. Let, in addition, $\hat{x} \in X$ be given which is thought of being an approximation of the true solution x^*. We define the Tikhonov functional by

$$J_{\alpha,\delta}(x) \;=\; \|K(x) - y^\delta\|_Y^2 \;+\; \alpha\|x - \hat{x}\|_X^2\,, \quad x \in \mathcal{D}(K)\,. \tag{4.2}$$

In the first subsection, we will discuss briefly the question of existence of minima of the Tikhonov functional and stability with respect to perturbation of y^δ. This part needs some knowledge of the weak topology in Hilbert spaces. We have collected the needed results in Section A.9 of the Appendix for the convenience of the reader. One can easily drop this subsection if one is not familiar (or interested) with this part of functional analysis. The further analysis is quite independent of this subsection.

4.2.1 Existence of Solutions and Stability

We recall the following definition (see remark following Definition A.75 of the Appendix A.9).

Definition 4.4 *Let X be a Hilbert space. A sequence (x_n) in X is said to converge* weakly *to $x \in X$ if $\lim_{n\to\infty}(x_n, z)_X = (x, z)_X$ for all $z \in X$.*

If Y is a second Hilbert space and $K : X \supset \mathcal{D}(K) \to Y$ is a (nonlinear) mapping then K is called weak-to-weak *continuous, if K maps weakly convergent sequences in $\mathcal{D}(K)$ into weakly convergent sequences in Y.*

We will need the following two results (see Corollary A.78 and part (d) of Theorem A.76).

Theorem 4.5 *Let X be a Hilbert space.*

(a) *Every bounded sequence (x_n) in X contains a weak accumulation point; that is, a weakly convergent subsequence.*

(b) *Every convex and closed set $U \subset X$ is also weakly closed; that is, if the sequence (x_n) in U converges weakly to some $x \in X$ then necessarily $x \in U$.*

The norm function $\|\cdot\|_X$ fails to be weakly continuous but has the following property which is sometimes called "weak lower semi-continuity".

Lemma 4.6 *Let X be a Hilbert space and let the sequence (x_n) converge weakly to x. Then*

$$\liminf_{n\to\infty} \|x_n - z\|_X \ \geq\ \|x - z\|_X \quad \text{for all } z \in X.$$

Proof: This follows from the formula

$$\|x_n - z\|_X^2 \ -\ \|x - z\|_X^2 \ =\ 2\,\mathrm{Re}(x_n - x, x - z)_X \ +\ \|x_n - x\|_X^2$$
$$\geq\ 2\,\mathrm{Re}(x_n - x, x - z)_X$$

because the right hand side of the inequality tends to zero as n tends to infinity. $\square$

Now we are able to prove the existence of minima of the Tikhonov functional under appropriate assumptions.

Theorem 4.7 *Let X and Y be Hilbert spaces, $y^\delta \in Y$, and $K : X \supset \mathcal{D}(K) \to Y$ weak-to-weak continuous with convex and closed domain of definition $\mathcal{D}(K)$. Then there exists a global minimum of $J_{\alpha,\delta}$, defined in (4.2), on $\mathcal{D}(K)$ for all $\alpha > 0$.*

Proof: Let $x_n \in \mathcal{D}(K)$ be a minimizing sequence; that is, $J_{\alpha,\delta}(x_n) \to J^* := \inf\{J_{\alpha,\delta}(x) : x \in \mathcal{D}(K)\}$ as $n \to \infty$. From

$$\alpha\|x_n - \hat{x}\|_X^2 \ \leq\ J_{\alpha,\delta}(x_n) \ \leq\ J^* + 1$$

for sufficiently large n, we conclude that the sequence (x_n) is bounded. By part (a) of Theorem 4.5, there exists a subsequence—which we also denote by (x_n)— which converges weakly to some $\overline{x}$. Also, $\overline{x} \in \mathcal{D}(K)$ by part (b) of this theorem. Furthermore, $K(x_n)$ converges weakly to $K(\overline{x})$ by the assumption on K. Lemma 4.6 implies that $\liminf_{n\to\infty} \|x_n - \hat{x}\|_X \geq \|\overline{x} - \hat{x}\|_X$ and $\liminf_{n\to\infty} \|K(x_n) - y^\delta\|_Y \geq \|K(\overline{x}) - y^\delta\|_Y$. Therefore, for any $\varepsilon > 0$ there exists $N \in \mathbb{N}$ such that for all $n \geq N$

$$
\begin{aligned}
J_{\alpha,\delta}(\overline{x}) \ &=\ \alpha\|\overline{x} - \hat{x}\|_X^2 \ +\ \|K(\overline{x}) - y^\delta\|_Y^2 \\
&\leq\ \alpha\|x_n - \hat{x}\|_X^2 \ +\ \|K(x_n) - y^\delta\|_Y^2 \ +\ \varepsilon \ =\ J_{\alpha,\delta}(x_n) \ +\ \varepsilon \\
&\leq\ J^* + 2\varepsilon.
\end{aligned}
$$

This holds for all $\varepsilon > 0$, thus $J^* \leq J_{\alpha,\delta}(\overline{x}) \leq J^*$ which proves that $\overline{x}$ is a minimum of $J_{\alpha,\delta}$. $\square$

In the same way one proves stability.

Theorem 4.8 *Let the assumptions of the previous theorem hold, and in addition, (y_n) be a sequence with $y_n \to y^\delta$ as $n \to \infty$. Let $x_n \in \mathcal{D}(K)$ be a minimum of $J_n(x) := \alpha\|x - \hat{x}\|_X^2 + \|K(x) - y_n\|_Y^2$ on $\mathcal{D}(K)$. Then there exist weak accumulation points of the sequence (x_n), and every weak accumulation point $\overline{x}$ is a minimum of $J_{\alpha,\delta}$.*

If in addition, K is weak-to-norm continuous, then every weak accumulation point $\overline{x}$ of (x_n) is also an accumulation point with respect to the norm.

Proof: The estimate

$$
\begin{aligned}
\alpha\|x_n - \hat{x}\|_X^2 + \|K(x_n) - y_n\|_Y^2 &= J_n(x_n) \le J_n(\hat{x}) = \|K(\hat{x}) - y_n\|_Y^2 \\
&\le \left(\|K(\hat{x}) - y^\delta\|_Y + \|y^\delta - y_n\|_Y\right)^2 \\
&\le \left(\|K(\hat{x}) - y^\delta\|_Y + 1\right)^2
\end{aligned}
$$

for sufficiently large n implies again that the sequence (x_n) is bounded. Thus, it contains weak accumulation points by Theorem 4.5. Let $\overline{x} \in \mathcal{D}(K)$ be a weak accumulation point, without loss of generality let x_n itself converge weakly to $\overline{x}$. Then $K(x_n) - y_n$ converges weakly to $K(\overline{x}) - y^\delta$ and thus as before $\liminf_{n\to\infty} \|x_n - \hat{x}\|_X \ge \|\overline{x} - \hat{x}\|_X$ and $\liminf_{n\to\infty} \|K(x_n) - y_n\|_Y \ge \|K(\overline{x}) - y^\delta\|_Y$. Set $J_n^* = J_n(x_n)$. Then, for any $\varepsilon > 0$, there exists $N \in \mathbb{N}$ such that for all $n \ge N$ and all $x \in \mathcal{D}(K)$

$$
\begin{aligned}
J_{\alpha,\delta}(\overline{x}) &\le \alpha\|x_n - \hat{x}\|_X^2 + \|K(x_n) - y_n\|_Y^2 + \varepsilon = J_n^* + \varepsilon \qquad (4.3)\\
&\le J_n(x) + \varepsilon = \alpha\|x - \hat{x}\|_X^2 + \|K(x) - y_n\|_Y^2 + \varepsilon.
\end{aligned}
$$

Now we let n tend to infinity. This yields

$$
J_{\alpha,\delta}(\overline{x}) \le \liminf_{n\to\infty} J_n^* + \varepsilon \le \liminf_{n\to\infty} J_n^* + \varepsilon \le J_{\alpha,\delta}(x) + \varepsilon.
$$

Letting ε tend to zero proves the optimality of $\overline{x}$ and also the convergence of J_n^* to $J_{\alpha,\delta}(\overline{x})$.

Let now K be weak-to-norm continuous. From

$$
\alpha\left(\|x_n - \hat{x}\|_X^2 - \|\overline{x} - \hat{x}\|_X^2\right) = J_n^* - J_{\alpha,\delta}(\overline{x}) + \|K(\overline{x}) - y^\delta\|_Y^2 - \|K(x_n) - y_n\|_Y^2
$$

and the convergence of $K(x_n)$ to $K(\overline{x})$, we conclude that the right-hand side converges to zero and thus $\|x_n - \hat{x}\|_X \to \|\overline{x} - \hat{x}\|_X$. Finally, the binomial formula yields

$$
\begin{aligned}
\|x_n - \overline{x}\|_X^2 &= \|(x_n - \hat{x}) - (\overline{x} - \hat{x})\|_X^2 \\
&= \|x_n - \hat{x}\|_X^2 + \|\overline{x} - \hat{x}\|_X^2 - 2\,\mathrm{Re}(x_n - \hat{x}, \overline{x} - \hat{x})_X
\end{aligned}
$$

which tends to zero. $\square$

4.2.2   Source Conditions And Convergence Rates

We make the following general assumption.

Assumption 4.9 *(a)* $\mathcal{D}(K)$ *is open, and* $x^* \in \mathcal{D}(K) \cap B(\hat{x}, \rho)$ *is a solution of* $K(x) = y^*$ *in some ball* $B(\hat{x}, \rho)$*, and* K *is continuously Fréchet-differentiable on* $\mathcal{D}(K) \cap B(\hat{x}, \rho)$,

(b) $K'(x^*)$ *is compact from* X *into* Y.

(c) The Tikhonov functional $J_{\alpha, \delta}$ *possesses global minima* $x^{\alpha, \delta} \in \mathcal{D}(K)$ *on* $\mathcal{D}(K)$ *for all* $\alpha, \delta > 0$*; that is,*

$$\alpha \|x^{\alpha, \delta} - \hat{x}\|_X^2 + \|K(x^{\alpha, \delta}) - y^\delta\|_Y^2 \ \leq\ \alpha \|x - \hat{x}\|_X^2 + \|K(x) - y^\delta\|_Y^2 \quad (4.4)$$

for all $x \in \mathcal{D}(K)$.

We refer to the previous subsection where we showed part (c) of this assumption under appropriate smoothness assumptions on K.

Now we are able to formulate the condition which corresponds to the "source condition" in the linear case. We will include the more general form with index functions and introduce the following notion.

Definition 4.10 *Any monotonically increasing and continuous function* φ : $[0, \delta_{max}] \to \mathbb{R}$ *(for some* $\delta_{max} > 0$*) with* $\varphi(0) = 0$ *is called an* index function.

The most prominent examples for index functions are $\varphi(t) = t^\sigma$ for any $\sigma > 0$ but also $\varphi(t) = -1/\ln t$ for $0 < t < 1$ (and $\varphi(0) = 0$). By calculating the second derivative, one observes that the latter one is concave on $[0, 1/e^2]$ and the first class is concave whenever $\sigma \leq 1$. The linear case $\varphi(t) = \beta t$ for $t \geq 0$ is particularly important.

Assumption 4.11 *(Source condition) Let* $\mathcal{D}(K)$ *be open and* $x^* \in \mathcal{D}(K) \cap B(\hat{x}, \rho)$ *be a solution of* $K(x) = y^*$ *in some ball* $B(\hat{x}, \rho)$ *and let* K *be differentiable on* $\mathcal{D}(K) \cap B(\hat{x}, \rho)$*. Furthermore, let* $\varphi : [0, \delta_{max}] \to \mathbb{R}$ *be a concave index function with* $\delta_{max} \geq \|K'(x^*)\|_{\mathcal{L}(X, Y)}$.

(i) Let K' *be locally Lipschitz continuous; that is, there exists* $\gamma > 0$ *with*

$$\|K'(x) - K'(x^*)\|_{\mathcal{L}(X, Y)} \ \leq\ \gamma \|x - x^*\|_X \quad \text{for all } x \in B(x^*, \rho) \cap \mathcal{D}(K),$$

(ii) and there exists $w \in X$ *with*

$$x^* - \hat{x} \ =\ \varphi\big([(K'(x^*)^*(K'(x^*)]^{1/2}\big)w \quad \text{and} \quad \gamma \|w\|_X < 1.$$

We note that for a linear compact operator $A : X \to Y$ (in the present case $A := K'(x^*)$) the operator $\varphi([A^*A]^{1/2})$ from X into itself is defined as in (A.47) by a singular system $\{\mu_j, x_j, y_j : j \in J\}$ for A, see Appendix A.6, Theorem A.57, where J is finite or $J = \mathbb{N}$, namely,

$$\varphi([A^*A]^{1/2})z \ =\ \sum_{j \in J} \varphi(\mu_j)\,(z, x_j)_X\, x_j\,, \quad z \in X\,.$$

In the special case that $\varphi(t) = t^\sigma$ the condition reads as $x^* - \hat{x} = (A^*A)^{\sigma/2})w$ which is just the source condition of the linear case (for $\hat{x} = 0$). Again, for $\sigma = 1$ the ranges $\mathcal{R}\big([(K'(x^*)^*(K'(x^*)]^{1/2}\big)$ and $\mathcal{R}\big(K'(x^*)^*\big)$ coincide which is seen from the singular system $\{\mu_j, x_j, y_j : j \in J\}$. Therefore, in this linear case part (ii) takes the form $x^* - \hat{x} = K'(x^*)^*v$ for some $v \in Y$ with $\gamma \|v\|_Y < 1$.

In the past decade, a different kind of source conditions has been developed which does not need the derivative of K. It can be generalized to a wider class of Tikhonov functionals with non-differentiable K acting between Banach spaces. We refer to Subsection 4.2.4 for a short glimpse on these extensions.

Assumption 4.12 *(Variational Source condition) Let $\mathcal{D}(K)$ be open and $x^* \in \mathcal{D}(K) \cap B(\hat{x}, \rho)$ be a solution of $K(x) = y^*$ in some ball $B(\hat{x}, \rho)$ and $\varphi : [0, \delta_{max}] \to \mathbb{R}$ be a concave index function with $\delta_{max} \geq \sup\{\|K(x^*) - K(x)\|_Y : x \in \mathcal{D}(K) \cap B(\hat{x}, \rho)\}$. Furthermore, there exists a constant $\beta > 0$ such that*

$$\beta \|x^* - x\|_X^2 \leq \|x - \hat{x}\|_X^2 - \|x^* - \hat{x}\|_X^2 + \varphi\big(\|K(x^*) - K(x)\|_Y\big)$$

for all $x \in B(\hat{x}, \rho) \cap \mathcal{D}(K)$.

This assumption is also known as a *variational inequality* (see, e.g., [245]). We prefer the notion of variational source condition as, e.g., in [96], because it takes the role of the source condition. We now show a relationship between Assumption 4.11 and Assumption 4.12.

Theorem 4.13 *Let $\mathcal{D}(K)$ be open and $x^* \in \mathcal{D}(K) \cap B(\hat{x}, \rho)$ be a solution of $K(x) = y^*$ with $\rho \in (0, 1/2)$ and $\varphi : [0, \delta_{max}] \to \mathbb{R}$ be a concave index function with $\delta_{max} \geq \|K'(x^*)\|_{\mathcal{L}(X,Y)}$ and $\delta_{max} \geq \sup\{\|K(x^*) - K(x)\|_Y : x \in \mathcal{D}(K) \cap B(\hat{x}, \rho)\}$.*

(a) *The variational source condition of Assumption 4.12 is equivalent to the following condition:*
   There exists $0 \leq \sigma < 1$ such that

$$2\,\mathrm{Re}(x^* - \hat{x}, x^* - x)_X \leq \sigma \|x^* - x\|_X^2 + \varphi\big(\|K(x^*) - K(x)\|_Y\big) \quad (4.5)$$

   for all $x \in B(\hat{x}, \rho) \cap \mathcal{D}(K)$.

(b) *Let Assumption 4.11 hold and, in addition, $\varphi(t) = t$ for all t or $\|K'(x^*)(x - x^*)\|_Y \leq \eta\big(\|K(x) - K(x^*)\|_Y\big)$ for all $x \in B(\hat{x}, \rho) \cap \mathcal{D}(K)$ where η is another concave index function. Then also Assumption 4.12 holds with some index function $\tilde{\varphi}$ which is linear if φ is linear.*

(c) *Let Assumption 4.12 hold for $\varphi(t) = \beta t$ for all $t \geq 0$. Then $x^* - \hat{x} \in \mathcal{R}\big(K'(x^*)^*\big) = \mathcal{R}\big([K'(x^*)^*K'(x^*)]^{1/2}\big)$.*

Proof: (a) From the elementary equation

$$\|x - \hat{x}\|_X^2 - \|x^* - \hat{x}\|_X^2 = 2\,\mathrm{Re}(x^* - \hat{x}, x - x^*)_X + \|x - x^*\|_X^2$$

we observe that the variational source condition is equivalent to

$$\beta\|x - x^*\|_X^2 \leq 2\operatorname{Re}(x^* - \hat{x}, x - x^*)_X + \|x - x^*\|_X^2 + \varphi(\|K(x^*) - K(x)\|_Y),$$

that is,

$$2\operatorname{Re}(x^* - \hat{x}, x^* - x)_X \leq (1 - \beta)\|x - x^*\|_X^2 + \varphi(\|K(x^*) - K(x)\|_Y),$$

which has the desired form with $\sigma = 1 - \beta$ if $\beta \leq 1$ and $\sigma = 0$ if $\beta > 1$.

(b) Set $A := K'(x^*)$ for abbreviation. Then, for $x \in B(\hat{x}, \rho) \cap \mathcal{D}(K)$,

$$\begin{aligned}
\operatorname{Re}(x^* - \hat{x}, x^* - x)_X &= \operatorname{Re}\big(\varphi([A^*A]^{1/2})w, x^* - x\big)_X \\
&= \operatorname{Re}\big(w, \varphi([A^*A]^{1/2})(x^* - x)\big)_X \\
&\leq \|w\|_X \big\|\varphi([A^*A]^{1/2})(x^* - x)\big\|_X.
\end{aligned}$$

Now we use the fact that for any concave index function

$$\big\|\varphi([A^*A]^{1/2})z\big\|_X \leq \varphi(\|Az\|_Y) \quad \text{for all } z \in X \text{ with } \|z\|_X \leq 1.$$

For a proof, we refer to Lemma A.73 of the Appendix. Therefore,

$$\operatorname{Re}(x^* - \hat{x}, x^* - x)_X \leq \|w\|_X \varphi(\|K'(x^*)(x^* - x)\|_Y). \tag{4.6}$$

If $\|K'(x^*)(x - x^*)\|_Y \leq \eta(\|K(x) - K(x^*)\|_Y)$ for all $x \in B(\hat{x}, \rho) \cap \mathcal{D}(K)$ then

$$\operatorname{Re}(x^* - \hat{x}, x^* - x)_X \leq \|w\|_X \varphi\big(\eta(\|K(x) - K(x^*)\|_Y)\big)$$

because of the monotonicity of φ. This proves the estimate (4.5) with $\sigma = 0$ and $\tilde{\varphi} = 2\|w\|_X \varphi \circ \eta$. Note that the composition of two concave index functions is again a concave index function.

If $\varphi(t) = t$ for all t we use the estimate

$$\|K(x) - K(x^*) - K'(x^*)(x - x^*)\|_Y \leq \frac{\gamma}{2}\|x^* - x\|_X^2$$

for all $x \in B(\hat{x}, \rho) \cap \mathcal{D}(K)$ (see Lemma A.63 of the Appendix A.7), thus $\|K'(x^*)(x - x^*)\|_Y \leq \|K(x) - K(x^*)\|_Y + \frac{\gamma}{2}\|x^* - x\|_X^2$ and thus from (4.6)

$$2\operatorname{Re}(x^* - \hat{x}, x^* - x)_X \leq 2\|w\|_X \|K(x^*) - K(x))\|_Y + \|w\|_X \gamma \|x^* - x\|_X^2$$

for all $x \in B(\hat{x}, \rho) \cap \mathcal{D}(K)$ which proves the estimate (4.5) with $\sigma = \gamma\|w\|_X < 1$ and $\tilde{\varphi}(t) = 2\|w\|_X t$ for $t \geq 0$.

(c) Let Assumption 4.12 hold for $\varphi(t) = \beta t$ for all $t \geq 0$. For any fixed $z \in X$ and $t \in \mathbb{K}$ sufficiently small such that $x := x^* - tz \in B(\hat{x}, \rho) \cap \mathcal{D}(K)$ we substitute x into (4.5) for $\varphi(t) = \beta t$ which yields

$$\begin{aligned}
2\operatorname{Re}\big[t\,(x^* - \hat{x}, z)_X\big] &\leq |t|^2 \sigma \|z\|_X^2 + \beta \|K(x^* - tz) - K(x^*)\|_Y \\
&\leq |t|^2 \sigma \|z\|_X^2 + |t|\beta \|K'(x^*)z\|_Y + \frac{\beta\gamma}{2}|t|^2 \|z\|_X^2.
\end{aligned}$$

Dividing by $|t|$ and letting t tend to zero yields[2] $\left|(x^* - \hat{x}, z)_X\right| \leq \frac{\beta}{2}\|Az\|_Y$ for all $z \in X$ where again $A = K'(x^*)$. We show that $x^* - \hat{x}$ belongs to the range of A^*. First we note that $x^* - \hat{x}$ is orthogonal to the nullspace $\mathcal{N}(A)$ of A. We choose a singular system $\{\mu_j, x_j, y_j : j \in J\}$ for A, see Appendix A.6, Theorem A.57, where J is finite or $J = \mathbb{N}$ and expand $x^* - \hat{x}$ in the form $x^* - \hat{x} = \sum_{j \in J} \gamma_j x_j$. (The component in the nullspace $\mathcal{N}(A)$ vanishes because $x^* - \hat{x}$ is orthogonal to $\mathcal{N}(A)$.) We set $J_n = J$ if J is finite and $J_n = \{1, \ldots, n\}$ if $J = \mathbb{N}$. For $z = \sum_{j \in J_n} \frac{\gamma_j}{\mu_j^2} x_j$ the inequality $\left|(x^* - \hat{x}, z)_X\right| \leq \frac{\beta}{2}\|Az\|$ is equivalent to

$$\left(\sum_{j \in J_n} \frac{\gamma_j^2}{\mu_j^2}\right)^2 \leq \frac{\beta^2}{4} \sum_{j \in J_n} \frac{\gamma_j^2}{\mu_j^2};$$

that is, $\sum_{j \in J_n} \frac{\gamma_j^2}{\mu_j^2} \leq \frac{\beta^2}{4}$. This proves that $w := \sum_{j \in J} \frac{\gamma_j}{\mu_j} y_j \in Y$ is well-defined. Finally, $A^* w = x^* - \hat{x}$. $\quad\square$

Under the Assumptions 4.9 and 4.12, we are able to prove convergence and also rates of convergence as in the linear theory. As we know from the linear theory there are (at least) two strategies to choose the regularization parameter α. To achieve the rate $\mathcal{O}(\sqrt{\delta})$, we should choose $\alpha = \alpha(\delta)$ to be proportional to δ (a priori choice) or such that the "discrepancy" $\|K(x^{\alpha(\delta),\delta}) - y^\delta\|_Y$ to be proportional to δ (a posteriori choice). For nonlinear operators, K essentially the same arguments as for linear operators (substitute $x = \hat{x}$ and $x = x^*$ into (4.4)) show that

$$\limsup_{\alpha \to 0} \|K(x^{\alpha,\delta}) - y^\delta\|_Y \leq \delta \quad \text{and} \quad \lim_{\alpha \to \infty} \|K(x^{\alpha,\delta}) - y^\delta\|_Y = \|K(\hat{x}) - y^\delta\|_Y$$

for any choice of minimizers $x^{\alpha,\delta}$. However, for nonlinear operators, the mapping $\alpha \mapsto \|K(x^{\alpha,\delta}) - y^\delta\|_Y$ is not necessarily continuous (see, e.g, [219] for a discussion of this topic). Therefore, the discrepancy principle is not well-defined unless more restrictive assumptions are made. In the following, we just take the possibility to choose the regularization parameter by the discrepancy principle as an assumption (see also [245], Section 4.1.2).

Theorem 4.14 *Let Assumptions 4.9 and 4.12 hold and let $\rho > 2\|x^* - \hat{x}\|$.*

(a) Let $\alpha = \alpha(\delta)$ be chosen such that

$$c_- \frac{\delta^2}{\varphi(\delta)} \leq \alpha(\delta) \leq c_+ \frac{\delta^2}{\varphi(\delta)} \quad \text{for all } \delta > 0$$

where $c_+ \geq c_- > 0$ are independent of δ (a priori choice),

(b) or assume that there exists $r_+ > r_- \geq 1$ and $\alpha(\delta) > 0$ such that

$$r_- \delta \leq \|K(x^{\alpha(\delta),\delta}) - y^\delta\|_Y \leq r_+ \delta \quad \text{for all } \delta > 0$$

(a posteriori choice) where $x^{\alpha(\delta),\delta}$ denotes a minimum of the Tikhonov functional $J_{\alpha(\delta),\delta}$ on $\mathcal{D}(K)$.

---

[2]Note that the phases of $t/|t|$ can be chosen arbitrarily!

Then $x^{\alpha(\delta),\delta} \in B(\hat{x}, \rho)$ for sufficiently small δ and

$$\|x^{\alpha(\delta),\delta} - x^*\|_X^2 = \mathcal{O}(\varphi(\delta)) \quad \text{and} \quad \|K(x^{\alpha(\delta),\delta}) - y^*\|_Y = \mathcal{O}(\delta), \quad \delta \to 0.$$

Proof: We show first that $x^{\alpha(\delta),\delta} \in B(\hat{x}, \rho)$ for sufficiently small δ. From (4.4) for $x = x^*$, we conclude that

$$\|K(x^{\alpha,\delta}) - y^\delta\|_Y^2 + \alpha\|x^{\alpha,\delta} - \hat{x}\|_X^2 \leq \delta^2 + \alpha\|x^* - \hat{x}\|_X^2 \tag{4.7a}$$

$$\leq \delta^2 + \alpha\frac{\rho^2}{4}. \tag{4.7b}$$

If $c_- \frac{\delta^2}{\varphi(\delta)} \leq \alpha(\delta) \leq c_+ \frac{\delta^2}{\varphi(\delta)}$ we conclude that $\|x^{\alpha(\delta),\delta} - \hat{x}\|_X^2 \leq \frac{\delta^2}{\alpha(\delta)} + \frac{\rho^2}{4} \leq \frac{\varphi(\delta)}{c_-} + \frac{\rho^2}{4} \leq \rho^2$ for sufficiently small δ.
If the discrepancy principle holds, then from (4.7b),

$$r_-^2 \delta^2 + \alpha(\delta)\|x^{\alpha(\delta),\delta} - \hat{x}\|_X^2 \leq \delta^2 + \alpha(\delta)\frac{\rho^2}{4},$$

and thus $\alpha(\delta)\|x^{\alpha(\delta),\delta} - \hat{x}\|_X^2 \leq \alpha(\delta)\frac{\rho^2}{4}$ because $r_- \geq 1$. This shows $\|x^{\alpha(\delta),\delta} - \hat{x}\|_X \leq \rho$ and ends the first part of the proof.
To show the error estimates, we use the variational source condition of Assumption 4.12 and (4.7a).

$$\|K(x^{\alpha,\delta}) - y^\delta\|_Y^2 + \alpha\beta\|x^{\alpha,\delta} - x^*\|_X^2 \leq \|K(x^{\alpha,\delta}) - y^\delta\|_Y^2$$
$$+ \alpha\|x^{\alpha,\delta} - \hat{x}\|_X^2 - \alpha\|x^* - \hat{x}\|_X^2 + \alpha\varphi(\|K(x^*) - K(x^{\alpha,\delta})\|_Y)$$
$$\leq \delta^2 + \alpha\varphi(\|K(x^*) - K(x^{\alpha,\delta})\|_Y)$$
$$\leq \delta^2 + \alpha\varphi(\|y^\delta - K(x^{\alpha,\delta})\|_Y + \delta). \tag{4.8}$$

Let first $\alpha(\delta)$ be chosen according to the discrepancy principle. Then

$$(r_-^2 - 1)\delta^2 + \alpha(\delta)\beta\|x^{\alpha(\delta),\delta} - x^*\|_X^2 \leq \alpha(\delta)\varphi((r_+ + 1)\delta)$$
$$\leq \alpha(\delta)(1 + r_+)\varphi(\delta)$$

where we have used that $\varphi(s\delta) \leq s\varphi(\delta)$ for all $s \geq 1$ (see Lemma A.73). This proves the assertion for $\|x^{\alpha(\delta),\delta} - x^*\|_X$ after division by $\alpha(\delta)$ and dropping the first term on the left hand side. The estimate for $\|K(x^{\alpha(\delta),\delta}) - y^*\|_Y$ follows obviously from the triangle inequality because $\|K(x^{\alpha(\delta),\delta}) - y^*\|_Y \leq \|K(x^{\alpha(\delta),\delta}) - y^\delta\|_Y + \delta \leq (r_+ + 1)\delta$.
Let now $c_- \frac{\delta^2}{\varphi(\delta)} \leq \alpha(\delta) \leq c_+ \frac{\delta^2}{\varphi(\delta)}$. Substituting this into (4.8) and dropping the second term on the left hand side yields

$$\|K(x^{\alpha(\delta),\delta}) - y^\delta\|_Y^2 \leq \delta^2 + \frac{c_+\delta^2}{\varphi(\delta)}\varphi(\|y^\delta - K(x^{\alpha(\delta),\delta})\|_Y + \delta).$$

Now we set $t = \|K(x^{\alpha(\delta),\delta}) - y^\delta\|_Y/\delta$ for abbreviation. Then the previous estimate reads as

$$t^2 \leq 1 + \frac{c_+}{\varphi(\delta)}\varphi((1+t)\delta) \leq 1 + c_+(1+t) = 1 + c_+ + c_+t,$$

where we used again $\varphi((1+t)\delta) \leq (1+t)\varphi(\delta)$. Completing the square yields $t \leq \frac{c_+}{2} + \sqrt{1 + c_+ + \frac{c_+^2}{4}}$; that is, $\|K(x^{\alpha(\delta),\delta}) - y^\delta\|_Y \leq c\delta$ for some $c > 0$. Now we substitute this and the bounds of $\alpha(\delta)$ into (4.8) again and drop the first term on the left hand side which yields

$$c_- \frac{\delta^2}{\varphi(\delta)} \beta \|x^{\alpha(\delta),\delta} - x^*\|_X^2 \leq \delta^2 + c_+ \frac{\delta^2}{\varphi(\delta)} \varphi((1+c)\delta) \leq \delta^2 + c_+(1+c)\delta^2$$

which yields $c_-\beta\|x^{\alpha(\delta),\delta} - x^*\|_X^2 \leq [1 + c_+(1+c)]\varphi(\delta)$ and ends the proof of the theorem. $\quad\square$

By Theorem 4.13, the special case $\varphi(t) = t$ corresponds to the source condition $x^* - \hat{x} \in \mathcal{R}((A^*A)^{1/2}) = \mathcal{R}(A^*)$ and leads to the order $\mathcal{O}(\sqrt{\delta})$ just as in the linear case. The cases $\varphi(t) = t^\sigma$ for $\sigma > 1$ are not covered by the previous theorem because these index functions are not concave anymore.

For proving the analogue of Theorem 2.12 to get the optimal order of convergence up to $\mathcal{O}(\delta^{2/3})$, we have to use the classical source condition of Assumption 4.11, which is the obvious extension of the one in the linear case, see Theorem 2.12. A variational source condition for this case is not available. We follow the approach in [92] (see also [260] for the original proof).

Theorem 4.15 *Let Assumptions 4.9 and 4.11 hold with $\rho > 2\|x^* - \hat{x}\|_X$ and (ii) modified in the way that $x^* - \hat{x} = (K'(x^*)^*K'(x^*))^{\sigma/2}v \in \mathcal{R}((K'(x^*)^*K'(x^*))^{\sigma/2})$ for some $\sigma \in [1, 2]$ and $v \in X$ such that $\gamma \|(K'(x^*)^*K'(x^*))^{-1/2}(x^* - \hat{x})\|_X < 1$ where γ denotes the Lipschitz constant from part (i) of Assumption 4.11. We choose $\alpha(\delta)$ such that*

$$c_- \delta^{2/(\sigma+1)} \leq \alpha(\delta) \leq c_+ \delta^{2/(\sigma+1)}.$$

Then

$$\|x^{\alpha(\delta),\delta} - x^*\|_X = \mathcal{O}(\delta^{\sigma/(\sigma+1)}), \quad \delta \to 0,$$
$$\|K(x^{\alpha(\delta),\delta}) - y^*\|_Y = \mathcal{O}(\delta^{\sigma/(\sigma+1)}), \quad \delta \to 0.$$

Proof: We leave $\alpha > 0$ arbitrary til the end of the proof. We set $A := K'(x^*)$ and choose a singular system $\{\mu_j, x_j, y_j : j \in J\}$ for A, see Appendix A.6, Theorem A.57, where J is finite or $J = \mathbb{N}$. Then we write $x^* - \hat{x}$ as

$$x^* - \hat{x} = (A^*A)^{\sigma/2}v = \sum_{j \in J}\mu_j^\sigma v_j x_j = A^*w \quad \text{with} \quad w = \sum_{j \in J}\mu_j^{\sigma-1}v_j y_j$$

where $v_j = (v, x_j)_X$ are the expansion coefficients of v. Then $\gamma\|w\|_Y < 1$. We define $z_\alpha \in X$ by

$$z_\alpha = x^* - \alpha(A^*A + \alpha I)^{-1}A^*w; \quad\quad (4.9)$$

that is,

$$z_\alpha = x^* - \alpha\sum_{j \in J}\frac{\mu_j^\sigma}{\mu_j^2 + \alpha}v_j x_j, \quad \text{thus} \quad \|z_\alpha - x^*\|_X^2 = \alpha^2\sum_{j \in J}\left(\frac{\mu_j^\sigma}{\mu_j^2 + \alpha}\right)^2 |v_j|^2.$$

Later, we will also need the form

$$A(z_\alpha - x^*) + \alpha w = \alpha \sum_{j \in J} \left(\mu_j^{\sigma-1} - \frac{\mu_j^{\sigma+1}}{\mu_j^2 + \alpha} \right) v_j\, y_j = \alpha^2 \sum_{j \in J} \frac{\mu_j^{\sigma-1}}{\mu_j^2 + \alpha} v_j\, y_j$$

with

$$\|A(z_\alpha - x^*) + \alpha w\|_Y^2 = \alpha^4 \sum_{j \in J} \left(\frac{\mu_j^{\sigma-1}}{\mu_j^2 + \alpha} \right)^2 |v_j|^2 .$$

With the elementary estimate (see Problem 4.3)

$$\frac{\mu^t}{\mu^2 + \alpha} \leq c_t\, \alpha^{t/2-1}, \quad \mu \geq 0,$$

for $t = \sigma$ and $t = \sigma - 1$, respectively, (where c_t depends on t only) we obtain

$$\|z_\alpha - x^*\|_X^2 \leq c_\sigma^2\, \alpha^\sigma , \tag{4.10a}$$
$$\|A(z_\alpha - x^*) + \alpha w\|_Y^2 \leq c_{\sigma-1}^2\, \alpha^{\sigma+1} . \tag{4.10b}$$

In particular, z_α converges to x^* as $\alpha \to 0$ and is, therefore, in $B(\hat{x}, \rho) \cap \mathcal{D}(K)$ for sufficiently small α.

We use the optimality of $x^{\alpha,\delta}$; that is (4.4), for $x = z_\alpha$ to obtain

$$\|K(x^{\alpha,\delta}) - y^\delta\|_Y^2 + \alpha\|x^{\alpha,\delta} - \hat{x}\|_X^2 \leq \|K(z_\alpha) - y^\delta\|_Y^2 + \alpha\|z_\alpha - \hat{x}\|_X^2 .$$

With

$$\|x^{\alpha,\delta} - \hat{x}\|_X^2 = \|x^* - \hat{x}\|_X^2 + 2\,\mathrm{Re}\big(x^{\alpha,\delta} - x^*, x^* - \hat{x}\big)_X + \|x^{\alpha,\delta} - x^*\|_X^2$$
$$= \|x^* - \hat{x}\|_X^2 + 2\,\mathrm{Re}\big(A(x^{\alpha,\delta} - x^*), w\big)_Y + \|x^{\alpha,\delta} - x^*\|_X^2 ,$$
$$\|z_\alpha - \hat{x}\|_X^2 = \|x^* - \hat{x}\|_X^2 + 2\,\mathrm{Re}\big(A(z_\alpha - x^*), w\big)_Y + \|z_\alpha - x^*\|_X^2$$

we obtain

$$\|K(x^{\alpha,\delta}) - y^\delta\|_Y^2 + 2\alpha\,\mathrm{Re}\big(w, A(x^{\alpha,\delta} - x^*)\big)_Y + \alpha\|x^{\alpha,\delta} - x^*\|_X^2$$
$$\leq \|K(z_\alpha) - y^\delta\|_Y^2 + 2\alpha\,\mathrm{Re}\big(w, A(z_\alpha - x^*)\big)_Y + \alpha\|z_\alpha - x^*\|_X^2 ,$$

and thus

$$\|K(x^{\alpha,\delta}) - y^\delta + \alpha w\|_Y^2 + \alpha\|x^{\alpha,\delta} - x^*\|_X^2$$
$$\leq \alpha^2\|w\|_Y^2 + 2\alpha\,\mathrm{Re}\big(w, K(x^{\alpha,\delta}) - y^\delta - A(x^{\alpha,\delta} - x^*)\big)_Y$$
$$+ \|K(z_\alpha) - y^\delta\|_Y^2 + 2\alpha\,\mathrm{Re}\big(w, A(z_\alpha - x^*)\big)_Y + \alpha\|z_\alpha - x^*\|_X^2 .$$

Now we use

$$K(x^{\alpha,\delta}) = K(x^*) + A(x^{\alpha,\delta} - x^*) + r^{\alpha,\delta} = y^* + A(x^{\alpha,\delta} - x^*) + r^{\alpha,\delta} \quad \text{and}$$
$$K(z_\alpha) = K(x^*) + A(z_\alpha - x^*) + s_\alpha = y^* + A(z_\alpha - x^*) + s_\alpha$$

with $\|r^{\alpha,\delta}\|_Y \leq \frac{\gamma}{2}\|x^{\alpha,\delta} - x^*\|_X^2$ and $\|s_\alpha\|_Y \leq \frac{\gamma}{2}\|z_\alpha - x^*\|_X^2$ and obtain

$$
\begin{aligned}
&\|K(x^{\alpha,\delta}) - y^\delta + \alpha w\|_Y^2 \; + \; \alpha\|x^{\alpha,\delta} - x^*\|_X^2 \\
&\leq \; \alpha^2\|w\|_Y^2 \; + \; 2\alpha\operatorname{Re}(w, y^* - y^\delta)_Y \; + \; 2\alpha\operatorname{Re}(w, r^{\alpha,\delta})_Y \\
&\quad + \|y^* - y^\delta + A(z_\alpha - x^*) + s_\alpha\|_Y^2 \; + \; 2\alpha\operatorname{Re}(w, A(z_\alpha - x^*))_Y \\
&\quad + \alpha\|z_\alpha - x^*\|_X^2 \\
&= \; \alpha^2\|w\|_Y^2 \; + \; 2\alpha\operatorname{Re}(w, y^* - y^\delta)_Y \; + \; 2\alpha\operatorname{Re}(w, r^{\alpha,\delta})_Y \\
&\quad + \|y^* - y^\delta\|_Y^2 \; + \; 2\operatorname{Re}(A(z_\alpha - x^*) + s_\alpha, y^* - y^\delta)_Y \\
&\quad + \|A(z_\alpha - x^*) + s_\alpha\|_Y^2 + 2\alpha\operatorname{Re}(w, A(z_\alpha - x^*))_Y + \alpha\|z_\alpha - x^*\|_X^2 \\
&= \; 2\operatorname{Re}(A(z_\alpha - x^*) + \alpha w, y^* - y^\delta)_Y \; + \; 2\operatorname{Re}(s_\alpha, y^* - y^\delta)_Y \\
&\quad + \alpha^2\|w\|_Y^2 + 2\alpha\operatorname{Re}(w, r^{\alpha,\delta})_Y \; + \; \|A(z_\alpha - x^*) + s_\alpha\|_Y^2 \\
&\quad + 2\alpha\operatorname{Re}(w, A(z_\alpha - x^*) + s_\alpha)_Y \; - \; 2\alpha\operatorname{Re}(w, s_\alpha)_Y \\
&\quad + \alpha\|z_\alpha - x^*\|_X^2 \\
&= \; 2\operatorname{Re}(A(z_\alpha - x^*) + \alpha w, y^* - y^\delta)_Y \; + \; 2\operatorname{Re}(s_\alpha, y^* - y^\delta)_Y \\
&\quad + 2\alpha\operatorname{Re}(w, r^{\alpha,\delta})_Y \; + \; \|A(z_\alpha - x^*) + \alpha w + s_\alpha\|_Y^2 \\
&\quad - 2\alpha\operatorname{Re}(w, s_\alpha)_Y \; + \; \alpha\|z_\alpha - x^*\|_X^2 \\
&\leq \; 2\delta\|A(z_\alpha - x^*) + \alpha w\|_Y \; + \; \gamma\delta\|z_\alpha - x^*\|_X^2 \\
&\quad + \alpha\gamma\|w\|_Y\|x^{\alpha,\delta} - x^*\|_X^2 \; + \; 2\|A(z_\alpha - x^*) + \alpha w\|_Y^2 \\
&\quad + \frac{\gamma^2}{2}\|z_\alpha - x^*\|_X^4 \; + \; \alpha\gamma\|w\|_Y\|z_\alpha - x^*\|_X^2 \; + \; \alpha\|z_\alpha - x^*\|_X^2 .
\end{aligned}
$$

Now we use that $\gamma\|w\|_Y < 1$ and thus

$$
\begin{aligned}
&\|K(x^{\alpha,\delta}) - y^\delta + \alpha w\|_Y^2 \; + \; \alpha(1 - \gamma\|w\|_Y)\|x^{\alpha,\delta} - x^*\|_X^2 \\
&\leq \; 2\delta\|A(z_\alpha - x^*) + \alpha w\|_Y \; + \; 2\|A(z_\alpha - x^*) + \alpha w\|_Y^2 \\
&\quad + \frac{\gamma^2}{2}\|z_\alpha - x^*\|_X^4 \; + \; (\gamma\delta + \alpha\gamma\|w\|_Y + \alpha)\|z_\alpha - x^*\|_X^2 .
\end{aligned}
$$

So far, we have not used the definition of z_α. We substitute the estimates (4.10a), (4.10b) and arrive at

$$
\begin{aligned}
&\|K(x^{\alpha,\delta}) - y^\delta + \alpha w\|_Y^2 \; + \; \alpha(1 - \gamma\|w\|_Y)\|x^{\alpha,\delta} - x^*\|_X^2 \\
&\leq \; c\left[\delta\alpha^{(\sigma+1)/2} + \alpha^{\sigma+1} + \alpha^{2\sigma} + \delta\alpha^\sigma\right]
\end{aligned}
$$

for some $c > 0$. Dropping one of the terms on the left hand side yields

$$
\begin{aligned}
\|K(x^{\alpha,\delta}) - y^\delta\|_Y^2 \; &\leq \; 2\|K(x^{\alpha,\delta}) - y^\delta + \alpha w\|_Y^2 \; + \; 2\alpha^2\|w\|_Y^2 \\
&\leq \; c\left[\delta\alpha^{(\sigma+1)/2} + \alpha^{\sigma+1} + \alpha^{2\sigma} + \delta\alpha^\sigma + \alpha^2\right], \\
\|x^{\alpha,\delta} - x^*\|_X^2 \; &\leq \; c\left[\delta\alpha^{(\sigma-1)/2} + \alpha^\sigma + \alpha^{2\sigma-1} + \delta\alpha^{\sigma-1}\right]
\end{aligned}
$$

for some $c > 0$. The choice $c_- \delta^{2/(\sigma+1)} \leq \alpha(\delta) \leq c_+ \delta^{2/(\sigma+1)}$ yields the desired result (note that $1 \leq \sigma \leq 2$). ☐

We note that the modified defect $\|K(x^{\alpha(\delta),\delta}) - y^* + \alpha(\delta)w\|_Y$ satisfies $\|K(x^{\alpha(\delta),\delta}) - y^* + \alpha(\delta)w\|_Y \leq c\delta$ (see Problem 4.4).

In the next subsection, we apply the result to the Tikhonov regularization of a parameter identification problem.

4.2.3   A Parameter-Identification Problem

Let $f \in L^2(0,1)$ be given. It is the aim to determine the parameter function $c \in L^2(0,1)$, $c \geq 0$ on $(0,1)$, in the boundary value problem

$$-u''(t) + c(t)\,u(t) = f(t),\ 0 < t < 1,\quad u(0) = u(1) = 0,\qquad (4.11)$$

from perturbed data $u^\delta(t)$. We recall the Sobolev spaces $H^p(0,1)$ from (1.24) as the spaces

$$H^p(0,1) = \left\{ u \in C^{p-1}[0,1] : u^{(p-1)}(t) = \alpha + \int_0^t \psi(s)\,ds\,,\ \alpha \in \mathbb{R}\,,\ \psi \in L^2(0,1) \right\}$$

and set $u^{(p)} := \psi$ for the pth derivative. Note that $H^p(0,1) \subset C[0,1]$ for $p \geq 1$ by definition. Then the differential equation of (4.11) for $u \in H^2(0,1)$ is understood in the L^2−sense. It is an easy exercise (see Problem 4.7) to show that $\|u\|_\infty \leq \|u'\|_{L^2(0,1)}$ for all $u \in H^1(0,1)$ with $u(0) = 0$. First, we consider the direct problem and show that the boundary value problem is equivalent to an integral equation of the second kind.

Lemma 4.16 Let $f \in L^2(0,1)$ and $c \in L^2_+(0,1) := \{c \in L^2(0,1) : c \geq 0 \text{ almost everywhere on } (0,1)\}$.

(a) If $u \in H^2(0,1)$ solves (4.11) then u solves the integral equation

$$u(t) + \int_0^1 g(t,s)\,c(s)\,u(s)\,ds = \int_0^1 g(t,s)\,f(s)\,ds,\quad t \in [0,1],\qquad (4.12)$$

where $g(t,s) = \begin{cases} s(1-t), & 0 \leq s \leq t \leq 1, \\ t(1-s), & 0 \leq t \leq s \leq 1. \end{cases}$

(b) We define the integral operator $G : L^2(0,1) \to L^2(0,1)$ by

$$(Gv)(t) = \int_0^1 g(t,s)\,v(s)\,ds = (1-t)\int_0^t s\,v(s)\,ds + t\int_t^1 (1-s)\,v(s)\,ds,$$

$t \in (0,1)$, $v \in L^2(0,1)$. The operator G is bounded from $L^2(0,1)$ into $H^2(0,1)$ and $(Gv)'' = -v$.

(c) If $u \in C[0,1]$ is a solution of (4.12); that is, of $u + G(cu) = Gf$, then $u \in H^2(0,1)$ and u is a solution of (4.11). Note that the right-hand side of (4.12) is continuous because $Gf \in H^2(0,1)$.

Proof: (a) Let $u \in H^2(0,1)$ solve (4.11) and set $h = f - cu$. Then $h \in L^2(0,1)$ (because u is continuous) and $-u'' = h$. Integrating this equation twice and using the boundary conditions $u(0) = u(1) = 0$ yields the assertion (see Problem 4.5).

(b) Let $v \in L^2(0,1)$ and set

$$u(t) \;=\; (Gv)(t) \;=\; (1-t)\int_0^t s\,v(s)\,ds \;+\; t\int_t^1 (1-s)\,v(s)\,ds\,, \quad t \in [0,1]\,,$$

$$\psi(t) \;=\; -\int_0^t v(s)\,ds \;+\; \int_0^1 (1-s)\,v(s)\,ds\,, \quad t \in [0,1]\,.$$

Then it is easy to see that $u(t) = \int_0^t \psi(s)\,ds$. Therefore, $u \in H^1(0,1)$ and $u' = \psi$. From the definitions of ψ and $H^2(0,1)$, we observe that $u \in H^2(0,1)$ and $v = -u''$.

(c) Let now $u \in C[0,1]$ be a solution of (4.12) and set again $h = f - cu$. Then again $h \in L^2(0,1)$, and u has the representation $u = Gh$. By part (b), we conclude that $u \in H^2(0,1)$ and $-u'' = h = f - cu$.   □

Theorem 4.17 *The integral equation (4.12) and the boundary value problem (4.11) are uniquely solvable for all $f \in L^2(0,1)$ and $c \in L^2_+(0,1)$. Furthermore, there exists $\gamma > 0$ (independent of f and c) such that $\|u\|_{H^2(0,1)} \leq \gamma\,(1 + \|c\|_{L^2(0,1)})\,\|f\|_{L^2(0,1)}$.*

Proof: By the previous lemma we have to study the integral equation

$$u \;+\; G(cu) \;=\; Gf \tag{4.13}$$

with the integral operator G with kernel g from the previous lemma. The operator $T : u \mapsto G(cu)$ is bounded from $C[0,1]$ into $H^2(0,1)$ and thus compact from $C[0,1]$ into itself (see again Problem 4.5). Now we use the following result from linear functional analysis (see Theorem A.36 of the Appendix A.3): If the homogeneous linear equation $u + Tu = 0$ with the compact operator T from $C[0,1]$ into itself admits *only* the trivial solution $u = 0$ then the inhomogeneous equation $u + Tu = h$ is uniquely solvable for all $h \in C[0,1]$, and the solution depends continuously on h. In other words, if $I + T$ is one-to-one then also onto and $I + T$ is boundedly invertible. Therefore, we have to show injectivity of $I + T$ in $C[0,1]$. Let $u \in C[0,1]$ solve (4.13) for $h = 0$. Then $u \in H^2(0,1)$ and u solves (4.11) for $f = 0$ by the previous lemma. Multiplication of (4.11) by $u(t)$ and integration yields

$$0 \;=\; \int_0^1 \left[-u''(t) + c(t)\,u(t)\right] u(t)\,dt \;=\; \int_0^1 \left[u'(t)^2 + c(t)\,u(t)^2\right] dt\,,$$

where we used partial integration and the fact that u vanishes at the boundary of $[0,1]$. Since $c \geq 0$ we conclude that u' vanishes on $[0,1]$. Therefore, u is constant and thus zero because of the boundary conditions. Therefore, $I + T$ is one-to-one and thus invertible. This shows that (4.11) is uniquely solvable in $H^2(0,1)$ for every $f, c \in L^2(0,1)$ with $c \geq 0$ almost everywhere on $(0,1)$. In order to derive the explicit estimate for $\|u\|_{H^2(0,1)}$ we observe first that any solution $u \in H^2(0,1)$ of (4.11) satisfies $\|u'\|^2_{L^2(0,1)} \leq \|f\|_{L^2(0,1)}\|u\|_{L^2(0,1)}$. Indeed, this follows by multiplication of the differential equation by $u(t)$ and integration:

$$-\int_0^1 u''(t)\, u(t)\, dt \; + \; \int_0^1 c(t)\, u(t)^2\, dt \; = \; \int_0^1 f(t)\, u(t)\, dt \; \leq \; \|f\|_{L^2(0,1)}\|u\|_{L^2(0,1)}\,.$$

Partial integration and the assumption $c(t) \geq 0$ yields the estimate $\|u'\|^2_{L^2(0,1)} \leq \|f\|_{L^2(0,1)}\|u\|_{L^2(0,1)}$. With $\|u\|_\infty \leq \|u'\|_{L^2(0,1)}$ and $\|u\|_{L^2(0,1)} \leq \|u\|_\infty$ this implies that $\|u\|_\infty \leq \|f\|_{L^2(0,1)}$. Therefore

$$
\begin{aligned}
\|u\|_{H^2(0,1)} &= \|G(f - cu)\|_{H^2(0,1)} \leq \|G\|_{\mathcal{L}(L^2(0,1),H^2(0,1))}\, \|f - cu\|_{L^2(0,1)} \\
&\leq \|G\|_{\mathcal{L}(L^2(0,1),H^2(0,1))}\, \big[\|f\|_{L^2(0,1)} + \|c\|_{L^2(0,1)}\|u\|_\infty\big] \\
&\leq \big(1 + \|c\|_{L^2(0,1)}\big)\, \|G\|_{\mathcal{L}(L^2(0,1),H^2(0,1))}\, \|f\|_{L^2(0,1)}\,.
\end{aligned}
$$

$\square$

We can even show existence and uniqueness for c from a small open neighborhood of $L^2_+(0,1) = \{c \in L^2(0,1) : c \geq 0 \text{ on } (0,1)\}$.

Corollary 4.18 *There exists $\delta > 0$ such that the boundary value problem (4.11) is uniquely solvable for all $f \in L^2(0,1)$ and $c \in U_\delta$ where*

$$U_\delta := \left\{ c = c_1 + h \in L^2(0,1) : \begin{array}{c} c_1 \in L^2_+(0,1),\ h \in L^2(0,1) \\ (1 + \|c_1\|_{L^2(0,1)})\|h\|_{L^2(0,1)} < \delta \end{array} \right\}.$$

Furthermore, there exists $\gamma > 0$ such that $\|u\|_{H^2(0,1)} \leq \gamma\,(1+\|c\|_{L^2(0,1)})\|f\|_{L^2(0,1)}$ for all $f \in L^2(0,1)$ and $c \in U_\delta$.

Proof: Let $K_{c_1} : L^2(0,1) \to H^2(0,1)$ be the operator $f \mapsto u$ where u is the solution of (4.11) for $c_1 \in L^2_+(0,1)$. Let $c = c_1 + h \in U_\delta$. We consider the fixed point equation $\tilde{u} + K_{c_1}(h\tilde{u}) = K_{c_1}f$ for $\tilde{u} \in H^2(0,1)$. We have for $v \in H^2(0,1)$ that

$$
\begin{aligned}
\|K_{c_1}(hv)\|_{H^2(0,1)} &\leq \gamma(1 + \|c_1\|_{L^2(0,1)})\,\|hv\|_{L^2(0,1)} \\
&\leq \gamma(1 + \|c_1\|_{L^2(0,1)})\,\|h\|_{L^2(0,1)}\,\|v\|_\infty \leq \gamma\,\delta\|v\|_{H^2(0,1)}\,.
\end{aligned}
$$

For $\delta < 1/\gamma$ we observe that $v \mapsto K_{c_1}(hv)$ is a contraction and, by the Contraction Theorem A.31, $\tilde{u} + K_{c_1}(h\tilde{u}) = K_{c_1}f$ has a unique solution $\tilde{u} \in H^2(0,1)$

and

$$\|\tilde{u}\|_{H^2(0,1)} \leq \frac{1}{1-\delta\gamma}\|K_{c_1}f\|_{H^2(0,1)} \leq \gamma\frac{1+\|c_1\|_{L^2(0,1)}}{1-\delta\gamma}\|f\|_{L^2(0,1)}$$

$$\leq \gamma\frac{1+\|c_1+h\|_{L^2(0,1)}+\delta}{1-\delta\gamma}\|f\|_{L^2(0,1)}$$

$$\leq \gamma\frac{1+\delta}{1-\delta\gamma}\left(1+\|c\|_{L^2(0,1)}\right)\|f\|_{L^2(0,1)}$$

because $\|h\|_{L^2(0,1)} \leq \delta$. Finally, we note that the equation $\tilde{u} = K_{c_1}f - K_{c_1}(h\tilde{u})$ is equivalent to $-\tilde{u}'' + (c_1+h)\tilde{u} = f$. $\square$

We note that the set U_δ is an open set containing $L^2_+(0,1)$ (see Problem 4.5). Therefore, K can be extended to the set U_δ. As a next step towards the inverse problem, we show that the nonlinear mapping $K : c \mapsto u$ is continuous and even differentiable.

Theorem 4.19 *Let $U_\delta \supset L^2_+(0,1)$ be as in the previous corollary and let $K : U_\delta \to H^2(0,1)$ defined by $K(c) = u$ where $u \in H^2(0,1)$ solves the boundary value problem (4.11). Then K is continuous and even differentiable in every $c \in U_\delta$. The derivative is given by $K'(c)h = v$ where $v \in H^2(0,1)$ is the solution of the boundary value problem*

$$-v''(t) + c(t)v(t) = -h(t)u(t), \quad 0 < t < 1, \quad v(0) = v(1) = 0, \qquad (4.14)$$

and $u \in H^2(0,1)$ is the solution of (4.11) for c; that is, $u = K(c)$.

Proof: Let $h \in L^2(0,1)$ such that $c+h \in U_\delta$ and let $\tilde{u} = K(c+h)$. Then u and $\tilde{u}$ satisfy

$$-\tilde{u}''+(c+h)\tilde{u} = f, \quad -u''+(c+h)u = f+hu, \quad \tilde{u}(0) = \tilde{u}(1) = u(0) = u(1) = 0,$$

respectively. We subtract both equations which yields $-(\tilde{u}-u)''+(c+h)(\tilde{u}-u) = -hu$. The stability estimate yields

$$\|\tilde{u} - u\|_{H^2(0,1)} \leq \gamma\left(1+\|c+h\|_{L^2(0,1)}\right)\|hu\|_{L^2(0,1)}$$

$$\leq \gamma\left(1+\|c\|_{L^2(0,1)}+\delta\right)\|h\|_{L^2(0,1)}\|u\|_\infty$$

which proves continuity (even Lipschitz continuity on bounded sets for c).

For the differentiability, we just subtract the equations for u and v from the one for $\tilde{u}$

$$-(\tilde{u} - u - v)'' + (c+h)(\tilde{u} - u - v) = -hv, \qquad (4.15)$$

with homogeneous boundary conditions. The stability estimate yields

$$\|\tilde{u} - u - v\|_{H^2(0,1)} \leq \gamma\left(1+\|c+h\|_{L^2(0,1)}\right)\|hv\|_{L^2(0,1)}$$

$$\leq \gamma\left(1+\|c\|_{L^2(0,1)}+\delta\right)\|h\|_{L^2(0,1)}\|v\|_\infty.$$

The stability estimate applied to v yields

$$
\begin{aligned}
\|v\|_\infty \le \|v\|_{H^2(0,1)} &\le \gamma\left(1 + \|c\|_{L^2(0,1)}\right)\|hu\|_{L^2(0,1)} \\
&\le \gamma\left(1 + \|c\|_{L^2(0,1)}\right)\|h\|_{L^2(0,1)}\|u\|_\infty
\end{aligned}
$$

which altogether ends up to

$$
\|\tilde{u} - u - v\|_{H^2(0,1)} \le \gamma^2\left(1 + \|c\|_{L^2(0,1)} + \delta\right)^2 \|h\|_{L^2(0,1)}^2 \|u\|_\infty .
$$

This proves differentiability.   □

In the following, we consider the parameter-to-solution map K as a mapping from $U_\delta \subset L^2(0,1)$ into $L^2(0,1)$ instead of $H^2(0,1)$. Then, of course, K is also differentiable with respect to this space with the same derivative.

From the theory, we know that for Assumption 4.11, we need the adjoint of $K'(c)$.

Lemma 4.20 *The adjoint operator $K'(c)^* : L^2(0,1) \to L^2(0,1)$ is given by $K'(c)^* w = -u\,y$ where $u = K(c)$, and $y \in H^2(0,1)$ solves the following boundary value problem (the "adjoint problem"):*

$$
-y''(t) + c(t)\,y(t) = w(t),\ 0 < t < 1,\quad y(0) = y(1) = 0, \qquad (4.16)
$$

Proof: Let $h, w \in L^2(0,1)$ and v the solution of (4.14) for $c = c^*$ and y the solution of (4.16). By partial integration we compute

$$
\begin{aligned}
\left(K'(c)h, w\right)_{L^2(0,1)} &= \int_0^1 v(t)\left[-y''(t) + c(t)\,y(t)\right] dt \\
&= \int_0^1 \left[-v''(t) + c(t)\,v(t)\right] y(t)\, dt = -\int_0^1 h(t)\,u(t)\,y(t)\, dt \\
&= -(h, uy)_{L^2(0,1)} .
\end{aligned}
$$

This proves the assertion.   □

Now we can formulate condition (ii) of Assumption 4.11 for a linear index function φ:

The existence of $w \in L^2(0,1)$ with $c^* - \hat{c} = K'(c^*)^* w$ is equivalent to the existence of $y \in H^2(0,1)$ with $y(0) = y(1) = 0$ and $c^* - \hat{c} = -u^* y$. Therefore, the condition is equivalent to

$$
\frac{c^* - \hat{c}}{u^*} \in H^2(0,1) \cap H_0^1(0,1) .
$$

This includes smoothness of $c^* - \hat{c}$ as well as a sufficiently strong boundary condition because also u^* vanishes at the boundary of $[0,1]$.

4.2.4   A Glimpse on Extensions to Banach Spaces

Recalling the classical Tikhonov functional $J_{\alpha,\delta}$ in Hilbert spaces from (4.2) we observe that the first term measures the misfit in the equation while the second part serves as a penalty term. The error $x^{\alpha,\delta} - x^*$ in the solution is measured in a third metric. In many cases, the canonical space for the unknown quantity x is only a Banach space rather than a Hilbert space. For example, in parameter identification problems as in the previous subsection the canonical space for the parameters are L^∞–spaces rather than L^2–spaces. The Hilbert space setting in the previous subsection only works because the pointwise multiplication is continuous as a mapping from $L^2 \times H^2$ into L^2. For more general partial differential equation this is not always true. For an elaborate motivation why to use Banach space settings, we refer to Chapter I of the excellent monograph [245] by Schuster, Kaltenbacher, Hofmann, and Kazimierski.

In this subsection, we will get the flavor of some aspects of this theory. Let X and Y be Banach spaces, $K : X \supset \mathcal{D}(K) \to Y$ a (nonlinear) operator with the domain of definition $\mathcal{D}(K) \subset X$ where $\mathcal{D}(K)$ is again convex and closed, and let $x^* \in \mathcal{D}(K)$ be the exact solution of $K(x) = y^*$ for some y^*. In the following we fix this pair x^*, y^*. As before, y^* is perturbed by $y^\delta \in Y$ such that $\|y^\delta - y^*\|_Y \le \delta$ for all $\delta \in (0, \delta_{max})$ for some $\delta_{max} > 0$. We note that we measure the error in the data with respect to the Banach space norm. The penalty term $\|x - \hat{x}\|_X^2$ is now replaced by any convex and continuous function $\Omega : X \to [0, \infty]$ where $\mathcal{D}(\Omega) := \{x \in X : \Omega(x) < \infty\}$ is not empty and, even more, $\mathcal{D}(K) \cap \mathcal{D}(\Omega) \neq \emptyset$. Further assumptions on Ω and K are needed to ensure the existence of minima of the Tikhonov functional

$$J_{\alpha,\delta}(x) := \|K(x) - y^\delta\|_Y^p + \alpha\,\Omega(x), \quad x \in \mathcal{D}(K) \cap \mathcal{D}(\Omega). \qquad (4.17)$$

Here $p > 1$ is a fixed parameter. Instead of posing assumptions concerning the weak topology, we just make the same assumption as at the beginning of Subsection 4.2.2.

Assumption 4.21 *The Tikhonov functional $J_{\alpha,\delta}$ possesses global minima $x^{\alpha,\delta} \in \mathcal{D}(K) \cap \mathcal{D}(\Omega)$ on $\mathcal{D}(K) \cap \mathcal{D}(\Omega)$ for all $\alpha, \delta > 0$; that is,*

$$\|K(x^{\alpha,\delta}) - y^\delta\|_Y^p + \alpha\,\Omega(x^{\alpha,\delta}) \le \|K(x) - y^\delta\|_Y^p + \alpha\,\Omega(x) \qquad (4.18)$$

for all $x \in \mathcal{D}(K) \cap \mathcal{D}(\Omega)$.

It remains to specify the metric in which we measure the error in x. As we will see in a moment the norm in X is not always the best possibility. To have more flexibility, we take any "measure function" $E(x)$ which measures the distance of x to x^*. We only require that $E(x) \ge 0$ for all $x \in X$ and $E(x^*) = 0$. Then the variational source condition of Assumption 4.12 is generalized into the following form.

Assumption 4.22 *(Variational Source condition)* Let $x^* \in \mathcal{D}(K) \cap \mathcal{D}(\Omega)$ be a solution of $K(x) = y^*$. Furthermore, let $\delta_{max} = \sup\{\|K(x^*) - K(x)\|_Y : x \in \mathcal{D}(K) \cap \mathcal{D}(\Omega)\}$ and $\varphi : [0, \delta_{max}) \to \mathbb{R}$ be a concave index function *(see Definition 4.10)*, and, for some constants $\beta > 0$ and $\rho > 0$ let the following estimate hold.

$$\beta E(x) \leq \Omega(x) - \Omega(x^*) + \varphi(\|K(x^*) - K(x)\|_Y)$$

for all $x \in \mathcal{M}_\rho := \{x \in \mathcal{D}(K) \cap \mathcal{D}(\Omega) : \Omega(x) \leq \Omega(x^*) + \rho\}$.

Theorem 4.23 *Let Assumptions 4.21 and 4.22 hold.*

(a) Let $\alpha = \alpha(\delta)$ be chosen such that

$$c_- \frac{\delta^p}{\varphi(\delta)} \leq \alpha(\delta) \leq c_+ \frac{\delta^p}{\varphi(\delta)} \quad \text{for all } \delta \in (0, \delta_1) \qquad (4.19)$$

where $c_+ \geq c_- \geq \frac{1}{\rho}$ are independent of δ and where $\delta_1 \in (0, \delta_{max})$ is chosen such that $\varphi(\delta_1) \leq \rho c_-$ (a priori choice),

(b) or assume that there exists $r_+ > r_- \geq 1$ and $\alpha(\delta) > 0$ such that

$$r_- \delta \leq \|K(x^{\alpha(\delta),\delta}) - y^\delta\|_Y \leq r_+ \delta \quad \text{for all } \delta \in (0, \delta_1) \qquad (4.20)$$

where $\delta_1 \in (0, \delta_{max})$ is arbitrary (a posteriori choice).

Then $x^{\alpha(\delta),\delta} \in \mathcal{M}_\rho$ and

$$E(x^{\alpha(\delta),\delta}) = \mathcal{O}(\varphi(\delta)) \quad \text{and} \quad \|K(x^{\alpha(\delta),\delta}) - y^*\|_Y = \mathcal{O}(\delta), \quad \delta \to 0.$$

Proof: We follow almost exactly the proof of Theorem 4.14. Substituting $x = x^*$ into (4.18) yields

$$\|K(x^{\alpha,\delta}) - y^\delta\|_Y^p + \alpha \Omega(x^{\alpha,\delta}) \leq \delta^p + \alpha \Omega(x^*). \qquad (4.21)$$

First we show that $x^{\alpha(\delta),\delta} \in \mathcal{M}_\rho$. If $\alpha(\delta)$ is chosen as in (4.19) then

$$\Omega(x^{\alpha(\delta),\delta}) \leq \frac{\delta^p}{\alpha(\delta)} + \Omega(x^*) \leq \frac{1}{c_-} \varphi(\delta) + \Omega(x^*) \leq \frac{1}{c_-} \varphi(\delta_1) + \Omega(x^*)$$

which shows $x^{\alpha(\delta),\delta} \in \mathcal{M}_\rho$ by the choice of δ_1. If $\alpha(\delta)$ is chosen by the discrepancy principle (4.20) then again from (4.21)

$$(r_-^p - 1) \delta^p + \alpha(\delta) \Omega(x^{\alpha(\delta),\delta}) \leq \alpha(\delta) \Omega(x^*)$$

and thus $\Omega(x^{\alpha(\delta),\delta}) \leq \Omega(x^*)$ because $r_- \geq 1$. Therefore, $x^{\alpha(\delta),\delta} \in \mathcal{M}_0 \subset \mathcal{M}_\rho$. Now we show the rates of convergence. Applying the variational source condition we conclude from (4.21) that

$$\|K(x^{\alpha,\delta}) - y^\delta\|_Y^p + \alpha\beta E(x^{\alpha,\delta}) \leq \delta^p + \alpha \varphi(\|K(x^{\alpha,\delta}) - y^*\|_Y)$$

$$\leq \delta^p + \alpha\,\varphi\big(\|K(x^{\alpha,\delta}) - y^\delta\|_Y + \delta\big). \tag{4.22}$$

If $\alpha = \alpha(\delta)$ is chosen according to the discrepancy principle (4.20) then

$$(r_-^p - 1)\delta^p + \alpha(\delta)\,\beta\,E(x^{\alpha(\delta),\delta}) \leq \alpha(\delta)\,\varphi\big((r_+ + 1)\delta\big)$$

and thus $\beta\,E(x^{\alpha(\delta),\delta}) \leq \varphi\big((r_+ + 1)\delta\big)$. Now we use the elementary estimate $\varphi(s\delta) \leq s\,\varphi(\delta)$ for all $s \geq 1$ and $\delta \geq 0$ (see Lemma A.73 of Appendix A.8). Therefore,

$$\beta\,E(x^{\alpha(\delta),\delta}) \leq (r_+ + 1)\,\varphi(\delta).$$

This proves the estimate for $E(x^{\alpha(\delta),\delta})$. The estimate for $\|K(x^{\alpha(\delta),\delta}) - y^*\|_Y$ follows obviously from the discrepancy inequality and the triangle inequality.

Let now $\alpha = \alpha(\delta)$ be given by (4.19). From (4.22), we obtain, using the upper estimate of $\alpha(\delta)$,

$$\|K(x^{\alpha(\delta),\delta}) - y^\delta\|_Y^p \leq \delta^p + c_+ \frac{\delta^p}{\varphi(\delta)}\,\varphi\big(\|K(x^{\alpha,\delta}) - y^\delta\|_Y + \delta\big).$$

We set $t = \|K(x^{\alpha(\delta),\delta}) - y^\delta\|_Y/\delta$ for abbreviation. Then the previous formula takes the form

$$t^p \leq 1 + c_+ \frac{\varphi\big(\delta(t+1)\big)}{\varphi(\delta)} \leq 1 + c_+(t+1) = (1 + c_+) + c_+ t$$

where we used the estimate $\varphi(s\delta) \leq s\,\varphi(\delta)$ for $s \geq 1$ again. Choose $c > 0$ with $c\,(c^{p-1} - c_+) > 1 + c_+$. Then $t \leq c$. Indeed, if $t > c$ then $t^p - c_+ t = t\,(t^{p-1} - c_+) > c\,(c^{p-1} - c_+) > 1 + c_+$, a contradiction. This proves that $\|K(x^{\alpha(\delta),\delta}) - y^\delta\|_Y \leq c\delta$. Now we substitute this into the right-hand side of (4.22), which yields

$$\beta\,E(x^{\alpha(\delta),\delta}) \leq \frac{\delta^p}{\alpha(\delta)} + \varphi\big((c+1)\delta\big) \leq \frac{1}{c_-}\,\varphi(\delta) + (c+1)\,\varphi(\delta).$$

This ends the proof. □

We note that in the case of Hilbert spaces X and Y and $\Omega(x) = \|x - \hat{x}\|_X^2$ and $E(x) = \|x - x^*\|_X^2$ Assumption 4.22 and Theorem 4.23 reduce to Assumption 4.12 and Theorem 4.14, respectively.

Before we continue with the general theory, we apply this theorem to the special situation to determine a *sparse* approximation of the linear problem $Kx = y^\delta$. By sparse we mean that the solution $x^* \in X$ can be expressed by only finitely many elements of a given basis of X. Therefore, let $\tilde{X}$ be a Banach space having a *Schauder basis* $\{b_j : j \in \mathbb{N}\}$ with $\|b_j\|_{\tilde{X}} = 1$ for all $j \in \mathbb{N}$; that is, every element $x \in \tilde{X}$ has a unique representation as $x = \sum_{j\in\mathbb{N}} x_j b_j$ where, of course, the convergence is understood in the norm of $\tilde{X}$. We define the subspace $X \subset \tilde{X}$ by

$$X = \left\{ x = \sum_{j\in\mathbb{N}} x_j b_j : \sum_{j\in\mathbb{N}} |x_j| < \infty \right\}$$

with the norm $\|\sum_{j\in\mathbb{N}} x_j b_j\|_X := \sum_{j\in\mathbb{N}} |x_j|$ for $x \in X$. Then X is bound-edly imbedded in $\tilde{X}$ because $\|x\|_{\tilde{X}} = \|\sum_{j\in\mathbb{N}} x_j b_j\|_{\tilde{X}} \leq \sum_{j\in\mathbb{N}} |x_j| \|b_j\|_{\tilde{X}} = \sum_{j\in\mathbb{N}} |x_j| = \|x\|_X$ for $x \in X$. Obviously, the space X is norm-isomorphic to the space ℓ^1 of sequences (x_j) such that $\sum_{j=1}^{\infty} |x_j|$ converge, equipped with the canonical norm $\|x\|_{\ell^1} = \sum_{j=1}^{\infty} |x_j|$ for $x = (x_j)$. As the Schauder basis of ℓ^1 we take $\{e^{(j)} : j = 1, 2, \ldots\} \subset \ell^1$ where $e^{(j)} \in \ell^1$ is defined as $e_k^{(j)} = 0$ for $k \neq j$ and $e_j^{(j)} = 1$. Therefore, we can take directly ℓ^1 as the space X. We note that the dual of ℓ^1 is just $(\ell^1)^* = \ell^{\infty}$, the space of bounded sequences with the sup-norm and the dual pairing[3] $\langle y, x \rangle_{\ell^{\infty}, \ell^1} = \sum_{j\in\mathbb{N}} y_j x_j$ for $y \in \ell^{\infty}$ and $x \in \ell^1$. Also we note that ℓ^1 itself is the dual of the space c_0 of sequences converging to zeros (see Example A.21). Therefore, by Theorem A.77 of Appendix A.9 the unit ball in ℓ^1 is weak* compact which is an important ingredient to prove existence of minimizers of the Tikhonov functional.

With these introductory remarks, we are able to show the following result where we followed the presentation in [96].

Theorem 4.24 *Let Y be a Banach space and $K : \ell^1 \to Y$ be a linear bounded operator such that $\langle \mu, K e^{(j)} \rangle_{Y^*, Y} \to 0$ as $j \to \infty$ for all $\mu \in Y^*$ where $\langle \mu, y \rangle_{Y^*, Y}$ denotes the application of $\mu \in Y^*$ to $y \in Y$. Let $Kx^* = y^*$ and $y^{\delta} \in Y$ with $\|y^{\delta} - y^*\|_Y \leq \delta$.*

(a) Let $x^{\alpha,\delta} \in \ell^1$ be a minimizer of

$$J_{\alpha,\delta}(x) = \|Kx - y^{\delta}\|_Y^p + \alpha \|x\|_{\ell^1}, \quad x \in \ell^1.$$

Then $x^{\alpha,\delta} \in \ell^1$ is sparse; that is, the number of non-vanishing components $x_j^{\alpha,\delta} \neq 0$ is finite.

(b) For every $j \in \mathbb{N}$ let there exists $f_j \in Y^*$ with $e^{(j)} = K^* f_j$ where $K^* : Y^* \to (\ell^1)^* = \ell^{\infty}$ is the dual operator corresponding to K. Define the function $\varphi : [0, \infty) \to \mathbb{R}$ as

$$\varphi(t) := 2 \inf_{n\in\mathbb{N}} \left[\gamma_n t + \sum_{j>n} |x_j^*| \right] \quad where \quad \gamma_n = \sup_{s_j \in \{0,1,-1\}} \left\| \sum_{j=1}^{n} s_j f_j \right\|_{Y^*}.$$

Then φ is a concave index function. With the choices (4.19) or (4.20) of $\alpha = \alpha(\delta)$ the following convergence rates hold:

$$\|x^{\alpha(\delta),\delta} - x^*\|_{\ell^1} = \mathcal{O}(\varphi(\delta)), \quad \|Kx^{\alpha(\delta),\delta} - y^*\|_Y = \mathcal{O}(\delta)$$

as δ tends to zero.

(c) If $x^* \in \ell^1$ is such that $\sum_{j=1}^{\infty} \gamma_j^{\sigma} |x_j^*| < \infty$ for some $\sigma > 0$ then we have

$$\|x^{\alpha(\delta),\delta} - x^*\|_{\ell^1} = \mathcal{O}(\delta^{\sigma/(1+\sigma)}), \quad \delta \to 0.$$

---

[3] Note that we denote the dual pairing by $\langle \ell, x \rangle_{X^*, X} = \ell(x)$ for $\ell \in X^*$ and $x \in X$. The mapping $(\ell, x) \mapsto \langle \ell, x \rangle_{X^*, X}$ is bilinear.

If x^* is sparse or if (γ_n) is bounded then we have $\|x^{\alpha(\delta),\delta} - x^*\|_{\ell^1} = \mathcal{O}(\delta)$ as δ tends to zero.

Proof: (a) Set $z = Kx^{\alpha,\delta} - y^\delta$ for abbreviation. The optimality of $x^{\alpha,\delta}$ reads as

$$\|z + Kh\|_Y^p - \|z\|_Y^p \geq -\alpha\big[\|x^{\alpha,\delta} + h\|_{\ell^1} - \|x^{\alpha,\delta}\|_{\ell^1}\big] \text{ for all } h \in \ell^1 . \qquad (4.23)$$

Define the sets $\mathcal{A}, \mathcal{B} \subset \mathbb{R} \times Y$ as follows:

$$\begin{aligned}
\mathcal{A} &= \big\{(r,y) \in \mathbb{R} \times Y : r > \|z + y\|_Y^p - \|z\|_Y^p\big\}, \\
\mathcal{B} &= \big\{(r, Kh) \in \mathbb{R} \times Y : h \in \ell^1, \ r \leq -\alpha\big[\|x^{\alpha,\delta} + h\|_{\ell^1} - \|x^{\alpha,\delta}\|_{\ell^1}\big]\big\}.
\end{aligned}$$

Then it is not difficult to show that $\mathcal{A}$ and $\mathcal{B}$ are convex, $\mathcal{A}$ is open, and $\mathcal{A} \cap \mathcal{B} = \emptyset$ because of (4.23). Now we apply the separation theorem for convex sets (see Theorem A.69 of Appendix A.8). There exists $(s, \mu) \in \mathbb{R} \times Y^*$ and $\gamma \in \mathbb{R}$ such that $(s, \mu) \neq (0,0)$ and

$$s\,r + \langle \mu, y \rangle_{Y^*,Y} \geq \gamma \geq s\,r' + \langle \mu, Kh \rangle_{Y^*,Y} \text{ for all } (r,y) \in \mathcal{A} \text{ and } (r', Kh) \in \mathcal{B} .$$

Letting r tend to infinity while keeping the other variables constant yields $s \geq 0$. It is $s \neq 0$ because otherwise we would have $\langle \mu, y \rangle_{Y^*,Y} \geq \gamma$ for all $y \in Y$ (set $r := \|z + y\|_Y^p - \|z\|_Y^p + 1$) which would yield that also μ vanishes[4], a contradiction. Therefore, $s > 0$, and without loss of generality, $s = 1$. Now we set $y = 0$ and fix any $h \in \ell^1$ and let r tend to zero from above and set $r' = -\alpha\big[\|x^{\alpha,\delta} + h\|_{\ell^1} - \|x^{\alpha,\delta}\|_{\ell^1}\big]$. This yields the inequality

$$0 \geq -\alpha\big[\|x^{\alpha,\delta} + h\|_{\ell^1} - \|x^{\alpha,\delta}\|_{\ell^1}\big] + \langle \mu, Kh \rangle_{Y^*,Y}$$

for all $h \in \ell^1$. For any $t \in \mathbb{R}$ and $k \in \mathbb{N}$, we set $h = te^{(k)}$ and arrive at

$$\alpha\,\big|x_k^{\alpha,\delta} + t\big| - \big|x_k^{\alpha,\delta}\big| - t\,\langle \mu, Ke^{(k)} \rangle_{Y^*,Y} \geq 0 .$$

For fixed k with $x_k^{\alpha,\delta} \neq 0$ we choose $|t|$ so small such that $\text{sign}\big(x_k^{\alpha,\delta} + t\big) = \text{sign}\big(x_k^{\alpha,\delta}\big)$. Then the previous inequality reads as

$$t\big[\alpha\,\text{sign}\big(x_k^{\alpha,\delta}\big) - \langle \mu, Ke^{(k)} \rangle_{Y^*,Y}\big] \geq 0 .$$

Choosing $t > 0$ and $t < 0$ yields $\alpha\,\text{sign}\big(x_k^{\alpha,\delta}\big) = \langle \mu, Ke^{(k)} \rangle_{Y^*,Y}$; that is, $\big|\langle \mu, Ke^{(k)} \rangle_{Y^*,Y}\big| = \alpha$. This holds for every $k \in J := \big\{k \in \mathbb{N} : x_k^{\alpha,\delta} \neq 0\big\}$. This implies that J is finite because of $\langle \mu, Ke^{(k)} \rangle_{Y^*,Y} \to 0$ as $k \to \infty$.

(b) We apply Theorem 4.23 and have to verify Assumptions 4.21 and 4.22 for the special case $\mathcal{D}(K) = \mathcal{D}(\Omega) = X = \ell^1$, $E(x) = \|x - x^*\|_{\ell^1}$, and $\Omega(x) = \|x\|_{\ell^1}$ for $x \in X = \ell^1$. First we show that φ is a concave index function. Indeed, φ is continuous, monotonic, and concave as the infimum of affine functions

---

[4]the reader should prove this himself.

(see Problem 4.6). Furthermore, $\varphi(0) = 2\inf\limits_{n\in\mathbb{N}}\sum_{j>n}|x_j^*| = 0$ which shows that φ is a concave index function. Assumption 4.21; that is, existence of a minimum of $J_{\alpha,\delta}$ can be shown using results on weak- and weak$*$-topologies. (The assumption that $\langle y^*, Ae^{(k)}\rangle_{Y^*,Y}$ tends to zero for all y^* is equivalent to the weak$*$-weak continuity of K. Then one uses that $\|\cdot\|_Y^p$ and $\|\cdot\|_{\ell^1}$ are lower weak semi-continuous and lower weak$*$ semi-continuous, respectively.) We do not carry out this part but refer to, e.g., [96]. Assumption 4.22 is therefore equivalent to (with $\beta = 1$)

$$\|x - x^*\|_{\ell^1} - \|x\|_{\ell^1} + \|x^*\|_{\ell^1} \leq \varphi(\|K(x - x^*)\|_Y) \quad \text{for all } x \in \ell^1. \quad (4.24)$$

To prove this we have for any $n \in \mathbb{N}$

$$\sum_{j=1}^{n}|x_j - x_j^*| = \sum_{j=1}^{n} s_j\langle e^{(j)}, x - x^*\rangle_{\ell^\infty,\ell^1} = \sum_{j=1}^{n} s_j\langle K^* f_j, x - x^*\rangle_{\ell^\infty,\ell^1}$$

$$= \sum_{j=1}^{n} s_j\langle f_j, K(x - x^*)\rangle_{Y^*,Y} \leq \left\|\sum_{j=1}^{n} s_j f_j\right\|_{Y^*} \|K(x - x^*)\|_Y$$

$$\leq \gamma_n \|K(x - x^*)\|_Y$$

where $s_j = \text{sign}(x_j - x_j^*) \in \{0, 1, -1\}$. Therefore, with $|x_j| \geq |x_j^*| - |x_j - x_j^*|$,

$$\sum_{j=1}^{n}\left[|x_j - x_j^*| - |x_j| + |x_j^*|\right] \leq 2\sum_{j=1}^{n}|x_j - x_j^*| \leq 2\gamma_n\|K(x - x^*)\|_Y.$$

Furthermore,

$$\sum_{j>n}\left[|x_j - x_j^*| - |x_j| + |x_j^*|\right] \leq 2\sum_{j>n}|x_j^*|;$$

that is,

$$\|x - x^*\|_{\ell^1} - \|x\|_{\ell^1} + \|x^*\|_{\ell^1} \leq 2\left[\gamma_n\|K(x - x^*)\|_Y + \sum_{j>n}|x_j^*|\right] \quad (4.25)$$

This shows (4.24) since this estimate holds for all $n \in \mathbb{N}$. Therefore, all of the assumptions of Theorem 4.23 are satisfied, and the error estimates are shown.

(c) We estimate $\varphi(t)$. First we observe that (γ_n) is monotonically increasing. If γ_n is bounded then it converges to some finite $\gamma \in \mathbb{R}$. Therefore, we let n tend to infinity in the definition of φ and arrive at $\varphi(t) = 2\gamma t$. We consider now the case that γ_n tends to infinity. First we estimate

$$\gamma_n t + \sum_{j>n}|x_j^*| \leq \gamma_n t + \frac{1}{\gamma_{n+1}^\sigma}\sum_{j>n}\gamma_j^\sigma|x_j^*| \leq \gamma_n t + \frac{c}{\gamma_{n+1}^\sigma}$$

with $c = \sum_{j=1}^{\infty}\gamma_j^\sigma|x_j^*|$. For sufficiently small t, the index

$$n(t) = \max\left\{n \in \mathbb{N} : \gamma_n \leq \frac{1}{t^{1/(1+\sigma)}}\right\}$$

is well-defined and finite. Then $\gamma_{n(t)} \leq t^{-1/(1+\sigma)}$ and $\gamma_{n(t)+1} > t^{-1/(1+\sigma)}$. Therefore,

$$\varphi(t) \leq \gamma_{n(t)}\, t + \frac{c}{\gamma_{n(t)+1}^{\sigma}} \leq (1+c)\, t^{\sigma/(1+\sigma)}.$$

Finally, if x^* is sparse then there exists $n \in \mathbb{N}$ such that $x_j^* = 0$ for all $j > n$. For that n the series $\sum_{j>n} |x_j^*|$ in (4.25) vanishes which shows the result for the linear index function $\varphi(t) = 2\gamma_n t$. $\qquad\square$

Remark: The reciprocals $1/\gamma_n$ play the role of the singular values in the case of Hilbert spaces $X = \ell^2$ and Y with a singular system $\{\mu_j, e^{(j)}, g^{(j)} : j \in J\}$. Indeed, the assumption $e^{(j)} = K^* f_j$ is satisfied with $f_j = \frac{1}{\mu_j} g^{(j)}$ and for γ_n one has the form

$$\gamma_n^2 = \sup_{s_j \in \{0,1,-1\}} \left\| \sum_{j=1}^{n} s_j f_j \right\|_Y^2 = \sup_{s_j \in \{0,1,-1\}} \sum_{j=1}^{n} \frac{s_j^2}{\mu_j^2} = \sum_{j=1}^{n} \frac{1}{\mu_j^2}$$

The condition that $\sum_{j=1}^{\infty} \gamma_j^{\sigma} |x_j^*|$ converges corresponds to the source condition $x^* \in \mathcal{R}\big((A^*A)^{\sigma/2}\big)$.

We go now back to the general case studied in Theorem 4.23. We wish to carry over Theorem 4.13 which proves the variational source condition from the classical one of Assumption 4.11. The essential formula used in the proof of part (a) of Theorem 4.13 was based on the binomial formula; that is,

$$\|x^* - x\|_X^2 = \|x - \hat{x}\|_X^2 - \|x^* - \hat{x}\|_X^2 - 2\,(x - x^*, x^* - \hat{x})_X.$$

If we denote the penalty term by $\Omega(x)$; that is, $\Omega(x) = \|x - \hat{x}\|_X^2$ then we can write this formula as

$$\|x^* - x\|_X^2 = B^{\Omega}(x, x^*) := \Omega(x) - \Omega(x^*) - \langle \Omega'(x^*), x - x^* \rangle_{X^*,X}$$

where $\Omega'(x^*) : X \to \mathbb{R}$ is the Fréchet derivative of Ω at x^* and $\langle \ell, z \rangle_{X^*,X}$ is the application of $\ell \in X^*$ to $z \in X$. The function $B^{\Omega}(x, x^*)$ is the famous *Bregman distance* corresponding to the function Ω. This can be extended to Banach spaces because for convex and differentiable functions Ω from a Banach space X into $\mathbb{R}$ the function

$$B^{\Omega}(x, x^*) := \Omega(x) - \Omega(x^*) - \langle \Omega'(x^*), x - x^* \rangle_{X^*,X}, \quad x \in X,$$

is nonnegative on X (see Lemma A.70 of Appendix A.8). If Ω is even strictly convex, then $B^{\Omega}(x, x^*) = 0 \Leftrightarrow x = x^*$. If $\Omega : X \to \mathbb{R}$ is convex and only continuous then the *subdifferential* $\partial\Omega(x^*) \subset X^*$; that is, set of *subgradients* is non-empty (see Lemma A.72 of Appendix A.8). We recall that $\partial\Omega(x^*) \subset X^*$ is the set of all $\ell \in X^*$ with

$$\Omega(x) - \Omega(x^*) - \langle \ell, x - x^* \rangle_{X^*,X} \geq 0 \quad \text{for all } x \in X.$$

If Ω is convex and differentiable then $\partial\Omega(x^*) = \{\Omega'(x^*)\}$ (Lemma A.72). Therefore, we formulate the following definition:

Definition 4.25 *Let X be a normed space, $A \subset X$ convex and open, and Ω : $A \to \mathbb{R}$ convex and continuous with subdifferential $\partial\Omega(x^*)$ at some $x^* \in A$. For $\ell \in \partial\Omega(x^*)$ the* Bregman distance *is defined as*

$$B_\ell^\Omega(x, x^*) := \Omega(x) - \Omega(x^*) - \langle \ell, x - x^* \rangle_{X^*, X}, \quad x \in A.$$

We note that the Bregman distance depends on the function Ω. It measures the defect of the function with its linearization at x^*. With the Bregman distance as $E(x)$ we have an analogue of Theorem 4.13.

Lemma 4.26 *Let $x^* \in \mathcal{D}(K) \cap \mathcal{D}(\Omega)$ be a solution of $K(x) = y^*$ and let φ : $[0, \infty) \to \mathbb{R}$ be a concave index function. Furthermore, let $\Omega : X \to \mathbb{R}$ be convex and continuous and $\ell \in \partial\Omega(x^*)$ and $E(x) = B_\ell^\Omega(x, x^*)$.*

(a) Then, for this particular choice of $E(x)$, the variational source condition of Assumption 4.22 is equivalent to the following condition: There exists $\rho > 0$ and $0 \le \sigma < 1$ such that

$$\langle \ell, x^* - x \rangle_{X^*, X} \le \sigma B_\ell^\Omega(x, x^*) + \varphi(\|K(x^*) - K(x)\|_Y) \qquad (4.26)$$

for all $x \in \mathcal{M}_\rho$.

(b) Assume that there exist $w \in Y^*$ and a concave index function η such that $\ell = K'(x^*)^* w$ and

$$\|K'(x^*)(x - x^*)\|_Y \le \eta(\|K(x^*) - K(x)\|_Y) \quad \text{for all } x \in \mathcal{M}_\rho.$$

Then Assumption 4.22 holds with $\varphi(t) = \|w\|_{Y^*} \eta(t)$ for $t \ge 0$.

Proof: (a) This follows directly from the definitions of $E(x)$ and $B_\ell^\Omega(x, x^*)$. Indeed, the estimate in Assumption 4.22 reads as

$$\beta B_\ell^\Omega(x, x^*) \le \Omega(x) - \Omega(x^*) + \varphi(\|K(x^*) - K(x)\|_Y);$$

that is,

$$\beta B_\ell^\Omega(x, x^*) \le B_\ell^\Omega(x, x^*) + \langle \ell, x - x^* \rangle_{X^*, X} + \varphi(\|K(x^*) - K(x)\|_Y)$$

which is equivalent to

$$\langle \ell, x^* - x \rangle_{X^*, X} \le (1 - \beta) B_\ell^\Omega(x, x^*) + \varphi(\|K(x^*) - K(x)\|_Y).$$

This proves part (a) with $\sigma = 1 - \beta$ if $\beta \le 1$ and $\sigma = 0$ otherwise.
(b) If $\ell = K'(x^*)^* w$ then

$$\begin{aligned}
\langle \ell, x^* - x \rangle_{X^*, X} &= \langle K'(x^*)^* w, x^* - x \rangle_{X^*, X} = \langle w, K'(x^*)(x^* - x) \rangle_{Y^*, Y} \\
&\le \|w\|_{Y^*} \|K'(x^*)(x^* - x)\|_Y \le \|w\|_{Y^*} \eta(\|K(x^*) - K(x)\|_Y).
\end{aligned}$$

This proves the condition of (a) with $\sigma = 0$; that is, $\beta = 1$. $\square$

Combining Theorem 4.23 with these particular choices, we have the following theorem:

Theorem 4.27 *Let Assumption 4.21 hold and let $x^* \in \mathcal{D}(K) \cap \mathcal{D}(\Omega)$ be a solution of $K(x) = y^*$. Furthermore, let $\delta_{max} = \sup\{\|K(x^*) - K(x)\|_Y : x \in \mathcal{D}(K) \cap \mathcal{D}(\Omega)\}$ and $\varphi : [0, \delta_{max}] \to \mathbb{R}$ be a concave index function, and for some constants $\rho > 0$ and $0 \leq \sigma < 1$, let the source condition (4.26) hold for some $\ell \in \partial\Omega(x^*)$. Let $\alpha = \alpha(\delta)$ be chosen according to (4.19) or (4.20). Then we have the error estimates*

$$B_\ell^\Omega(x^{\alpha(\delta),\delta}, x^*) = \mathcal{O}(\varphi(\delta)) \quad and \quad \|K(x^{\alpha(\delta),\delta}) - y^*\|_Y = \mathcal{O}(\delta)$$

as δ tends to zero.

As a particular and obviously important example, we now take

$$\Omega(x) := \|x - \hat{x}\|_X^p, \quad x \in X,$$

for some $p > 1$. Then one would like to characterize the Bregman distance — or, at least, construct lower bounds of $B_\ell^\Omega(x, x^*)$ in terms of $\|x - x^*\|_X^p$. This leads to the concept of p−convex Banach spaces.

Definition 4.28 *A Banach space X is called p−convex for some $p > 0$ if there exists $c > 0$ such that*

$$\|x + y\|_X^p - \|x\|_X^p - \langle \ell_x, y \rangle_{X^*,X} \geq c\|y\|_X^p \quad for\ all\ \ell_x \in \partial(\|\cdot\|_X^p)(x)$$

and $x, y \in X$.[5]

In other words, for p−convex spaces the Bregman distance $B_\ell^\Omega(x, z)$ corresponding to $\Omega(x) = \|x - \hat{x}\|_X^p$ can be bounded below by $c\|x - z\|_X^p$ for all $x, z \in X$. Therefore, if the assumptions of the previous Theorem 4.27 holds one has the rate $\|x^{\alpha(\delta),\delta} - x^*\|_X^p = \mathcal{O}(\varphi(\delta))$.

As a particular example, we show that $L^p(D)$ are p−convex for all $p > 2$.

Lemma 4.29 *Let $p > 1$ and $D \subset \mathbb{R}^n$ open and*

$$f(x) = \|x\|_{L^p(D)}^p = \int_D |x(t)|^p \, dt \quad for\ x \in L^p(D).$$

(a) Then f is differentiable and

$$f'(x)y = p \int_D y(t) |x(t)|^{p-1} \operatorname{sign} x(t) \, dt \quad for\ x, y \in L^p(D).$$

(b) Let $p > 2$. Then there exists $c_p > 0$ with

$$f(x + y) - f(x) - f'(x)y \geq c_p \|y\|_{L^p(D)}^p \quad for\ all\ x, y \in L^p(D).$$

---

[5]Actually, the classical definition uses the dual mapping instead of the subdifferential. However, by Asplund's theorem (see [48]) they are equivalent.

Proof: (a) First we observe that the integral for $f'(x)y$ exists by Hölder's inequality. Indeed, set $r = \frac{p}{p-1}$ and $s = p$ then $\frac{1}{r} + \frac{1}{s} = 1$ and thus

$$\int_D |x(t)|^{p-1}|y(t)|\, dt \;\leq\; \left(\int_D |x(t)|^{r(p-1)}dt\right)^{1/r}\left(\int_D |y(t)|^s dt\right)^{1/s}$$

$$= \left(\int_D |x(t)|^p dt\right)^{(p-1)/p}\left(\int_D |y(t)|^p dt\right)^{1/p}$$

$$= \|x\|_{L^p(D)}^{p-1}\|y\|_{L^p(D)}\,.$$

We use the following elementary estimate. There exist constants $c_+ > 0$ and $c_p \geq 0$ with $c_p = 0$ for $p \leq 2$ and $c_p > 0$ for $p > 2$ such that for all $z \in \mathbb{R}$

$$c_p|z|^p \;\leq\; |1+z|^p - 1 - pz \;\leq\; \begin{cases} c_+\,|z|^p & \text{if } p \leq 2 \text{ or } |z| \geq \frac{1}{2}, \\ c_+\,|z|^2 & \text{if } p > 2 \text{ and } |z| \leq \frac{1}{2}. \end{cases} \tag{4.27}$$

We give a proof of this estimate in Lemma A.74 for the convenience of the reader.

Therefore, for any $x, y \in \mathbb{R}$ with $x \neq 0$, we have

$$|x+y|^p - |x|^p - py\,|x|^{p-1}\operatorname{sign} x \;=\; |x|^p\left[\left|1+\frac{y}{x}\right|^p - 1 - p\frac{y}{x}\right] \;\geq\; 0$$

and

$$|x+y|^p - |x|^p - py\,|x|^{p-1}\operatorname{sign} x \;=\; |x|^p\left[\left|1+\frac{y}{x}\right|^p - 1 - p\frac{y}{x}\right]$$

$$\leq\; \begin{cases} c_+\,|x|^p|y/x|^p = c_+\,|y|^p & \text{if } p \leq 2 \text{ or } 2|y| \geq |x|, \\ c_+\,|x|^p|y/x|^2 = c_+\,|x|^{p-2}|y|^2 & \text{if } p > 2 \text{ and } 2|y| \leq |x|. \end{cases}$$

This holds obviously also for $x = 0$. Now we apply this to $x(t)$ and $y(t)$ with $x, y \in L^p(D)$. This shows already that

$$\int_D \left[|x(t)+y(t)|^p - |x(t)|^p - py(t)\,|x(t)|^{p-1}\operatorname{sign} x(t)\right] dt \;\geq\; 0\,.$$

Next we show that

$$\int_D \left[|x(t)+y(t)|^p - |x(t)|^p - py(t)\,|x(t)|^{p-1}\operatorname{sign} x(t)\right] dt \;\leq\; c\,\|y\|_{L^p(D)}^{\min\{2,p\}}$$

for $\|y\|_{L^p(D)} \leq 1$. This would finish the proof of part (a) because $p > 1$. For proving this estimate we define $T := \{t \in D : 2|y(t)| \geq |x(t)|\}$. Then

$$\int_T \left[|x(t)+y(t)|^p - |x(t)|^p - py(t)\,|x(t)|^{p-1}\operatorname{sign} x(t)\right] dt \;\leq\; c_+\int_T |y(t)|^p dt$$

and

$$\int_{D\setminus T} \left[|x(t) + y(t)|^p - |x(t)|^p - p\,y(t)\,|x(t)|^{p-1} \operatorname{sign} x(t) \right] dt$$

$$\leq \begin{cases} c_+ \displaystyle\int_{D\setminus T} |y(t)|^p dt & \text{if } p \leq 2, \\[2ex] c_+ \displaystyle\int_{D\setminus T} |x(t)|^{p-2} |y(t)|^2 dt & \text{if } p > 2. \end{cases}$$

If $p \leq 2$ we just add the two estimates and have shown the estimate

$$\int_D \left[|x(t) + y(t)|^p - |x(t)|^p - p\,y(t)\,|x(t)|^{p-1} \operatorname{sign} x(t) \right] dt \ \leq \ c_+ \|y\|^p_{L^p(D)}.$$

If $p > 2$ we apply Hölder's inequality to the integral $\int_{D\setminus T} |x(t)|^{p-2} |y(t)|^2 dt$.
Indeed, set $r = \frac{p}{p-2}$ and $s = \frac{p}{2}$ then $\frac{1}{r} + \frac{1}{s} = 1$ and thus

$$\int_{D\setminus T} |x(t)|^{p-2} |y(t)|^2 dt \ \leq \ \left(\int_{D\setminus T} |x(t)|^{r(p-2)} dt \right)^{1/r} \left(\int_{D\setminus T} |y(t)|^{2s} dt \right)^{1/s}$$

$$= \left(\int_{D\setminus T} |x(t)|^p dt \right)^{(p-2)/p} \left(\int_{D\setminus T} |y(t)|^p dt \right)^{2/p}$$

$$\leq \ \|x\|^{p-2}_{L^p(D)} \|y\|^2_{L^p(D)}.$$

Therefore,

$$\int_D \left[|x(t) + y(t)|^p - |x(t)|^p - p\,y(t)\,|x(t)|^{p-1} \operatorname{sign} x(t) \right] dt$$

$$\leq \ c_+ \|y\|^p_{L^p(D)} + c_+ \|x\|^{p-2}_{L^p(D)} \|y\|^2_{L^p(D)}.$$

For $p > 2$ and $\|y\|_{L^p(D)} \leq 1$ we have $\|y\|^p_{L^p(D)} \leq \|y\|^2_{L^p(D)}$ and thus

$$\int_D \left[|x(t) + y(t)|^p - |x(t)|^p - p\,y(t)\,|x(t)|^{p-1} \operatorname{sign} x(t) \right] dt \ \leq \ c \|y\|^2_{L^p(D)}.$$

This proves part (a).

(b) By part (a) we have to show that

$$\int_D \left[|x(t) + y(t)|^p - |x(t)|^p - p\,y(t)\,|x(t)|^{p-1} \operatorname{sign} x(t) - c_p \left| y(t) \right|^p \right] dt \ \geq \ 0$$

for all $x, y \in L^p(D)$. We show that the integrand is non-negative. Indeed, we have

$$|x + y|^p - |x|^p - p\, y\, |x|^{p-1} \operatorname{sign} x - c_p|y|^p \;=\; |x|^p \left[\left|1 + \frac{y}{x}\right|^p - 1 - p\frac{y}{x} - c_p \left|\frac{y}{x}\right|^p \right]$$

and this is nonnegative by (4.27).   □

By the same method, it can be shown that also the spaces ℓ^p of sequences and the Sobolev spaces $W^{m,p}(D)$ are p−convex for $p \geq 2$ and any $m \in \mathbb{N}_0$.

4.3   The Nonlinear Landweber Iteration

As a general criticism towards Tikhonov's method in the nonlinear case, we mention the disadvantage that the convergence results hold only for the *global* minima of the Tikhonov functional $J_{\alpha,\delta}$ which is, in general, a non-convex function. In the best of all cases, the global minima can only be computed by iterative methods[6]. Therefore, it seems to be natural to solve the nonlinear equation $K(x) = y^\delta$ directly by an iterative scheme such as Newton-type methods or Landweber methods. In this section, we present the simplest of such an algorithm and follow the presentation of the paper [126] and also the monograph [149].

In this section, let again X and Y be Hilbert spaces and $K : X \supset \mathcal{D}(K) \to Y$ a continuously Fréchet differentiable mapping from the open domain of definition $\mathcal{D}(K)$ and let $\hat{x} \in \mathcal{D}(K)$.

First we recall the nonlinear Tikhonov functional

$$J_{\alpha,\delta}(x) \;=\; \|K(x) - y^\delta\|_Y^2 \;+\; \alpha\|x - \hat{x}\|_X^2 \,, \quad x \in \mathcal{D}(K) \,,$$

for $\alpha, \delta \geq 0$. This functional is differentiable.

Lemma 4.30 $J_{\alpha,\delta}$ *is differentiable in every* $x^* \in \mathcal{D}(K)$ *and*

$$J_{\alpha,\delta}'(x^*)h \;=\; 2\operatorname{Re}\big(K(x^*) - y^\delta, K'(x^*)h\big)_Y \;+\; 2\alpha \operatorname{Re}(x^* - \hat{x}, h)_X \,, \quad h \in X \,.$$

Proof: We have $K(x^*+h) = K(x^*)+K'(x^*)h+r(h)$ with $\|r(h)\|_Y/\|h\|_X \to 0$ for $h \to 0$. Using the binomial formula $\|a + b\|^2 = \|a\|^2 + 2\operatorname{Re}(a, b) + \|b\|^2$ we have

$$
\begin{aligned}
J_{\alpha,\delta}(x^* + h) \;&=\; \|K(x^* + h) - y^\delta\|_Y^2 \;+\; \alpha\|x^* + h - \hat{x}\|_X^2 \\
&=\; \|(K(x^*) - y^\delta) + (K'(x^*)h + r(h))\|_Y^2 \;+\; \alpha\|(x^* - \hat{x}) + h\|_X^2 \\
&=\; \|K(x^*) - y^\delta\|_Y^2 + 2\operatorname{Re}\big(K(x^*) - y^\delta, K'(x^*)h + r(h)\big)_Y \\
&\quad + \|K'(x^*)h + r(h)\|_Y^2 \;+\; \alpha\|x^* - \hat{x}\|_X^2 \;+\; 2\alpha \operatorname{Re}(x^* - \hat{x}, h)_X \\
&\quad + \alpha\|h\|_X^2 \\
&=\; J_{\alpha,\delta}(x^*) + 2\operatorname{Re}\big(K(x^*) - y^\delta, K'(x^*)h\big)_Y + 2\alpha \operatorname{Re}(x^* - \hat{x}, h)_X + \tilde{r}(h)
\end{aligned}
$$

---

[6] Even this is only possible in very special cases. Usually, one must be satisfied with critical points of $J_{\alpha,\delta}$

with

$$\tilde{r}(h) \;=\; 2\operatorname{Re}\big(K(x^*) - y^\delta, r(h)\big)_Y \;+\; \|K'(x^*)h + r(h)\|_Y^2 \;+\; \alpha\|h\|_X^2 ,$$

thus $\lim_{h \to 0} \tilde{r}(h)/\|h\|_X = 0$.   □

Lemma 4.31 *If $x^* \in \mathcal{D}(K)$ is a local minimum of $J_{\alpha,\delta}$ then $J'_{\alpha,\delta}(x^*)h = 0$ for all $h \in X$, thus $K'(x^*)^*\big(K(x^*) - y^\delta\big) + \alpha\,(x^* - \hat{x}) = 0$.*

Proof: Let $h \neq 0$ fixed. For sufficiently small $t > 0$ the point $x^* + th$ belongs to $\mathcal{D}(K)$ because $\mathcal{D}(K)$ is open. Therefore, for sufficiently small $t > 0$ the optimality of x^* implies $\big[J_{\alpha,\delta}(x^* + th) - J_{\alpha,\delta}(x^*)\big]/\|th\|_X \geq 0$, thus

$$0 \;\leq\; \frac{J_{\alpha,\delta}(x^* + th) - J_{\alpha,\delta}(x^*) - J'_{\alpha,\delta}(x^*)(th)}{\|th\|_X} \;+\; \frac{J'_{\alpha,\delta}(x^*)h}{\|h\|_X}.$$

The inequality $J'_{\alpha,\delta}(x^*)h \geq 0$ follows as t tends to zero. This implies the assertion since this holds for all h.   □

The equation

$$K'(x^*)^*\big(K(x^*) - y^\delta\big) + \alpha\,(x^* - \hat{x}) \;=\; 0 \qquad (4.28)$$

reduces to the well known normal equation (2.16) in the case that K is linear and $\hat{x} = 0$ because the derivative is given by $K'(x^*) = K$. However, in general, this equation is nonlinear, and one has to apply an iterative method for solving this equation. Instead of solving the well posed nonlinear equation (4.28), one can directly use an iterative scheme to solve—and regularize—the unregularized equation $K'(x^*)^*\big(K(x^*) - y^\delta\big) = 0$, which can be written as

$$x^* \;=\; x^* \;-\; a\,K'(x^*)^*\big(K(x^*) - y^\delta\big)$$

with an arbitrary number $a > 0$. This is a fixed point equation, and it is natural to solve it iteratively by the fixed point iteration: Fix $\hat{x} \in \mathcal{D}(K)$. Set $x_0^\delta = \hat{x}$ and

$$x_{k+1}^\delta \;=\; x_k^\delta \;-\; a\,K'(x_k^\delta)^*\big(K(x_k^\delta) - y^\delta\big), \quad k = 0, 1, 2, \ldots$$

This is called the *nonlinear Landweber iteration* because it reduces to the well known Landweber iteration in the linear case, see Sections 2.3 and 2.6.

At the moment, it is not clear that it is well-defined; that is, that all of the iterates lie in the domain of definition $\mathcal{D}(K)$.

We choose $\rho > 0$ and $a > 0$ such that $a\|K'(x)\|_{\mathcal{L}(X,Y)}^2 \leq 1$ for all $x \in B(\hat{x}, \rho)$. Then we scale the equation $K(x) = y^\delta$; that is, we replace it by the equivalent equation $\tilde{K}(x) = \tilde{y}^\delta$ with $\tilde{K} = \sqrt{a}\,K$ and $\tilde{y}^\delta = \sqrt{a}\,y^\delta$. Then $\|\tilde{K}'(x)\|_{\mathcal{L}(X,Y)}^2 \leq 1$ for all $x \in B(\hat{x}, \rho)$, and the Landweber iteration takes the form

$$x_{k+1}^\delta \;=\; x_k^\delta \;-\; \tilde{K}'(x_k^\delta)^*\big(\tilde{K}(x_k^\delta) - \tilde{y}^\delta\big), \quad k = 0, 1, 2, \ldots$$

Therefore, we assume from now on that $a = 1$ and $\|K'(x)\|_{\mathcal{L}(X,Y)} \leq 1$ for all $x \in B(\hat{x}, \rho)$. The Landweber iteration thus takes the form

$$x_0^\delta = \hat{x}, \quad x_{k+1}^\delta = x_k^\delta - K'(x_k^\delta)^* \big(K(x_k^\delta) - y^\delta \big), \quad k = 0, 1, 2, \ldots \quad (4.29)$$

The following, rather strong, assumption is called the "tangential cone condition" and will ensure that the Landweber iteration is well-defined and will also provide convergence.

Assumption 4.32 *(Tangential Cone Condition)*
Let $B(\hat{x}, \rho) \subset \mathcal{D}(K)$ be some ball, K differentiable in $B(\hat{x}, \rho)$ with $\|K'(x)\|_{\mathcal{L}(X,Y)} \leq 1$ for all $x \in B(\hat{x}, \rho)$. Furthermore, let K' be Lipschitz continuous on $B(\hat{x}, \rho)$, and there exists $\eta < \frac{1}{2}$ with

$$\big\| K(\tilde{x}) - K(x) - K'(x)(\tilde{x} - x) \big\|_Y \leq \eta \, \|K(\tilde{x}) - K(x)\|_Y \quad (4.30)$$

for all $x, \tilde{x} \in B(\hat{x}, \rho)$.

We compare this estimate with the estimate

$$\big\| K(\tilde{x}) - K(x) - K'(x)(\tilde{x} - x) \big\|_Y \leq c \|\tilde{x} - x\|_X^2,$$

which holds for Lipschitz-continuously differentiable functions (see Lemma A.63). If there exists c' with $\|\tilde{x} - x\|_X \leq c' \|K(\tilde{x}) - K(x)\|_Y$ then (4.30) follows. However, such an estimate does not hold for ill-posed problems (see Definition 4.1 and Problem 4.1). The condition (4.30) is very strong but in some cases it can be verified (see Problem 4.7). First we draw some conclusions from this condition. The first one is a basic inequality which will be used quite often in the following.

Corollary 4.33 *Under Assumption 4.32, the following holds:*

$$\frac{1}{1 + \eta} \|K'(x)(\tilde{x} - x)\|_Y \leq \|K(\tilde{x}) - K(x)\|_Y \leq \frac{1}{1 - \eta} \|K'(x)(\tilde{x} - x)\|_Y \quad (4.31)$$

for all $x, \tilde{x} \in B(\hat{x}, \rho)$.

Proof: The right estimate follows from

$$\begin{aligned}
\|K(\tilde{x}) - K(x)\|_Y &\leq \|K(\tilde{x}) - K(x) - K'(x)(\tilde{x} - x)\|_Y + \|K'(x)(\tilde{x} - x)\|_Y \\
&\leq \eta \|K(\tilde{x}) - K(x)\|_Y + \|K'(x)(\tilde{x} - x)\|_Y,
\end{aligned}$$

and the left from

$$\begin{aligned}
\|K(\tilde{x}) - K(x)\|_Y &= \big\| K'(x)(\tilde{x} - x) - \big[K'(x)(\tilde{x} - x) - K(\tilde{x}) + K(x) \big] \big\|_Y \\
&\geq \|K'(x)(\tilde{x} - x)\|_Y - \eta \|K(\tilde{x}) - K(x)\|_Y.
\end{aligned}$$

$\square$

With this lemma, we try to justify the notion "tangential cone condition" and define the convex cones $C_\pm(x)$ in $\mathbb{R} \times X$ with vertex at $(0, x)$ as

$$C_\pm(x) = \left\{ (r, \tilde{x}) \in \mathbb{R} \times X : r \geq \frac{1}{1 \pm \eta} \| K'(x)(x - \tilde{x}) \|_Y \right\}.$$

Then (4.31) can be formulated as $\big(\| K(\tilde{x}) - K(x) \|_Y, \tilde{x} \big) \in C_+(x) \setminus \operatorname{int} C_-(x)$ for every $x, \tilde{x} \in B(\hat{x}, \rho)$. Therefore, for every $x \in B(\hat{x}, \rho)$ the graph of $\tilde{x} \mapsto \| K(\tilde{x}) - K(x) \|_Y$ lies between the cones $C_+(x)$ and $C_-(x)$ which are build by the tangent space at x.

We draw two conclusions from this corollary. Under Assumption 4.32, the reverse of Theorem 4.3 holds, and there exists a unique minimum norm solution with respect to $\hat{x}$.

Definition 4.34 *Let* $K : X \supset \mathcal{D}(K) \to Y$ *be a nonlinear mapping,* $y^* \in Y$, *and* $\hat{x} \in X$. *A point* $x^* \in \mathcal{D}(K)$ *with* $K(x^*) = y^*$ *is called* minimum norm solution *with respect to* $\hat{x}$ *if* $\| x^* - \hat{x} \|_X \leq \| x - \hat{x} \|_X$ *for all solutions* $x \in \mathcal{D}(K)$ *of* $K(x) = y^*$.

Lemma 4.35 *Let Assumption 4.32 hold for some ball* $B(\hat{x}, \rho)$ *and let* $x^* \in B(\hat{x}, \rho)$ *with* $K(x^*) = y^*$.

(a) *Let the linear equation* $K'(x^*)h = 0$ *be ill-posed in the sense of Definition 4.1. Then the nonlinear equation* $K(x) = y$ *is locally ill-posed in* x^*.

(b) x^* *is a minimum norm solution with respect to* $\hat{x}$ *if, and only if,* $x^* - \hat{x} \perp \mathcal{N}\big(K'(x^*) \big)$

(c) *There exists a unique minimum norm solution* $x^\dagger$ *of* $K(x) = y^*$ *with respect to* $\hat{x}$.

Proof: (a) Let $r \in \big(0, \rho - \| x^* - \hat{x} \|_X \big)$. Since the linear equation is ill-posed there exists a sequence $h_n \in X$ with $\| h_n \|_X = 1$ and $\| K'(x^*)h_n \|_Y \to 0$ for $n \to \infty$. We set $x_n = x^* + rh_n$. For $\tilde{x} = x^*$ and $x = x_n$ in (4.31) it follows that $K(x_n) \to K(x^*)$ and $\| x_n - x^* \|_X = r$.

(b) For the characterization of a minimum norm solution it suffices to compare $\| x^* - \hat{x} \|_X$ with $\| \tilde{x} - \hat{x} \|_X$ for solutions $\tilde{x} \in B(\hat{x}, \rho)$ of $K(\tilde{x}) = y^*$. The estimate (4.31) for $x = x^*$ implies that $\tilde{x} \in B(\hat{x}, \rho)$ is a solution of $K(\tilde{x}) = y^*$ if, and only if, $K'(x^*)(x^* - \tilde{x}) = 0$; that is, if $x^* - \tilde{x} \in \mathcal{N}\big(K'(x^*) \big)$. The point x^* is a minimum norm solution with respect to $\hat{x}$ if, and only if, $\| x^* - \hat{x} \|_X \leq \| \tilde{x} - \hat{x} \|_X$ for all solutions $\tilde{x} \in B(\hat{x}, \rho)$ of $K(\tilde{x}) = y^*$; that is, for all $\tilde{x}$ with $x^* - \tilde{x} \in \mathcal{N}\big(K'(x^*) \big)$. Replacing $x^* - \tilde{x}$ by z shows that x^* is a minimum norm solution with respect to $\hat{x}$ if, and only if, 0 is the best approximation of $x^* - \hat{x}$ in $\mathcal{N}\big(K'(x^*) \big)$ which is equivalent to $x^* - \hat{x} \perp \mathcal{N}\big(K'(x^*) \big)$.

(c) The subspace $\mathcal{N}\big(K'(x^*) \big)$ is closed. Therefore, there exists a unique best approximation p of $x^* - \hat{x}$ in $\mathcal{N}\big(K'(x^*) \big)$.[7] Then $x^\dagger = x^* - p \in x^* + \mathcal{N}\big(K'(x^*) \big)$

---

[7]This follows from a general theorem on best approximations in Hilbert spaces.

is the best approximation at $\hat{x}$ in $x^* + \mathcal{N}(K'(x^*))$. Also, $K(x^\dagger) = y^*$ because $x^* - x^\dagger = p \in \mathcal{N}(K'(x^*))$. Finally, $\|x^\dagger - \hat{x}\|_X \le \|x^* - \hat{x}\|_X < \rho$, thus $x^\dagger \in B(\hat{x}, \rho)$. $\qquad\square$

The following result proves that the Landweber iteration is well-defined, and it motivates a stopping rule.

Theorem 4.36 *Let Assumption 4.32 hold. Let $\|y^\delta - y^*\|_Y \le \delta$ and $x^* \in B(\hat{x}, \rho/2)$ with $K(x^*) = y^*$. Furthermore, let $r > \frac{2(1+\eta)}{1-2\eta}$ and assume that there exists $k_* \in \mathbb{N}$ with $x_k^\delta \in B(x^*, \rho/2)$ for all $k = 0, \ldots, k_* - 1$ (where x_k^δ are defined by (4.29)) such that*

$$\|K(x_k^\delta) - y^\delta\|_Y \ge r\delta \quad \text{for all } k = 0, \ldots, k_* - 1. \tag{4.32}$$

Then the following holds.

(a) *$\|x_{k+1}^\delta - x^*\|_X \le \|x_k^\delta - x^*\|_X$ for all $k = 0, \ldots, k_* - 1$. In particular, $\|x_k^\delta - x^*\|_X \le \|x_0^\delta - x^*\|_X = \|\hat{x} - x^*\|_X < \rho/2$ for all $k = 0, \ldots, k_*$. Therefore, all x_k^δ belong to $B(x^*, \rho/2) \subset B(\hat{x}, \rho)$ for $k = 0, \ldots, k_*$.*

(b) *For all $\ell \in \{0, \ldots, k_* - 1\}$ it holds that*

$$\left(1 - 2\eta - \frac{2(1+\eta)}{r}\right) \sum_{k=\ell}^{k_*-1} \|K(x_k^\delta) - y^\delta\|_Y^2 \le \|x_\ell^\delta - x^*\|_X^2 - \|x_{k_*}^\delta - x^*\|_X^2. \tag{4.33}$$

(c) *If $\delta = 0$ then (4.32) holds for all $k \in \mathbb{N}$, and (we write x_k instead of x_k^0)*

$$\sum_{k=0}^{\infty} \|K(x_k) - y^*\|_Y^2 \le \frac{\|\hat{x} - x^*\|_X^2}{1 - 2\eta}. \tag{4.34}$$

In particular, $K(x_k)$ converges to y^*.

Proof: Since $\|K'(x_k^\delta)^*\|_{\mathcal{L}(Y,X)} \le 1$ we have for $k = 0, \ldots, k_* - 1$

$$\|x_{k+1}^\delta - x^*\|_X^2 - \|x_k^\delta - x^*\|_X^2$$
$$= 2 \operatorname{Re}(x_k^\delta - x^*, x_{k+1}^\delta - x_k^\delta)_X + \|x_{k+1}^\delta - x_k^\delta\|_X^2$$
$$= 2 \operatorname{Re}(x_k^\delta - x^*, K'(x_k^\delta)^*(y^\delta - K(x_k^\delta)))_X + \|K'(x_k^\delta)^*(y^\delta - K(x_k^\delta))\|_X^2$$
$$\le 2 \operatorname{Re}(K'(x_k^\delta)(x_k^\delta - x^*), y^\delta - K(x_k^\delta))_X + \|y^\delta - K(x_k^\delta)\|_Y^2$$
$$= 2 \operatorname{Re}(K'(x_k^\delta)(x_k^\delta - x^*) - K(x_k^\delta) + y^\delta, y^\delta - K(x_k^\delta))_X - \|y^\delta - K(x_k^\delta)\|_Y^2$$
$$\le \left[2\eta\|K(x_k^\delta) - y^*\|_Y + 2\delta - \|K(x_k^\delta) - y^\delta\|_Y\right]\|K(x_k^\delta) - y^\delta\|_Y$$
$$\le \left[(2\eta - 1)\|K(x_k^\delta) - y^\delta\|_Y + 2\delta + 2\eta\delta\right]\|K(x_k^\delta) - y^\delta\|_Y$$
$$= \left[2\delta(1+\eta) - (1-2\eta)\|K(x_k^\delta) - y^\delta\|_Y\right]\|K(x_k^\delta) - y^\delta\|_Y \tag{4.35}$$
$$= \frac{1}{r}\left[2(1+\eta)\delta r - (1-2\eta)r\|K(x_k^\delta) - y^\delta\|_Y\right]\|K(x_k^\delta) - y^\delta\|_Y$$
$$\le \frac{2(1+\eta) - (1-2\eta)r}{r}\|K(x_k^\delta) - y^\delta\|_Y^2$$

where we used (4.32) in the last estimate. This implies (a), because $2(1 + \eta) - (1 - 2\eta)r < 0$ by the choice of r.

With $\alpha = \left[(1 - 2\eta)r - 2(1 + \eta)\right]/r > 0$ we just have shown the estimate

$$\alpha \, \|K(x_k^\delta) - y^\delta\|_Y^2 \leq \|x_k^\delta - x^*\|_X^2 - \|x_{k+1}^\delta - x^*\|_X^2 \,, \quad k = 0, \ldots, k_* - 1 \,,$$

and thus for $\ell \in \{0, \ldots, k_* - 1\}$:

$$\alpha \sum_{k=\ell}^{k_*-1} \|K(x_k^\delta) - y^\delta\|_Y^2 \leq \|x_\ell^\delta - x^*\|_X^2 - \|x_{k_*}^\delta - x^*\|_X^2 \,.$$

(c) We consider the above estimate up to line (4.35) for $\delta = 0$; that is,

$$(1 - 2\eta) \, \|K(x_k) - y^*\|_Y^2 \leq \|x_k - x^*\|_X^2 - \|x_{k+1} - x^*\|_X^2 \,,$$

and thus for all $m \in \mathbb{N}$

$$(1 - 2\eta) \sum_{k=0}^{m-1} \|K(x_k) - y^*\|_Y^2 \leq \|x_0 - x^*\|_X^2 - \|x_m - x^*\|_X^2 \leq \|x_0 - x^*\|_X^2 \,.$$

Therefore the series converges which yields the desired estimate. □

Let now $\delta > 0$ and again $r > \frac{2(1+\eta)}{1-2\eta}$. The condition (4.32) can not hold for every k_*. Indeed, otherwise the sequence $\left(\|x_k - x^*\|_X\right)$ would be a monotonically decreasing and bounded sequence, thus convergent. From (4.33) for $\ell = k_* - 1$ it would follows that

$$\left(1 - 2\eta - \frac{2(1 + \eta)}{r}\right) \|K(x_{k_*-1}^\delta) - y^\delta\|_Y^2 \leq \|x_{k_*-1}^\delta - x^*\|_X^2 - \|x_{k_*}^\delta - x^*\|_X^2 \longrightarrow 0$$

for $k_* \to \infty$, a contradiction. Therefore, the following stopping rule is well-defined.

Stopping rule: Let $\delta > 0$ and $r > \frac{2(1+\eta)}{1-2\eta}$. We define $k_* = k_*(\delta) \in \mathbb{N}$ as the uniquely determined number such that

$$\|K(x_k^\delta) - y^\delta\|_Y \geq r\delta > \|K(x_{k_*}^\delta) - y^\delta\|_Y \quad \text{for all } k = 0, \ldots, k_* - 1. \quad (4.36)$$

With this stopping rule, we can show convergence as $\delta \to 0$. First we note that for any fixed $\delta > 0$ and $k \leq k_*(\delta)$, the mapping $y^\delta \mapsto x_k^\delta$ is continuous. Indeed, in every of the first k steps of the algorithm only continuous operations of the Landweber iteration are performed.

We first consider the case of no noise; that is, $\delta = 0$.

Theorem 4.37 *Let Assumption 4.32 hold and let $\rho > 0$ and $x^* \in B(\hat{x}, \rho/2)$ with $K(x^*) = y^*$. The Landweber iteration is well defined for $\delta = 0$ (that is, the iterates x_k belong to $\mathcal{D}(K)$) and the sequence (x_k) converges to a solution $\tilde{x} \in B[x^*, \rho/2] \subset B(\hat{x}, \rho)$ of $K(x) = y^*$.*

Proof: We have seen already in Theorem 4.36 that $x_k \in B(x^*, \rho/2)$ for all k. We now show that (x_k) is a Cauchy sequence. Let $\ell \leq m$ be fixed. Determine k with $\ell \leq k \leq m$ and

$$\|K(x_k) - y^*\|_Y \ \leq \ \|K(x_i) - y^*\|_Y \quad \text{for all } \ell \leq i \leq m.$$

Because of $\|x_\ell - x_m\|^2 \leq 2\|x_\ell - x_k\|^2 + 2\|x_k - x_m\|^2$ we estimate both terms separately and use the formula $\|u - v\|^2 + \|u\|^2 - \|v\|^2 = 2\,\mathrm{Re}(u, u - v)$ and the left estimate of (4.31).

$$\|x_k - x_m\|_X^2 + \|x_k - x^*\|_X^2 - \|x_m - x^*\|_X^2$$

$$= 2\,\mathrm{Re}\big(x_k - x^*, x_k - x_m\big)_X \ = \ 2 \sum_{i=k}^{m-1} \mathrm{Re}\big(x_k - x^*, x_i - x_{i+1}\big)_X$$

$$= 2 \sum_{i=k}^{m-1} \mathrm{Re}\big(x_k - x^*, K'(x_i)^*(K(x_i) - y^*)\big)_X$$

$$= 2 \sum_{i=k}^{m-1} \mathrm{Re}\big(K'(x_i)(x_k - x^*), K(x_i) - y^*\big)_Y$$

$$\leq 2 \sum_{i=k}^{m-1} \|K(x_i) - y^*\|_Y \, \|K'(x_i)(x_k - x^*)\|_Y$$

$$\leq 2 \sum_{i=k}^{m-1} \|K(x_i) - y^*\|_Y \big[\|K'(x_i)(x_i - x^*)\|_Y + \|K'(x_i)(x_i - x_k)\|_Y\big]$$

$$\leq 2(1 + \eta) \sum_{i=k}^{m-1} \|K(x_i) - y^*\|_Y \big[\|K(x_i) - K(x^*)\|_Y + \|K(x_i) - K(x_k)\|_Y\big]$$

$$\leq 2(1 + \eta) \sum_{i=k}^{m-1} \|K(x_i) - y^*\|_Y \big[\|K(x_i) - y^*)\|_Y + \|K(x_i) - y^*\|_Y + \|y^* - K(x_k)\|_Y\big]$$

$$\leq 2(1 + \eta) \sum_{i=k}^{m-1} \|K(x_i) - y^*\|_Y \big[\|K(x_i) - y^*\|_Y + \|K(x_i) - y^*\|_Y + \|y^* - K(x_i)\|_Y\big]$$

$$= 6(1 + \eta) \sum_{i=k}^{m-1} \|K(x_i) - y^*\|_Y^2 .$$

Analogously, we estimate

$$\|x_k - x_\ell\|_X^2 + \|x_k - x^*\|_X^2 - \|x_\ell - x^*\|_X^2 \ \leq \ 6(1 + \eta) \sum_{i=\ell}^{k-1} \|K(x_i) - y^*\|_Y^2$$

and thus

$$
\begin{aligned}
\|x_\ell - x_m\|_X^2 &\leq 2\|x_\ell - x_k\|_X^2 + 2\|x_k - x_m\|_X^2 \\
&\leq 2\|x_\ell - x^*\|_X^2 - 2\|x_k - x^*\|_X^2 + 12(1+\eta) \sum_{i=\ell}^{k-1} \|K(x_i) - y^*\|_Y^2 \\
&\quad + 2\|x_m - x^*\|_X^2 - 2\|x_k - x^*\|_X^2 + 12(1+\eta) \sum_{i=k}^{m-1} \|K(x_i) - y^*\|_Y^2 \\
&= 2\|x_\ell - x^*\|_X^2 + 2\|x_m - x^*\|_X^2 - 4\|x_k - x^*\|_X^2 \\
&\quad + 12(1+\eta) \sum_{i=\ell}^{m-1} \|K(x_i) - y^*\|_Y^2 .
\end{aligned}
$$

Since both, the sequence $\big(\|x_\ell - x^*\|_X\big)_\ell$ and also the series $\sum_{i=0}^\infty \|K(x_i) - y^*\|_Y^2$, converge by Theorem 4.36 we conclude that also $\|x_\ell - x_m\|_X^2$ converges to zero as $\ell \to \infty$. Therefore, (x_m) is a Cauchy sequence and thus convergent; that is, $x_m \to \tilde{x}$ for some $\tilde{x} \in B[x^*, \rho/2]$. The continuity of K implies that $K(\tilde{x}) = y^*$.   □

Now we prove convergence of $x_{k_*(\delta)}^\delta$ as $\delta \to 0$ if $k_*(\delta)$ is determined by the stopping rule.

Theorem 4.38 *Let Assumption 4.32 hold and let $\rho > 0$ and $x^* \in B(\hat{x}, \rho/2)$ with $K(x^*) = y^*$ and $r > \frac{2(1+\eta)}{1-2\eta}$. Then the sequence stops by the stopping rule (4.36) and defines $k_*(\delta)$ for all $\delta > 0$. Then $\lim_{\delta \to 0} x_{k_*(\delta)}^\delta = \tilde{x}$ where $\tilde{x} \in B[x^*, \rho/2]$ is the limit of the sequence for $\delta = 0$ – which exists by the previous theorem and is a solution of $K(\tilde{x}) = y^*$.*

Proof: Let (x_k) be the sequence of the Landweber iteration for $\delta = 0$; that is, $x_k = x_k^0$, and $\tilde{x} = \lim_{k \to \infty} x_k$. Furthermore, let (δ_n) be a sequence which converges to zero. Set for abbreviation $k_n = k_*(\delta_n)$. We distinguish between two cases.

Case 1: The sequence (k_n) of natural numbers has finite accumulation points. Let $k \in \mathbb{N}$ be the smallest accumulation point. Then there exists $I \subset \mathbb{N}$ of infinite cardinality such that $k_n = k$ for all $n \in I$. By the definition of k_n, we have

$$\|K(x_k^{\delta_n}) - y^{\delta_n}\|_Y < r\,\delta_n \quad \text{for all } n \in I .$$

Since for this fixed k, the iteration $x_k^{\delta_n}$ depends continuously on y^{δ_n}, and since y^{δ_n} converges to y^*, we conclude for $n \in I$, $n \to \infty$, that $x_k^{\delta_n} \to x_k$ (this is the k-th iteration of the sequence for $\delta = 0$) and $K(x_k) = y^*$. Landweber's iteration for $\delta = 0$ implies that then $x_m = x_k$ for all $m \geq k$; that is, the sequence is constant for $m \geq k$. In particular, $x_m = \tilde{x}$ for all $m \geq k$. The same argument holds for any other accumulation point $\tilde{k} \geq k$ and thus $\lim_{n \to \infty, n \in \tilde{I}} x_{k_*(\delta_n)}^{\delta_n} = x_{\tilde{k}} = \tilde{x}$.

Case 2: The sequence $(k_n)_n$ tends to infinity. Without loss of generality, we assume that k_n converges monotonically to infinity. Let $n > m$. We apply Theorem 4.36 for $\tilde{x}$ instead of x^* and get

$$\|x_{k_n}^{\delta_n} - \tilde{x}\|_X \leq \|x_{k_{n-1}}^{\delta_n} - \tilde{x}\|_X \leq \cdots \leq \|x_{k_m}^{\delta_n} - \tilde{x}\|_X$$
$$\leq \|x_{k_m}^{\delta_n} - x_{k_m}\|_X + \|x_{k_m} - \tilde{x}\|_X .$$

Let $\varepsilon > 0$. Choose m such that $\|x_{k_m} - \tilde{x}\|_X \leq \varepsilon/2$. For this fixed m, the sequence $x_{k_m}^{\delta_n}$ converges to x_{k_m} for $n \to \infty$. Therefore, we can find n_0 with $\|x_{k_m}^{\delta_n} - x_{k_m}\|_X \leq \varepsilon/2$ for all $n \geq n_0$. For these $n \geq n_0$, we have $\|x_{k_n}^{\delta_n} - \tilde{x}\|_X \leq \varepsilon$. Therefore, we have shown convergence of every sequence $\delta_n \to 0$ to the same limit $\tilde{x}$. This ends the proof. □

Before we prove a result on the order of convergence, we formulate a condition under which the Landweber iteration converges to the minimum norm solution.

Lemma 4.39 *Let Assumption 4.32 hold and let $x^* \in B(\hat{x}, \rho/2)$ be the minimum norm solution of $K(x) = y^*$ with respect to $\hat{x}$. (It exists and is unique by Lemma 4.35.) Let, furthermore, $\mathcal{N}(K'(x^*)) \subset \mathcal{N}(K'(x))$ for all $x \in B(\hat{x}, \rho/2)$. Then the sequences (x_k) and $\left(x_{k(\delta_n)}^{\delta_n}\right)_n$ of the Landweber iteration converge to the minimum norm solution x^* of $K(x) = y^*$.*

Proof: Convergence of the sequences has been shown in Theorems 4.37 and 4.38 already. Let $\tilde{x} = \lim_{k\to\infty} x_k$. Then also $\tilde{x} = \lim_{\delta\to 0} x_{k(\delta)}^{\delta}$ by Theorem 4.38. We show by induction that $x_k - x^* \perp \mathcal{N}(K'(x^*))$ for all $k = 0, 1, \ldots$. For $k = 0$ this is true because $x_0 = \hat{x}$, the minimum property of x^*, and part (b) of Lemma 4.35. Let the assertion be true for k and let $z \in \mathcal{N}(K'(x^*))$. Then $z \in \mathcal{N}(K'(x_k))$, thus

$$(x_{k+1} - x^*, z)_X = (x_k - x^*, z)_X + \big(K(x_k) - y^*, K'(x_k)z\big)_Y = 0 .$$

Since the orthogonal complement of $\mathcal{N}(K'(x^*))$ is closed we conclude that also $\tilde{x} - x^* \perp \mathcal{N}(K'(x^*))$. Furthermore, $\tilde{x} - x^* \in \mathcal{N}(K'(x^*))$ because of (4.31) for $x = x^*$. Therefore, $\tilde{x} = x^*$. □

We will now prove a rate of convergence. This part will be a bit more technical. It is not surprising that we need the source condition of Assumption 4.11 just as in the linear case or for the nonlinear Tikhonov regularization. Unfortunately, Assumptions 4.32 and 4.11 are not sufficient, and we have to strengthen them.

Assumption 4.40 *Let Assumption 4.32 hold and $x^* \in B(\hat{x}, \rho)$ be the minimum norm solution of $K(x) = y^*$ with respect to $\hat{x}$. (It exists by Lemma 4.35). Furthermore, let $K'(x^*)$ compact and there exists $C > 0$ and a family $\{R_x : x \in B(\hat{x}, \rho)\}$ of linear bounded operators $R_x : Y \to Y$ with*

$$K'(x) = R_x K'(x^*) \quad \text{and} \quad \|R_x - I\|_{\mathcal{L}(Y,Y)} \leq C \|x - x^*\|_X \quad \text{for } x \in B(\hat{x}, \rho) .$$

In the linear case, both conditions are satisfied for $R_x = I$ because $K'(x) = K$ for all x. Under this additional assumption, we can partially sharpen the tangential cone condition of Assumption 4.32.

Lemma 4.41 *Under Assumption 4.40, we have $\mathcal{N}(K'(x^*)) \subset \mathcal{N}(K'(x))$ for all $x \in B(\hat{x}, \rho)$ and*

$$\|K(x) - K(x^*) - K'(x^*)(x - x^*)\|_Y \leq \frac{C}{2} \|x - x^*\|_X \|K'(x^*)(x - x^*)\|_Y, \qquad (4.37)$$

$$\|K(x) - K(x^*) - K'(x^*)(x - x^*)\|_Y \leq \frac{C}{2 - \rho C} \|x - x^*\|_X \|K(x) - K(x^*)\|_Y \qquad (4.38)$$

for all $x \in B(\hat{x}, \rho)$. Therefore, if we choose ρ such that $\eta := \frac{C\rho}{2 - \rho C} < 1/2$ then condition (4.30) is satisfied for $x = x^*$.

Proof: The first assertion is obvious. For the second, we write we

$$K(x) - K(x^*) - K'(x^*)(x - x^*)$$

$$= \int_0^1 \left[\frac{d}{dt} K(tx + (1 - t)x^*) - K'(x^*)(x - x^*) \right] dt$$

$$= \int_0^1 \left[K'(tx + (1 - t)x^*) - K'(x^*) \right](x - x^*) \, dt,$$

thus

$$\|K(x) - K(x^*) - K'(x^*)(x - x^*)\|_Y$$

$$\leq \int_0^1 \left\| \left[K'(tx + (1 - t)x^*) - K'(x^*) \right](x - x^*) \right\|_Y dt$$

$$\leq \int_0^1 \|R_{tx+(1-t)x^*} - I\|_{\mathcal{L}(Y,Y)} \, dt \, \|K'(x^*)(x - x^*)\|_Y$$

$$\leq C \int_0^1 t \, dt \, \|x - x^*\|_X \|K'(x^*)(x - x^*)\|_Y$$

$$= \frac{C}{2} \|x - x^*\|_X \|K'(x^*)(x - x^*)\|_Y$$

which proves (4.37). We continue and estimate

$$\left\| K(x) - K(x^*) - K'(x^*)(x - x^*) \right\|_Y$$

$$\leq \frac{C}{2} \|x - x^*\|_X \left\| K'(x^*)(x - x^*) + K(x^*) - K(x) \right\|_Y$$

$$+ \frac{C}{2} \|x - x^*\|_X \|K(x^*) - K(x)\|_Y$$

and thus

$$\left(2-C\|x-x^*\|_X\right)\left\|K(x)-K(x^*)-K'(x^*)(x-x^*)\right\|_Y \leq C\,\|x-x^*\|_X\,\|K(x^*)-K(x)\|_Y.$$

This proves (4.38) because $\|x - x^*\|_X \leq \rho$.   □

Under this Assumption 4.40 (and $x^* \in B(\hat{x}, \rho/2)$), we know from Lemma 4.39 that $x_{k(\delta)}^\delta$ converges to x^* as $\delta \to 0$. For the proof of the order of convergence, we need the following elementary estimates:

Lemma 4.42

$$\frac{1-(1-x^2)^k}{x} \leq \sqrt{k} \quad \text{for all } k \in \mathbb{N} \text{ and } x \in [0,1]\,, \tag{4.39}$$

$$x\,(1-x^2)^k \leq \frac{1}{\sqrt{k+1}} \quad \text{for all } k \in \mathbb{N} \text{ and } x \in [0,1]\,, \tag{4.40}$$

$$x^2\,(1-x^2)^k \leq \frac{1}{k+1} \quad \text{for all } k \in \mathbb{N} \text{ and } x \in [0,1]\,, \tag{4.41}$$

for every $\alpha \in (0,1]$ there exists $c(\alpha) > 0$ with

$$\sum_{j=0}^{k-1}(j+1)^{-\alpha}(k-j)^{-3/2} \leq \frac{c(\alpha)}{(k+1)^\alpha} \quad \text{for all } k \in \mathbb{N}. \tag{4.42}$$

Proof: To show (4.39) let first $0 < x \leq 1/\sqrt{k}$. Bernoulli's inequality implies $(1-x^2)^k \geq 1-kx^2$, thus $\frac{1-(1-x^2)^k}{x} \leq \frac{1-(1-kx^2)}{x} = kx \leq \sqrt{k}$. Let now $x \geq 1/\sqrt{k}$. Then $\frac{1-(1-x^2)^k}{x} \leq \frac{1}{x} \leq \sqrt{k}$. For $x = 0$ the estimate follows by l'Hospital's rule. The proofs of estimates (4.40) and (4.41) are elementary and left to the reader (see also the proof of Theorem 2.8).

Proof of (4.42):   Define the function $f(x) = (x + 1)^{-\alpha}(k - x)^{-3/2}$ for $-1 < x < k$. By the explicit computation of the second derivative, one obtains that $f''(x) \geq 0$ for all $-1 < x < k$.[8] Taylor's formula yields for $j < k$

$$f(x) = f(j) + f'(j)(x-j) + \frac{1}{2}f''(z_{j,x})(x-j)^2 \geq f(j) + f'(j)(x-j),$$

for $j - 1/2 \leq x \leq j + 1/2$ with some intermediate point $z_{j,x}$, thus

$$\int_{j-1/2}^{j+1/2} f(x)\,dx \geq \int_{j-1/2}^{j+1/2} \left[f(j) + f'(j)(x-j)\right]dx = f(j),$$

---

[8]Therefore, f is convex.

because the integral $\int\limits_{j-1/2}^{j+1/2} (x-j)\,dx$ vanishes by the symmetry of the intervals.
Therefore,

$$\sum_{j=0}^{k-1} (j+1)^{-1/2}(k-j)^{-3/2} = \sum_{j=0}^{k-1} f(j) \leq \sum_{j=0}^{k-1} \int\limits_{j-1/2}^{j+1/2} f(x)\,dx = \int\limits_{-1/2}^{k-1/2} f(x)\,dx$$

$$= \int\limits_{-1/2}^{k/2} f(x)\,dx + \int\limits_{k/2}^{k-1/2} f(x)\,dx\,.$$

For the first integral, we have

$$\int\limits_{-1/2}^{k/2} f(x)\,dx \leq \left(\frac{2}{k}\right)^{3/2} \int\limits_{-1/2}^{k/2} (x+1)^{-\alpha}\,dx\,.$$

If $\alpha < 1$ then

$$\int\limits_{-1/2}^{k/2} f(x)\,dx \leq \left(\frac{2}{k}\right)^{3/2} \frac{1}{1-\alpha} \left(\frac{k}{2}+1\right)^{1-\alpha} \leq \tilde{c}(\alpha)\,k^{-\alpha-1/2} \leq \frac{c(\alpha)}{(k+1)^\alpha}\,.$$

If $\alpha = 1$ then

$$\int\limits_{-1/2}^{k/2} f(x)\,dx \leq \left(\frac{2}{k}\right)^{3/2} \ln(k/2) \leq \frac{c(1)}{k+1}\,.$$

For the remaining integral, we estimate

$$\int\limits_{k/2}^{k-1/2} f(x)\,dx \leq \left(\frac{2}{k}\right)^\alpha \int\limits_{k/2}^{k-1/2} (k-x)^{-3/2}\,dx = 2\left(\frac{2}{k}\right)^\alpha (k-x)^{-1/2}\Big|_{k/2}^{k-1/2}$$

$$\leq 2\left(\frac{2}{k}\right)^\alpha \sqrt{2}\,.$$

This finishes the proof.   □

(4.39), (4.40), and (4.41) directly imply

Corollary 4.43 *Let $A : X \to Y$ be a linear and compact operator with adjoint A^* and let $\|A\|_{\mathcal{L}(X,Y)} \leq 1$. Then*

(a) $\|(I - A^*A)^k A^*\|_{\mathcal{L}(Y,X)} \leq (k+1)^{-1/2}$,

(b) $\left\|(I - A A^*)^k A A^*\right\|_{\mathcal{L}(Y,Y)} \le (k+1)^{-1}$,

(c) $\left\|\sum_{j=0}^{k-1}(I - A^* A)^j A^*\right\|_{\mathcal{L}(Y,X)} \le \sqrt{k}$.

Proof: Let $\{\mu_i, x_i, y_i : i \in I\}$ be a singular system for A, see Appendix A.6, Theorem A.57. For $y = y_0 + \sum_{i \in I} \alpha_i\, y_i \in Y$ with $A^* y_0 = 0$, we have

$$\left\|(I - A^* A)^k A^* y\right\|_X^2 = \sum_{i \in I}\left[(1 - \mu_i^2)^k \mu_i\right]^2 \alpha_i^2 \le \frac{1}{k+1}\sum_{i \in I}\alpha_i^2 \le \frac{1}{k+1}\|y\|_Y^2$$

because of (4.40). The part (b) follows in the way from (4.41). Finally, we observe that

$$\sum_{j=0}^{k-1}(I - A^* A)^j A^* y = \sum_{j=0}^{k-1}\sum_{i \in I}\left[(1 - \mu_i^2)^j \mu_i\right]\alpha_i\, y_i = \sum_{i \in I}\frac{1 - (1 - \mu_i^2)^k}{\mu_i}\alpha_i\, y_i$$

and thus by (4.39)

$$\left\|\sum_{j=0}^{k-1}(I - A^* A)^j A^* y\right\|_Y^2 \le k\sum_{i \in I}\alpha_i^2 \le k\,\|y\|_Y^2.$$

□

Now we are able to prove the main theorem on the order of convergence. The proof is lengthy.

Theorem 4.44 *Let Assumption 4.40 hold, and let $x^* \in B(\hat{x}, \rho)$ be the minimum norm solution of $K(x) = y^*$ with respect to $\hat{x}$. Define the constant*

$$c_* = 2\left(1 + \frac{1 - 2\eta}{\eta\,(5 - 4\eta)}\right)$$

where $\eta < 1/2$ is the constant of Assumption 4.32. Furthermore, let the source condition hold; that is, there exists $w \in Y$ such that $x^* - \hat{x} = K'(x^*)^* w$ and

$$\frac{9}{2}C\,c_*^2\,\max\{c(1/2), c(1)\}\,\|w\|_Y \le 1,$$

where C is the constant of Assumption 4.40 and $c(\alpha)$ are the constants of (4.42) for $\alpha = 1/2$ and $\alpha = 1$. Then there exists a constant $c > 0$ such that

$$\|x_{k_*(\delta)}^\delta - x^*\|_X \le c\,\|w\|_Y^{1/2}\,\delta^{1/2}, \qquad \|K(x_{k_*(\delta)}^\delta) - y^*\|_Y \le (1 + r)\,\delta. \quad (4.43)$$

Again, $k_*(\delta)$ is the iteration index determined by the stopping rule (4.36).

Proof: The second estimate follows immediately from the property of $k_*(\delta)$ and the triangle inequality. Set for abbreviation $e_k = x_k^\delta - x^*$ and $A = K'(x^*)$. Then $K'(x_k^\delta) = R_{x_k^\delta} A$ because of Assumption 4.40 and thus

$$
\begin{aligned}
e_{k+1} &= e_k - K'(x_k^\delta)^* \big(K(x_k^\delta) - y^\delta \big) \\
&= e_k - A^* \big(K(x_k^\delta) - y^\delta \big) + A^* (I - R_{x_k^\delta}^*) \big(K(x_k^\delta) - y^\delta \big) \\
&= (I - A^* A) e_k - A^* \big(K(x_k^\delta) - y^\delta - K'(x^*) e_k \big) \\
&\quad + A^* (I - R_{x_k^\delta}^*) \big(K(x_k^\delta) - y^\delta \big) \\
&= (I - A^* A) e_k + A^* (y^\delta - y^*) - A^* \big[K(x_k^\delta) - K(x^*) - K'(x^*)(x_k^\delta - x^*) \big] \\
&\quad + A^* (I - R_{x_k^\delta}^*) \big(K(x_k^\delta) - y^\delta \big) \\
&= (I - A^* A) e_k + A^* (y^\delta - y^*) + A^* z_k , \quad k = 0, \ldots, k_*(\delta) - 1 , \quad (4.44)
\end{aligned}
$$

with

$$
z_k = (I - R_{x_k^\delta}^*) \big(K(x_k^\delta) - y^\delta \big) - \big[K(x_k^\delta) - K(x^*) - K'(x^*)(x_k^\delta - x^*) \big] . \quad (4.45)
$$

This is a recursion formula for e_k. It is solved in terms of $y^\delta - y^*$ and z_j by

$$
\begin{aligned}
e_k &= (I - A^* A)^k e_0 + \sum_{j=0}^{k-1} (I - A^* A)^j A^* \big[(y^\delta - y^*) + z_{k-j-1} \big] \\
&= -(I - A^* A)^k A^* w + \left[\sum_{j=0}^{k-1} (I - A^* A)^j A^* \right] (y^\delta - y^*) \\
&\quad + \sum_{j=0}^{k-1} (I - A^* A)^j A^* z_{k-j-1} \quad (4.46)
\end{aligned}
$$

for $k = 0, \ldots, k_*(\delta)$. (Proof by induction with respect to k.) Furthermore, because $A(I - A^* A)^j = (I - A A^*)^j A$,

$$
\begin{aligned}
Ae_k \;=\;& -(I - AA^*)^k AA^* w + \sum_{j=0}^{k-1}(I - AA^*)^j AA^*(y^\delta - y^*) \\[4pt]
& + \sum_{j=0}^{k-1}(I - AA^*)^j AA^* z_{k-j-1} \\[4pt]
=\;& -(I - AA^*)^k AA^* w - \sum_{j=0}^{k-1}\big[(I - AA^*)^{j+1} - [(I - AA^*)^j]\big](y^\delta - y^*) \\[4pt]
& + \sum_{j=0}^{k-1}(I - AA^*)^j AA^* z_{k-j-1} \\[4pt]
=\;& -(I - AA^*)^k AA^* w + \big[I - (I - AA^*)^k\big](y^\delta - y^*) \\[4pt]
& + \sum_{j=0}^{k-1}(I - AA^*)^j AA^* z_{k-j-1} \tag{4.47}
\end{aligned}
$$

for $k = 0, \ldots, k_*(\delta)$. Now we consider z_k from (4.45) for $k \in \{0, \ldots, k_*(\delta) - 1\}$ and estimate both terms of z_k separately. Since $r > 2$ and $\|K(x_k^\delta) - y^\delta\|_Y \geq r\delta > 2\delta$ we have

$$
\|K(x_k^\delta) - y^\delta\|_Y \;\leq\; \|K(x_k^\delta) - y^*\|_Y + \delta \;\leq\; \|K(x_k^\delta) - y^*\|_Y + \tfrac{1}{2}\|K(x_k^\delta) - y^\delta\|_Y
$$

and thus $\|K(x_k^\delta) - y^\delta\|_Y \leq 2\,\|K(x_k^\delta) - y^*\|_Y$. With (4.31) and because $\eta < 1/2$ we have

$$
\big\|(I - R_{x_k^\delta}^*)\big(K(x_k^\delta) - y^\delta\big)\big\|_Y \leq \|I - R_{x_k^\delta}^*\|_{\mathcal{L}(Y,Y)}\,\|K(x_k^\delta) - y^\delta\|_Y \;\leq\; 4C\,\|e_k\|_X\,\|Ae_k\|_Y.
$$

For second term of z_k we use (4.37) for $x = x_k^\delta$:

$$
\big\|K(x_k^\delta) - K(x^*) - K'(x^*)(x_k^\delta - x^*)\big\|_Y \;\leq\; \frac{C}{2}\,\|e_k\|_X\,\|Ae_k\|_Y
$$

and thus $\|z_k\|_Y \leq \tfrac{9}{2}C\,\|e_k\|_X\,\|Ae_k\|_Y$.

Now we estimate δ from above. Because of the stopping rule (4.36) and (4.31) we have for $k = 0, \ldots, k_*(\delta) - 1$

$$
r\delta \;\leq\; \|K(x_k^\delta) - y^\delta\|_Y \;\leq\; \|K(x_k^\delta) - K(x^*)\|_Y + \delta \;\leq\; \frac{1}{1-\eta}\|Ae_k\|_Y + \delta,
$$

and thus, because $r - 1 > \frac{2(1+\eta)}{1-2\eta} - 1 = \frac{1+4\eta}{1-2\eta}$,

$$
\delta \;\leq\; \frac{1 - 2\eta}{(1 + 4\eta)(1 - \eta)}\,\|Ae_k\|_Y, \qquad k = 0, \ldots, k_*(\delta) - 1. \tag{4.48}
$$

Now we show

$$\|e_j\|_X \leq c_*\|w\|_Y \frac{1}{\sqrt{j+1}}, \quad \|Ae_j\|_Y \leq c_*\|w\|_Y \frac{1}{j+1}, \qquad (4.49)$$

and

$$\delta \leq \frac{2(1-2\eta)}{\eta(5-4\eta)} \frac{\|w\|_Y}{k+1}. \qquad (4.50)$$

for $j = 0,\ldots,k$ and every $k \leq k_*(\delta) - 1$. We show these three estimates by induction with respect to $k < k_*(\delta)$. These are true for $k = 0$ because $\|e_0\|_X \leq \|w\|_Y$ and $\|Ae_0\|_Y \leq \|w\|_Y$ and by (4.48) and $\frac{1-2\eta}{(1+4\eta)(1-\eta)} \leq \frac{2(1-2\eta)}{\eta(5-4\eta)}$. Let now (4.49) and (4.50) hold for $k - 1$. From the above representation (4.46) of e_k, parts (a) and (c) of Corollary 4.43, the assumption of induction, and (4.42) for $\alpha = 1/2$ we conclude that

$$\begin{aligned}
\|e_k\|_X &\leq \frac{\|w\|_Y}{\sqrt{k+1}} + \sqrt{k}\,\delta + \frac{9}{2}C\sum_{j=0}^{k-1} \frac{1}{\sqrt{j+1}} \|e_{k-j-1}\|_X \|Ae_{k-j-1}\|_Y \\
&\leq \frac{\|w\|_Y}{\sqrt{k+1}} + \sqrt{k}\,\delta + \frac{9}{2}C c_*^2 \|w\|_Y^2 \sum_{j=0}^{k-1} \frac{1}{\sqrt{j+1}} \frac{1}{\sqrt{k-j}} \frac{1}{k-j} \\
&\leq \frac{\|w\|_Y}{\sqrt{k+1}} + \sqrt{k}\,\delta + \frac{9}{2}C c_*^2 \|w\|_Y^2 \frac{c(1/2)}{\sqrt{k+1}} \\
&\leq \sqrt{k+1}\,\delta + 2\frac{\|w\|_Y}{\sqrt{k+1}}
\end{aligned}$$

by assumption on $\|w\|_Y$. Analogously, it follows for $\|Ae_k\|_Y$ with (4.47) and part (b) of Corollary 4.43 and (4.42) for $\alpha = 1$ with constant $c(1)$:

$$\begin{aligned}
\|Ae_k\|_Y &\leq \frac{\|w\|_Y}{k+1} + \delta + \frac{9}{2}C\sum_{j=0}^{k-1} \frac{1}{j+1} \|e_{k-j-1}\|_X \|Ae_{k-j-1}\|_Y \\
&\leq \frac{\|w\|_Y}{k+1} + \delta + \frac{9}{2}C c_*^2 \|w\|_Y^2 \sum_{j=0}^{k-1} \frac{1}{j+1} \frac{1}{\sqrt{k-j}} \frac{1}{k-j} \\
&\leq \frac{\|w\|_Y}{k+1} + \delta + \frac{9}{2}C c_*^2 \|w\|_Y^2 \frac{c(1)}{k+1} \\
&\leq \delta + 2\frac{\|w\|_Y}{k+1}.
\end{aligned}$$

Next we substitute this estimate of $\|Ae_k\|_Y$ into the estimate (4.48) for δ and obtain

$$\delta \leq \frac{1-2\eta}{(1+4\eta)(1-\eta)} \|Ae_k\|_Y \leq \frac{1-2\eta}{(1+4\eta)(1-\eta)} \left[\delta + 2\frac{\|w\|_Y}{k+1}\right].$$

Since $1 - \frac{1-2\eta}{(1+4\eta)(1-\eta)} = \frac{\eta(5-4\eta)}{(1+4\eta)(1-\eta)}$ it follows estimate (4.50) for k; that is,

$$\delta \leq \frac{2(1-2\eta)}{\eta(5-4\eta)} \frac{\|w\|_Y}{k+1}.$$

We substitute this into the estimates of $\|e_k\|_X$ and $\|Ae_k\|_Y$:

$$\|e_k\|_X \leq 2\left[1 + \frac{1 - 2\eta}{\eta(5 - 4\eta)}\right]\frac{\|w\|_Y}{\sqrt{k+1}} = c_*\frac{\|w\|_Y}{\sqrt{k+1}},$$

$$\|Ae_k\|_Y \leq 2\left[1 + \frac{1 - 2\eta}{\eta(5 - 4\eta)}\right]\frac{\|w\|_Y}{k+1} = c_*\frac{\|w\|_Y}{k+1}.$$

This shows (4.49) for k. Therefore, (4.49) and (4.50) are proven for all $0 \leq j, k < k_*(\delta)$.

In the case $k_*(\delta) \geq 1$, we take $k = k_*(\delta) - 1$ in (4.50) and obtain

$$k_*(\delta) \leq \frac{2(1 - 2\eta)}{\eta(5 - 4\eta)}\frac{\|w\|_Y}{\delta}. \tag{4.51}$$

From the estimate $\|K(x_k^\delta) - y^\delta\|_Y \leq 2\|K(x_k^\delta) - K(x^*)\|_Y \leq \frac{2}{1-\eta}\|Ae_k\|_Y \leq 4\|Ae_k\|_Y$ for $k < k_*(\delta)$ it follows that

$$\|K(x_k^\delta) - y^\delta\|_Y \leq 4c_*\|w\|_Y\frac{1}{k+1}, \quad k = 0, \ldots, k_*(\delta) - 1. \tag{4.52}$$

Now we come to the final part of the proof. We write e_k from (4.46) in der form

$$e_k = -(I - A^*A)^k A^* w + \left[\sum_{j=0}^{k-1}(I - A^*A)^j A^*\right](y^\delta - y^*)$$

$$+ \sum_{j=0}^{k-1}(I - A^*A)^j A^* z_{k-j-1}$$

$$= -A^*(I - AA^*)^k w + \left[\sum_{j=0}^{k-1}(I - A^*A)^j A^*\right](y^\delta - y^*)$$

$$+ A^* \sum_{j=0}^{k-1}(I - AA^*)^j z_{k-j-1}$$

$$= A^* w_k + \left[\sum_{j=0}^{k-1}(I - A^*A)^j A^*\right](y^\delta - y^*) \tag{4.53}$$

for $k = 0, \ldots, k_*(\delta)$ with

$$w_k = -(I - AA^*)^k w + \sum_{j=0}^{k-1}(I - AA^*)^j z_{k-j-1}.$$

Since $\|(I - AA^*)^j\|_{\mathcal{L}(Y,Y)} \leq 1$ for all j, we conclude

$$
\begin{aligned}
\|w_k\|_Y &\leq \|w\|_Y + \sum_{j=0}^{k-1} \|z_{k-j-1}\|_Y \leq \|w\|_Y + \frac{9}{2}C\sum_{j=0}^{k-1} \|e_{k-j-1}\|_X \|Ae_{k-j-1}\|_Y \\
&\leq \|w\|_Y + \frac{9}{2}C c_*^2 \|w\|_Y^2 \sum_{j=0}^{k-1} \frac{1}{\sqrt{k-j}} \frac{1}{k-j} \leq \|w\|_Y + \frac{9}{2}C c_*^2 \|w\|_Y^2 \sum_{j=1}^{\infty} \frac{1}{j^{3/2}} \\
&\leq \|w\|_Y \left[1 + \frac{9}{2}C c_*^2 \sum_{j=1}^{\infty} \frac{1}{j^{3/2}} \right]
\end{aligned}
$$

for all $k = 0, \ldots, k_*(\delta)$. Here we used without loss of generality that $\|w\|_Y \leq 1$. The bound on the right-hand side is independent of δ. Similarly, we estimate $\|AA^*w_k\|_Y$. Indeed, with

$$
AA^*w_k = Ae_k - \left[A\sum_{j=0}^{k-1}(I - A^*A)^j A^* \right](y^\delta - y^*)
$$

from (4.53) and

$$
\begin{aligned}
A\sum_{j=0}^{k-1}(I - A^*A)^j A^* &= \sum_{j=0}^{k-1}(I - AA^*)^j AA^* \\
&= -\sum_{j=0}^{k-1}(I - AA^*)^{j+1} + \sum_{j=0}^{k-1}(I - AA^*)^j = I - (I - AA^*)^k
\end{aligned}
$$

we estimate

$$
\begin{aligned}
\|AA^*w_k\|_Y &\leq \|Ae_k\|_Y + \left\| A\sum_{j=0}^{k-1}(I - A^*A)^j A^* \right\|_{\mathcal{L}(Y,Y)} \|y^\delta - y^*\|_Y \\
&\leq \|Ae_k\|_Y + \left\| I - (I - AA^*)^k \right\|_Y \delta \leq \|Ae_k\|_Y + \delta
\end{aligned}
$$

for $k = 0, \ldots, k_*(\delta)$. For $k = k_*(\delta)$ the estimate (4.31) and the stopping rule imply that

$$
\begin{aligned}
\|AA^*w_{k_*(\delta)}\|_Y &\leq (1+\eta)\|K(x^\delta_{k_*(\delta)}) - y^*\|_Y + \delta \\
&\leq (1+\eta)\left[\|K(x^\delta_{k_*(\delta)}) - y^\delta\|_Y + \delta \right] + \delta \leq \left[(1+\eta)(1+r) + 1 \right]\delta.
\end{aligned}
$$

This implies

$$
\begin{aligned}
\|A^*w_{k_*(\delta)}\|_X^2 &= \left(A^*w_{k_*(\delta)}, A^*w_{k_*(\delta)} \right)_X = \left(AA^*w_{k_*(\delta)}, w_{k_*(\delta)} \right)_Y \\
&\leq \|AA^*w_{k_*(\delta)}\|_Y \|w_{k_*(\delta)}\|_Y \leq c_1 \delta \|w\|_Y
\end{aligned}
$$

with some constant c_1 which is independent of δ and w. Now we go back to (4.53) and get, using part (c) of Corollary 4.43 and (4.51) (in the case $k_*(\delta) \geq 1$, otherwise directly)

$$
\begin{aligned}
\|e_{k_*(\delta)}\|_X \;\leq&\; \sqrt{c_1\delta}\,\|w\|_Y \;+\; \left\|\sum_{j=0}^{k_*(\delta)-1}(I - A^*A)^j A^*\right\|_{\mathcal{L}(X,Y)}\delta \\
\leq&\; \sqrt{c_1\delta}\,\|w\|_Y \;+\; \sqrt{k_*(\delta)}\,\delta \\
\leq&\; \sqrt{c_1\delta}\,\|w\|_Y \;+\; \sqrt{\frac{2(1-2\eta)}{\eta(5-4\eta)}}\,\|w\|_Y^{1/2}\sqrt{\delta} \;\leq\; c\,\|w\|_Y^{1/2}\sqrt{\delta}.
\end{aligned}
$$

This, finally, ends the proof. $\square$

The Landweber iteration is only one member of the large class of iterative regularization methods. In particular, Newton-type methods, combined with various forms of regularization, have been investigated in the past and are subject of current research. We refer to the monograph [149] of Kaltenbacher, Neubauer, and Scherzer.

4.4 Problems

4.1 Let $K : X \supset \mathcal{D}(K) \to K$ be a mapping with $\|\tilde{x}-x\|_X \leq c\|K(\tilde{x})-K(x)\|_Y$ for all $x, \tilde{x} \in B(x^*, \rho)$ where c is independent of $\tilde{x}$ and x. Set $y^* = K(x^*)$. Show that the equation $K(x) = y^*$ is not locally ill-posed in the sense of Definition 4.1.

4.2 Define the integral operator (auto-convolution) K from $L^2(0,1)$ into itself by

$$
K(x)(t) \;=\; \int_0^t x(t-s)\,x(s)\,ds, \quad t \in (0,1).
$$

(a) Show that K is well-defined; that is, that $K(x) \in L^2(0,1)$ for every $x \in L^2(0,1)$. *Remark:* K is even well-defined as a mapping from $L^2(0,1)$ into $C[0,1]$. One try to prove this.

(b) Show that $K(x) = y$ is locally ill-posed in the sense of Definition 4.1 in every $x^* \in L^2(0,1)$ with $x^*(t) \geq 0$ for almost all $t \in (0,1)$.

Hint: For any $r > 0$ and $n \in \mathbb{N}$ define $x_n \in L^2(0,1)$ by

$$
x_n(t) \;=\; \begin{cases} x^*(t), & 0 < t < 1 - 1/n, \\ x^*(t) + r\sqrt{n}, & 1 - 1/n < t < 1. \end{cases}
$$

4.3 Show that for every $t \in [0,2]$ there exists $c_t > 0$ such that

$$
\frac{\mu^t}{\mu^2 + \alpha} \;\leq\; c_t \alpha^{t/2-1} \quad \text{for all } \mu, \alpha > 0.
$$

4.4 Show that $\|K(x^{\alpha(\delta),\delta}) - y^\delta + \alpha w\|_Y = \mathcal{O}(\delta)$ as $\delta \to 0$ for the choice $c_- \delta^{2/(\sigma+1)} \le \alpha(\delta) \le c_+ \delta^{2/(\sigma+1)}$ and $x^{\alpha,\delta}$ and w as in Theorem 4.15.

4.5 (a) Show that $\|u\|_\infty \le \|u'\|_{L^2(0,1)}$ for all $u \in H^1(0,1)$ with $u(0) = 0$.

(b) Show that $H^2(0,1)$ is compactly imbedded in $C[0,1]$.

(c) Show that the solution $u \in H^2(0,1)$ of $u'' = -g$ in $(0,1)$ and $u(0) = u(1) = 0$ is given by $u(t) = \int_0^1 G(t,s)g(s)ds$ where $G(t,s)$ is defined in Lemma 4.16.

4.6 Let the affine functions $f_n : \mathbb{R}_{\ge 0} \to \mathbb{R}$ for $n \in \mathbb{N}$ be given by $f_n(t) = \gamma_n t + \eta_n$, $t \ge 0$, where $\gamma_n, \eta_n > 0$. Show that the function $\varphi(t) := \inf_{n \in \mathbb{N}} f_n(t)$, $t \ge 0$, is continuous, monotonic, and concave.

4.7 (a) Show that the set U_δ from Corollary 4.18 is open in $L^2(0,1)$.

(b) Let $K : U_\delta \to H^2(0,1)$ be the operator from Theorem 4.19; that is, $c \mapsto u$ where $u \in H^2(0,1)$ solves the boundary value problem (4.11). Show that this K satisfies the Tangential Cone Condition of Assumption 4.32.

Hint: Modify the differential equation (4.15) such that $u - \tilde{u}$ appears on the right hand side and continue as in the proof of Theorem 4.19.

Chapter 5

Inverse Eigenvalue Problems

5.1 Introduction

Inverse eigenvalue problems are not only interesting in their own right, but also have important practical applications. We recall the fundamental paper by Kac [148]. Other applications appear in parameter identification problems for parabolic or hyperbolic differential equations—as we study in Section 5.6 for a model problem—(see also [167, 187, 255]) or in grating theory ([156]).

We study the Sturm–Liouville eigenvalue problem in canonical form. The direct problem is to determine the eigenvalues λ and the corresponding eigenfunctions $u \neq 0$ such that

$$-\frac{d^2 u(x)}{dx^2} + q(x)\, u(x) = \lambda\, u(x), \quad 0 \leq x \leq 1, \tag{5.1a}$$

$$u(0) = 0 \quad \text{and} \quad hu'(1) + Hu(1) = 0, \tag{5.1b}$$

where $q \in L^2(0,1)$ and $h, H \in \mathbb{R}$ with $h^2 + H^2 > 0$ are given. In this chapter, we assume that all functions are real-valued. In some applications, e.g., in grating theory, complex-valued functions q are also of practical importance. Essentially, all of the results of this chapter hold also for complex-valued q and are proven mainly by the same arguments. We refer to the remarks at the end of each section.

The eigenvalue problem (5.1a), (5.1b) is a special case of the more general eigenvalue problem to determine $\rho \in \mathbb{R}$ and non-vanishing w such that

$$\frac{d}{dt}\left(p(t)\,\frac{dw(t)}{dt}\right) + \left[\rho\, r(t) - g(t)\right] w(t) = 0, \quad t \in [a,b], \tag{5.2a}$$

$$\alpha_a w'(a) + \beta_a\, w(a) = 0, \quad \alpha_b w'(b) + \beta_b w(b) = 0. \tag{5.2b}$$

© Springer Nature Switzerland AG 2021
A. Kirsch, *An Introduction to the Mathematical Theory of Inverse Problems*,
Applied Mathematical Sciences 120,
https://doi.org/10.1007/978-3-030-63343-1_5

Here p, r, and g are given functions with $p(t) > 0$ and $r(t) > 0$ for $t \in [a, b]$, and $\alpha_a, \alpha_b, \beta_a, \beta_b \in \mathbb{R}$ are constants with $\alpha_a^2 + \beta_a^2 > 0$ and $\alpha_b^2 + \beta_b^2 > 0$. If we assume, however, that $g \in C[a, b]$ and $p, r \in C^2[a, b]$, then the *Liouville transformation* reduces the eigenvalue problem (5.2a), (5.2b) to the canonical form (5.1a), (5.1b). In particular, we define

$$\sigma(t) := \sqrt{\frac{r(t)}{p(t)}}, \quad f(t) := [p(t)\,r(t)]^{1/4}, \quad L := \int_a^b \sigma(s)\,ds, \tag{5.3}$$

the monotonic function $x : [a, b] \to [0, 1]$ by

$$x(t) := \frac{1}{L} \int_a^t \sigma(s)\,ds, \quad t \in [a, b], \tag{5.4}$$

and the new function $u : [0, 1] \to \mathbb{R}$ by $u(x) := f(t(x))\,w(t(x))$, $x \in [0, 1]$, where $t = t(x)$ denotes the inverse of $x = x(t)$. Elementary calculations show that u satisfies the differential equation (5.1a) with $\lambda = L^2 \rho$ and

$$q(x) = L^2 \left[\frac{g(t)}{r(t)} + \frac{f(t)}{r(t)} \left(\frac{p(t)f'(t)}{f(t)^2} \right)' \right]_{t=t(x)}. \tag{5.5}$$

Also, it is easily checked that the boundary conditions (5.2b) are mapped into the boundary conditions

$$h_0 u'(0) + H_0 u(0) = 0 \quad \text{and} \quad h_1 u'(1) + H_1 u(1) = 0 \tag{5.6}$$

with $h_0 = \alpha_a \sigma(a)/(L\,f(a))$ and $H_0 = \beta_a/f(a) - \alpha_a f'(a)/f(a)^2$ and, analogously, h_1, H_1 with a replaced by b.

In this chapter, we restrict ourselves to the study of the canonical Sturm–Liouville eigenvalue problem (5.1a), (5.1b). In the first part, we study the case $h = 0$ in some detail. At the end of Section 5.3, we briefly discuss the case where $h = 1$. In Section 5.3, we prove that there exists a countable number of eigenvalues λ_n of this problem and also prove an asymptotic formula. Because q is real-valued, the problem is self-adjoint, and the existence of a countable number of eigenvalues follows from the general spectral theorem of functional analysis (see Appendix A.6, Theorem A.53). Because this general theorem provides only the information that the eigenvalues tend to infinity, we need other tools to obtain more information about the rate of convergence. The basic ingredient in the proof of the asymptotic formula is the asymptotic behavior of the fundamental system of the differential equation (5.1a) as $|\lambda|$ tends to infinity. Although all of the data and the eigenvalues are real-valued, we use results from complex analysis, in particular, Rouché's theorem. This makes it necessary to allow the parameter λ in the fundamental system to be complex-valued. The existence of a fundamental solution and its asymptotics is the subject of the next section.

Section 5.5 is devoted to the corresponding inverse problem: Given the eigenvalues λ_n, determine the function q. In Section 5.6, we demonstrate how inverse spectral problems arise in a parameter identification problem for a parabolic initial value problem. Section 5.7, finally, studies numerical procedures for recovering q that have been suggested by Rundell and others (see [186, 233, 234]).

We finish this section with a "negative" result, as seen in Example 5.1.

Example 5.1

Let λ be an eigenvalue and u a corresponding eigenfunction of

$$-u''(x) + q(x)\, u(x) = \lambda\, u(x), \ 0 < x < 1, \quad u(0) = 0, \ u(1) = 0\,.$$

Then λ is also an eigenvalue with corresponding eigenfunction $v(x) := u(1-x)$ of the eigenvalue problem

$$-v''(x) + \tilde{q}(x)\, v(x) = \lambda\, v(x), \ 0 < x < 1, \quad v(0) = 0, \ v(1) = 0\,,$$

where $\tilde{q}(x) := q(1-x)$.

This example shows that it is generally impossible to recover the function q unless more information is available. We will see that q can be recovered uniquely, provided we know that it is an even function with respect to $1/2$ or if we know a second spectrum; that is, a spectrum for a boundary condition different from $u(1) = 0$.

5.2 Construction of a Fundamental System

It is well-known from the theory of linear ordinary differential equations that the following initial value problems are uniquely solvable for every fixed (real- or complex-valued) $q \in C[0,1]$ and every given $\lambda \in \mathbb{C}$:

$$-u_1'' + q(x)\, u_1 = \lambda\, u_1\,, \ 0 < x < 1, \quad u_1(0) = 1\,, \ u_1'(0) = 0 \tag{5.7a}$$

$$-u_2'' + q(x)\, u_2 = \lambda\, u_2\,, \ 0 < x < 1, \quad u_2(0) = 0\,, \ u_2'(0) = 1\,. \tag{5.7b}$$

Uniqueness and existence for $q \in L^2(0,1)$ is shown in Theorem 5.4 below. The set of functions $\{u_1, u_2\}$ is called a *fundamental system* of the differential equation $-u'' + q\, u = \lambda\, u$ in $(0,1)$. The functions u_1 and u_2 are linearly independent because the *Wronskian determinant* is one

$$[u_1, u_2] := \det \begin{bmatrix} u_1 & u_2 \\ u_1' & u_2' \end{bmatrix} = u_1 u_2' - u_1' u_2 = 1\,. \tag{5.8}$$

This is seen from

$$\frac{d}{dx}[u_1, u_2] = u_1 u_2'' - u_1'' u_2 = u_1\,(q - \lambda)\, u_2 - u_2\,(q - \lambda)\, u_1 = 0$$

and $[u_1, u_2](0) = 1$. The functions u_1 and u_2 depend on λ and q. We express this dependence often by $u_j = u_j(\cdot, \lambda, q)$, $j = 1, 2$. For $q \in L^2(0,1)$, the solution is not twice continuously differentiable anymore but is only an element of the *Sobolev space*

$$H^2(0,1) := \left\{ u \in C^1[0,1] : u'(x) = \alpha + \int_0^x v(t)\, dt, \ \alpha \in \mathbb{C}, \ v \in L^2(0,1) \right\},$$

see (1.24). We write u'' for v and observe that $u'' \in L^2(0,1)$. The most important example is when $q = 0$. In this case, we can solve (5.7a) and (5.7b) explicitly and have the following:

Example 5.2

Let $q = 0$. Then the solutions of (5.7a) and (5.7b) are given by

$$u_1(x,\lambda,0) = \cos(\sqrt{\lambda}\,x) \quad \text{and} \quad u_2(x,\lambda,0) = \frac{\sin(\sqrt{\lambda}\,x)}{\sqrt{\lambda}}, \qquad (5.9)$$

respectively. An arbitrary branch of the square root can be taken because $s \mapsto \cos(sx)$ and $s \mapsto \sin(sx)/s$ are even functions.

We will see that the fundamental solution for any function $q \in L^2(0,1)$ behaves as (5.9) as $|\lambda|$ tends to infinity. For the proof of the next theorem, we need the following technical lemma.

Lemma 5.3 *Let $q \in L^2(0,1)$ and $k, \tilde{k} \in C[0,1]$ such that there exists $\mu > 0$ with $|k(\tau)| \leq \exp(\mu\tau)$ and $|\tilde{k}(\tau)| \leq \exp(\mu\tau)$ for all $\tau \in [0,1]$. Let $K, \tilde{K} : C[0,1] \to C[0,1]$ be the Volterra integral operators with kernels $k(x-t)\,q(t)$ and $\tilde{k}(x-t)\,q(t)$, respectively; that is,*

$$(K\phi)(x) = \int_0^x k(x-t)\,q(t)\,x(t)\, dt, \quad 0 \leq x \leq 1,$$

and analogously for $\tilde{K}$. Then the following estimate holds:

$$|(\tilde{K}\, K^{n-1}\phi)(x)| \leq \|\phi\|_\infty \frac{1}{n!}\, \hat{q}(x)^n\, e^{\mu x}, \quad 0 \leq x \leq 1, \qquad (5.10)$$

for all $\phi \in C[0,1]$ and all $n \in \mathbb{N}$. Here, $\hat{q}(x) := \int_0^x |q(t)|\, dt$. If $\phi \in C[0,1]$ satisfies also the estimate $|\phi(\tau)| \leq \exp(\mu\tau)$ for all $\tau \in [0,1]$, then we have

$$|(\tilde{K}\, K^{n-1}\phi)(x)| \leq \frac{1}{n!}\, \hat{q}(x)^n\, e^{\mu x}, \quad 0 \leq x \leq 1, \qquad (5.11)$$

for all $n \in \mathbb{N}$.

Proof: We prove the estimates by induction with respect to n.

For $n = 1$, we estimate

$$\left|(\tilde{K}\phi)(x)\right| = \left|\int\limits_0^x \tilde{k}(x - t)\, q(t)\, \phi(t)\, dt\right|$$

$$\leq \|\phi\|_\infty \int\limits_0^x e^{\mu(x-t)}\, |q(t)|\, dt \leq \|\phi\|_\infty\, e^{\mu x}\, \hat{q}(x)\,.$$

Now we assume the validity of (5.10) for n. Because it holds also for $K = \tilde{K}$, we estimate

$$\left|(\tilde{K}\, K^n \phi)(x)\right| \leq \int\limits_0^x e^{\mu(x-t)}\, |q(t)|\, |(K^n \phi)(t)|\, dt$$

$$\leq \|\phi\|_\infty\, \frac{1}{n!}\, e^{\mu x} \int\limits_0^x |q(t)|\, \hat{q}(t)^n\, dt\,.$$

We compute the last integral by

$$\int\limits_0^x |q(t)|\, \hat{q}(t)^n\, dt = \int\limits_0^x \hat{q}'(t)\, \hat{q}(t)^n\, dt = \frac{1}{n+1} \int\limits_0^x \frac{d}{dt}\left(\hat{q}(t)^{n+1}\right) dt$$

$$= \frac{1}{n+1}\, \hat{q}(x)^{n+1}\,.$$

This proves the estimate (5.10) for $n + 1$.

For estimate (5.11), we only change the initial step $n = 1$ into

$$\left|(\tilde{K}\phi)(x)\right| \leq \int\limits_0^x e^{\mu(x-t)}\, e^{\mu t}\, |q(t)|\, dt \leq e^{\mu x}\, \hat{q}(x)\,.$$

The remaining part is proven by the same arguments.   □

Now we prove the equivalence of the initial value problems for $u_j,\ j = 1, 2$, to Volterra integral equations.

Theorem 5.4 *Let $q \in L^2(0,1)$ and $\lambda \in \mathbb{C}$. Then we have*

(a) $u_1, u_2 \in H^2(0,1)$ are solutions of (5.7a) and (5.7b), respectively, if and only if $u_1, u_2 \in C[0,1]$ solve the Volterra integral equations:

$$u_1(x) = \cos(\sqrt{\lambda}\, x) + \int\limits_0^x \frac{\sin\sqrt{\lambda}(x - t)}{\sqrt{\lambda}}\, q(t)\, u_1(t)\, dt, \qquad (5.12a)$$

$$u_2(x) = \frac{\sin(\sqrt{\lambda}\, x)}{\sqrt{\lambda}} + \int\limits_0^x \frac{\sin\sqrt{\lambda}(x - t)}{\sqrt{\lambda}}\, q(t)\, u_2(t)\, dt\,, \qquad (5.12b)$$

respectively, for $0 \leq x \leq 1$.

(b) *The integral equations (5.12a) and (5.12b) and the initial value problems (5.7a) and (5.7b) are uniquely solvable. The solutions can be represented by a Neumann series. Let K denote the integral operator*

$$(K\phi)(x) := \int_0^x \frac{\sin\sqrt{\lambda}(x-t)}{\sqrt{\lambda}}\, q(t)\,\phi(t)\,dt\,, \quad x \in [0,1]\,, \qquad (5.13)$$

and define

$$C(x) := \cos(\sqrt{\lambda}x) \quad and \quad S(x) := \frac{\sin(\sqrt{\lambda}x)}{\sqrt{\lambda}}. \qquad (5.14)$$

Then

$$u_1 = \sum_{n=0}^{\infty} K^n\, C \quad and \quad u_2 = \sum_{n=0}^{\infty} K^n\, S. \qquad (5.15)$$

The series converge uniformly with respect to $(x,\lambda,q) \in [0,1] \times \Lambda \times Q$ for all bounded sets $\Lambda \subset \mathbb{C}$ and $Q \subset L^2(0,1)$.

Proof: (a) We use the following version of partial integration for $f,g \in H^2(0,1)$:

$$\int_a^b \left[f''(t)\, g(t) - f(t)\, g''(t) \right] dt \;=\; \left[f'(t)\, g(t) - f(t)\, g'(t) \right] \Big|_a^b. \qquad (5.16)$$

We restrict ourselves to the proof for u_1. Let u_1 be a solution of (5.7a). Then

$$\int_0^x S(x-t)\, q(t)\, u_1(t)\, dt \;=\; \int_0^x S(x-t)\left[\lambda\, u_1(t) + u_1''(t) \right] dt$$

$$= \int_0^x u_1(t) \underbrace{\left[\lambda\, S(x-t) + S''(x-t) \right]}_{=0} dt$$

$$+ \left[u_1'(t)\, S(x-t) + u_1(t)\, S'(x-t) \right] \Big|_{t=0}^{t=x}$$

$$= u_1(x) - \cos(\sqrt{\lambda}x)\,.$$

On the other hand, let $u_1 \in C[0,1]$ be a solution of the integral equation (5.12a). The operator $A : L^2(0,1) \to L^2(0,1)$ defined by $(A\phi)(x) = \int_0^x S(x-t)\,\phi(t)\,dt$, $x \in (0,1)$, is bounded, and it is easily seen that $(A\phi)'' + \lambda(A\phi) = \phi$ for $\phi \in C[0,1]$. Therefore, A is even bounded from $L^2(0,1)$ into $H^2(0,1)$. Writing (5.12a) in the form $u_1 = C + A(qu_1)$ yields that $u_1 \in H^2(0,1)$ and $u_1'' = -\lambda C + qu_1 - \lambda A(qu_1) = qu_1 - \lambda u_1$. This proves the assertion because the initial conditions are obviously satisfied.

(b) We observe that all of the functions $k(\tau) = \cos(\sqrt{\lambda}\tau)$, $k(\tau) = \sin(\sqrt{\lambda}\tau)$, and $k(\tau) = \sin(\sqrt{\lambda}\tau)/\sqrt{\lambda}$ for $\tau \in [0,1]$ satisfy the estimate $|k(\tau)| \le \exp(\mu\,\tau)$

with $\mu = |\operatorname{Im} \sqrt{\lambda}|$. This is obvious for the first two functions. For the third, it follows from

$$\left| \frac{\sin(\sqrt{\lambda}\tau)}{\sqrt{\lambda}} \right| \leq \int_0^\tau |\cos(\sqrt{\lambda}s)|\, ds \;=\; \int_0^\tau \cosh(\mu s)\, ds \;\leq\; \cosh(\mu\tau) \;\leq\; e^{\mu\tau}.$$

We have to study the integral operator K with kernel $k(x-t)q(t)$, where $k(\tau) = \sin(\sqrt{\lambda}\tau)/\sqrt{\lambda}$. We apply Lemma 5.3 with $\tilde{K} = K$. Estimate (5.10) yields

$$\|K^n\|_\infty \;\leq\; \frac{\hat{q}(1)^n}{n!}\, e^\mu \;<\; 1$$

for sufficiently large n uniformly for $q \in Q$ and $\lambda \in \Lambda$. Therefore, the Neumann series converges (see Appendix A.3, Theorem A.31), and part (b) is proven. □

The integral representation of the previous theorem yields the following asymptotic behavior of the fundamental system by comparing the case for arbitrary q with the case of $q = 0$.

Theorem 5.5 *Let $q \in L^2(0,1)$, $\lambda \in \mathbb{C}$, and u_1, u_2 be the fundamental system; that is, the solutions of the initial value problems (5.7a) and (5.7b), respectively. Then we have for all $x \in [0,1]$:*

$$\left| u_1(x) - \cos(\sqrt{\lambda}x) \right| \;\leq\; \frac{1}{|\sqrt{\lambda}|}\, \exp\!\left(|\operatorname{Im}\sqrt{\lambda}|\, x + \int_0^x |q(t)|\, dt \right) \qquad (5.17a)$$

$$\left| u_2(x) - \frac{\sin(\sqrt{\lambda}x)}{\sqrt{\lambda}} \right| \;\leq\; \frac{1}{|\lambda|}\, \exp\!\left(|\operatorname{Im}\sqrt{\lambda}|\, x + \int_0^x |q(t)|\, dt \right) \qquad (5.17b)$$

$$\left| u_1'(x) + \sqrt{\lambda}\sin(\sqrt{\lambda}x) \right| \;\leq\; \exp\!\left(|\operatorname{Im}\sqrt{\lambda}|\, x + \int_0^x |q(t)|\, dt \right) \qquad (5.17c)$$

$$\left| u_2'(x) - \cos(\sqrt{\lambda}x) \right| \;\leq\; \frac{1}{|\sqrt{\lambda}|}\, \exp\!\left(|\operatorname{Im}\sqrt{\lambda}|\, x + \int_0^x |q(t)|\, dt \right) \qquad (5.17d)$$

Proof: Again, we use the Neumann series and define $C(\tau) := \cos(\sqrt{\lambda}\tau)$ and $S(\tau) := \sin(\sqrt{\lambda}\tau)/\sqrt{\lambda}$.   Let K be the integral operator with kernel $q(t)\sin\!\big(\sqrt{\lambda}(x-t)\big)/\sqrt{\lambda}$. Then

$$\left| u_1(x) - \cos(\sqrt{\lambda}x) \right| \;\leq\; \sum_{n=1}^\infty |(K^n C)(x)|.$$

Now we set $\tilde{k}(\tau) = \sin(\sqrt{\lambda}\tau)$ and $k(\tau) = \sin(\sqrt{\lambda}\tau)/\sqrt{\lambda}$ and denote by $\tilde{K}$ and K the Volterra integral operators with kernels $\tilde{k}(x-t)$ and $k(x-t)$, respectively.

Then $K^n = \frac{1}{\sqrt{\lambda}} \tilde{K} K^{n-1}$ and, by Lemma 5.3, part (b), we conclude that

$$|(K^n C)(x)| \leq \frac{1}{|\sqrt{\lambda}| \, n!} \left(\int_0^x |q(t)| \, dt \right)^n \exp(|\operatorname{Im} \sqrt{\lambda}| \, x)$$

for $n \geq 1$. Summation now yields the desired estimate:

$$|u_1(x) - \cos(\sqrt{\lambda} x)| \leq \frac{1}{|\sqrt{\lambda}|} \exp\left(|\operatorname{Im} \sqrt{\lambda}| \, x + \int_0^x |q(t)| \, dt \right).$$

Because $|S(x)| \leq \frac{1}{|\sqrt{\lambda}|} \exp(|\operatorname{Im} \sqrt{\lambda}| \, x)$, the same arguments prove the estimate (5.17b). Differentiation of the integral equations (5.12a) and (5.12b) yields

$$u_1'(x) + \sqrt{\lambda} \sin(\sqrt{\lambda} x) = \int_0^x \cos \sqrt{\lambda} (x - t) \, q(t) \, u_1(t) \, dt,$$

$$u_2'(x) - \cos(\sqrt{\lambda} x) = \int_0^x \cos \sqrt{\lambda} (x - t) \, q(t) \, u_2(t) \, dt.$$

With K as before and $\tilde{K}$ defined as the operator with kernel $q(t) \cos \sqrt{\lambda} (x - t)$. Then

$$u_1'(x) + \sqrt{\lambda} \sin(\sqrt{\lambda} x) = \tilde{K} \sum_{n=0}^{\infty} K^n C,$$

$$u_2'(x) - \cos(\sqrt{\lambda} x) = \tilde{K} \sum_{n=0}^{\infty} K^n S,$$

and we use Lemma 5.3, estimate (5.11), again. Summation yields the estimates (5.17c) and (5.17d).   □

In the next section, we need the fact that the eigenfunctions are continuously differentiable with respect to q and λ. We remind the reader of the concept of Fréchet differentiability (F-differentiability) of an operator between Banach spaces X and Y (see Appendix A.7, Definition A.60). Here we consider the mapping $(\lambda, q) \mapsto u_j(\cdot, \lambda, q)$ from $\mathbb{C} \times L^2(0, 1)$ into $C[0, 1]$ for $j = 1, 2$. We denote these mappings by u_j again and prove the following theorem:

Theorem 5.6 *Let $u_j : \mathbb{C} \times L^2(0, 1) \to C[0, 1]$, $j = 1, 2$, be the solution operator of (5.7a) and (5.7b), respectively. Then we have the following:*

(a) u_j is continuous.

(b) u_j *is continuously F-differentiable for every* $(\hat{\lambda}, \hat{q}) \in \mathbb{C} \times L^2(0,1)$ *with partial derivatives*

$$\frac{\partial}{\partial \lambda} u_j(\cdot, \hat{\lambda}, \hat{q}) \;=\; u_{j,\lambda}(\cdot, \hat{\lambda}, \hat{q}) \tag{5.18a}$$

and

$$\frac{\partial}{\partial q} u_j(\cdot, \hat{\lambda}, \hat{q})(q) \;=\; u_{j,q}(\cdot, \hat{\lambda}, \hat{q}), \tag{5.18b}$$

where $u_{j,\lambda}(\cdot, \hat{\lambda}, \hat{q})$ *and* $u_{j,q}(\cdot, \hat{\lambda}, \hat{q})$ *are solutions of the following initial boundary value problems for* $j = 1, 2$:

$$\begin{aligned}
-u_{j,\lambda}'' + (\hat{q} - \hat{\lambda}) u_{j,\lambda} &= u_j(\cdot, \hat{\lambda}, \hat{q}) \quad \text{in } (0,1) \\
u_{j,\lambda}(0) &= 0, \quad u_{j,\lambda}'(0) = 0, \\
-u_{j,q}'' + (\hat{q} - \hat{\lambda}) u_{j,q} &= -q\, u_j(\cdot, \hat{\lambda}, \hat{q}) \quad \text{in } (0,1), \\
u_{j,q}(0) &= 0, \quad u_{j,q}'(0) = 0.
\end{aligned} \tag{5.19}$$

(c) Furthermore, for all $x \in [0,1]$ *we have*

$$\int_0^x u_j(t)^2 dt \;=\; [u_{j,\lambda}, u_j](x), \quad j = 1, 2, \tag{5.20a}$$

$$\int_0^x u_1(t)\, u_2(t)\, dt \;=\; [u_{1,\lambda}, u_2](x) \;=\; [u_{2,\lambda}, u_1](x), \tag{5.20b}$$

$$-\int_0^x q(t)\, u_j(t)^2 dt \;=\; [u_{j,q}, u_j](x), \quad j = 1, 2, \tag{5.20c}$$

$$-\int_0^x q(t)\, u_1(t)\, u_2(t)\, dt \;=\; [u_{1,q}, u_2](x) \;=\; [u_{2,q}, u_1](x), \tag{5.20d}$$

where $[u, v]$ *denotes the Wronskian determinant from (5.8).*

Proof: (a), (b): Continuity and differentiability of u_j follow from the integral equations (5.12a) and (5.12b) because the kernel and the right-hand sides depend continuously and differentiably on λ and q. It remains to show the representation of the derivatives in (b). Let $u = u_j$, $j = 1$ or 2. Then

$$\begin{aligned}
-u''(\cdot, \hat{\lambda} + \varepsilon) + (\hat{q} - \hat{\lambda} - \varepsilon)\, u(\cdot, \hat{\lambda} + \varepsilon) &= 0, \\
-u''(\cdot, \hat{\lambda}) + (\hat{q} - \hat{\lambda})\, u(\cdot, \hat{\lambda}) &= 0;
\end{aligned}$$

thus

$$-\frac{1}{\varepsilon}\big[u(\cdot, \hat{\lambda} + \varepsilon) - u(\cdot, \hat{\lambda})\big]'' + (\hat{q} - \hat{\lambda})\frac{1}{\varepsilon}\big[u(\cdot, \hat{\lambda} + \varepsilon) - u(\cdot, \hat{\lambda})\big] = u(\cdot, \hat{\lambda} + \varepsilon).$$

Furthermore, the homogeneous initial conditions are satisfied for the difference quotient. The right-hand side converges uniformly to $u(\cdot, \hat{\lambda})$ as $\varepsilon \to 0$. Therefore, the difference quotient converges to u_λ uniformly in x. The same arguments yield the result for the derivative with respect to q.

(c) Multiplication of the differential equation for $u_{j,\lambda}$ by u_j and the differential equation for u_j by $u_{j,\lambda}$ and subtraction yields

$$
\begin{aligned}
u_j^2(x) &= u_j''(x)\, u_{j,\lambda}(x) - u_{j,\lambda}''(x)\, u_j(x) \\
&= \frac{d}{dx}\left(u_j'(x)\, u_{j,\lambda}(x) - u_{j,\lambda}'(x)\, u_j(x) \right).
\end{aligned}
$$

Integration of this equation and the homogeneous boundary conditions yield the first equation of (5.20a). The proofs for the remaining equations use the same arguments and are left to the reader.   □

At no place in this section have we used the assumption that q is real-valued. Therefore, the assertions of Theorems 5.4, 5.5, and 5.6 also hold for complex-valued q.

5.3   Asymptotics of the Eigenvalues and Eigenfunctions

We first restrict ourselves to the Dirichlet problem; that is, the eigenvalue problem

$$
-u''(x) + q(x)\, u(x) = \lambda\, u(x), \ 0 < x < 1, \quad u(0) = u(1) = 0. \tag{5.21}
$$

We refer to the end of this section for different boundary conditions. Again, let $q \in L^2(0,1)$ be real-valued. We observe that $\lambda \in \mathbb{C}$ is an eigenvalue of this problem if and only if λ is a zero of the function

$$
f(\lambda) := u_2(1, \lambda, q).
$$

Again, $u_2 = u_2(\cdot, \lambda, q)$ denotes the solution of the differential equation $-u_2'' + q\, u_2 = \lambda\, u_2$ in $(0,1)$ with initial conditions $u_2(0) = 0$ and $u_2'(0) = 1$. If $u_2(1, \lambda, q) = 0$, then $u = u_2(\cdot, \lambda, q)$ is an eigenfunction corresponding to the eigenvalue λ, normalized such that $u'(0) = 1$. There are different ways to normalize the eigenfunctions. Later we will sometimes normalize them such that the L^2-norms are one; that is, use $g = u/\|u\|_{L^2}$ instead of u.

The function f plays exactly the role of the well-known characteristic polynomial for matrices and is, therefore, called the *characteristic function* of the eigenvalue problem. Theorem 5.6 implies that f is differentiable; that is, analytic in all of $\mathbb{C}$. This observation makes it possible to use tools from complex analysis. First, we summarize well-known facts about eigenvalues and eigenfunctions for the Sturm–Liouville problem.

Theorem 5.7 *Let* $q \in L^2(0,1)$ *be real-valued. Then*

(a) *All eigenvalues λ are real.*

(b) *There exists a countable number of real eigenvalues λ_j, $j \in \mathbb{N}$, which tend to infinity as $j \to \infty$. The corresponding eigenfunctions $g_j \in C[0,1]$, normalized by $\|g_j\|_{L^2} = 1$, form a complete orthonormal system in $L^2(0,1)$.*

(c) *The geometric and algebraic multiplicities of the eigenvalues λ_j are one; that is, the eigenspaces are one-dimensional and the zeros of the characteristic function f are simple.*

(d) *Let the eigenvalues be ordered as $\lambda_1 < \lambda_2 < \lambda_3 < \cdots$. The eigenfunction g_j corresponding to λ_j has exactly $j - 1$ zeros in $(0,1)$.*

(e) *Let q be even with respect to $1/2$; that is, $q(1-x) = q(x)$ for all $x \in [0,1]$. Then g_j is even with respect to $1/2$ for odd j and odd with respect to $1/2$ for even j.*

Proof: (a) and (b) follow from the fact that the boundary value problem is self-adjoint. We refer to Problem 5.1 for a repetition of the proof (see also Theorems A.52 and A.53).

(c) Let λ be an eigenvalue and u, v be two corresponding eigenfunctions. Choose α, β with $\alpha^2 + \beta^2 > 0$, such that $\alpha\, u'(0) = \beta\, v'(0)$. The function $w := \alpha u - \beta v$ solves the differential equation and $w(0) = w'(0) = 0$; that is, w vanishes identically. Therefore, u and v are linearly dependent.

We apply Theorem 5.6, part (c), to show that λ is a simple zero of f. Because $u_2(1, \lambda, q) = 0$, we have from (5.20a) for $j = 2$ that

$$f'(\lambda) \;=\; \frac{\partial}{\partial\lambda} u_2(1,\lambda,q) \;=\; u_{2,\lambda}(1,\lambda,q)$$

$$\;=\; \frac{1}{u_2'(1,\lambda,q)} \int\limits_0^1 u_2(x,\lambda,q)^2 \, dx \;\neq\; 0. \tag{5.22}$$

This proves part (c).

(d) First we note that every g_j has only a finite number of zeros. Otherwise, they would accumulate at some point $x \in [0,1]$, and it is not difficult to show that g_j and also g_j' vanish at x. This would imply that g_j vanishes identically.

We fix $j \in \mathbb{N}$ and define the function $h : [0,1] \times [0,1] \to \mathbb{R}$ by $h(t,x) = u_j(x; tq)$. Here, $u_j(\cdot; tq)$ is the j-th eigenfunction u_j corresponding to tq instead of q and normalized such that $u_j'(0) = 1$. Then h is continuously differentiable and $h(t,0) = h(t,1) = 0$ and every zero of $h(t,\cdot)$ is simple. This holds for every t. By part (a) of Lemma 5.8 below, the number of zeros of $h(t,\cdot)$ is constant with respect to t. Therefore, $u_j(\cdot,q) = h(1,\cdot)$ has exactly the same number of zeros as $h(0,x) = u_j(x,0) = \sqrt{2}\sin(j\pi x)$ which is $j - 1$. The normalization to $g_j = u_j(\cdot,q)/\|u_j(\cdot,q)\|_{L^2}$ does not change the number of zeros.

(e) Again, it is sufficient to prove that $v_j = u_2(\cdot,\lambda_j,q)$ is even (or odd) for odd (or even) j. First, we note that also $\tilde{v}_j(x) := v_j(1-x)$ is an eigenfunction.

Since the eigenspace is one-dimensional, we conclude that there exists $\rho_j \in \mathbb{R}$ with $\tilde{v}_j(x) = v_j(1 - x) = \rho_j v_j(x)$ for all x. For $x = 1/2$ this implies that $(1 - \rho_j)v_j(1/2) = 0$ and also by differentiation $(1 + \rho_j)v'_j(1/2) = 0$ for all j. Since $v_j(1/2)$ and $v'_j(1/2)$ cannot vanish simultaneously, we conclude that $\rho_j \in \{+1, -1\}$ and even $\rho_j = \rho_j v'_j(0) = -v'_j(1)$. From (5.22), we conclude that $\text{sign} f'(\lambda_j) = \text{sign} v'_j(1) = -\text{sign} \rho_j$. Since λ_j are the subsequent zeros of f, we conclude that $f'(\lambda_j)f'(\lambda_{j+1}) < 0$ for all j; that is, $\rho_j = \sigma(-1)^{j+1}$, $j \in \mathbb{N}$, for some $\sigma \in \{+1, -1\}$. The first eigenfunction v_1 has no zero by part (d) which yields $\sigma = 1$. This ends the proof.   □

The first part of the following technical result has been used in the previous proof, the second part will be needed below.

Lemma 5.8   *(a) Let $h : [0, 1] \times [0, 1] \to \mathbb{R}$ be continuously differentiable such that $h(t, \cdot)$ has finitely many zeros in $[0, 1]$ and all are simple for every $t \in [0, 1]$. Then the number $m(t)$ of zeros of $h(t, \cdot)$ is constant with respect to t.*

(b) Let $z \in \mathbb{C}$ with $|z - n\pi| \geq \pi/4$ for all $n \in \mathbb{Z}$. Then

$$\exp(|\text{Im} z|) < 4 |\sin z|.$$

Proof: (a) It suffices to show that $t \mapsto m(t)$ is continuous. Fix $\hat{t} \in [0, 1]$ and let $\hat{x}_j$, $j = 1, \ldots, m(\hat{t})$, be the zeros of $h(\hat{t}, \cdot)$. Because $\partial h(\hat{t}, \hat{x}_j)/\partial x \neq 0$ there exist intervals $T = (\hat{t} - \delta, \hat{t} + \delta) \cap [0, 1]$ and $J_j = (\hat{x}_j - \delta, \hat{x}_j + \delta) \cap [0, 1]$ with $\partial h(t, x)/\partial x \neq 0$ for all $t \in T$ and $x \in \bigcup_j J_j$. Therefore, for every $t \in T$, the function $h(t, \cdot)$ has at most one zero in every J_j. On the other hand, by the implicit function theorem (applicable because $\partial h(\hat{t}, \hat{x}_j)/\partial x \neq 0$), for every $t \in T$ there exists at least one zero of $h(t, \cdot)$ in every J_j, where T and J_j are possibly made smaller. Outside of $\bigcup J_j$ there are no zeros of $h(\hat{t}, \cdot)$, and thus by making perhaps T smaller again, no zeros of $h(t, \cdot)$ either for all $t \in T$. This shows that $m(t) = m(\hat{t})$ for all $t \in T$ which shows that $t \mapsto m(t)$ is continuous, thus constant.

(b) Let $\psi(z) = \exp |z_2|/|\sin z|$ for $z = z_1 + iz_2$, $z_1, z_2 \in \mathbb{R}$ with $z_1 \notin \{n\pi : n \in \mathbb{Z}\}$. We consider two cases:

1st case: $|z_2| > \ln 2/2$. Then

$$\psi(z) = \frac{2 e^{|z_2|}}{|e^{iz_1 - z_2} - e^{-iz_1 + z_2}|} \leq \frac{2 e^{|z_2|}}{e^{|z_2|} - e^{-|z_2|}} = \frac{2}{1 - e^{-2|z_2|}} < 4$$

because $\exp(-2|z_2|) < 1/2$.

2nd case: $|z_2| \leq \ln 2/2$. From $|z - n\pi| \geq \pi/4$ for all n, we conclude that $|z_1 - n\pi|^2 \geq \pi^2/16 - z_2^2 \geq \pi^2/16 - (\ln 2)^2/4 \geq \pi^2/64$; thus $|\sin z_1| \geq \sin \frac{\pi}{8}$. With $|\text{Re} \sin z| = |\sin z_1| |\cosh z_2| \geq |\sin z_1|$, we conclude that

$$\psi(z) \leq \frac{e^{|z_2|}}{|\text{Re} \sin z|} \leq \frac{\sqrt{2}}{|\sin z_1|} \leq \frac{\sqrt{2}}{|\sin \frac{\pi}{8}|} < 4.   □$$

Now we prove the "counting lemma," a first crude asymptotic formula for the eigenvalues. As the essential tool in the proof, we use the theorem of Rouché from complex analysis (see [2]), which we state for the convenience of the reader: *Let $U \subset \mathbb{C}$ be a domain and the functions F and G be analytic in $\mathbb{C}$ and $|F(z) - G(z)| < |G(z)|$ for all $z \in \partial U$. Then F and G have the same number of zeros in U.*

Lemma 5.9 *Let $q \in L^2(0,1)$ and $N > 2\exp(\|q\|_{L^1})$ be an integer. Then*

(a) *The characteristic function $f(\lambda) := u_2(1, \lambda, q)$ has exactly N zeros in the half-plane*

$$H := \{\lambda \in \mathbb{C} : \operatorname{Re}\lambda < (N + 1/2)^2\pi^2\}. \tag{5.23}$$

(b) *For every $m > N$ there exists exactly one zero of f in the set*

$$U_m := \{\lambda \in \mathbb{C} : |\sqrt{\lambda} - m\pi| < \pi/2\}. \tag{5.24}$$

Here we take the branch with $\operatorname{Re}\sqrt{\lambda} \geq 0$.

(c) *There are no other zeros of f in $\mathbb{C}$.*

Proof: We are going to apply Rouché's theorem to the function $F(z) = f(z^2) = u_2(1, z^2, q)$ and the corresponding function G of the eigenvalue problem for $q = 0$; that is, $G(z) := \sin z/z$. For U, we take one of the sets W_m or V_R defined by

$$W_m := \{z \in \mathbb{C} : |z - m\pi| < \pi/2\},$$
$$V_R := \{z \in \mathbb{C} : |\operatorname{Re} z| < (N + 1/2)\pi, \ |\operatorname{Im} z| < R\}$$

for fixed $R > (N + 1/2)\pi$ and want to apply Lemma 5.8:

(i) First let $z \in \partial W_m$: For $n \in \mathbb{Z}$, $n \neq m$, we have $|z - n\pi| \geq |m - n|\pi - |z - m\pi| \geq \pi - \pi/2 > \pi/4$. For $n = m$, we observe that $|z - m\pi| = \pi/2 > \pi/4$. Therefore, we can apply Lemma 5.8 for $z \in \partial W_m$. Furthermore, we note the estimate $|z| \geq m\pi - |z - m\pi| = (m - 1/2)\pi > N\pi > 2N$ for all $z \in \partial W_m$.

(ii) Let $z \in \partial V_R$, $n \in \mathbb{Z}$. Then $|\operatorname{Re} z| = (N+1)\pi$ or $|\operatorname{Im} z| = R$. In either case, we estimate $|z - n\pi|^2 = (\operatorname{Re} z - n\pi)^2 + (\operatorname{Im} z)^2 \geq \pi^2/4 > \pi^2/16$. Therefore, we can apply Lemma 5.8 for $z \in \partial V_R$. Furthermore, we have the estimate $|z| \geq (N + 1/2)\pi > 2N$ for all $z \in \partial V_R$.

Application of Theorem 5.5 and Lemma 5.8 yields the following estimate for all $z \in \partial V_R \cup \partial W_m$:

$$\left| F(z) - \frac{\sin z}{z} \right| \leq \frac{1}{|z|^2} \exp(|\operatorname{Im} z| + \|q\|_{L^1}) \leq \frac{4|\sin z|}{|z|^2} \frac{N}{2}$$
$$= \frac{2N}{|z|} \left| \frac{\sin z}{z} \right| < \left| \frac{\sin z}{z} \right|.$$

Therefore, F and $G(z) := \sin z/z$ have the same number of zeros in V_R and every W_m. Because the zeros of G are $\pm n\pi$, $n = 1, 2, \ldots$, we conclude that G

has exactly $2N$ zeros in V_R and exactly one zero in every W_m. By the theorem of Rouché, this also holds for F.

Now we show that F has no zero outside of $V_R \cup \bigcup_{m>N} W_m$. Again, we apply Lemma 5.8: Let $z \notin V_R \cup \bigcup_{m>N} W_m$. From $z \notin V_R$, we conclude that $|z| = \sqrt{(\mathrm{Re}\, z)^2 + (\mathrm{Im}\, z)^2} \geq (N+1/2)\pi$. For $n > N$, we have that $|z-n\pi| > \pi/2$ because $z \notin W_n$. For $n \leq N$, we conclude that $|z - n\pi| \geq |z| - n\pi \geq (N + 1/2 - n)\pi \geq \pi/2$. We apply Theorem 5.5 and Lemma 5.8 again and use the second triangle inequality. This yields

$$
\begin{aligned}
|F(z)| &\geq \left| \frac{\sin z}{z} \right| - \frac{1}{|z|^2} \exp\big(|\mathrm{Im}\, z| + \|q\|_{L^1}\big) \\
&\geq \left| \frac{\sin z}{z} \right| \left[1 - \frac{4\exp\big(\|q\|_{L^1}\big)}{|z|} \right] \\
&\geq \left| \frac{\sin z}{z} \right| \left[1 - \frac{2N}{|z|} \right] > 0
\end{aligned}
$$

because $|z| \geq (N + 1/2)\pi > 2N$. Therefore, we have shown that f has exactly one zero in every U_m, $m > N$, and N zeros in the set

$$
H_R := \big\{ \lambda \in \mathbb{C} : 0 < \mathrm{Re}\,\sqrt{\lambda} < (N + 1/2)\pi,\ \big|\mathrm{Im}\,\sqrt{\lambda}\big| < R \big\}
$$

and no other zeros. It remains to show that $H_R \subset H$. For $\lambda = |\lambda| \exp(i\theta) \in H_R$, we conclude that $\mathrm{Re}\,\sqrt{\lambda} = \sqrt{|\lambda|}\cos\frac{\theta}{2} < (N+1/2)\pi$; thus $\mathrm{Re}\,\lambda = |\lambda|\cos\big(2\frac{\theta}{2}\big) \leq |\lambda|\cos^2\frac{\theta}{2} < (N+1/2)^2\pi^2$. □

This lemma proves again the existence of infinitely many eigenvalues. The arguments are not changed for the case of complex-valued functions q. In this case, the general spectral theory is not applicable anymore because the boundary value problem is not self-adjoint. This lemma also provides more information about the eigenvalue distribution, even for the real-valued case. First, we order the eigenvalues in the form

$$
\lambda_1 < \lambda_2 < \lambda_3 < \cdots .
$$

Lemma 5.9 implies that

$$
\sqrt{\lambda_n} = n\pi + \mathcal{O}(1); \quad \text{that is,} \quad \lambda_n = n^2\pi^2 + \mathcal{O}(n). \tag{5.25}
$$

For the treatment of the inverse problem, it is necessary to improve this formula. It is our aim to prove that

$$
\lambda_n = n^2\pi^2 + \int_0^1 q(t)\, dt + \tilde{\lambda}_n \quad \text{where} \quad \sum_{n=1}^{\infty} |\tilde{\lambda}_n|^2 < \infty . \tag{5.26}
$$

There are several methods to prove (5.26). We follow the treatment in [218]. The key is to apply the fundamental theorem of calculus to the function $t \mapsto \lambda_n(tq)$

for $t \in [0, 1]$, thus connecting the eigenvalues λ_n corresponding to q with the eigenvalues $n^2\pi^2$ corresponding to $q = 0$ by the parameter t. For this approach, we need the differentiability of the eigenvalues with respect to q.

For fixed $n \in \mathbb{N}$, the function $q \mapsto \lambda_n(q)$ from $L^2(0, 1)$ into $\mathbb{C}$ is well-defined and Fréchet differentiable by the following theorem.

Theorem 5.10 *For every $n \in \mathbb{N}$, the mapping $q \mapsto \lambda_n(q)$ from $L^2(0, 1)$ into $\mathbb{C}$ is continuously Fréchet differentiable for every $\hat{q} \in L^2(0, 1)$ and*

$$\lambda_n'(\hat{q})q \;=\; \int_0^1 g_n(x, \hat{q})^2 \, q(x) \, dx\,, \quad q \in L^2(0,1)\,. \tag{5.27}$$

Here,

$$g_n(x, \hat{q}) \;:=\; \frac{u_2(x, \hat{\lambda}_n, \hat{q})}{\left\| u_2(\cdot, \hat{\lambda}_n, \hat{q}) \right\|_{L^2}}$$

denotes the L^2-normalized eigenfunction corresponding to $\hat{\lambda}_n := \lambda_n(\hat{q})$. Note that the integral is well-defined because $u_2(\cdot, \hat{\lambda}_n, \hat{q}) \in H^2(0, 1) \subset C[0, 1]$.

Proof: We observe that $u_2(1, \hat{\lambda}_n, \hat{q}) = 0$ and apply the implicit function theorem to the equation

$$u_2(1, \lambda, q) = 0$$

in a neighborhood of $(\hat{\lambda}_n, \hat{q})$. This is possible because the zero $\hat{\lambda}_n$ of $u_2(1, \cdot, \hat{q})$ is simple by Lemma 5.7. The implicit function theorem (see Appendix A.7, Theorem A.66) yields the existence of a unique function $\lambda_n = \lambda_n(q)$ such that $u_2(1, \lambda_n(q), q) = 0$ for all q in a neighborhood of $\hat{q}$; we know this already. But it also implies that the function λ_n is continuously differentiable with respect to q and

$$0 \;=\; \frac{\partial}{\partial \lambda} u_2(1, \hat{\lambda}_n, \hat{q}) \, \lambda_n'(\hat{q})q \;+\; \frac{\partial}{\partial q} u_2(1, \hat{\lambda}_n, \hat{q})q\,;$$

that is, $u_{2,\lambda}(1)\,\lambda_n'(\hat{q})q + u_{2,q}(1) = 0$. With Theorem 5.6, part (c), we conclude that

$$\lambda_n'(\hat{q})q \;=\; -\frac{u_{2,q}(1)}{u_{2,\lambda}(1)} \;=\; -\frac{u_{2,q}(1)\, u_2'(1)}{u_{2,\lambda}(1)\, u_2'(1)}$$

$$\;=\; -\frac{[u_{2,q}, u_2](1)}{[u_{2,\lambda}, u_2](1)} \;=\; \frac{\int_0^1 q(x)\, u_2(x)^2 dx}{\int_0^1 u_2(x)^2 dx} \;=\; \int_0^1 g_n(x, \hat{q})^2 q(x)\, dx\,,$$

where we have dropped the arguments $\hat{\lambda}$ and $\hat{q}$. $\square$

Now we are ready to formulate and prove the main theorem which follows:

Theorem 5.11 *Let $Q \subset L^2(0,1)$ be bounded, $q \in Q$, and $\lambda_n \in \mathbb{C}$ the corresponding eigenvalues. Then we have*

$$\lambda_n = n^2\pi^2 + \int_0^1 q(t)\, dt - \int_0^1 q(t)\cos(2n\pi t)\, dt + \mathcal{O}(1/n) \qquad (5.28)$$

for $n \to \infty$ uniformly for $q \in Q$. Furthermore, the corresponding eigenfunctions g_n, normalized to $\|g_n\|_{L^2} = 1$, have the following asymptotic behavior:

$$g_n(x) = \sqrt{2}\sin(n\pi x) + \mathcal{O}(1/n) \quad \text{and} \qquad (5.29a)$$
$$g_n'(x) = \sqrt{2}\,n\pi\cos(n\pi x) + \mathcal{O}(1) \qquad (5.29b)$$

as $n \to \infty$ uniformly for $x \in [0,1]$ and $q \in Q$.

We observe that the second integral on the right-hand side of (5.28) is the nth Fourier coefficient a_n of q with respect to $\{\cos(2\pi n t) : n = 0, 1, 2, \ldots, \}$. From Fourier theory, it is known that a_n converges to zero, and even more: Bessel's inequality (A.7) yields that $\sum_{n=0}^{\infty} |a_n|^2 < \infty$; that is, (5.26) is satisfied. If q is smooth enough, e.g., continuously differentiable, then a_n tends to zero faster than $1/n$. In that case, this term is absorbed in the $\mathcal{O}(1/n)$ expression.

Proof: We split the proof into four parts:

(a) First, we show that $g_n(x) = \sqrt{2}\sin(\sqrt{\lambda_n}x) + \mathcal{O}(1/n)$ uniformly for $(x,q) \in [0,1] \times Q$. By Lemma 5.9, we know that $\sqrt{\lambda_n} = n\pi + \mathcal{O}(1)$, and thus by Theorem 5.5

$$u_2(x, \lambda_n) = \frac{\sin(\sqrt{\lambda_n}x)}{\sqrt{\lambda_n}} + \mathcal{O}(1/n^2).$$

With the formula $2\int_0^1 \sin^2(\alpha t)\, dt = 1 - \sin(2\alpha)/(2\alpha)$, we compute

$$\int_0^1 u_2(t, \lambda_n)^2 dt = \frac{1}{\lambda_n} \int_0^1 \sin^2\left(\sqrt{\lambda_n}t\right) dt + \mathcal{O}(1/n^3)$$
$$= \frac{1}{2\lambda_n}\left[1 - \frac{\sin(2\sqrt{\lambda_n})}{2\sqrt{\lambda_n}}\right] + \mathcal{O}(1/n^3)$$
$$= \frac{1}{2\lambda_n}[1 + \mathcal{O}(1/n)].$$

Therefore, we have

$$g_n(x) = \frac{u_2(x, \lambda_n)}{\sqrt{\int_0^1 u_2(t, \lambda_n)^2 dt}} = \sqrt{2}\sin(\sqrt{\lambda_n}x) + \mathcal{O}(1/n).$$

(b) Now we show that $\sqrt{\lambda_n} = n\pi + \mathcal{O}(1/n)$ and $g_n(x) = \sqrt{2}\sin(n\pi x) + \mathcal{O}(1/n)$.

We apply the fundamental theorem of calculus and use Theorem 5.10

$$\lambda_n - n^2\pi^2 \;=\; \lambda_n(q) - \lambda_n(0) \;=\; \int_0^1 \frac{d}{dt}\lambda_n(tq)\,dt \tag{5.30}$$

$$=\; \int_0^1 \lambda_n'(tq)q\,dt \;=\; \int_0^1\int_0^1 g_n(x,tq)^2 q(x)\,dx\,dt \;=\; \mathcal{O}(1)\,.$$

This yields $\sqrt{\lambda_n} = n\pi + \mathcal{O}(1/n)$ and, with part (a), the asymptotic form $g_n(x) = \sqrt{2}\,\sin(n\pi x) + \mathcal{O}(1/n)$.

(c) Now the asymptotics of the eigenvalues follow easily from (5.30) by the observation that

$$g_n(x,tq)^2 \;=\; 2\sin^2(n\pi x) + \mathcal{O}(1/n) \;=\; 1 - \cos(2n\pi x) + \mathcal{O}(1/n)\,,$$

uniformly for $t \in [0,1]$ and $q \in Q$.

(d) Similarly, we have for the derivatives

$$g_n'(x) \;=\; \frac{u_2'(x,\lambda_n)}{\sqrt{\int_0^1 u_2(t,\lambda_n)^2 dt}} \;=\; \frac{\sqrt{2}\,\sqrt{\lambda_n}\,\cos(\sqrt{\lambda_n}x) + \mathcal{O}(1)}{\sqrt{1+\mathcal{O}(1/n)}}$$

$$=\; \sqrt{2}\,n\pi\,\cos(n\pi x) + \mathcal{O}(1)\,. \qquad\qquad \square$$

Example 5.12
We illustrate Theorem 5.11 by the following two numerical examples:

(a) Let $q_1(x) = \exp\!\big(\sin(2\pi x)\big)$, $x \in [0,1]$. Then q_1 is analytic and periodic with period 1.
    Plots of the characteristic functions $\lambda \mapsto f(\lambda)$ for q_1 and $q = 0$; that is, $\lambda \mapsto \sin\sqrt{\lambda}/\sqrt{\lambda}$ are shown in Figure 5.1.

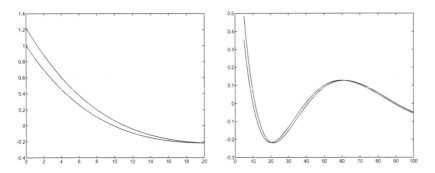

Figure 5.1: Characteristic functions of q, q_1, respectively, on $[0,20]$ and $[5,100]$

(b) Let $q_2(x) = -5\,x$ for $0 \le x \le 0.4$ and $q_2(x) = 4$ for $0.4 < x \le 1$. The function q_2 is not continuous.

Plots of the characteristic functions $\lambda \mapsto f(\lambda)$ for q_2 and $q = 0$ are shown in Figure 5.2.

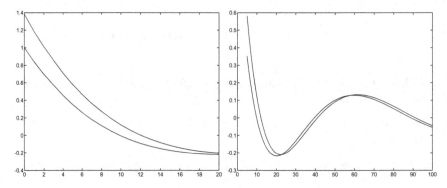

Figure 5.2: Characteristic functions of q, q_2, respectively, on $[0, 20]$ and $[5, 100]$

The Fourier coefficients of q_1 converge to zero of exponential order. The following table shows the eigenvalues λ_n corresponding to q_1, the eigenvalues $n^2\pi^2$ corresponding to $q = 0$ and the difference

$$c_n := \lambda_n - n^2\pi^2 - \int_0^1 q(x)\,dx \quad \text{for } n = 1, \dots, 10:$$

λ_n	$n^2\pi^2$	c_n
11.1	9.9	$-2.04 * 10^{-2}$
40.9	39.5	$1.49 * 10^{-1}$
90.1	88.8	$2.73 * 10^{-3}$
159.2	157.9	$-1.91 * 10^{-3}$
248.0	246.7	$7.74 * 10^{-4}$
356.6	354.3	$4.58 * 10^{-4}$
484.9	483.6	$4.58 * 10^{-4}$
632.9	631.7	$4.07 * 10^{-4}$
800.7	799.4	$3.90 * 10^{-4}$
988.2	987.0	$3.83 * 10^{-4}$

We clearly observe the rapid convergence.

Because q_2 is not continuous, the Fourier coefficients converge to zero only slowly. Again, we list the eigenvalues λ_n for q_2, the eigenvalues $n^2\pi^2$ corresponding to $q = 0$, and the differences

$$c_n := \lambda_n - n^2\pi^2 - \int_0^1 q(x)\,dx \quad \text{and}$$

$$d_n := \lambda_n - n^2\pi^2 - \int_0^1 q(x)\,dx + \int_0^1 q(x)\cos(2\pi nx)\,dx$$

for $n = 1, \ldots, 10$:

λ_n	$n^2\pi^2$	c_n	d_n
12.1	9.9	$1.86 * 10^{-1}$	$-1.46 * 10^{-1}$
41.1	39.5	$-3.87 * 10^{-1}$	$8.86 * 10^{-2}$
91.1	88.8	$3.14 * 10^{-1}$	$2.13 * 10^{-2}$
159.8	157.9	$1.61 * 10^{-1}$	$-6.70 * 10^{-3}$
248.8	246.7	$2.07 * 10^{-2}$	$2.07 * 10^{-2}$
357.4	354.3	$8.29 * 10^{-2}$	$-4.24 * 10^{-3}$
484.5	483.6	$-1.25 * 10^{-1}$	$6.17 * 10^{-3}$
633.8	631.7	$1.16 * 10^{-1}$	$3.91 * 10^{-3}$
801.4	799.4	$-6.66 * 10^{-2}$	$-1.38 * 10^{-3}$
989.0	987.0	$5.43 * 10^{-3}$	$5.43 * 10^{-3}$

Now we sketch the modifications necessary for Sturm–Liouville eigenvalue problems of the type

$$-u''(x) + q(x)\, u(x) = \lambda\, u(x), \quad 0 < x < 1, \tag{5.31a}$$

$$u(0) = 0, \quad u'(1) + Hu(1) = 0. \tag{5.31b}$$

Now the eigenvalues are zeros of the characteristic function

$$f(\lambda) = u_2'(1, \lambda, q) + H u_2(1, \lambda, q), \quad \lambda \in \mathbb{C}. \tag{5.32}$$

For the special case, where $q = 0$, we have $u_2(x, \lambda, 0) = \sin(\sqrt{\lambda}x)/\sqrt{\lambda}$. The characteristic function for this case is then given by

$$g(\lambda) = \cos\sqrt{\lambda} + H\frac{\sin\sqrt{\lambda}}{\sqrt{\lambda}}.$$

The zeros of f for $q = 0$ and $H = 0$ are $\lambda_n = (n + 1/2)^2\pi^2$, $n = 0, 1, 2, \ldots$ If $H \neq 0$, one has to solve the transcendental equation $z \cot z + H = 0$. One can show (see Problem 5.2) by an application of the implicit function theorem in $\mathbb{R}^2$ that the eigenvalues for $q = 0$ behave as

$$\lambda_n = (n + 1/2)^2\pi^2 + 2H + \mathcal{O}(1/n).$$

Lemma 5.7 is also valid because the boundary value problem is again self-adjoint. The Counting Lemma 5.9 now takes the following form:

Lemma 5.13 *Let $q \in L^2(0, 1)$ and $N > 2 \exp\big(\|q\|_{L^1}\big)(1 + |H|)$ be an integer. Then we have*

(a) *The mapping $f(\lambda) := u_2'(1, \lambda, q) + H u_2(1, \lambda, q)$ has exactly N zeros in the half-plane*

$$H := \big\{\lambda \in \mathbb{C} : \operatorname{Re}\lambda < N^2\pi^2\big\}.$$

(b) f has exactly one zero in every set

$$U_m := \left\{ \lambda \in \mathbb{C} : \left| \sqrt{\lambda} - (m - 1/2)\pi \right| < \pi/2 \right\}$$

provided $m > N$.

(c) There are no other zeros of f in $\mathbb{C}$.

For the proof, we refer to Problem 5.3. We can apply the implicit function theorem to the equation

$$u_2'(1, \lambda_n(q), q) + H u_2(1, \lambda_n(q), q) = 0$$

because the zeros are again simple. Differentiating this equation with respect to q yields

$$\left[u_{2,\lambda}'(1, \hat{\lambda}_n, \hat{q}) + H u_{2,\lambda}(1, \hat{\lambda}_n, \hat{q}) \right] \lambda_n'(\hat{q}) q$$
$$+ \; u_{2,q}'(1, \hat{\lambda}_n, \hat{q}) + H u_{2,q}(1, \hat{\lambda}_n, \hat{q}) = 0.$$

Theorem 5.6 yields

$$\int_0^1 u_2(t)^2 dt = u_{2,\lambda}(1) \underbrace{u_2'(1)}_{=-H u_2(1)} - u_{2,\lambda}'(1) \, u_2(1)$$

$$= -u_2(1) \left[u_{2,\lambda}'(1) + H \, u_{2,\lambda}(1) \right]$$

where again we have dropped the arguments $\hat{\lambda}_n$ and $\hat{q}$. Analogously, we compute

$$-\int_0^1 q(t) u_2(t)^2 dt = -u_2(1) \left[u_{2,q}'(1) + H \, u_{2,q}(1) \right]$$

and thus

$$\lambda_n'(\hat{q}) q = -\frac{u_{2,q}'(1) + H \, u_{2,q}(1)}{u_{2,\lambda}'(1) + H \, u_{2,\lambda}(1)} = \frac{\int_0^1 q(t) u_2(t)^2 dt}{\int_0^1 u_2(t)^2 dt}.$$

This has the same form as before. We continue as in the case of the Dirichlet boundary condition and arrive at Theorem 5.14.

Theorem 5.14 *Let $Q \subset L^2(0,1)$ be bounded, $q \in Q$, and $H \in \mathbb{R}$. The eigenvalues λ_n have the asymptotic form*

$$\lambda_n = \left(n + \frac{1}{2} \right)^2 \pi^2 + 2H + \int_0^1 q(t) \, dt - \int_0^1 q(t) \cos(2n+1)\pi t \, dt + \mathcal{O}(1/n) \quad (5.33)$$

as n *tends to infinity, uniformly in* $q \in Q$. *For the* L^2-*normalized eigenfunctions, we have*

$$
\begin{aligned}
g_n(x) &= \sqrt{2}\,\sin(n+1/2)\pi x \,+\, \mathcal{O}(1/n) \quad and \quad &\text{(5.34a)}\\
g_n'(x) &= \sqrt{2}\,(n+1/2)\,\pi\,\cos(n+1/2)\pi x \,+\, \mathcal{O}(1) &\text{(5.34b)}
\end{aligned}
$$

uniformly for $x \in [0,1]$ *and* $q \in Q$.

As mentioned at the beginning of this section, there are other ways to prove the asymptotic formulas for the eigenvalues and eigenfunctions that avoid Lemma 5.9 and the differentiability of λ_n with respect to q. But the proof in, e.g., [276], seems to yield only the asymptotic behavior

$$
\lambda_n = m_n^2 \pi^2 + \int_0^1 q(t)\,dt + \mathcal{O}(1/n)
$$

instead of (5.28). Here, (m_n) denotes some sequence of natural numbers.

Before we turn to the inverse problem, we make some remarks concerning the case where q is complex-valued. Now the eigenvalue problems are no longer self-adjoint, and the general spectral theory is not applicable anymore. With respect to Lemma 5.7, it is still easy to show that the eigenfunctions corresponding to different eigenvalues are linearly independent and that the geometric multiplicities are still one. The Counting Lemma 5.9 is valid without restrictions. From this, we observe also that the algebraic multiplicities of λ_n are one, at least for $n > N$. Thus, the remaining arguments of this section are valid if we restrict ourselves to the eigenvalues λ_n with $n > N$. Therefore, the asymptotic formulas (5.28), (5.29a), (5.29b), (5.33), (5.34a), and (5.34b) hold equally well for complex-valued q.

5.4   Some Hyperbolic Problems

As a preparation for the following sections, in particular, Sections 5.5 and 5.7, we study some initial value problems for the two-dimensional linear hyperbolic partial differential equation

$$
\frac{\partial^2 W(x,t)}{\partial x^2} - \frac{\partial^2 W(x,t)}{\partial t^2} + a(x,t)\,W(x,t) = 0,
$$

where the coefficient a has the special form $a(x,t) = p(t) - q(x)$. It is well-known that the method of characteristics reduces initial value problems to Volterra integral equations of the second kind, which can be studied in spaces of continuous functions. This approach naturally leads to solution concepts for nonsmooth coefficients and boundary data. We summarize the results in three theorems. In each of them, we formulate first the results for the case of smooth coefficients and then for the nonsmooth case. We remark that it is not our aim to relax the solution concept to the weakest possible case but rather to relax the assumptions only to the extent that are needed in Sections 5.5 and 5.7 and in Subsection 7.6.3.

Although most of the problems—at least for smooth data—are subjects of elementary courses on partial differential equations, we include the complete proofs for the convenience of the reader.

Before we begin with the statements of the theorems, we recall some function spaces

$$C_{00}[0,1] := \{f \in C[0,1] : f(0) = 0\},$$

$$H^1(0,1) := \left\{f \in C[0,1] : f(x) = \alpha + \int_0^x g(t)\, dt, \ \alpha \in \mathbb{R}, \ g \in L^2(0,1)\right\},$$

$$H_{00}^1(0,1) := H^1(0,1) \cap C_{00}[0,1]$$

and equip them with their canonical norms

$$\|f\|_\infty := \max_{0 \le x \le 1} |f(x)| \quad \text{in } C_{00}[0,1],$$

$$\|f\|_{H^1} := \sqrt{\|f\|_{L^2}^2 + \|f'\|_{L^2}^2} \quad \text{in } H^1(0,1) \text{ and } H_{00}^1(0,1).$$

The notations $C_{00}[0,1]$ and $H_{00}^1(0,1)$ should indicate that the boundary condition is set only at $x = 0$. By $\|\cdot\|_{C^j}$ for $j \ge 1$ we denote the canonical norm in $C^j[0,1]$.

Furthermore, we define the triangular regions $\Delta_0 \subset \mathbb{R}^2$ and $\Delta \subset \mathbb{R}^2$ by

$$\Delta_0 := \{(x,t) \in \mathbb{R}^2 : 0 < t < x < 1\}, \tag{5.35a}$$

$$\Delta := \{(x,t) \in \mathbb{R}^2 : |t| < x < 1\}, \tag{5.35b}$$

respectively. We begin with an initial value problem, sometimes called the *Goursat problem*.

Theorem 5.15 *(a) Let $p, q \in C[0,1]$ and $f \in C^2[0,1]$ with $f(0) = 0$. Then there exists a unique solution $W \in C^2(\overline{\Delta_0})$ of the following hyperbolic initial value problem:*

$$\frac{\partial^2 W(x,t)}{\partial x^2} - \frac{\partial^2 W(x,t)}{\partial t^2} + (p(t) - q(x))\, W(x,t) = 0 \quad \text{in } \Delta_0, \tag{5.36a}$$

$$W(x,x) = f(x), \quad 0 \le x \le 1, \tag{5.36b}$$

$$W(x,0) = 0, \quad 0 \le x \le 1. \tag{5.36c}$$

(b) The solution operator $(p, q, f) \mapsto W$ has an extension to a bounded operator from $L^2(0,1) \times L^2(0,1) \times C_{00}[0,1]$ into $C(\overline{\Delta_0})$.

(c) The operator $(p, q, f) \mapsto (W(1,\cdot), W_x(1,\cdot))$ has an extension to a bounded operator from $L^2(0,1) \times L^2(0,1) \times H_{00}^1(0,1)$ into $H^1(0,1) \times L^2(0,1)$. Here and in the following, we denote by W_x the partial derivative with respect to x.

Proof: (a) First, we extend the problem to the larger region Δ and study the problem

$$\frac{\partial^2 W(x,t)}{\partial x^2} - \frac{\partial^2 W(x,t)}{\partial t^2} + a(x,t)\,W(x,t) = 0 \quad \text{in } \Delta, \tag{5.37a}$$

$$W(x,x) = f(x), \quad 0 \le x \le 1, \tag{5.37b}$$

$$W(x,-x) = -f(x), \quad 0 \le x \le 1, \tag{5.37c}$$

where we have extended $p(t) - q(x)$ to $a(x,t) := p(|t|) - q(x)$ for $(x,t) \in \Delta$. To treat problem (5.37a)–(5.37c), we make the change of variables

$$x = \xi + \eta, \quad t = \xi - \eta.$$

Then $(x,t) \in \Delta$ if and only if $(\xi, \eta) \in D$, where

$$D := \{(\xi, \eta) \in (0,1) \times (0,1) : \eta + \xi < 1\}. \tag{5.38}$$

We set $w(\xi, \eta) := W(\xi+\eta, \xi-\eta)$ for $(\xi, \eta) \in D$. Then W solves problem (5.37a)–(5.37c) if and only if w solves the hyperbolic problem

$$\frac{\partial^2 w(\xi, \eta)}{\partial \xi\, \partial \eta} = \underbrace{-a(\xi + \eta, \xi - \eta)}_{=:\tilde{a}(\xi,\eta)}\, w(\xi, \eta), \quad (\xi, \eta) \in D, \tag{5.39a}$$

$$w(\xi, 0) = f(\xi) \quad \text{for } \xi \in [0,1], \tag{5.39b}$$

$$w(0, \eta) = -f(\eta) \quad \text{for } \eta \in [0,1]. \tag{5.39c}$$

Now let w be a solution of (5.39a)–(5.39c). We integrate the differential equation twice and use the initial conditions. Then w solves the integral equation

$$w(\xi, \eta) = \int_0^\eta \int_0^\xi \tilde{a}(\xi', \eta')\, w(\xi', \eta')\, d\xi' d\eta' - f(\eta) + f(\xi), \tag{5.40}$$

for $(\xi, \eta) \in D$. This is a Volterra integral equation in two dimensions. We use the standard method to solve this equation by successive iteration in $C(\overline{D})$ where we assume only $p, q \in L^2(0,1)$, and thus $\tilde{a} \in L^2(D)$. Let A be the Volterra integral operator defined by the integral on the right-hand side of (5.40). By induction with respect to $n \in \mathbb{N}$, it can easily be seen (compare with the proof of Theorem 5.18 below for a similar, but more complicated estimate) that

$$|(A^n w)(\xi, \eta)| \le \|w\|_\infty \|\tilde{a}\|_{L^2}^n \frac{1}{n!} (\xi \eta)^{n/2}, \quad n = 1, 2, \dots;$$

thus $\|A^n w\|_\infty \le \|w\|_\infty \|\tilde{a}\|_{L^2}^n \frac{1}{n!}$. Therefore, $\|A^n\|_{\mathcal{L}(C[0,1])} < 1$ for sufficiently large n, and the Neumann series converges (see Appendix A.3, Theorem A.31).

This proves that there exists a unique solution $w \in C(\overline{D})$ of (5.40). From our arguments, uniqueness also holds for (5.37a)–(5.37c).

Now we prove that the solution $w \in C(\overline{D})$ is even in $C^2(\overline{D})$. Obviously, from (5.40) and the differentiability of f, we conclude that w is differentiable with partial derivative (with respect to ξ, the derivative with respect to η is seen analogously)

$$
\begin{aligned}
w_\xi(\xi, \eta) &= \int_0^\eta \left[q(\xi + \eta') - p(|\xi - \eta'|) \right] w(\xi, \eta')\, d\eta' + f'(\xi) \\
&= \int_\xi^{\xi+\eta} q(s)\, w(\xi, s - \xi)\, ds - \int_{\xi-\eta}^\xi p(|s|)\, w(\xi, \xi - s)\, ds + f'(\xi).
\end{aligned}
$$

This second form can be differentiated again. Thus $w \in C^2(\overline{D})$, and we have shown that W is the unique solution of (5.37a)–(5.37c).

Because $a(x, \cdot)$ is an even function and the initial data are odd functions with respect to t, we conclude from the uniqueness result that the solution $W(x, \cdot)$ is also odd. In particular, this implies that $W(x, 0) = 0$ for all $x \in [0, 1]$, which proves that W solves problem (5.36a)–(5.36c) and finishes part (a).

Part (b) follows immediately from the integral equation (5.40) because the integral operator $A : C(\overline{D}) \to C(\overline{D})$ depends continuously on the kernel $\tilde{a} \in L^2(D)$.

For part (c), we observe that

$$
\begin{aligned}
W(1, 2\xi - 1) &= w(\xi, 1 - \xi), \quad \text{thus} \\
\frac{d}{d\xi} W(1, 2\xi - 1) &= \frac{d}{d\xi} w(\xi, 1 - \xi) = w_\xi(\xi, 1 - \xi) - w_\eta(\xi, 1 - \xi) \quad \text{and} \\
W_x(1, 2\xi - 1) &= \frac{1}{2} w_\xi(\xi, 1 - \xi) + \frac{1}{2} w_\eta(\xi, 1 - \xi).
\end{aligned}
$$

We have computed w_ξ already above and have

$$
w_\xi(\xi, 1 - \xi) = \int_\xi^1 q(s)\, w(\xi, s - \xi)\, ds - \int_{2\xi-1}^\xi p(|s|)\, w(\xi, \xi - s)\, ds + f'(\xi)
$$

which is in $L^2(0, 1)$ for $p, q \in L^2(0, 1)$ and $f \in H_{00}^1(0, 1)$. An analogous formula holds for w_η and shows that $W(t, \cdot) \in H^1(0, 1)$ and $W_x(1, \cdot) \in L^2(0, 1)$ for $p, q \in L^2(0, 1)$ and $f \in H_{00}^1(0, 1)$. This ends the proof. $\square$

Remark 5.16 (a) If $p, q \in L^2(0, 1)$ and $f \in C[0, 1]$ with $f(0) = 0$, we call the solution $W \in C(\overline{\Delta_0})$, given by

$$
W(x, t) = w\left(\frac{1}{2}(x + t), \frac{1}{2}(x - t) \right),
$$

where $w \in C(\overline{D})$ solves the integral equation (5.40), the weak solution of the Goursat problem (5.36a)–(5.36c). We observe that for every weak solution W there exist sequences (p_n), (q_n) in $C[0,1]$ and (f_n) in $C^2[0,1]$ with $f_n(0) = 0$ and $\|p_n - p\|_{L^2} \to 0$, $\|q_n - q\|_{L^2} \to 0$ and $\|f_n - f\|_\infty \to 0$ such that the solutions $W_n \in C^2(\overline{\Delta_0})$ of (5.36a)–(5.36c) corresponding to p_n, q_n, and f_n converge uniformly to W.

(b) We observe from the integral equation (5.40) that w has a decomposition into $w(\xi, \eta) = w_1(\xi, \eta) - f(\eta) + f(\xi)$ where $w_1 \in C^1(D)$ even if only $p, q \in L^2(0,1)$. This transforms into $W(x,t) = W_1(x,t) - f\left(\frac{1}{2}(x-t)\right) + f\left(\frac{1}{2}(x+t)\right)$ with $W_1 \in C^1(\Delta)$.

For the special case $p = q = 0$, the integral equation (5.40) reduces to the well-known solution formula

$$W(x,t) = f\left(\frac{1}{2}(x+t)\right) - f\left(\frac{1}{2}(x-t)\right).$$

The next theorem studies a Cauchy problem for the same hyperbolic differential equation.

Theorem 5.17 (a) Let $f \in C^2[0,1]$, $g \in C^1[0,1]$ with $f(0) = f''(0) = g(0) = 0$, and $p, q \in C[0,1]$ and $F \in C^1(\overline{\Delta_0})$. Then there exists a unique solution $W \in C^2(\overline{\Delta_0})$ of the Cauchy problem

$$\frac{\partial^2 W(x,t)}{\partial x^2} - \frac{\partial^2 W(x,t)}{\partial t^2} + \big(p(t) - q(x)\big)\, W(x,t) = F(x,t) \quad \text{in } \Delta_0, \quad (5.41a)$$

$$W(1,t) = f(t) \quad \text{for } 0 \le t \le 1, \quad (5.41b)$$

$$\frac{\partial}{\partial x} W(1,t) = g(t) \quad \text{for } 0 \le t \le 1. \quad (5.41c)$$

(b) Furthermore, the solution operator $(p, q, F, f, g) \mapsto W$ has an extension to a bounded operator from $L^2(0,1) \times L^2(0,1) \times L^2(\Delta_0) \times H^1_{00}(0,1) \times L^2(0,1)$ into $C(\overline{\Delta_0})$.

Proof: As in the proof of Theorem 5.15, we set $a(x,t) := p(|t|) - q(x)$ for $(x,t) \in \Delta$ and extend F to an even function on Δ by $F(x,-t) = F(x,t)$ for $(x,t) \in \Delta_0$. We also extend f and g to odd functions on $[-1,1]$ by $f(-t) = -f(t)$ and $g(-t) = -g(t)$ for $t \in [0,1]$. Then $F \in C^1\big(\overline{\Delta} \setminus ([0,1] \times \{0\})\big) \cap C(\overline{\Delta})$, $f \in C^2[-1,1]$, and $g \in C^1[-1,1]$. We again make the change of variables

$$x = \xi + \eta, \quad t = \xi - \eta, \quad w(\xi,\eta) = W(\xi + \eta, \xi - \eta) \quad \text{for } (\xi, \eta) \in D,$$

where D is given by (5.38). Then W solves (5.41a)–(5.41c) if and only if w solves

$$\frac{\partial^2 w(\xi, \eta)}{\partial \xi\, \partial \eta} = \tilde{a}(\xi, \eta)\, w(\xi, \eta) + \tilde{F}(\xi, \eta), \quad (\xi, \eta) \in D,$$

where $\tilde{F}(\xi,\eta) = F(\xi+\eta, \xi-\eta)$ and $\tilde{a}(\xi,\eta) = -a(\xi+\eta,\xi-\eta) = q(\xi+\eta) - p(|\xi-\eta|)$. The Cauchy conditions (5.41b) and (5.41c) transform into

$$w(\xi, 1-\xi) = f(2\xi - 1) \quad \text{and} \quad w_\xi(\xi, 1-\xi) + w_\eta(\xi, 1-\xi) = 2\,g(2\xi - 1)$$

for $0 \le \xi \le 1$. Differentiating the first equation and solving for w_ξ and w_η yields

$$w_\xi(\xi, 1-\xi) = g(2\xi - 1) + f'(2\xi - 1) \quad \text{and} \quad w_\eta(\xi, 1-\xi) = g(2\xi - 1) - f'(2\xi - 1)$$

for $0 \le \xi \le 1$. Integration of the differential equation with respect to ξ from ξ to $1 - \eta$ yields

$$\frac{\partial w(\xi,\eta)}{\partial \eta} = -\int_\xi^{1-\eta} \left[\tilde{a}(\xi',\eta)\,w(\xi',\eta) + \tilde{F}(\xi',\eta)\right] d\xi' + g(1 - 2\eta) - f'(1 - 2\eta).$$

Now we integrate this equation with respect to η from η to $1 - \xi$ and arrive at

$$w(\xi,\eta) = \int_\eta^{1-\xi} \int_\xi^{1-\eta'} \left[\tilde{a}(\xi',\eta')\,w(\xi',\eta') + \tilde{F}(\xi',\eta')\right] d\xi'\, d\eta' \tag{5.42}$$

$$- \int_\eta^{1-\xi} g(1 - 2\eta')\, d\eta' \;+\; \frac{1}{2} f(2\xi - 1) \;+\; \frac{1}{2} f(1 - 2\eta)$$

for $(\xi,\eta) \in D$. This is again a Volterra integral equation in two variables. The solution $w \in C(\overline{D})$ is in $C^2(\overline{D})$. Indeed, since it is obviously differentiable we take the derivative with respect to η and arrive at the formula above and, after substitution of $\tilde{a}(\xi',\eta)$ and making the change of variables $s = \xi' - \eta$ and $s = \xi' + \eta$, respectively, at the representation

$$\frac{\partial w(\xi,\eta)}{\partial \eta} = \int_{\xi-\eta}^{1-2\eta} p(|s|)\, w(s + \eta, \eta)\, ds - \int_{\xi+\eta}^{1} \left[q(s)\, w(s - \eta, \eta)\right] ds$$

$$- \int_\xi^{1-\eta} \tilde{F}(\xi',\eta)\, d\xi' \;+\; g(1 - 2\eta) \;-\; f'(1 - 2\eta).$$

An analogous formula holds for the derivative with respect to ξ. We can differentiate again because the function $\psi(\xi,\eta) = \int_\xi^{1-\eta} \tilde{F}(\xi',\eta)\, d\xi'$ is differentiable in D – although $\tilde{F}$ is not differentiable at the line $\xi = \eta$. The reader should try to prove this. If only $p, q \in L^2(0,1)$, $F \in L^2(\Delta_0)$, $f \in H_{00}^1(0,1)$, and $g \in L^2(0,1)$ then (5.42) defines the weak solution.

Let A denote the integral operator

$$(Aw)(\xi,\eta) = \int_\eta^{1-\xi} \int_\xi^{1-\eta'} \tilde{a}(\xi',\eta')\, w(\xi',\eta')\, d\xi'\, d\eta', \quad (\xi,\eta) \in D.$$

By induction, it is easily seen that

$$|(A^n w)(\xi, \eta)| \leq \|w\|_\infty \|\tilde{a}\|_{L^2}^n \frac{1}{\sqrt{(2n)!}} (1 - \xi - \eta)^n$$

for all $(\xi, \eta) \in D$ and $n \in \mathbb{N}$; thus

$$\|A^n w\|_\infty \leq \|w\|_\infty \|\tilde{a}\|_{L^2}^n \frac{1}{\sqrt{(2n)!}}$$

for all $n \in \mathbb{N}$. For sufficiently large n, we conclude that $\|A^n\|_{\mathcal{L}(C[0,1])} < 1$, which again implies that (5.42) is uniquely solvable in $C(\overline{D})$ for any $p, q, g \in L^2(0,1)$, $F \in L^2(\Delta_0)$, and $f \in H_{00}^1(0,1)$. □

For the special case $p = q = 0$ and $F = 0$, the integral equation (5.42) reduces to the well-known d'Alembert formula

$$W(x,t) = -\frac{1}{2} \int\limits_{t-(1-x)}^{t+(1-x)} g(\tau)\,d\tau + \frac{1}{2} f\big(t + (1-x)\big) + \frac{1}{2} f\big(t - (1-x)\big).$$

Finally, the third theorem studies a quite unusual coupled system for a pair (W, r) of functions. We treat this system with the same methods as above.

Theorem 5.18 *(a) Let $q \in C[0,1]$, $F \in C^1(\overline{\Delta_0})$, $f \in C^2[0,1]$, and $g \in C^1[0,1]$ such that $f(0) = f''(0) = g(0) = 0$. Then there exists a unique pair of functions $(W, r) \in C^2(\overline{\Delta_0}) \times C^1[0,1]$ with*

$$\frac{\partial^2 W(x,t)}{\partial x^2} - \frac{\partial^2 W(x,t)}{\partial t^2} - q(x)\,W(x,t) = F(x,t)\,r(x) \quad in\ \Delta_0, \quad (5.43a)$$

$$W(x,x) = \frac{1}{2} \int\limits_0^x r(s)\,ds, \quad 0 \leq x \leq 1, \quad (5.43b)$$

$$W(x,0) = 0, \quad 0 \leq x \leq 1, \quad (5.43c)$$

and

$$W(1,t) = f(t) \quad and \quad \frac{\partial}{\partial x} W(1,t) = g(t) \quad for\ all\ t \in [0,1]. \quad (5.43d)$$

(b) Furthermore, the solution operator $(q, F, f, g) \mapsto (W, r)$ has an extension to a bounded operator from $L^2(0,1) \times C(\overline{\Delta_0}) \times H_{00}^1(0,1) \times L^2[0,1]$ into $C(\overline{\Delta_0}) \times L^2(0,1)$.

Proof: (a) We apply the same arguments as in the proofs of Theorems 5.15 and 5.17. We extend $F(x, \cdot)$ to an even function and f and g to odd functions. We again make the change of variables $x = \xi + \eta$ and $t = \xi - \eta$ and set $\tilde{F}(\xi, \eta) = F(\xi + \eta, \xi - \eta)$. In Theorem 5.17 (for $p = 0$), we have shown that the solution W of the Cauchy problem (5.43a) and (5.43d) is equivalent to the integral equation

$$
w(\xi, \eta) = \int_{\eta}^{1-\xi} \int_{\xi}^{1-\eta'} \left[q(\xi' + \eta') \, w(\xi', \eta') + \tilde{F}(\xi', \eta') \, r(\xi' + \eta') \right] d\xi' \, d\eta'
$$

$$
- \int_{\eta}^{1-\xi} g(1 - 2\eta') \, d\eta' + \frac{1}{2} f(2\xi - 1) + \frac{1}{2} f(1 - 2\eta) \qquad (5.44a)
$$

for $w(\xi, \eta) = W(\xi + \eta, \xi - \eta)$ (see equation (5.42)). From this and the initial condition (5.43b), we derive a second integral equation. We set $\eta = 0$ in (5.44a), differentiate, and substitute (5.43b). This yields the following Volterra equation after an obvious change of variables:

$$
\frac{1}{2} r(x) = - \int_{x}^{1} \left[q(s) \, w(x, s - x) + r(s) \, \tilde{F}(x, s - x) \right] ds
$$

$$
+ g(2x - 1) + f'(2x - 1). \qquad (5.44b)
$$

Assume that there exists a solution $(w, r) \in C(\overline{D}) \times L^2(0, 1)$ of (5.44a) and (5.44b). From (5.44b), we observe that r is continuous on $[0, 1]$, and thus by (5.44a), $w \in C^1(\overline{D})$ and thus also $r \in C^1[0, 1]$ because the function $\psi(x) = \int_x^1 r(s) \tilde{F}(x, s - x) \, ds$ is differentiable on $[0, 1]$. Therefore, the right-hand side of (5.43a) is differentiable and we conclude as in the previous theorem that $w \in C^2(\overline{D})$. Furthermore, $\frac{d}{dx} W(x, x) = \frac{d}{dx} w(x, 0) = \frac{1}{2} r(x)$. Now, because $F(x, \cdot)$ is even and f and g are odd functions, we conclude that $W(x, \cdot)$ is also odd. In particular, $W(x, 0) = 0$ for all $x \in [0, 1]$. This implies $W(0, 0) = 0$ and thus $W(x, x) = \frac{1}{2} \int_0^x r(s) \, ds$. Therefore, we have shown that every solution of equations (5.44a) and (5.44b) satisfies (5.43a)–(5.43d) and vice versa.

Now we sketch the proof that the system (5.44a), (5.44b) is uniquely solvable for $(w, r) \in C(\overline{D}) \times L^2(0, 1)$ for given $q \in L^2(0, 1)$, $F \in C(\overline{\Delta_0})$, $f \in H_{00}^1(0, 1)$, and $g \in L^2(0, 1)$. This would also include the proof of part (b). We write this system in the form $(w, r) = A(w, r) + b$ in the product space $C(\overline{D}) \times L^2(0, 1)$ which we equip with the norm $\|(w, r)\|_{\infty, L^2} = \max\{\|w\|_\infty, \|r\|_{L^2(0,1)}\}$. To apply the fixed point theorem we define first the constant $c := 2 \left[\|q\|_{L^2(0,1)} + \|F\|_\infty \right]$ and, for given $(w, r) \in C(\overline{D}) \times L^2(0, 1)$, the functions $(w_n, r_n) = A^n(w, r)$. By induction we prove the following estimates

$$
|w_n(\xi, \eta)| \leq \|(w, r)\|_{\infty, L^2} \frac{c^n}{\sqrt{n!}} (1 - \xi - \eta)^{n/2}, \quad (\xi, \eta) \in D,
$$

$$
|r_n(x)| \leq \|(w, r)\|_{\infty, L^2} \frac{c^n}{\sqrt{n!}} (1 - x)^{n/2}, \quad x \in (0, 1),
$$

for all $n = 1, 2, \ldots$.

We use the elementary integral $\int_\eta^{1-\xi} \int_\xi^{1-\eta'} (1 - \xi' - \eta')^n \, d\xi' \, d\eta' = \frac{1}{(n+1)(n+2)} (1 - \xi - \eta)^{n+2}$ for $n = 0, 1, \ldots$ and set $\tilde{q}(\xi, \eta) = q(\xi + \eta)$ and $\tilde{r}(\xi, \eta) = r(\xi + \eta)$ for abbreviation. We note that $\|\tilde{q}\|_{L^2(D)}^2 = \int_D |q(\xi + \eta)|^2 d(\xi, \eta) = \int_0^1 t \, |q(t)|^2 dt \leq \|q\|_{L^2(0,1)}^2$ and analogously $\|\tilde{r}\|_{L^2(D)} \leq \|r\|_{L^2(D)}$.

For $n = 1$ we have by the Cauchy–Schwarz inequality

$$
\begin{aligned}
|w_1(\xi, \eta)| &\leq \|w\|_\infty \int_\eta^{1-\xi} \int_\xi^{1-\eta'} |\tilde{q}(\xi, \eta)| \, d\xi' \, d\eta' + \|F\|_\infty \int_\eta^{1-\xi} \int_\xi^{1-\eta'} |\tilde{r}(\xi, \eta)| \, d\xi' \, d\eta' \\
&\leq \|(w, r)\|_{\infty, L^2} \big[\|\tilde{q}\|_{L^2(D)} + \|F\|_\infty \big] \sqrt{\int_\eta^{1-\xi} \int_\xi^{1-\eta'} d\xi' \, d\eta'} \\
&\leq c \, \|(w, r)\|_{\infty, L^2} \frac{1}{\sqrt{2}} (1 - \xi - \eta) \leq c \, \|(w, r)\|_{\infty, L^2} \sqrt{1 - \xi - \eta}, \\
|r(x)| &\leq 2 \big[\|w\|_\infty \|q\|_{L^2(0,1)} + \|r\|_{L^2(0,1)} \|F\|_\infty \big] \sqrt{1 - x} \\
&\leq 2 \|(w, r)\|_{\infty, L^2} \big[\|q\|_{L^2(0,1)} + \|F\|_\infty \big] \sqrt{1 - x} \\
&\leq c \, \|(w, r)\|_{\infty, L^2} \sqrt{1 - x}.
\end{aligned}
$$

The step from n to $n+1$ is proven in just the same way. Therefore, $\|A^n\|_{\infty, L^2} \leq \frac{c^n}{\sqrt{n!}}$ which tends to zero as n tends to infinity. Application of Theorem A.31 yields that (5.44a), (5.44b) has a unique solution in $C(\overline{D}) \times L^2(0, 1)$ for all $q \in L^2(0, 1)$, $F \in C(\overline{\Delta_0})$, $f \in H^1_{00}(0, 1)$, and $g \in L^2(0, 1)$. □

5.5 The Inverse Problem

Now we study the inverse spectral problem. This is, given the eigenvalues λ_n of the Sturm–Liouville eigenvalue problem

$$-u''(x) + q(x) \, u(x) = \lambda \, u(x), \; 0 < x < 1, \quad u(0) = 0, \; u(1) = 0, \qquad (5.45)$$

determine the function q. We saw in Example 5.1 that the knowledge of the spectrum $\{\lambda_n : n \in \mathbb{N}\}$ is, in general, not sufficient to determine q uniquely. We need more information, such as a second spectrum μ_n of an eigenvalue problem of the form

$$-v''(x) + q(x) \, v(x) = \mu \, v(x), \quad v(0) = 0, \; v'(1) + Hv(1) = 0, \qquad (5.46)$$

or some knowledge about the eigenfunctions.

The basic tool in the uniqueness proof for this inverse problem is the use of the *Gelfand–Levitan–Marchenko integral operator* (see [101]). This integral operator maps solutions of initial value problems for the equation $-u'' + qu = \lambda u$ onto solutions for the equation $-u'' + pu = \lambda u$ and, most importantly, does not depend on λ. It turns out that the kernel of this operator is the solution for the hyperbolic boundary value problem that was studied in the previous section.

Theorem 5.19 *Let $p, q \in L^2(0,1)$, $\lambda \in \mathbb{C}$, and $u, v \in H^2(0,1)$ be solutions of*

$$-u''(x) + q(x)\,u(x) = \lambda\,u(x), \quad 0 < x < 1, \qquad u(0) = 0, \tag{5.47a}$$

$$-v''(x) + p(x)\,v(x) = \lambda\,v(x), \quad 0 < x < 1, \qquad v(0) = 0, \tag{5.47b}$$

such that $u'(0) = v'(0)$. Also let $K \in C(\overline{\Delta_0})$ be the weak solution of the Goursat problem

$$\frac{\partial^2 K(x,t)}{\partial x^2} - \frac{\partial^2 K(x,t)}{\partial t^2} + \big(p(t) - q(x)\big)\,K(x,t) = 0 \quad in\ \Delta_0, \tag{5.48a}$$

$$K(x,0) = 0, \quad 0 \le x \le 1, \tag{5.48b}$$

$$K(x,x) = \frac{1}{2}\int_0^x \big(q(s) - p(s)\big)\,ds, \quad 0 \le x \le 1, \tag{5.48c}$$

where the triangular region Δ_0 is again defined by

$$\Delta_0 := \big\{ (x,t) \in \mathbb{R}^2 : 0 < t < x < 1 \big\}. \tag{5.49}$$

Then we have

$$u(x) = v(x) + \int_0^x K(x,t)\,v(t)\,dt, \quad 0 \le x \le 1. \tag{5.50}$$

We remark that Theorem 5.15 with $f(x) = \frac{1}{2}\int_0^x \big(q(s) - p(s)\big)\,ds$ implies that this Goursat problem is uniquely solvable in the weak sense.

Proof: First, let $p, q \in C[0,1]$. Then $K \in C^2(\overline{\Delta_0})$ by Theorem 5.15. Define w by the right-hand side of (5.50); that is,

$$w(x) := v(x) + \int_0^x K(x,t)\,v(t)\,dt \quad \text{for } 0 \le x \le 1.$$

Then $w(0) = v(0) = 0 = u(0)$ and w is differentiable with

$$w'(x) = v'(x) + K(x,x)v(x) + \int_0^x K_x(x,t)\,v(t)\,dt, \quad 0 < x < 1.$$

Again, we denote by K_x, K_t, etc., the partial derivatives. For $x = 0$, we have $w'(0) = v'(0) = u'(0)$. Furthermore,

$$
\begin{aligned}
w''(x) &= v''(x) + v(x)\frac{d}{dx}K(x,x) + K(x,x)\,v'(x) \\
&\quad + K_x(x,x)\,v(x) + \int_0^x K_{xx}(x,t)\,v(t)\,dt \\
&= \left[p(x) - \lambda + \frac{d}{dx}K(x,x) + K_x(x,x)\right]v(x) + K(x,x)\,v'(x) \\
&\quad + \int_0^x \left[(q(x) - p(t))K(x,t)v(t) + K_{tt}(x,t)\,v(t)\right]dt\,.
\end{aligned}
$$

Partial integration yields

$$
\begin{aligned}
& \int_0^x K_{tt}(x,t)\,v(t)\,dt \\
&= \int_0^x K(x,t)\,v''(t)\,dt + \left[K_t(x,t)\,v(t) - K(x,t)\,v'(t)\right]_{t=0}^{t=x} \\
&= \int_0^x \left(p(t) - \lambda\right)K(x,t)\,v(t)\,dt + K_t(x,x)\,v(x) - K(x,x)\,v'(x)\,.
\end{aligned}
$$

Therefore, we have

$$
\begin{aligned}
w''(x) &= \left[p(x) - \lambda + \underbrace{\frac{d}{dx}K(x,x) + K_x(x,x) + K_t(x,x)}_{=2\frac{d}{dx}K(x,x)=q(x)-p(x)}\right]v(x) \\
&\quad + \left(q(x) - \lambda\right)\int_0^x K(x,t)\,v(t)\,dt \\
&= \left(q(x) - \lambda\right)\left[v(x) + \int_0^x K(x,t)v(t)\,dt\right] = \left(q(x) - \lambda\right)w(x)\,;
\end{aligned}
$$

that is, w solves the same initial value problem as u. The Picard–Lindelöf uniqueness theorem for initial boundary value problems yields $w = u$. Thus, we have proven the theorem for smooth functions p and q.

Now let $p, q \in L^2(0,1)$. Then we choose functions $(p_n), (q_n)$ in $C[0,1]$ with $p_n \to p$ and $q_n \to q$ in $L^2(0,1)$, respectively. Let K_n be the solution of (5.48a)–

(5.48c) for p_n and q_n. We have already shown that

$$u_n(x) = v_n(x) + \int_0^x K_n(x,t) v_n(t)\, dt, \quad 0 \le x \le 1,$$

for all $n \in \mathbb{N}$, where u_n and v_n solve (5.47a) and (5.47b), respectively, with $u_n'(0) = v_n'(0) = u'(0) = v'(0)$. From the continuous dependence results of Theorems 5.6 and 5.15(b), the functions u_n, v_n, and K_n converge uniformly to u, v, and K, respectively. This proves the assertion of the theorem for $p, q \in L^2(0,1)$. $\square$

As an example, we take $p = 0$ and $v(x) = \sin(\sqrt{\lambda}x)/\sqrt{\lambda}$ and have the following result:

Example 5.20
Let u be a solution of

$$-u''(x) + q(x)\, u(x) = \lambda\, u(x), \quad u(0) = 0, \quad u'(0) = 1, \tag{5.51}$$

for given $q \in L^2(0,1)$. Then we have the representation

$$u(x) = \frac{\sin\sqrt{\lambda}\, x}{\sqrt{\lambda}} + \int_0^x K(x,t) \frac{\sin\sqrt{\lambda}\, t}{\sqrt{\lambda}}\, dt, \quad 0 \le x \le 1, \tag{5.52}$$

where the kernel K solves the following Goursat problem in the weak sense:

$$K_{xx}(x,t) - K_{tt}(x,t) - q(x)\, K(x,t) = 0 \quad \text{in } \Delta_0, \tag{5.53a}$$

$$K(x,0) = 0, \quad 0 \le x \le 1, \tag{5.53b}$$

$$K(x,x) = \frac{1}{2} \int_0^x q(s)\, ds, \quad 0 \le x \le 1. \tag{5.53c}$$

This example has an application that is interesting in itself but that we also need in Section 5.7 and later in Subsection 7.6.3

Theorem 5.21 *Let λ_n be the eigenvalues of one of the eigenvalue problems (5.45) or (5.46) where again $q \in L^2(0,1)$. Then the set of functions $\{\sin(\sqrt{\lambda_n}\cdot) : n \in \mathbb{N}\}$ is complete in $L^2(0,1)$. This means that $\int_0^1 h(x) \sin\sqrt{\lambda_n}x\, dx = 0$ for all $n \in \mathbb{N}$ implies that $h = 0$.*

Proof: Let $T : L^2(0,1) \to L^2(0,1)$ be the Volterra integral operator of the second kind with kernel K; that is,

$$(Tv)(x) := v(x) + \int_0^x K(x,t) v(t)\, dt, \quad x \in (0,1), \ v \in L^2(0,1),$$

where K solves the Goursat problem (5.53a)–(5.53c) in the weak sense. Then we know that T is an isomorphism from $L^2(0,1)$ onto itself. Define $v_n(x) := \sin\sqrt{\lambda_n}x$ for $x \in [0,1]$, $n \in \mathbb{N}$. Let u_n be the eigenfunction corresponding to λ_n, normalized to $u_n'(0) = 1$. By the preceding example,

$$u_n = \frac{1}{\sqrt{\lambda_n}} T v_n \quad \text{or} \quad v_n = \sqrt{\lambda_n}\, T^{-1}u_n\,.$$

Now, if $\int_0^1 h(x)\, v_n(x)\, dx = 0$ for all $n \in \mathbb{N}$, then

$$0 = \int_0^1 h(x)\, T^{-1}u_n(x)\, dx = \int_0^1 u_n(x)\, (T^*)^{-1}h(x)\, dx \quad \text{for all } n \in \mathbb{N},$$

where T^* denotes the L^2-adjoint of T. Because $\{u_n/\|u_n\|_{L^2} : n \in \mathbb{N}\}$ is complete in $L^2(0,1)$ by Lemma 5.7, we conclude that $(T^*)^{-1}h = 0$ and thus $h = 0$. □

Now we can prove the main uniqueness theorem.

Theorem 5.22 *Let $H \in \mathbb{R}$, $p, q \in L^2(0,1)$, and $\lambda_n(p)$, $\lambda_n(q)$ be the eigenvalues of the eigenvalue problem*

$$-u'' + r\,u = \lambda\,u \text{ in } (0,1), \quad u(0) = 0,\ u(1) = 0\,,$$

corresponding to $r = p$ and $r = q$, respectively. Furthermore, let $\mu_n(p)$ and $\mu_n(q)$ be the eigenvalues of

$$-u'' + r\,u = \mu\,u \text{ in } (0,1), \quad u(0) = 0,\ u'(1) + Hu(1) = 0\,,$$

corresponding to $r = p$ and $r = q$, respectively.
If $\lambda_n(p) = \lambda_n(q)$ and $\mu_n(p) = \mu_n(q)$ for all $n \in \mathbb{N}$, then $p = q$.

Proof: From the asymptotics of the eigenvalues (Theorem 5.11), we conclude that

$$\lambda_n(p) = n^2\pi^2 + \int_0^1 p(t)\, dt + o(1), \quad n \to \infty,$$

$$\lambda_n(q) = n^2\pi^2 + \int_0^1 q(t)\, dt + o(1), \quad n \to \infty,$$

and thus

$$\int_0^1 (p(t) - q(t))\, dt = \lim_{n\to\infty} (\lambda_n(p) - \lambda_n(q)) = 0\,. \tag{5.54}$$

Now let K be the weak solution of the Goursat problem (5.48a)–(5.48c). Then K depends only on p and q and is independent of the eigenvalues $\lambda_n := \lambda_n(p) =$

$\lambda_n(q)$ and $\mu_n := \mu_n(p) = \mu_n(q)$. Furthermore, from (5.54), we conclude that $K(1,1) = 0$.

Now let u_n, v_n be the eigenfunctions corresponding to $\lambda_n(q)$ and $\lambda_n(p)$, respectively; that is, solutions of the differential equations

$$-u_n''(x) + q(x)\, u_n(x) = \lambda_n\, u_n(x), \quad -v_n''(x) + p(x)\, v_n(x) = \lambda_n\, v_n(x)$$

for $0 < x < 1$ with homogeneous Dirichlet boundary conditions on both sides. Furthermore, we assume that they are normalized by $u_n'(0) = v_n'(0) = 1$. Then Theorem 5.19 is applicable and yields the relationship

$$u_n(x) \; = \; v_n(x) \; + \; \int_0^x K(x,t)\, v_n(t)\, dt \quad \text{for } x \in [0,1], \tag{5.55}$$

and all $n \in \mathbb{N}$. For $x = 1$, the boundary conditions yield

$$0 \; = \; \int_0^1 K(1,t)\, v_n(t)\, dt \quad \text{for all } n \in \mathbb{N}. \tag{5.56}$$

Now we use the fact that the set $\{v_n/\|v_n\|_{L^2} : n \in \mathbb{N}\}$ forms a complete orthonormal system in $L^2(0,1)$. From this, $K(1,t) = 0$ for all $t \in [0,1]$ follows.

Now let $\tilde{u}_n$ and $\tilde{v}_n$ be eigenfunctions corresponding to μ_n and q and p, respectively, with the normalization $\tilde{u}_n'(0) = \tilde{v}_n'(0) = 1$. Again, Theorem 5.19 is applicable and yields the relationship (5.55) for $\tilde{u}_n$ and $\tilde{v}_n$ instead of u_n und v_n, respectively. Assume for the moment that K is differentiable. Then we can differentiate this equation, set $x = 1$, and arrive at

$$0 \; = \; \tilde{u}_n'(1) \; - \; \tilde{v}_n'(1) \; + \; H\big[\tilde{u}_n(1) - \tilde{v}_n(1)\big]$$

$$= \; \underbrace{K(1,1)}_{=0}\, \tilde{v}_n(1) \; + \; \int_0^1 \big[K_x(1,t) + H\,\underbrace{K(1,t)}_{=0}\big]\, \tilde{v}_n(t)\, dt.$$

We conclude that $\int_0^1 K_x(1,t)\, \tilde{v}_n(t)\, dt = 0$ for all $n \in \mathbb{N}$. From this, $K_x(1,t) = 0$ for all $t \in (0,1)$ follows because $\{\tilde{v}_n/\|\tilde{v}_n\|_{L^2}\}$ forms a complete orthonormal system. However, K is only a weak solution since $p, q \in L^2(0,1)$. From part (b) of Remark 5.16 we know that K has the form $K(x,t) = K_1(x,t) - f\big(\frac{1}{2}(x - t)\big) + f\big(\frac{1}{2}(x + t)\big)$ with $K \in C^1(\Delta_0)$ where in the present situation $f(x) = \frac{1}{2}\int_0^x (q(s) - p(s))\, ds$ is in $H_{00}^1(0,1)$. Then one can easily prove (approximation of f by C^1-functions) that $\psi \in C^1[0,1]$ where $\psi(x) = \int_0^x f\big(\frac{1}{2}(x \pm t)\big)\, \tilde{v}_n(t)\, dt$ and $\psi'(x) = f\big(\frac{1}{2}(x \pm x)\big)\, \tilde{v}_n(x) + \frac{1}{2}\int_0^x f'\big(\frac{1}{2}(x \pm t)\big)\, \tilde{v}_n(t)\, dt$. Therefore, one can argue as in the smooth case of K.

Now we apply the uniqueness part of Theorem 5.17 (in particular, the integral equation (5.42) for $f = g = 0$ and $F = 0$) which yields that K has to vanish identically. In particular, this means that $\cdot$

$$0 \; = \; K(x,x) \; = \; \frac{1}{2}\int_0^x (p(s) - q(s))\, ds \quad \text{for all } x \in (0,1).$$

Differentiating this equation yields that $p = q$.   □

We have seen in Example 5.1 that the knowledge of one spectrum for the Sturm–Liouville differential equation is not enough information to recover the function q uniquely. Instead of knowing the spectrum for a second pair of boundary conditions, we can use other kinds of information, as the following theorem shows:

Theorem 5.23 *Let $p, q \in L^2(0,1)$ with eigenvalues $\lambda_n(p)$, $\lambda_n(q)$, and eigenfunctions u_n and v_n, respectively, corresponding to Dirichlet boundary conditions $u(0) = 0$, $u(1) = 0$. Let the eigenvalues coincide; that is, $\lambda_n(p) = \lambda_n(q)$ for all $n \in \mathbb{N}$. Let one of the following assumptions also be satisfied:*

(a) Let p and q be even functions with respect to $1/2$; that is, $p(1-x) = p(x)$ and $q(1-x) = q(x)$ for all $x \in [0,1]$.

(b) Let the Neumann boundary values coincide; that is, let

$$\frac{u_n'(1)}{u_n'(0)} = \frac{v_n'(1)}{v_n'(0)} \quad \text{for all } n \in \mathbb{N}. \tag{5.57}$$

Then $p = q$.

Proof: (a) From Theorem 5.7, part (e), we know that the eigenfunctions u_n and v_n, again normalized by $u_n'(0) = v_n'(0) = 1$, are even with respect to $x = 1/2$ for odd n and odd for even n. In particular, $u_n'(1) = v_n'(1)$. This reduces the uniqueness question for part (a) to part (b).

(b) We follow the first part of the proof of Theorem 5.22. From (5.56), we again conclude that $K(1,t)$ vanishes for all $t \in (0,1)$. The additional assumption (5.57) yields that $u_n'(1) = v_n'(1)$. We differentiate (5.55), set $x = 1$, and arrive at $\int_0^1 K_x(1,t)v_n(t)\,dt = 0$ for all $n \in \mathbb{N}$. Again, this implies that $K_x(1,\cdot) = 0$, and the proof follows the same lines as the proof of Theorem 5.22.   □

5.6   A Parameter Identification Problem

This section and the next two chapters are devoted to the important field of parameter identification problems for partial differential equations. In Chapter 6, we study the problem of impedance tomography to determine the conductivity distribution from boundary measurements, while in Chapter 7, we study the inverse scattering problem to determine the refractive index of a medium from measurements of the scattered field. In the present section, we consider an application of the inverse Sturm–Liouville eigenvalue problem to the following parabolic initial boundary value problem. First, we formulate the direct problem:

Let $T > 0$ and $\Omega_T := (0,1) \times (0,T) \subset \mathbb{R}^2$, $q \in C[0,1]$ and $f \in C^2[0,T]$ be given with $f(0) = 0$ and $q(x) \geq 0$ for $x \in [0,1]$. Determine $U \in C(\overline{\Omega_T})$, which is twice continuously differentiable with respect to x and continuously differentiable with respect to t in Ω_T such that $\partial U/\partial x \in C\left(\overline{\Omega_T}\right)$ and

$$\frac{\partial U(x,t)}{\partial t} = \frac{\partial^2 U(x,t)}{\partial x^2} - q(x)\,U(x,t) \quad \text{in } \Omega_T, \tag{5.58a}$$

$$U(x,0) = 0, \quad x \in [0,1], \tag{5.58b}$$

$$U(0,t) = 0, \quad \frac{\partial}{\partial x}U(1,t) = f(t), \quad t \in (0,T). \tag{5.58c}$$

From the theory of parabolic initial boundary value problems, it is known that there exists a unique solution of this problem. We prove uniqueness and refer to [170] or (5.60) for the question of existence.

Theorem 5.24 *Let* $f = 0$. *Then* $U = 0$ *is the only solution of (5.58a)–(5.58c) in* Ω_T.

Proof: Multiply the differential equation (5.58a) by $U(x,t)$ and integrate with respect to x. This yields

$$\frac{1}{2}\frac{d}{dt}\int_0^1 U(x,t)^2\,dx = \int_0^1\left[\frac{\partial^2 U(x,t)}{\partial x^2}\,U(x,t) - q(x)\,U(x,t)^2\right]dx.$$

We integrate by parts and use the homogeneous boundary conditions:

$$\frac{1}{2}\frac{d}{dt}\int_0^1 U(x,t)^2\,dx = -\int_0^1\left[\left(\frac{\partial U(x,t)}{\partial x}\right)^2 + q(x)\,U(x,t)^2\right]dx \leq 0.$$

This implies that $t \mapsto \int_0^1 U(x,t)^2 dx$ is nonnegative and monotonically nonincreasing. From $\int_0^1 U(x,0)^2\,dx = 0$, we conclude that $\int_0^1 U(x,t)^2\,dx = 0$ for all t; that is, $U = 0$. $\quad\square$

Now we turn to the inverse problem. Let f be known and, in addition, $U(1,t)$ for all $0 < t \leq T$. The inverse problem is to determine the coefficient q.

In this section, we are only interested in the question if this provides sufficient information in principle to recover q uniquely; that is, we study the question of uniqueness of the inverse problem. It is our aim to prove the following theorem:

Theorem 5.25 *Let* U_1, U_2 *be solutions of (5.58a)–(5.58c) corresponding to* $q = q_1 \geq 0$ *and* $q = q_2 \geq 0$, *respectively, and to the same* $f \in C^2[0,T]$ *with* $f(0) = 0$ *and* $f'(0) \neq 0$. *Let* $U_1(1,t) = U_2(1,t)$ *for all* $t \in (0,T)$. *Then* $q_1 = q_2$ *on* $[0,1]$.

Proof: Let (q,U) be (q_1,U_1) or (q_2,U_2), respectively. Let λ_n and g_n, $n \in \mathbb{N}$, be the eigenvalues and eigenfunctions, respectively, of the Sturm–Liouville eigenvalue problem (5.46) for $H = 0$; that is,

$$-u''(x) + q(x)\,u(x) = \lambda\,u(x), \quad 0 < x < 1, \qquad u(0) = u'(1) = 0.$$

We assume that the eigenfunctions are normalized by $\|g_n\|_{L^2} = 1$ for all $n \in \mathbb{N}$. Furthermore, we can assume that $g_n(1) > 0$ for all $n \in \mathbb{N}$[1]. We know that $\{g_n : n \in \mathbb{N}\}$ forms a complete orthonormal system in $L^2(0,1)$. Theorem 5.14 implies the asymptotic behavior

$$\lambda_n = (n + 1/2)^2 \pi^2 + \hat{q} + \tilde{\lambda}_n \quad \text{with} \quad \sum_{n=1}^{\infty} \tilde{\lambda}_n^2 < \infty, \tag{5.59a}$$

$$g_n(x) = \sqrt{2} \sin(n + 1/2)\pi x + \mathcal{O}(1/n), \tag{5.59b}$$

where $\hat{q} = \int_0^1 q(x)\,dx$. In the first step, we derive a series expansion for the solution U of the initial boundary value problem (5.58a)–(5.58c). From the completeness of $\{g_n : n \in \mathbb{N}\}$, we have the Fourier expansion

$$U(x,t) = \sum_{n=1}^{\infty} a_n(t)\,g_n(x) \quad \text{with} \quad a_n(t) = \int_0^1 U(x,t)\,g_n(x)\,dx, \quad n \in \mathbb{N},$$

where the convergence is understood in the $L^2(0,1)$-sense for every $t \in (0,T]$. We would like to substitute this into the differential equation and the initial and boundary conditions. Because for this formal procedure the interchanging of summation and differentiation is not justified, we suggest a different derivation of a_n. We differentiate a_n and use the partial differential equation (5.58a). This yields

$$
\begin{aligned}
a_n'(t) &= \int_0^1 \frac{\partial U(x,t)}{\partial t}\, g_n(x)\,dx = \int_0^1 \left[\frac{\partial^2 U(x,t)}{\partial x^2} - q(x)\,U(x,t) \right] g_n(x)\,dx \\[2mm]
&= \left[g_n(x) \frac{\partial U(x,t)}{\partial x} - U(x,t) g_n'(x) \right]_{x=0}^{x=1} \\[2mm]
&\quad + \int_0^1 U(x,t) \underbrace{\left[g_n''(x) - q(x)\,g_n(x) \right]}_{=-\lambda_n g_n(x)} dx \\[2mm]
&= f(t)\,g_n(1) - \lambda_n a_n(t).
\end{aligned}
$$

With the initial condition $a_n(0) = 0$, the solution is given by

$$a_n(t) = g_n(1) \int_0^t f(\tau)\, e^{-\lambda_n (t-\tau)}\,d\tau;$$

that is, the solution U of (5.58a)–(5.58c) takes the form

$$U(x,t) = \sum_{n=1}^{\infty} g_n(1)\, g_n(x) \int_0^t f(\tau)\, e^{-\lambda_n (t-\tau)}\,d\tau. \tag{5.60}$$

---

[1] $g_n(1) = 0$ is impossible because of $g_n'(1) = 0$

From partial integration, we observe that

$$\int_0^t f(\tau)\,e^{-\lambda_n(t-\tau)}d\tau \;=\; \frac{1}{\lambda_n}\,f(t) \;-\; \frac{1}{\lambda_n}\int_0^t f'(\tau)\,e^{-\lambda_n(t-\tau)}d\tau\,,$$

and this decays as $1/\lambda_n$. Using this and the asymptotic behavior (5.59a) and (5.59b), we conclude that the series (5.60) converges uniformly in $\overline{\Omega}_T$. For $x = 1$, the representation (5.60) reduces to

$$
\begin{aligned}
U(1,t) \;&=\; \sum_{n=1}^{\infty} g_n(1)^2 \int_0^t f(\tau)\,e^{-\lambda_n(t-\tau)}d\tau \\[2mm]
&=\; \int_0^t f(\tau) \underbrace{\sum_{n=1}^{\infty} g_n(1)^2\,e^{-\lambda_n(t-\tau)}}_{=:\,A(t-\tau)} \, d\tau\,, \quad t \in [0,T]\,.
\end{aligned}
$$

Changing the orders of integration and summation is justified by Lebesgue's theorem of dominated convergence. This is seen from the estimate

$$\sum_{n=1}^{\infty} g_n(1)^2\,e^{-\lambda_n s} \;\leq\; c\sum_{n=1}^{\infty} e^{-n^2\pi^2 s} \;\leq\; c\int_0^{\infty} e^{-\sigma^2\pi^2 s}d\sigma \;=\; \frac{c}{2\sqrt{\pi s}}$$

and the fact that the function $s \mapsto 1/\sqrt{s}$ is integrable in $(0,T]$.

Such a representation holds for $U_1(1,\cdot)$ and $U_2(1,\cdot)$ corresponding to q_1 and q_2, respectively. We denote the dependence on q_1 and q_2 by superscripts (1) and (2), respectively. From $U_1(1,\cdot) = U_2(1,\cdot)$, we conclude that

$$0 \;=\; \int_0^t f(\tau)\,[A^{(1)}(t-\tau)-A^{(2)}(t-\tau)]\,d\tau \;=\; \int_0^t f(t-\tau)\,[A^{(1)}(\tau)-A^{(2)}(\tau)]\,d\tau\,;$$

that is, the function $w := A^{(1)} - A^{(2)}$ solves the homogeneous Volterra integral equation of the *first kind* with kernel $f(t-\tau)$. We differentiate this equation twice and use $f(0) = 0$ and $f'(0) \neq 0$. This yields a Volterra equation of the *second kind* for w:

$$f'(0)\,w(t) \;+\; \int_0^t f''(t-s)\,w(s)\,ds \;=\; 0\,, \quad t \in [0,T]\,.$$

Because Volterra equations of the second kind are uniquely solvable (see Example A.32 of Appendix A.3), this yields $w(t) = 0$ for all t, that is

$$\sum_{n=1}^{\infty} [g_n^{(1)}(1)]^2\,e^{-\lambda_n^{(1)}t} \;=\; \sum_{n=1}^{\infty} [g_n^{(2)}(1)]^2\,e^{-\lambda_n^{(2)}t} \quad \text{for all } t \in (0,T)\,.$$

We note that $g_n^{(j)'}(1) > 0$ for $j = 1, 2$ by our normalization. Now we can apply a result from the theory of Dirichlet series (see Lemma 5.26) and conclude that $\lambda_n^{(1)} = \lambda_n^{(2)}$ and $g_n^{(1)'}(1) = g_n^{(2)'}(1)$ for all $n \in \mathbb{N}$. Applying the uniqueness result analogous to Theorem 5.23, part (b), for the boundary conditions $u(0) = 0$ and $u'(1) = 0$ (see Problem 5.5), we conclude that $q_1 = q_2$. □

It remains to prove the following lemma:

Lemma 5.26 *Let λ_n and μ_n be strictly increasing sequences that tend to infinity. Let the series*

$$\sum_{n=1}^{\infty} \alpha_n e^{-\lambda_n t} \quad and \quad \sum_{n=1}^{\infty} \beta_n e^{-\mu_n t}$$

converge for every $t \in (0, T]$ and uniformly on some interval $[\delta, T]$. Let the limits coincide, that is

$$\sum_{n=1}^{\infty} \alpha_n e^{-\lambda_n t} = \sum_{n=1}^{\infty} \beta_n e^{-\mu_n t} \quad for~all~t \in (0, T].$$

If we also assume that $\alpha_n \neq 0$ and $\beta_n \neq 0$ for all $n \in \mathbb{N}$, then $\alpha_n = \beta_n$ and $\lambda_n = \mu_n$ for all $n \in \mathbb{N}$.

Proof: Assume that $\lambda_1 \neq \mu_1$ or $\alpha_1 \neq \beta_1$. Without loss of generality, we can assume that $\mu_1 \geq \lambda_1$ (otherwise, interchange the roles of λ_n and μ_n). Define

$$C_n(t) := \alpha_n e^{-(\lambda_n - \lambda_1)t} - \beta_n e^{-(\mu_n - \lambda_1)t} \quad for~t \geq \delta.$$

By analytic continuation, we conclude that $\sum_{n=1}^{\infty} C_n(t) = 0$ for all $t \geq \delta$ and that the series converges uniformly on $[\delta, \infty)$. Because

$$C_1(t) = \alpha_1 - \beta_1 e^{-(\mu_1 - \lambda_1)t}$$

and $\alpha_1 \neq \beta_1$ or $\mu_1 > \lambda_1$ there exist $\epsilon > 0$ and $t_1 > \delta$ such that $|C_1(t)| \geq \epsilon$ for all $t \geq t_1$. Choose $n_0 \in \mathbb{N}$ with

$$\left| \sum_{n=1}^{n_0} C_n(t) \right| < \frac{\epsilon}{2} \quad for~all~t \geq t_1.$$

Then we conclude that

$$\left| \sum_{n=2}^{n_0} C_n(t) \right| = \left| C_1(t) - \sum_{n=1}^{n_0} C_n(t) \right| \geq |C_1(t)| - \left| \sum_{n=1}^{n_0} C_n(t) \right| \geq \frac{\epsilon}{2}$$

for all $t \geq t_1$. Now we let t tend to infinity. The first finite sum converges to zero, which is a contradiction. Therefore, we have shown that $\lambda_1 = \mu_1$ and $\alpha_1 = \beta_1$. Now we repeat the argument for $n = 2$, etc. This proves the lemma.
□

5.7   Numerical Reconstruction Techniques

In this section, we discuss some numerical algorithms for solving the inverse spectral problem which was suggested and tested by W. Rundell, P. Sacks, and others. We follow closely the papers [186, 233, 234].

From now on, we assume knowledge of eigenvalues λ_n and μ_n, $n \in \mathbb{N}$, of the Sturm–Liouville eigenvalue problems (5.45) or (5.46). It is our aim to determine the unknown function q. Usually, only a finite number of eigenvalues is known. Then one cannot expect to recover the total function q but only "some portion" of it (see (5.62)).

The first algorithm we discuss uses the concept of the characteristic function again. For simplicity, we describe the method only for the case where q is known to be an even function; that is, $q(1 - x) = q(x)$. Then we know that only one spectrum suffices to recover q (see Theorem 5.23).

Recalling the characteristic function $f(\lambda) = u_2(1, \lambda, q)$ for the problem (5.45), the inverse problem can be written as the problem of solving the equations

$$u_2(1, \lambda_n, q) = 0 \quad \text{for all } n \in \mathbb{N} \tag{5.61}$$

for the function q. If we know only a finite number, say λ_n for $n = 1, \ldots, N$, then we assume that q is of the form

$$q(x; a) = \sum_{n=1}^{N} a_n q_n(x), \quad x \in [0, 1], \tag{5.62}$$

for coefficients $a = (a_1, \ldots, a_N) \in \mathbb{R}^N$ and some given linear independent even functions q_n. If q is expected to be smooth and periodic, a good choice for q_n is $q_n(x) = \cos(2\pi(n - 1)x)$, $n = 1, \ldots, N$. Equation (5.61) then reduces to the finite nonlinear system $F(a) = 0$, where $F : \mathbb{R}^N \to \mathbb{R}^N$ is defined by

$$F_n(a) := u_2(1, \lambda_n, q(\cdot; a)) \quad \text{for } a \in \mathbb{R}^N \text{ and } n = 1, \ldots, N.$$

Therefore, all of the well-developed methods for solving systems of nonlinear equations can be used. For example, Newton's method

$$a^{(k+1)} = a^{(k)} - F'(a^{(k)})^{-1} F(a^{(k)}), \quad k = 0, 1, \ldots,$$

is known to be quadratically convergent if $F'(a)^{-1}$ is regular. As we know from Section 5.2, Theorem 5.6, the mapping F is continuously Fréchet differentiable for every $a \in \mathbb{R}^N$. The computation of the derivative is rather expensive, and in general, it is not known if $F'(a)$ is regular. In [186], it was proven that $F'(a)$ is regular for sufficiently small a and is of triangular form for $a = 0$. This observation leads to the *simplified Newton method* of the form

$$a^{(k+1)} = a^{(k)} - F'(0)^{-1} F(a^{(k)}), \quad k = 0, 1, \ldots.$$

For further aspects of this method, we refer to [186].

Before we describe a second algorithm, we observe that from the asymptotic form (5.28) of the eigenvalues, we have an estimate of $\hat{q} = \int_0^1 q(x)\,dx$. Writing the differential equation in the form

$$-u_n''(x) + \left(q(x) - \hat{q}\right) u_n(x) = \left(\lambda_n - \hat{q}\right) u_n(x), \quad 0 \le x \le 1,$$

we observe that we can assume without loss of generality that $\int_0^1 q(x)\,dx = 0$.

Now we describe an algorithm that follows the idea of the uniqueness Theorem 5.22. We allow $q \in L^2(0,1)$ to be arbitrary. The algorithm consists of two steps. First, we recover the Cauchy data $f = K(1,\cdot)$ and $g = K_x(1,\cdot)$ from the two sets of eigenvalues. Then we suggest Newton-type methods to compute q from these Cauchy data.

The starting point is Theorem 5.19 for the case $p = 0$. We have already formulated this special case in Example 5.20. Therefore, let (λ_n, u_n) be the eigenvalues and eigenfunctions of the eigenvalue problem (5.45) normalized such that $u_n'(0) = 1$. The eigenvalues λ_n are assumed to be known. From Example 5.20, we have the representation

$$u_n(x) = \frac{\sin\sqrt{\lambda_n}x}{\sqrt{\lambda_n}} + \int_0^x K(x,t) \frac{\sin\sqrt{\lambda_n}t}{\sqrt{\lambda_n}}\,dt, \quad 0 \le x \le 1, \qquad (5.63)$$

where K satisfies (5.53a)–(5.53c) with $K(1,1) = \frac{1}{2}\int_0^1 q(t)\,dt = 0$. From (5.63) for $x = 1$, we can compute $K(1,t)$ because, by Theorem 5.21, the functions $v_n(t) = \sin\sqrt{\lambda_n}t$ form a complete system in $L^2(0,1)$. When we know only a finite number $\lambda_1, \ldots, \lambda_N$ of eigenvalues, we suggest representing $K(1,\cdot)$ as a finite sum of the form

$$K(1,t) = \sum_{k=1}^{N} a_k \sin(k\pi t),$$

arriving at the finite linear system

$$\sum_{k=1}^{N} a_k \int_0^1 \sin(k\pi t)\,\sin\sqrt{\lambda_n}t\,dt = -\sin\sqrt{\lambda_n} \quad \text{for } n = 1, \ldots, N. \qquad (5.64)$$

The same arguments yield a set of equations for the second boundary condition $u'(1) + H\,u(1) = 0$ in the form

$$\sqrt{\mu_n}\,\cos\sqrt{\mu_n} + H\,\sin\sqrt{\mu_n} + \int_0^1 \left(K_x(1,t) + H\,K(1,t)\right)\sin\sqrt{\mu_n}t\,dt = 0,$$

where now μ_n are the corresponding known eigenvalues. The representation

$$K_x(1,t) + H\,K(1,t) = \sum_{k=1}^{N} b_k \sin(k\pi t)$$

leads to the system

$$\sum_{k=1}^{N} b_k \int_0^1 \sin(k\pi t)\, \sin\sqrt{\mu_n}t\, dt \;=\; -\sqrt{\mu_n}\,\cos\sqrt{\mu_n} \;-\; H\,\sin\sqrt{\mu_n} \qquad (5.65)$$

for $n = 1, \ldots, N$. Equations (5.64) and (5.65) are of the same form and we restrict ourselves to the discussion of (5.64). Asymptotically, the matrix $A \in \mathbb{R}^{N \times N}$ defined by $A_{kn} = \int_0^1 \sin(k\pi t)\,\sin\sqrt{\lambda_n}t\, dt$ is just $\frac{1}{2}I$. More precisely, from Parseval's identity (see (A.8) from Theorem A.15 of AppendixA.2)

$$\sum_{k=1}^{\infty} \left| \int_0^1 \psi(t)\,\sin(k\pi t)\, dt \right|^2 \;=\; \frac{1}{2} \int_0^1 |\psi(t)|^2 dt$$

we conclude that (set $\psi(t) = \sin\sqrt{\lambda_n}t - \sin(n\pi t)$ for some $n \in \mathbb{N}$)

$$\sum_{k=1}^{\infty} \left| \int_0^1 \sin(k\pi t)\,\left[\sin\sqrt{\lambda_n}t - \sin(n\pi t)\right] dt \right|^2$$

$$= \frac{1}{2} \int_0^1 \left|\sin\sqrt{\lambda_n}t - \sin(n\pi t)\right|^2 dt \;\leq\; \frac{1}{2}\left|\sqrt{\lambda_n} - n\pi\right|^2$$

where we used the mean value theorem. The estimate (5.30) yields $|\lambda_n - n^2\pi^2| \leq \tilde{c}\|q\|_\infty$ and thus

$$\left|\sqrt{\lambda_n} - n\pi\right| \;\leq\; \frac{c}{n}\|q\|_\infty\,,$$

where c is independent of q and n. From this, we conclude that

$$\sum_{k=1}^{\infty} \left| \int_0^1 \sin(k\pi t)\,\left[\sin\sqrt{\lambda_n}t - \sin(n\pi t)\right] dt \right|^2 \;\leq\; \frac{c^2}{2\,n^2}\|q\|_\infty^2\,.$$

The matrix A is thus diagonally dominant, and therefore, invertible for sufficiently small $\|q\|_\infty$. Numerical experiments have shown that also for "large" values of q the numerical solution of (5.65) does not cause any problems.

We are now facing the following inverse problem: Given (approximate values of) the Cauchy data $f = K(1,\cdot) \in H_{00}^1(0,1)$ and $g = K_x(1,\cdot) \in L^2(0,1)$, compute $q \in L^2(0,1)$ such that the solution of the Cauchy problem (5.41a)–(5.41c) for $p = 0$ and $F = 0$ assumes the boundary data $K(x,x) = \frac{1}{2}\int_0^x q(t)\, dt$ for $x \in [0,1]$. An alternative way of formulating the inverse problem is to start with the Goursat problem (5.53a)–(5.53c): Compute $q \in L^2(0,1)$ such that the solution of the initial value problem (5.53a)–(5.53c) has Cauchy data $f(t) = K(1,t)$ and $g(t) = K_x(1,t)$ for $t \in [0,1]$.

We have studied these coupled systems for K and q in Theorem 5.18. Here we apply it for the case where $F = 0$. It has been shown that the pair (K, r) solves the system

$$\frac{\partial^2 K(x,t)}{\partial x^2} - \frac{\partial^2 K(x,t)}{\partial t^2} - q(x)\,K(x,t) = 0 \quad \text{in } \Delta_0\,,$$

$$K(x,x) = \frac{1}{2}\int_0^x r(t)\,dt\,, \quad 0 \le x \le 1\,,$$

$$K(x,0) = 0\,, \quad 0 \le x \le 1\,,$$

and

$$K(1,t) = f(t) \quad \text{and} \quad \frac{\partial}{\partial x}K(1,t) = g(t) \quad \text{for all } t \in [0,1]$$

if and only if $w(\xi,\eta) = K(\xi + \eta, \xi - \eta)$ and r solve the system of integral equations (5.44a) and (5.44b). For this special choice of F, (5.44b) reduces to

$$\frac{1}{2}r(x) = -\int_x^1 q(s)\,K(s, 2x - s)\,ds + g(2x - 1) + f'(2x - 1)\,, \qquad (5.66)$$

where we have extended f and g to odd functions on $[-1, 1]$. Denote by $T(q)$ the expression on the right-hand side of (5.66). For the evaluation of $T(q)$, one has to solve the Cauchy problem (5.41a)–(5.41c) for $p = 0$. Note that the solution K; that is, the kernel $K(y, 2x - y)$ of the integral operator T, also depends on q. The operator T is therefore nonlinear!

The requirement $r = q$ leads to a fixed point equation $q = 2T(q)$ in $L^2(0,1)$. It was shown in [233] that there exists at most one fixed point $q \in L^\infty(0,1)$ of T. Even more, Rundell and Sachs proved that the projected operator $P_M T$ is a contraction on the ball $B_M := \{q \in L^\infty(0,1) : \|q\|_\infty \le M\}$ with respect to some weighted L^∞-norms. Here, P_M denotes the projection onto B_M defined by

$$(P_M q)(x) = \begin{cases} q(x), & |q(x)| \le M, \\ M\,\mathrm{sign}\,q(x), & |q(x)| > M. \end{cases}$$

Also, they showed the effectiveness of the iteration method $q^{(k+1)} = 2T(q^{(k)})$ by several numerical examples. We observe that for $q^{(0)} = 0$ the first iterate $q^{(1)}$ is simply $q^{(1)}(x) = 2\,g(2x - 1) + 2\,f'(2x - 1)$, $x \in [0,1]$. We refer to [233] for more details.

As suggested earlier, an alternative numerical procedure based on the kernel function K is to define the operator S from $L^2(0,1)$ into $H^1_{00}(0,1) \times L^2(0,1)$ by $S(q) = \big(K(1, \cdot), K_x(1, \cdot)\big)$, where K solves the Goursat problem (5.53a)–(5.53c) in the weak sense. This operator is well-defined and bounded by Theorem 5.15, part (c). If $f \in H^1_{00}(0,1)$ and $g \in L^2(0,1)$ are the given Cauchy values $K(1, \cdot)$

and $K_x(1, \cdot)$, respectively, then we have to solve the nonlinear equation $S(q) = (f, g)$. Newton's method does it by the iteration procedure

$$q^{(k+1)} = q^{(k)} - S'(q^{(k)})^{-1}[S(q^{(k)}) - (f, g)], \quad k = 0, 1, \ldots. \qquad (5.67)$$

For the implementation, one has to compute the Fréchet derivative of S. Using the Volterra equation (5.40) derived in the proof of Theorem 5.15, it is not difficult to prove that S is Fréchet differentiable and that $S'(q)r = (W(1, \cdot), W_x(1, \cdot))$, where W solves the inhomogeneous Goursat problem

$$W_{xx}(x, t) - W_{tt}(x, t) - q(x) W(x, t) = K(x, t) r(x) \quad \text{in } \Delta_0, \qquad (5.68a)$$

$$W(x, 0) = 0, \quad 0 \le x \le 1, \qquad (5.68b)$$

$$W(x, x) = \frac{1}{2} \int_0^x r(t) \, dt, \quad 0 \le x \le 1. \qquad (5.68c)$$

In part (b) of Theorem 5.18, we showed that $S'(q)$ is an isomorphism. We reformulate this result.

Theorem 5.27 *Let $q \in L^2(0, 1)$ and K be the weak solution of (5.53a)–(5.53c). For every $f \in H_{00}^1(0, 1)$ and $g \in L^2(0, 1)$, there exists a unique $r \in L^2(0, 1)$ and a weak solution W of (5.68a)–(5.68c) with $W(1, \cdot) = f$ and $W_x(1, \cdot) = g$ in the sense of (5.44a), (5.44b); that is, $S'(q)$ is an isomorphism.*

Implementing Newton's method is quite expensive because in every step one has to solve a coupled system of the form (5.68a)–(5.68c). Rundell and Sachs suggested a simplified Newton method of the form

$$q^{(k+1)} = q^{(k)} - S'(0)^{-1}[S(q^{(k)}) - (f, g)], \quad k = 0, 1, \ldots.$$

Because $S(0) = 0$, we can invert the linear operator $S'(0)$ analytically. In particular, we have $S'(0)r = (W(1, \cdot), W_x(1, \cdot))$, where W now solves

$$W_{xx}(x, t) - W_{tt}(x, t) = 0 \quad \text{in } \Delta_0,$$

$$W(x, 0) = 0, \quad \text{and} \quad W(x, x) = \frac{1}{2} \int_0^x r(t) \, dt, \quad 0 \le x \le 1,$$

because also $K = 0$. The solution W of the Cauchy problem

$$W_{xx}(x, t) - W_{tt}(x, t) = 0 \quad \text{in } \Delta_0,$$

$$W(1, t) = f(t), \quad \text{and} \quad W_x(1, t) = g(t), \quad 0 \le t \le 1,$$

is given by

$$W(x, t) = -\frac{1}{2} \int_{t-(1-x)}^{t+(1-x)} g(\tau) \, d\tau + \frac{1}{2} f(t + (1-x)) + \frac{1}{2} f(t - (1-x)),$$

where we have extended f and g to odd functions again. The solution r of $S'(0)r = (f, g)$ is, therefore, given by

$$r(x) \;=\; 2\frac{d}{dx}W(x,x) \;=\; 2f'(2x-1) \,+\, 2g(2x-1).$$

In this chapter, we have studied only one particular inverse eigenvalue problem. Similar theoretical results and constructive algorithms can be obtained for other inverse spectral problems; see [4, 16]. For an excellent overview, we refer to the lecture notes by W. Rundell [232].

5.8   Problems

5.1 Let $q, f \in C[0,1]$ and $q(x) \geq 0$ for all $x \in [0,1]$.

  (a) Show that the following boundary value problem on $[0,1]$ has at most one solution $u \in C^2[0,1]$:

$$-u''(x) + q(x)\,u(x) \;=\; f(x), \quad u(0) \;=\; u(1) \;=\; 0. \qquad (5.69)$$

  (b) Let v_1 and v_2 be the solutions of the following initial value problems on $[0,1]$:

$$\begin{aligned}
-v_1''(x) + q(x)\,v_1(x) &= 0, \quad v_1(0) = 0,\ v_1'(0) = 1, \\
-v_2''(x) + q(x)\,v_2(x) &= 0, \quad v_2(1) = 0,\ v_2'(1) = 1.
\end{aligned}$$

Show that the Wronskian $v_1'v_2 - v_2'v_1$ is constant. Define the following function for some $a \in \mathbb{R}$:

$$G(x,y) \;=\; \begin{cases} a\,v_1(x)\,v_2(y), & 0 \leq x \leq y \leq 1, \\ a\,v_2(x)\,v_1(y), & 0 \leq y < x \leq 1. \end{cases}$$

Determine $a \in \mathbb{R}$ such that

$$u(x) \;:=\; \int_0^1 G(x,y)\,f(y)\,dy, \quad x \in [0,1],$$

solves (5.69).

The function G is called *Green's function* of the boundary value problem (5.69).

  (c) Show that the eigenvalue problem

$$-u''(x) + q(x)\,u(x) \;=\; \lambda\,u(x), \quad u(0) \;=\; u(1) \;=\; 0,$$

is equivalent to the eigenvalue problem for the integral equation

$$\frac{1}{\lambda}u(x) \;=\; \int_0^1 G(x,y)\,u(y)\,dy, \quad x \in [0,1].$$

Prove Theorem 5.7, parts (a) and (b) by the general spectral theorem (Theorem A.53 of Appendix A.6).

(d) How can one treat the case of part (c) when q changes sign?

5.2 Let $H \in \mathbb{R}$. Prove that the transcendental equation $z \cot z + H = 0$ has a countable number of zeros z_n and that

$$z_n = (n + 1/2)\pi + \frac{H}{(n + 1/2)\pi} + \mathcal{O}(1/n^2).$$

From this,

$$z_n^2 = (n + 1/2)^2 \pi^2 + 2H + \mathcal{O}(1/n)$$

follows. *Hint:* Make the substitution $z = x + (n + 1/2)\pi$, set $\varepsilon = 1/(n + 1/2)\pi$, write $z \cot z + H = 0$ in the form $f(x, \varepsilon) = 0$, and apply the implicit function theorem.

5.3 Prove Lemma 5.13.

5.4 Let $q \in C[0,1]$ be real- or complex-valued and λ_n, g_n be the eigenvalues and L^2-normalized eigenfunctions, respectively, corresponding to q and boundary conditions $u(0) = 0$ and $hu'(1) + Hu(1) = 0$. Show by modifying the proof of Theorem 5.21 that $\{g_n : n \in \mathbb{N}\}$ is complete in $L^2(0,1)$. This gives—even for real q—a proof different from the one obtained by applying the general spectral theory.

5.5 Consider the eigenvalue problem on $[0,1]$:

$$-u''(x) + q(x)u(x) = \lambda u(x), \quad u(0) = u'(1) = 0.$$

By modifying the proof of Theorem 5.22, prove the following uniqueness result for the inverse problem: Let (λ_n, u_n) and (μ_n, v_n) be the eigenvalues and eigenfunctions corresponding to p and q, respectively. If $\lambda_n = \mu_n$ for all $n \in \mathbb{N}$ and

$$\frac{u_n(1)}{u'_n(0)} = \frac{v_n(1)}{v'_n(0)} \quad \text{for all } n \in \mathbb{N},$$

then p and q coincide.

Chapter 6

An Inverse Problem in Electrical Impedance Tomography

6.1 Introduction

Electrical impedance tomography (EIT) is a medical imaging technique in which an image of the conductivity (or permittivity) of part of the body is determined from electrical surface measurements. Typically, conducting electrodes are attached to the skin of the subject and small alternating currents are applied to some or all of the electrodes. The resulting electrical potentials are measured, and the process may be repeated for numerous different configurations of applied currents.

Applications of EIT as an imaging tool can be found in fields such as medicine (monitoring of the lung function or the detection of skin cancer or breast cancer), geophysics (locating of underground deposits, detection of leaks in underground storage tanks), or nondestructive testing (determination of cracks in materials).

To derive the EIT model, we start from the time-harmonic Maxwell system in the form

$$\operatorname{curl} H + (i\omega\varepsilon - \gamma)E = 0, \quad \operatorname{curl} E - i\omega\mu H = 0$$

in some domain which we take as a cylinder of the form $B \times \mathbb{R} \subset \mathbb{R}^3$ with bounded cross-section $B \subset \mathbb{R}^2$. Here, ω, ε, γ, and μ denote the frequency, electric permittivity, conductivity, and magnetic permeability, respectively, which are all assumed to be constant along the axis of the cylinder; that is, depend on x_1 and x_2 only. We note that the real parts $\operatorname{Re}\big[\exp(-i\omega t)\,E(x)\big]$ and $\operatorname{Re}\big[\exp(-i\omega t)\,H(x)\big]$ are the physically meaningful electric and magnetic field, respectively. For low frequencies ω (i.e., for small $(\omega\mu\gamma)\cdot L^2$ where L is a typical

© Springer Nature Switzerland AG 2021
A. Kirsch, *An Introduction to the Mathematical Theory of Inverse Problems*,
Applied Mathematical Sciences 120,
https://doi.org/10.1007/978-3-030-63343-1_6

length scale of B), one can show (see, e.g., [47]) that the Maxwell system is approximated by

$$\text{curl } H - \gamma E = 0, \quad \text{curl } E = 0.$$

The second equation yields the existence[1] of a scalar potential u such that $E = -\nabla u$. Substituting this into the first equation and taking the divergence yields $\text{div}(\gamma \nabla u) = 0$ in the cylinder. We restrict ourselves to the two-dimensional case and consider the conductivity equation

$$\text{div}(\gamma \nabla u) = 0 \quad \text{in } B. \tag{6.1}$$

There are several possibilities for modeling the attachment of the electrodes on the boundary ∂B of B. The simplest of these is the *continuum model* in which the potential $U = u|_{\partial B}$ and the boundary current distribution $f = \gamma \nabla u \cdot \nu = \gamma \, \partial u / \partial \nu$ are both given on the boundary ∂B. Here, $\nu = \nu(x)$ is the unit normal vector at $x \in \partial B$ directed into the exterior of B. First, we observe that,[2] by the divergence theorem,

$$0 = \int_B \text{div}(\gamma \nabla u) \, dx = \int_{\partial B} \gamma \frac{\partial u}{\partial \nu} \, d\ell = \int_{\partial B} f \, d\ell;$$

that is, the boundary current distribution f has zero mean. In practice, $f(x)$ is not known for all $x \in \partial B$. One actually knows the currents sent along wires attached to N discrete electrodes that in turn are attached to the boundary ∂B. Therefore, in the *gap model* one approximates f by assuming that f is constant at the surface of each electrode and zero in the gaps between the electrodes. An even better choice is the *complete model*. Suppose that f_j is the electric current sent through the wire attached to the jth electrode. At the surface S_j of this electrode, the current density satisfies

$$\int_{S_j} \gamma \frac{\partial u}{\partial \nu} \, d\ell = f_j.$$

In the gaps between the electrodes, we have

$$\gamma \frac{\partial u}{\partial \nu} = 0 \quad \text{in } \partial B \setminus \bigcup_j S_j.$$

If electrochemical effects at the contact of S_j with ∂B are taken into account, the Dirichlet boundary condition $u = U_j$ on S_j is replaced by

$$u + z_j \gamma \frac{\partial u}{\partial \nu} = U_j \quad \text{on } S_j,$$

where z_j denotes the surface impedance of the jth electrode. We refer to [20, 145, 146, 252] for a discussion of these electrode models and the well-posedness of the corresponding boundary value problems (for given γ).

---

[1] If the domain is simply connected.
[2] Provided γ, f, and u are smooth enough.

In the *inverse problem of EIT* the conductivity function γ is unknown and has to be determined from simultaneous measurements of the boundary voltages U and current densities f, respectively.

In this introductory chapter on EIT, we restrict ourselves to the continuum model as the simplest electrode model. We start with the precise mathematical formulation of the direct and the inverse problem and prove well-posedness of the direct problem: existence, uniqueness, and continuous dependence on both the boundary data f and the conductivity γ. Then we consider the inverse problem of EIT. The question of uniqueness is addressed, and we prove uniqueness of the inverse linearized problem. This problem is interesting also from an historical point of view because the proof, given in Calderón's fundamental paper [38], has influenced research on inverse medium problems monumentally. In the last section, we introduce a technique to determine the *support* of the contrast $\gamma - \gamma_1$ where γ_1 denotes the known background conductivity. This *factorization method* has been developed fairly recently—after publication of the first edition of this monograph—and is a prominent member of a whole class of newly developed methods subsumed under the name *Sampling Methods*.

6.2 The Direct Problem and the Neumann–Dirichlet Operator

Let $B \subset \mathbb{R}^2$ be a given bounded domain with boundary ∂B and $\gamma : B \to \mathbb{R}$ and $f : \partial B \to \mathbb{R}$ be given real-valued functions. The direct problem is to determine u such that

$$\mathrm{div}\,(\gamma \nabla u) \;=\; 0 \quad \text{in } B\,, \qquad \gamma \frac{\partial u}{\partial \nu} \;=\; f \quad \text{on } \partial B\,. \qquad (6.2)$$

Throughout this chapter, $\nu = \nu(x)$ again denotes the exterior unit normal vector at $x \in \partial B$. As mentioned in the introduction, we have to assume that $\int_{\partial B} f \, d\ell = 0$. Therefore, throughout this chapter, we make the following assumptions on B, γ, and f:

Assumption 6.1   *(a) $B \subset \mathbb{R}^2$ is a bounded Lipschitz domain[3] such that the exterior of B is connected.*

(b) $\gamma \in L^\infty(B)$, and there exists $\gamma_0 > 0$ such that $\gamma(x) \geq \gamma_0$ for almost all $x \in B$.

(c) $f \in L^2_\diamond(\partial B)$ where

$$L^2_\diamond(\partial B) \;=\; \left\{ f \in L^2(\partial B) : \int_{\partial B} f \, d\ell = 0 \right\}.$$

---

[3]For a definition see, e.g., [191, 161].

In this chapter and the following one on scattering theory, we have to use Sobolev spaces as the appropriate functions spaces for the solution. For a general theory on Sobolev spaces, we refer to the standard literature such as [1, 191]. For obvious reasons, we also refer to the monograph [161], Sections 4.1 and 5.1. Also, in Appendix A.5, we introduce and study the Sobolev space $H^1(B)$ for the particular case of B being the unit disc. At this place we recall only the very basic definition.

For any open and bounded set $B \subset \mathbb{R}^2$ the Sobolev space $H^1(B)$ is defined as the completion of $C^1(\overline{B})$ with respect to the norm

$$\|u\|_{H^1(B)} = \sqrt{\int_B \left[|\nabla u|^2 + |u|^2 \right] dx} \, .$$

An important property of Sobolev spaces for Lipschitz domains is the existence of traces; that is, for every $u \in H^1(B)$ the trace $u|_{\partial B}$ on ∂B is well-defined and represents an $L^2(\partial B)$-function[4] (Theorem 5.10 in [161]). Also, there exists $c_T > 0$ (independent of u) such that $\|u|_{\partial B}\|_{L^2(\partial B)} \leq c_T \|u\|_{H^1(B)}$ for all $u \in H^1(B)$; that is, the trace operator $u \mapsto u|_{\partial B}$ is bounded.

We note that the solution u of (6.2) is only unique up to an additive constant. Therefore, we normalize the solution $u \in H^1(B)$ such that it has vanishing mean on the boundary; that is, $u \in H^1_\diamond(B)$ where

$$H^1_\diamond(B) = \left\{ u \in H^1(B) : \int_{\partial B} u \, d\ell = 0 \right\}. \tag{6.3}$$

The formulation (6.2) of the boundary value problem has to be understood in the variational (or weak) sense. By multiplying the first equation of (6.2) with some test function ψ and using Green's first formula we arrive at

$$\begin{aligned} 0 &= \int_B \psi \operatorname{div}(\gamma \nabla u) \, dx = -\int_B \gamma \nabla \psi \cdot \nabla u \, dx + \int_{\partial B} \psi \gamma \nabla u \cdot \nu \, d\ell \\ &= -\int_B \gamma \nabla \psi \cdot \nabla u \, dx + \int_{\partial B} \psi f \, d\ell \, . \end{aligned}$$

Therefore, we *define* the variational solution $u \in H^1_\diamond(B)$ of (6.2) by the solution of

$$\int_B \gamma \nabla \psi \cdot \nabla u \, dx = \int_{\partial B} \psi f \, d\ell \quad \text{for all } \psi \in H^1_\diamond(B) \, . \tag{6.4}$$

Existence and uniqueness follows from the representation theorem due to Riesz (cf. Theorem A.23 of Appendix A.3).

Theorem 6.2 *Let Assumption 6.1 be satisfied. For every $f \in L^2_\diamond(\partial B)$ there exists a unique variational solution $u \in H^1_\diamond(B)$ of (6.2), that is, a solution*

---

[4]The trace is even more regular and belongs to the fractional Sobolev space $H^{1/2}(\partial B)$.

of the variational equation (6.4). Furthermore, there exists a constant $c > 0$ (independent of f) such that $\|u\|_{H^1(B)} \leq c\|f\|_{L^2(\partial B)}$. In other words: the operator $f \mapsto u$ from $L^2_\diamond(\partial B)$ to $H^1_\diamond(B)$ is bounded.

Proof: We define a new inner product in the space $H^1_\diamond(B)$ by

$$(u, v)_* = \int_B \gamma \, \nabla u \cdot \nabla v \, dx, \quad u, v \in H^1_\diamond(B).$$

The corresponding norm $\|u\|_* = \sqrt{(u, u)_*}$ is equivalent to the ordinary norm $\|\cdot\|_{H^1(B)}$ in $H^1_\diamond(B)$. This follows from Friedrich's inequality in the form (see Theorem A.50 for the case of B being the unit disc and [1, 191] for more general Lipschitz domains):

There exists $c_F > 0$ such that

$$\|v\|_{L^2(B)} \leq c_F \|\nabla v\|_{L^2(B)} \quad \text{for all } v \in H^1_\diamond(B). \tag{6.5}$$

Indeed, from this the equivalence follows inasmuch as

$$\frac{\gamma_0}{1 + c_F^2} \|v\|^2_{H^1(B)} \leq \gamma_0 \|\nabla v\|^2_{L^2(B)} \leq \|v\|^2_* \leq \|\gamma\|_\infty \|v\|^2_{H^1(B)} \tag{6.6}$$

for all $v \in H^1_\diamond(B)$. For fixed $f \in L^2_\diamond(\partial B)$ we can interpret the right-hand side of (6.4) as a linear functional F on the space $H^1_\diamond(B)$; that is, $F(\psi) = (f, \psi)_{L^2(\partial B)}$ for $\psi \in H^1_\diamond(B)$. This functional F is bounded by the inequality of Cauchy-Schwarz and the trace theorem (with constant $c_T > 0$) because

$$\begin{aligned} |F(\psi)| &\leq \|f\|_{L^2(\partial B)} \|\psi\|_{L^2(\partial B)} \leq c_T \|f\|_{L^2(\partial B)} \|\psi\|_{H^1(B)} \\ &\leq c\|f\|_{L^2(\partial B)} \|\psi\|_* \end{aligned}$$

with $c = c_T \sqrt{(1 + c_F^2)/\gamma_0}$. In particular, $\|F\|_{H^1_\diamond(B)^*} \leq c\|f\|_{L^2(\partial B)}$, and we can apply the representation theorem of Riesz in the Hilbert space $\left(H^1_\diamond(B), (\cdot, \cdot)_*\right)$: there exists a unique $u \in H^1_\diamond(B)$ with $(u, \psi)_* = F(\psi)$ for all $\psi \in H^1_\diamond(B)$. This is exactly the variational equation (6.4). Furthermore, $\|u\|_* = \|F\|_{H^1_\diamond(B)^*}$ and thus by (6.6),

$$\|u\|^2_{H^1(B)} \leq \frac{1 + c_F^2}{\gamma_0} \|u\|^2_* = \frac{1 + c_F^2}{\gamma_0} \|F\|^2_{H^1_\diamond(B)^*} \leq c^2 \frac{1 + c_F^2}{\gamma_0} \|f\|^2_{L^2(\partial B)};$$

that is, the operator $f \mapsto u$ is bounded from $L^2_\diamond(\partial B)$ into $H^1_\diamond(B)$. $\qquad\square$

This theorem implies the existence and boundedness of the Neumann–Dirichlet operator.

Definition 6.3

The Neumann–Dirichlet operator $\Lambda : L^2_\diamond(\partial B) \to L^2_\diamond(\partial B)$ *is defined by* $\Lambda f = u|_{\partial B}$, *where* $u \in H^1_\diamond(B)$ *is the uniquely determined variational solution of (6.2); that is, the solution of (6.4).*

Remark: This operator is bounded by the boundedness of the solution map $f \mapsto u$ from $L^2_\diamond(\partial B)$ to $H^1_\diamond(B)$ and the boundedness of the trace operator from $H^1_\diamond(B)$ to $L^2_\diamond(\partial B)$. It is even compact because the trace operator is compact from $H^1_\diamond(B)$ to $L^2_\diamond(\partial B)$. However, we do not make use of this latter property.

We show some properties of the Neumann–Dirichlet operator.

Theorem 6.4 *Let Assumption 6.1 be satisfied. Then the Neumann–Dirichlet map Λ is self-adjoint and positive; that is, $(\Lambda f, g)_{L^2(\partial B)} = (f, \Lambda g)_{L^2(\partial B)}$ and $(\Lambda f, f)_{L^2(\partial B)} > 0$ for all $f, g \in L^2_\diamond(\partial B)$, $f \neq 0$.*

Proof: This follows simply from the definition of Λ and Green's first identity. Let $u, v \in H^1_\diamond(B)$ be the solutions of (6.4) corresponding to boundary data f and g, respectively. Then, by (6.4) for the pair g, v and the choice $\psi = u$ (note that $u|_{\partial B} = \Lambda f$),

$$(\Lambda f, g)_{L^2(\partial B)} = \int_{\partial B} u\, g\, d\ell = \int_B \gamma\, \nabla u \cdot \nabla v\, dx,$$

and this term is symmetric with respect to u and v. For $f = g$ this also yields that Λ is positive.   □

In the following, we write Λ_γ to indicate the dependence on γ. The next interesting property is a monotonicity result.

Theorem 6.5 *Let $\gamma_1, \gamma_2 \in L^\infty(B)$ with $\gamma_1 \geq \gamma_2 \geq \gamma_0$ a.e. on B. Then $\Lambda_{\gamma_1} \leq \Lambda_{\gamma_2}$ in the sense that*

$$\left(\Lambda_{\gamma_1} f, f\right)_{L^2(\partial B)} \leq \left(\Lambda_{\gamma_2} f, f\right)_{L^2(\partial B)} \quad \text{for all } f \in L^2_\diamond(\partial B).$$

Proof: For fixed $f \in L^2_\diamond(\partial B)$ let $u_j \in H^1_\diamond(B)$ be the corresponding solution of (6.2) for γ_j, $j = 1, 2$. From (6.4) (for $\gamma = \gamma_2$, $u = u_2$, and $\psi = u_1 - u_2$) we conclude that

$$
\begin{aligned}
\left((\Lambda_{\gamma_1} - \Lambda_{\gamma_2})f, f\right)_{L^2(\partial B)} &= \int_{\partial B} (u_1 - u_2)\, f\, d\ell = \int_B \gamma_2 \left(\nabla u_1 - \nabla u_2\right) \cdot \nabla u_2\, dx \\
&= \frac{1}{2} \int_B \gamma_2 \left[|\nabla u_1|^2 - |\nabla u_2|^2 - |\nabla(u_1 - u_2)|^2\right] dx \\
&\leq \frac{1}{2} \int_B \gamma_2 |\nabla u_1|^2\, dx - \frac{1}{2} \int_B \gamma_2 |\nabla u_2|^2\, dx \\
&\leq \frac{1}{2} \int_B \gamma_1 |\nabla u_1|^2\, dx - \frac{1}{2} \int_B \gamma_2 |\nabla u_2|^2\, dx \\
&= \frac{1}{2} \left((\Lambda_{\gamma_1} - \Lambda_{\gamma_2})f, f\right)_{L^2(\partial B)}
\end{aligned}
$$

which proves the result.   □

6.3   The Inverse Problem

As described in the introduction, the problem of electrical impedance tomography is to determine (properties of) the conductivity distribution γ from all
—or at least a large number of—pairs $(f, u|_{\partial B})$. Because $u|_{\partial B} = \Lambda f$ we can
rephrase this problem as follows:

Inverse Problem: Determine the conductivity γ from the given Neumann–
Dirichlet operator $\Lambda_\gamma : L_\diamond^2(\partial B) \to L_\diamond^2(\partial B)$!

As we have seen already in the previous chapter (and this is typical for
studying inverse problems), an intensive investigation of the direct problem
has to precede the treatment of the inverse problem. In particular, we study
the dependence of the Neumann–Dirichlet map on γ. First, we show with an
example that the inverse problem of impedance tomography is ill-posed.

Example 6.6 Let $B = B(0, 1)$ be the unit disk, $\hat{q} > 0$ constant, and $R \in (0, 1)$.
We define $\gamma_R \in L^\infty(B)$ by

$$\gamma_R(x) = \begin{cases} 1, & R < |x| < 1, \\ 1 + \hat{q}, & |x| < R. \end{cases}$$

Because γ_R is piecewise constant the solution $u \in H_\diamond^1(B)$ is a *harmonic function*
in $B(0, 1) \setminus \{x : |x| = R\}$ (that is, $\Delta u = 0$ in $B(0, 1) \setminus \{x : |x| = R\}$) and satisfies
the jump conditions $u_- = u_+$ and $(1 + \hat{q}) \partial u / \partial r|_- = \partial u / \partial r|_+$ for $|x| = R$ where
$v|_\pm$ denotes the trace of v from the interior $(-)$ and exterior $(+)$ of $\{x : |x| = R\}$,
respectively. We refer to Problem 6.1 for a justification of this statement.

We solve the boundary value problem (6.2) by expanding the boundary data
$f \in L_\diamond^2(\partial B)$ and the solution u into Fourier series; that is, for

$$f(\varphi) = \sum_{n \neq 0} f_n e^{in\varphi}, \quad \varphi \in [0, 2\pi],$$

we make an ansatz for the solution of (6.2) in the form

$$u(r, \varphi) = \begin{cases} \sum_{n \neq 0} (b_n + c_n) \left(\frac{r}{R}\right)^{|n|} e^{in\varphi}, & r < R, \\ \sum_{n \neq 0} \left[b_n \left(\frac{r}{R}\right)^{|n|} + c_n \left(\frac{r}{R}\right)^{-|n|}\right] e^{in\varphi}, & r > R. \end{cases}$$

The ansatz already guarantees that u is continuous on the circle $r = R$. The
unknown coefficients b_n, c_n are determined from the conditions $(1 + \hat{q}) \partial u / \partial r|_- = \partial u / \partial r|_+$ for $r = R$ and $\partial u / \partial r = f$ for $r = 1$. This yields the set of equations

$$(1 + \hat{q}) (b_n + c_n) = b_n - c_n$$

and

$$b_n \frac{|n|}{R^{|n|}} - c_n |n| R^{|n|} = f_n$$

for all $n \neq 0$ which yields explicit formulas for b_n and c_n. Substituting this into the form of u and taking $r = 1$ yields

$$(\Lambda_{\gamma_R} f)(\varphi) \;=\; u(1, \varphi) \;=\; \sum_{n \neq 0} \frac{\alpha - R^{2|n|}}{\alpha + R^{2|n|}} \frac{f_n}{|n|} e^{in\varphi}, \quad \varphi \in [0, 2\pi], \qquad (6.7)$$

with $\alpha = 1 + 2/\hat{q}$. We observe that Λ_{γ_R} is a diagonal operator from $L_\diamond^2(0, 2\pi)$ into itself with eigenvalues that behave asymptotically as $1/|n|$. Therefore, the natural setting for Λ_{γ_R} is to consider it as an operator from the Sobolev space $H_\diamond^{-1/2}(0, 2\pi)$ of order $-1/2$ into the Sobolev space $H_\diamond^{1/2}(0, 2\pi)$ of order $1/2$; see Section A.4 of Appendix A. We prefer the setting in $L_\diamond^2(0, 2\pi)$ because the more general setting does not give any more insight with respect to the inverse problem.

Let Λ_1 be the operator with $\gamma = 1$ which is given by

$$(\Lambda_1 f)(\varphi) \;=\; \sum_{n \neq 0} \frac{f_n}{|n|} e^{in\varphi}, \quad \varphi \in [0, 2\pi].$$

We estimate the difference by

$$\|(\Lambda_{\gamma_R} - \Lambda_1) f\|_{L^2(0,2\pi)}^2 \;=\; 2\pi \sum_{n \neq 0} \left| \frac{\alpha - R^{2|n|}}{\alpha + R^{2|n|}} - 1 \right|^2 \frac{|f_n|^2}{n^2}$$

$$=\; 8\pi \sum_{n \neq 0} \frac{R^{4|n|}}{\left(\alpha + R^{2|n|}\right)^2} \frac{|f_n|^2}{n^2}$$

$$\leq\; \frac{4 R^4}{\alpha^2} \|f\|_{L^2(0,2\pi)}^2 \,;$$

that is,

$$\|\Lambda_{\gamma_R} - \Lambda_1\|_{\mathcal{L}(L^2(0,2\pi))} \;\leq\; \frac{2R^2}{\alpha} \;\leq\; 2R^2$$

because $\alpha \geq 1$. Therefore, we have convergence of Λ_{γ_R} to Λ_1 in the operator norm as R tends to zero. On the other hand, the difference $\|\gamma_R - 1\|_\infty = \hat{q}$ is constant and does not converge to zero as R tends to zero. This shows clearly that the inverse problem to determine γ_R from Λ is ill-posed.

One can argue that perhaps the sup-norm for γ is not appropriate to measure the error in γ. Our example, however, shows that even if we replace $\hat{q}$ by a constant $\hat{q}_R$ which depends on R such that $\lim_{R \to 0} \hat{q}_R = \infty$ we still have convergence of Λ_{γ_R} to Λ_1 in the operator norm as R tends to zero. Taking, for example, $\hat{q}_R = \hat{q}/R^3$, we observe that $\|\gamma_R - 1\|_{L^p(B)} \to \infty$ as R tends to zero for arbitrary $p \geq 1$, and the problem of impedance tomography is also ill-posed with respect to any L^p-norm.

A fundamental question for every inverse problem is the question of *unique-ness*: is the information—at least in principle—sufficient to determine the unknown quantity? Therefore, in electrical impedance tomography, we ask: does the knowledge of the Neumann–Dirichlet operator Λ determine the conductivity γ uniquely or is it possible that two different γ correspond to the same Λ?

In full generality, this fundamental question was not answered until 2006 by K. Astala and L. Päivärinta in [10]. We state the result without proof.

Theorem 6.7 *Let $\gamma_1, \gamma_2 \in L^\infty(B)$ with $\gamma_j(x) \geq \gamma_0$ for $j = 1, 2$ and almost all $x \in B$. We denote the corresponding Neumann–Dirichlet operators by Λ_1 and Λ_2, respectively. If $\Lambda_1 = \Lambda_2$ then $\gamma_1 = \gamma_2$ in B.*

Instead of proving this theorem which uses refined arguments from complex analysis, we consider the linearized problem. Therefore, writing $\Lambda(\gamma)$ instead of Λ_γ to indicate the dependence on γ, we consider the linear problem

$$\Lambda(\gamma) \,+\, \Lambda'(\gamma)q \;=\; \Lambda_{meas}\,, \tag{6.8}$$

where $\Lambda'(\gamma) : L^\infty(B) \to \mathcal{L}\big(L^2(\partial B)\big)$ denotes the Fréchet derivative of the nonlinear operator $\gamma \mapsto \Lambda(\gamma)$ from $L^\infty(B)$ to $\mathcal{L}\big(L_\diamond^2(\partial B)\big)$ at γ. Here, $\mathcal{L}\big(L_\diamond^2(\partial B)\big)$ denotes again the space of all linear and bounded operators from $L_\diamond^2(\partial B)$ into itself equipped with the operator norm. The right-hand side $\Lambda_{meas} \in \mathcal{L}\big(L^2(\partial B)\big)$ is given ("measured"), and the contrast $q \in L^\infty(B)$ has to be determined.

Theorem 6.8
Let $U \subset L^\infty(B)$ be given by $U = \big\{\gamma \in L^\infty(B) : \gamma \geq \gamma_0 \text{ a.e. on } B\big\}$.

(a) *The mapping $\gamma \mapsto \Lambda(\gamma)$ from U to $\mathcal{L}\big(L_\diamond^2(\partial B)\big)$ is Lipschitz continuous.*

(b) *The mapping $\gamma \mapsto \Lambda(\gamma)$ from U to $\mathcal{L}\big(L_\diamond^2(\partial B)\big)$ is Fréchet differentiable. The Fréchet derivative $\Lambda'(\gamma)$ at $\gamma \in U$ in the direction $q \in L^\infty(B)$ is given by $\big[\Lambda'(\gamma)q\big]f = v|_{\partial B}$ where $v \in H_\diamond^1(B)$ solves*

$$\operatorname{div}\big(\gamma \nabla v\big) = -\operatorname{div}\big(q \nabla u\big) \text{ in } B\,, \quad \gamma \frac{\partial v}{\partial \nu} = -q \frac{\partial u}{\partial \nu} \text{ on } \partial B\,, \tag{6.9}$$

and $u \in H_\diamond^1(B)$ solves (6.2) with data $\gamma \in U$ and $f \in L_\diamond^2(\partial B)$. The solution of (6.9) is again understood in the weak sense; that is,

$$\int_B \gamma \nabla \psi \cdot \nabla v \, dx \;=\; -\int_B q \nabla \psi \cdot \nabla u \, dx \quad \text{for all } \psi \in H_\diamond^1(B)\,. \tag{6.10}$$

Proof: (a) Let $\gamma_1, \gamma_2 \in U$, $f \in L_\diamond^2(\partial B)$, and $u_1, u_2 \in H_\diamond^1(B)$ be the corresponding weak solutions of (6.2). Taking the difference of (6.4) for the triples (γ_1, u_1, f) and (γ_2, u_2, f) yields

$$\int_B \gamma_1 \nabla(u_1 - u_2) \cdot \nabla \psi \, dx \;=\; \int_B (\gamma_2 - \gamma_1) \nabla u_2 \cdot \nabla \psi \, dx \quad \text{for all } \psi \in H_\diamond^1(B)\,.$$

With $\psi = u_1 - u_2$ and the lower bound $\gamma_0 \leq \gamma_1$ this yields

$$\gamma_0 \|\nabla(u_1 - u_2)\|_{L^2(B)}^2 \leq \int_B \gamma_1 |\nabla(u_1 - u_2)|^2 \, dx$$

$$= \int_B (\gamma_2 - \gamma_1) \nabla u_2 \cdot \nabla(u_1 - u_2) \, dx$$

$$\leq \|\gamma_1 - \gamma_2\|_\infty \|\nabla(u_1 - u_2)\|_{L^2(B)} \|\nabla u_2\|_{L^2(B)} \, ;$$

that is, there exists a constant $c_1 > 0$ (independent of γ_1, γ_2) with

$$\|\nabla(u_1 - u_2)\|_{L^2(B)} \leq \frac{1}{\gamma_0} \|\gamma_1 - \gamma_2\|_\infty \|\nabla u_2\|_{L^2(B)}$$

$$\leq c_1 \|\gamma_1 - \gamma_2\|_\infty \|f\|_{L^2(\partial B)} \, , \qquad (6.11)$$

where we use the boundedness of the mapping $f \mapsto u_2$ (see Theorem 6.2). Now we use the trace theorem and (6.6) to conclude that

$$\|\Lambda(\gamma_1)f - \Lambda(\gamma_2)f\|_{L^2(\partial B)} = \left\|(u_1 - u_2)\big|_{\partial B}\right\|_{L^2(\partial B)} \leq c_2 \|\gamma_1 - \gamma_2\|_\infty \|f\|_{L^2(\partial B)} \, ;$$

that is,

$$\|\Lambda(\gamma_1) - \Lambda(\gamma_2)\|_{\mathcal{L}(L^2(\partial B))} \leq c_2 \|\gamma_1 - \gamma_2\|_\infty$$

which proves part (a).

(b) Let $\gamma \in U$ and $q \in L^\infty(B)$ such that $\|q\|_\infty \leq \gamma_0/2$. Then $\gamma + q \geq \gamma_0/2$ a.e. on B. Let $u, u_q \in H_\diamond^1(B)$ correspond to γ and $\gamma + q$, respectively, and boundary data f. Subtraction of (6.4) for the triple (γ, u, f) and (6.10) from (6.4) for $(\gamma + q, u_q, f)$ yields

$$\int_B \gamma \nabla(u_q - u - v) \cdot \nabla\psi \, dx = \int_B q \nabla(u - u_q) \cdot \nabla\psi \, dx \quad \text{for all } \psi \in H_\diamond^1(B).$$

Taking $\psi = u_q - u - v$ yields as in part (a) an estimate of the form

$$\|\nabla(u_q - u - v)\|_{L^2(B)} \leq \frac{1}{\gamma_0} \|q\|_\infty \|\nabla(u - u_q)\|_{L^2(B)} \, .$$

Now we use (6.11) (with $u_1 = u$ and $u_2 = u_q$) to conclude that

$$\|\nabla(u_q - u - v)\|_{L^2(B)} \leq \frac{c_1}{\gamma_0} \|q\|_\infty^2 \|f\|_{L^2(\partial B)} \, .$$

Again by the trace theorem and (6.6) this yields

$$\|\Lambda(\gamma + q)f - \Lambda(\gamma)f - [\Lambda'(\gamma)q]f\|_{L^2(\partial B)} = \left\|(u_q - u - v)\big|_{\partial B}\right\|_{L^2(\partial B)}$$

$$\leq c \|q\|_\infty^2 \|f\|_{L^2(\partial B)} \, ,$$

which proves part (b). $\square$

We now show that, for any given *constant* background medium γ, the linearized inverse problem of electrical impedance tomography (6.8) has at most one solution; that is, the Fréchet derivative is one-to-one. As already mentioned in the introduction, this proof is due to Calderón (see [38]) and has "opened the door" to many uniqueness results in tomography and scattering theory. We come back to this method in the next chapter where we prove uniqueness of an inverse scattering problem by this method.

Theorem 6.9 *Let γ be constant. Then the Fréchet derivative $\Lambda'(\gamma) : L^\infty(B) \to \mathcal{L}\big(L_\diamond^2(\partial B)\big)$ is one-to-one.*

Proof: First we note that we can assume without loss of generality that $\gamma = 1$. Let $q \in L^\infty(B)$ such that $\Lambda'(\gamma)q = 0$; that is, $\big[\Lambda'(\gamma)q\big]f = 0$ for all $f \in L_\diamond^2(\partial B)$.

The proof consists of two parts. First, we show that q is orthogonal to all products of two gradients of harmonic functions. Then, in the second part, by choosing special harmonic functions we show that the Fourier transform of q vanishes.

Let $u_1 \in C^2(\overline{B})$ be any harmonic function; that is, $\Delta u_1 = 0$ in B. Define $f \in L_\diamond^2(\partial B)$ by $f = \partial u_1/\partial \nu$ on ∂B. Then u_1 is the solution of (6.2) with Neumann boundary data f. We denote by $v_1 \in H_\diamond^1(B)$ the corresponding solution of (6.10); that is,

$$\int_B \nabla \psi \cdot \nabla v_1 \, dx \; = \; -\int_B q \, \nabla \psi \cdot \nabla u_1 \, dx \quad \text{for all } \psi \in H_\diamond^1(B)\,.$$

Now we take a second arbitrary harmonic function $u_2 \in C^2(\overline{B})$ and set $\psi = u_2$ in the previous equation. This yields

$$\int_B q \, \nabla u_2 \cdot \nabla u_1 \, dx \; = \; -\int_B \nabla u_2 \cdot \nabla v_1 \, dx \; = \; -\int_{\partial B} v_1 \frac{\partial u_2}{\partial \nu} \, d\ell$$

by Green's first theorem. Now we note that $v_1|_{\partial B} = \big[\Lambda'(\gamma)q\big]f = 0$. Therefore, we conclude that the right-hand side vanishes; that is

$$\int_B q \, \nabla u_2 \cdot \nabla u_1 \, dx \; = \; 0 \quad \text{for all harmonic functions } u_1, u_2 \in C^2(\overline{B})\,. \quad (6.12)$$

So far, we considered real-valued functions u_1 and u_2. By taking the real and imaginary parts, we can also allow complex-valued harmonic functions for u_1 and u_2.

Now we fix any $y \in \mathbb{R}^2$ with $y \neq 0$. Let $y^\perp \in \mathbb{R}^2$ be a vector (unique up to sign) with $y \cdot y^\perp = 0$ and $|y| = |y^\perp|$. Define the complex vectors $z^\pm \in \mathbb{C}^2$ by $z^\pm = \frac{1}{2}\,(iy \pm y^\perp)$. Then one computes that $z^\pm \cdot z^\pm = \sum_{j=1}^2 (z_j^\pm)^2 = 0$ and $z^+ \cdot z^- = \sum_{j=1}^2 z_j^+ z_j^- = -\frac{1}{2}|y|^2$ and $z^+ + z^- = iy$. From this, we observe that the functions $u^{\pm}(x) = \exp(z^{\pm} \cdot x)$, $x \in \mathbb{R}^2$, are harmonic in all of $\mathbb{R}^2$. Therefore, substituting u^+ and u^- into (6.12) yields

$$0 = \int_B q \, \nabla u^+ \cdot \nabla u^- \, dx = z^+ \cdot z^- \int_B q(x) \, e^{(z^+ + z^-) \cdot x} \, dx = -\frac{1}{2}|y|^2 \int_B q(x) \, e^{iy \cdot x} \, dx.$$

From this, we conclude that the Fourier transform of q (extended by zero in the exterior of B) vanishes on $\mathbb{R}^2 \setminus \{0\}$, and thus also q itself. This ends the proof.
$\square$

6.4   The Factorization Method

In this section, we consider the full nonlinear problem but restrict ourselves to the more modest problem to determine only the shape of the region D, where γ differs from the known background medium which we assume to be homogeneous with conductivity 1.

We sharpen the assumption on γ of Assumption 6.1.

Assumption 6.10 *In addition to Assumption 6.1, let there exist finitely many domains D_j, $j = 1, \ldots, m$, such that $\overline{D_j} \subset B$ and $\overline{D_j} \cap \overline{D_k} = \emptyset$ for $j \neq k$ and such that the complement $B \setminus \overline{D}$ of the closure of the union $D = \bigcup_{j=1}^{m} D_j$ is connected. Every domain D_j is assumed to satisfy the exterior cone condition (see. e.g., [103]); that is, for every $z \in \partial D_j$ there exists a set C (part of a cone) of the form*

$$C = z + \left\{ x \in \mathbb{R}^2 : \hat{\theta} \cdot \frac{x}{|x|} > 1 - \delta, \ 0 < |x| < \varepsilon_0 \right\}$$

for some $\varepsilon_0, \delta > 0$ and $\hat{\theta} \in \mathbb{R}^2$ with $|\hat{\theta}| = 1$, such that $C \cap D_j = \emptyset$.

Furthermore, there exists $q_0 > 0$ such that $\gamma = 1$ on $B \setminus D$ and $\gamma \geq 1 + q_0$ on D. We define the contrast $q \in L^{\infty}(B)$ by $q = \gamma - 1$ and note that D is the support of q.

It is not difficult to show that every Lipschitz domain D satisfies the exterior cone condition (see Problem 6.6).

The **inverse problem** of this section is to determine the shape of D from the Neumann–Dirichlet operator Λ.

In the following, we use the *relative* data $\Lambda - \Lambda_1$ where $\Lambda_1 : L_\diamond^2(\partial B) \to L_\diamond^2(\partial B)$ corresponds to the known background medium; that is, to $\gamma = 1$. The information that $\Lambda - \Lambda_1$ does not vanish simply means that the background is perturbed by some contrast $q = \gamma - 1$. In the factorization method, we develop a criterion to decide whether or not a given point $z \in B$ belongs to D. The idea is then to take a fine grid in B and to check this criterion for every grid point z. This provides a pixel-based picture of D.

We recall that $\Lambda f = u|_{\partial B}$ and $\Lambda_1 f = u_1|_{\partial B}$, where $u, u_1 \in H^1_\diamond(B)$ solve

$$\int_B (1+q)\, \nabla u \cdot \nabla \psi \, dx \;=\; (f, \psi)_{L^2(\partial B)} \text{ for all } \psi \in H^1_\diamond(B)\,, \qquad (6.13)$$

$$\int_B \nabla u_1 \cdot \nabla \psi \, dx \;=\; (f, \psi)_{L^2(\partial B)} \quad \text{ for all } \psi \in H^1_\diamond(B)\,. \qquad (6.14)$$

For the difference, we have $(\Lambda_1 - \Lambda)f = (u_1 - u)|_{\partial B}$, and $u_1 - u \in H^1_\diamond(B)$ satisfies the variational equation

$$\int_B (1+q)\, \nabla (u_1 - u) \cdot \nabla \psi \, dx \;=\; \int_D q\, \nabla u_1 \cdot \nabla \psi \, dx \quad \text{ for all } \psi \in H^1_\diamond(B)\,. \qquad (6.15)$$

It is the aim to *factorize* the operator $\Lambda_1 - \Lambda$ in the form

$$\Lambda_1 - \Lambda \;=\; A^* T\, A\,,$$

where the operators $A : L^2_\diamond(\partial B) \to L^2(D)^2$ and $T : L^2(D)^2 \to L^2(D)^2$ are defined as follows:[5]

- $Af = \nabla u_1|_D$, where $u_1 \in H^1_\diamond(B)$ solves the variational equation (6.14), and

- $Th = q(h - \nabla w)$ where $w \in H^1_\diamond(B)$ solves the variational equation

$$\int_B (1+q)\, \nabla w \cdot \nabla \psi \, dx \;=\; \int_D q\, h \cdot \nabla \psi \, dx \quad \text{ for all } \psi \in H^1_\diamond(B)\,. \qquad (6.16)$$

We note that the solution w of (6.16) exists and is unique. This is seen by the representation theorem A.23 of Riesz because the right-hand side again defines a linear and bounded functional $F(\psi) = \int_D q\, h \cdot \nabla \psi \, dx$ on $H^1_\diamond(B)$. The left-hand side of (6.16) is again the inner product $(w, \psi)_*$. The classical interpretation of the variational equation (under the assumption that all functions are sufficiently smooth) can again be seen from Green's first theorem, applied in D and in $B \setminus D$. Indeed, in this case (6.16) is equivalent to

$$\begin{aligned}
0 \;=\; & \int_D \psi \,\operatorname{div}\big[(1+q)\, \nabla w - q\, h\big]\, dx \;-\; \int_{\partial D} \psi\, \nu \cdot \big[(1+q)\, \nabla w - q\, h\big]\, d\ell \\
& +\; \int_{B \setminus D} \psi\, \Delta w \, dx \;-\; \int_{\partial(B \setminus D)} \psi\, \frac{\partial w}{\partial \nu}\, d\ell
\end{aligned}$$

for all ψ. This yields

$$\operatorname{div}\big[(1+q)\, \nabla w - q\, h\big] = 0 \text{ in } D\,, \quad \Delta w = 0 \text{ in } B \setminus \overline{D}\,,$$

---

[5] Here, $L^2(D)^2$ denotes the space of vector-valued functions $D \to \mathbb{R}^2$ such that both components are in $L^2(D)$.

and

$$\nu \cdot \left[(1+q)\nabla w - q\,h\right]\big|_- = \left.\frac{\partial w}{\partial \nu}\right|_+ \quad \text{on } \partial D, \qquad \frac{\partial w}{\partial \nu} = 0 \text{ on } \partial B\,;$$

that is,

$$\nu \cdot \left[(1+q)\nabla w\right]\big|_- - \left.\frac{\partial w}{\partial \nu}\right|_+ = q|_-\,\nu \cdot h \text{ on } \partial D, \qquad \frac{\partial w}{\partial \nu} = 0 \text{ on } \partial B\,.$$

Theorem 6.11 *Let the operators* $A : L^2_\diamond(\partial B) \to L^2(D)^2$ *and* $T : L^2(D)^2 \to L^2(D)^2$ *be defined as above by (6.14) and (6.16), respectively. Then*

$$\Lambda_1 - \Lambda = A^* T A. \tag{6.17}$$

Proof: We define the auxiliary operator $\mathcal{H} : L^2(D)^2 \to L^2_\diamond(\partial B)$ by $\mathcal{H}h = w|_{\partial B}$ where $w \in H^1_\diamond(B)$ solves (6.16). Obviously, we conclude from (6.15) that $\Lambda_1 - \Lambda = \mathcal{H}A$.

We determine the adjoint $A^* : L^2(D)^2 \to L^2_\diamond(\partial B)$ of A and prove that $A^*h = v|_{\partial B}$ where $v \in H^1_\diamond(B)$ solves the variational equation

$$\int_B \nabla v \cdot \nabla \psi \, dx = \int_D h \cdot \nabla \psi \, dx \quad \text{for all } \psi \in H^1_\diamond(B) \tag{6.18}$$

and even for all $\psi \in H^1(B)$ because it obviously holds for constants. The solution v exists and is unique by the same arguments as above. Again, by applying Green's theorem we note that v is the variational solution of the boundary value problem

$$\Delta v = \begin{cases} \text{div } h & \text{in } D, \\ 0 & \text{in } B \setminus D, \end{cases} \qquad \frac{\partial v}{\partial \nu} = 0 \text{ on } \partial B, \tag{6.19a}$$

$$v|_- = v|_+ \text{ on } \partial D, \qquad \frac{\partial}{\partial \nu}v|_- - \frac{\partial}{\partial \nu}v|_+ = \nu \cdot h \text{ on } \partial D. \tag{6.19b}$$

To prove the representation of A^*h, we conclude from the definition of A, equation (6.18) for $\psi = u_1$, and (6.14) that

$$(Af, h)_{L^2(D)^2} = \int_D \nabla u_1 \cdot h \, dx = \int_B \nabla u_1 \cdot \nabla v \, dx = (f, v)_{L^2(\partial B)}\,,$$

and thus $v|_{\partial B}$ is indeed the value of the adjoint A^*h.

Now it remains to show that $\mathcal{H} = A^*T$. Let $h \in L^2(D)^2$ and $w \in H^1_\diamond(B)$ solve (6.16). Then $\mathcal{H}h = w|_{\partial B}$. We rewrite (6.16) as

$$\int_B \nabla w \cdot \nabla \psi \, dx = \int_D q\,(h - \nabla w) \cdot \nabla \psi \, dx \quad \text{for all } \psi \in H^1_\diamond(B). \tag{6.20}$$

The comparison with (6.18) yields $A^*\big(q(h - \nabla w)\big) = w|_{\partial B} = \mathcal{H}h$; that is, $A^*T = \mathcal{H}$. Substituting this into $\Lambda_1 - \Lambda = \mathcal{H}A$ yields the assertion. $\qquad\square$

Properties of the operators A and T are listed in the following theorem:

Theorem 6.12 *The operator* $A : L^2_\diamond(\partial B) \to L^2(D)^2$ *is compact, and the operator* $T : L^2(D)^2 \to L^2(D)^2$ *is self-adjoint and coercive*

$$(Th, h)_{L^2(D)^2} \geq c \|h\|^2_{L^2(D)^2} \quad \text{for all } h \in L^2(D)^2, \tag{6.21}$$

where $c = q_0\big(1 - q_0/(1 + q_0)\big) > 0.$

Proof: (i) For smooth functions $u_1 \in C^2(\overline{B})$ with $\Delta u_1 = 0$ in B and $\partial u_1/\partial \nu = f$ on ∂B the following representation formula holds (see [53] or Theorem 7.16 for the case of the three-dimensional Helmholtz equation).

$$
\begin{aligned}
u_1(x) &= \int\limits_{\partial B} \left[\Phi(x, y) \frac{\partial u_1(y)}{\partial \nu} - u_1(y) \frac{\partial}{\partial \nu(y)} \Phi(x, y) \right] d\ell(y) \\
&= \int\limits_{\partial B} \left[\Phi(x, y) f(y) - (\Lambda_1 f)(y) \frac{\partial}{\partial \nu(y)} \Phi(x, y) \right] d\ell(y), \quad x \in B,
\end{aligned}
$$

where Φ denotes the fundamental solution of the Laplace equation in $\mathbb{R}^2$; that is,

$$\Phi(x, y) = -\frac{1}{2\pi} \ln |x - y|, \quad x \neq y.$$

We can write ∇u_1 in D in the form $\nabla u_1\big|_D = K_1 f - K_2 \Lambda_1 f$ where the operators $K_1, K_2 : L^2_\diamond(\partial B) \to L^2(D)^2$, defined by

$$
\begin{aligned}
(K_1 f)(x) &= \nabla \int\limits_{\partial B} \Phi(x, y) f(y) \, d\ell(y), \quad x \in D, \\
(K_2 g)(x) &= \nabla \int\limits_{\partial B} g(y) \frac{\partial}{\partial \nu(y)} \Phi(x, y) \, d\ell(y), \quad x \in D,
\end{aligned}
$$

are compact as integral operators on bounded regions of integration with smooth kernels. (Note that $\overline{D} \subset B$.) The representation $A = K_1 - K_2 \Lambda_1$ holds by a density argument (see Theorem A.30). Therefore, also A is compact.

(ii) Let $h_1, h_2 \in L^2(D)^2$ with corresponding solutions $w_1, w_2 \in H^1_\diamond(B)$ of (6.16). Then, with (6.16) for h_2, w_2 and $\psi = w_1$:

$$
\begin{aligned}
(Th_1, h_2)_{L^2(D)^2} &= \int\limits_D q\,(h_1 - \nabla w_1) \cdot h_2 \, dx \\
&= \int\limits_D q\, h_1 \cdot h_2 \, dx - \int\limits_D q\, \nabla w_1 \cdot h_2 \, dx \\
&= \int\limits_D q\, h_1 \cdot h_2 \, dx - \int\limits_B (1 + q)\, \nabla w_1 \cdot \nabla w_2 \, dx.
\end{aligned}
$$

This expression is symmetric with respect to h_1 and h_2. Therefore, T is self-adjoint.

For $h \in L^2(D)^2$ and corresponding solution $w \in H_\circ^1(B)$ of (6.16), we conclude that

$$
\begin{aligned}
(Th, h)_{L^2(D)^2} &= \int_D q\,|h - \nabla w|^2\, dx \;+\; \int_D q\,(h - \nabla w) \cdot \nabla w\, dx \\[2mm]
&= \int_D q\,|h - \nabla w|^2\, dx \;+\; \int_B |\nabla w|^2\, dx \quad \text{(with the help of (6.20))} \\[2mm]
&\geq \int_D \left[q_0\,|h|^2 - 2\,q_0\, h \cdot \nabla w + (1 + q_0)\,|\nabla w|^2 \right] dx \\[2mm]
&= \int_D \left[\left| \sqrt{1 + q_0}\,\nabla w - \frac{q_0}{\sqrt{1 + q_0}}\, h \right|^2 + q_0 \left(1 - \frac{q_0}{1 + q_0} \right) |h|^2 \right] dx \\[2mm]
&\geq q_0 \left(1 - \frac{q_0}{1 + q_0} \right) \|h\|_{L^2(D)^2}^2 .
\end{aligned}
$$

$\square$

From this result and the factorization (6.17), we note that $\Lambda_1 - \Lambda$ is compact, self-adjoint (this follows already from Theorem 6.4), and nonnegative.

Now we derive the binary criterion on a point $z \in B$ to decide whether or not this point belongs to D. First, for every point $z \in B$ we define a particular function $G(\cdot, z)$ such that $\Delta G(\cdot, z) = 0$ in $B \setminus \{z\}$ and $\partial G(\cdot, z)/\partial \nu = 0$ on ∂B such that $G(x, z)$ becomes singular as x tends to z. We construct G from the Green's function N for Δ in B with respect to the Neumann boundary conditions.

We make an ansatz for N in the form $N(x, z) = \Phi(x, z) - \tilde{N}(x, z)$ where again

$$
\Phi(x, z) \;=\; -\frac{1}{2\pi}\, \ln|x - z|, \quad x \neq z,
$$

is the fundamental solution of the Laplace equation in $\mathbb{R}^2$ and determine $\tilde{N}(\cdot, z) \in H_\circ^1(B)$ as the unique solution of the Neumann problem

$$
\Delta \tilde{N}(\cdot, z) = 0 \text{ in } B \quad \text{and} \quad \frac{\partial \tilde{N}}{\partial \nu}(\cdot, z) = \frac{\partial \Phi}{\partial \nu}(\cdot, z) + \frac{1}{|\partial B|} \text{ on } \partial B.
$$

We note that the solution exists because $\int_{\partial B} \left[\partial \Phi(\cdot, z)/\partial \nu + 1/|\partial B| \right] d\ell = 0$. This is seen by Green's first theorem in the region $B \setminus B(z, \varepsilon)$:

$$
\begin{aligned}
\int_{\partial B} \frac{\partial \Phi}{\partial \nu}(\cdot, z)\, d\ell &= \int_{|x - z| = \varepsilon} \frac{\partial \Phi}{\partial \nu}(x, z)\, d\ell(x) \\[2mm]
&= -\frac{1}{2\pi} \int_{|x - z| = \varepsilon} \frac{x - z}{|x - z|^2} \cdot \frac{x - z}{|x - z|}\, d\ell(x) = -1 .
\end{aligned}
$$

Then $N = \Phi - \tilde{N}$ is the Green's function in B with respect to the Neumann boundary conditions; that is, N satisfies

$$\Delta N(\cdot, z) = 0 \text{ in } B \setminus \{z\} \quad \text{and} \quad \frac{\partial N}{\partial \nu}(\cdot, z) = -\frac{1}{|\partial B|} \text{ on } \partial B.$$

From the differentiable dependence of the solution $\tilde{N}(\cdot, z)$ on the parameter $z \in B$, we conclude that, for any fixed $a \in \mathbb{R}^2$ with $|a| = 1$, the function $G(\cdot, z) = a \cdot \nabla_z N(\cdot, z)$ satisfies

$$\Delta G(\cdot, z) = 0 \text{ in } B \setminus \{z\} \quad \text{and} \quad \frac{\partial G}{\partial \nu}(\cdot, z) = 0 \text{ on } \partial B. \tag{6.22}$$

The function $G(\cdot, z)$ has the following desired properties.

Lemma 6.13 *Let $z \in B$, $\varepsilon_0 > 0$, $\theta \in [0, 2\pi]$, and $\delta > 0$ be kept fixed. For $\varepsilon \in [0, \varepsilon_0)$ define the set (part of a cone, compare with Assumption 6.10 for $\hat{\theta} = \binom{\cos \theta}{\sin \theta}$)*

$$C_\varepsilon = z + \left\{ r \binom{\cos t}{\sin t} : \varepsilon < r < \varepsilon_0, \ |\theta - t| < \arccos(1 - \delta) \right\}$$

with vertex in z. Let ε_0 be so small such that $\overline{C_0} \subset B$. Then

$$\lim_{\varepsilon \to 0} \|G(\cdot, z)\|_{L^2(C_\varepsilon)} = \infty.$$

Proof: By the smoothness of $\tilde{N}(\cdot, z)$ it is sufficient to consider only the part $a \cdot \nabla_z \ln |x - z|$. Using polar coordinates for x with respect to z (i.e., $x = z + r \binom{\cos t}{\sin t}$), and the representation of a as $a = \binom{\cos s}{\sin s}$, we have with $\eta = \arccos(1 - \delta)$

$$\int_{C_\varepsilon} |a \cdot \nabla_z \ln |x - z||^2 dx = \int_{C_\varepsilon} \frac{((z - x) \cdot a)^2}{|x - z|^4} dx = \int_{\varepsilon}^{\varepsilon_0} \int_{\theta - \eta}^{\theta + \eta} \frac{r^2 \cos^2(s - t)}{r^4} r \, dt \, dr$$

$$= \underbrace{\int_{\theta - \eta}^{\theta + \eta} \cos^2(s - t) \, dt}_{= c} \int_{\varepsilon}^{\varepsilon_0} \frac{1}{r} dr = c \ln \frac{\varepsilon_0}{\varepsilon}.$$

Therefore,

$$\|G(\cdot, z)\|_{L^2(C_\varepsilon)} \geq \sqrt{c \ln \frac{\varepsilon_0}{\varepsilon}} - \|a \cdot \nabla_z \tilde{N}(\cdot, z)\|_{L^2(C_0)} \to \infty \quad \text{for } \varepsilon \to 0.$$

□

We observe that the functions $\phi_z(x) = G(\cdot, z)|_{\partial B}$ are traces of harmonic functions in $B \setminus \{z\}$ with vanishing normal derivatives on ∂B. Comparing this with the classical formulation (6.19a) (6.19b) of the adjoint A^* of A it seems to be plausible that the "source region" D of (6.19a), (6.19b) can be determined by moving the source point z in ϕ_z. This is confirmed in the following theorem.

Theorem 6.14 *Let Assumptions 6.10 hold and let $a \in \mathbb{R}^2$ with $|a| = 1$ be fixed. For every $z \in B$ define $\phi_z \in L^2_\diamond(\partial B)$ by*

$$\phi_z(x) = G(x, z) = a \cdot \nabla_z N(x, z), \quad x \in \partial B, \tag{6.23}$$

where N denotes the Green's function with respect to the Neumann boundary condition. Then

$$z \in D \iff \phi_z \in \mathcal{R}(A^*), \tag{6.24}$$

where $A^* : L^2(D)^2 \to L^2_\diamond(\partial B)$ is the adjoint of A, given by (6.18), and $\mathcal{R}(A^*)$ its range.

Proof: First let $z \in D$. Choose a disc $B[z, \varepsilon] = \{x \in \mathbb{R}^2 : |x - z| \leq \varepsilon\}$ with center z and radius $\varepsilon > 0$ such that $B[z, \varepsilon] \subset D$. Furthermore, choose a function $\varphi \in C^\infty(\mathbb{R}^2)$ such that $\varphi(x) = 0$ for $|x - z| \leq \varepsilon/2$ and $\varphi(x) = 1$ for $|x - z| \geq \varepsilon$ and set $w(x) = \varphi(x)G(x, z)$ for $x \in B$. Then $w \in H^1_\diamond(B)$ and $w = G(\cdot, z)$ in $B \setminus D$, thus $w|_{\partial B} = \phi_z$.

Next, we determine $u \in H^1_\diamond(D)$ as a solution of $\Delta u = \Delta w$ in D, $\partial u/\partial \nu = 0$ on ∂D; that is, in weak form

$$\int_D \nabla u \cdot \nabla \psi \, dx = \int_D \nabla w \cdot \nabla \psi \, dx - \int_{\partial D} \psi \frac{\partial}{\partial \nu} G(\cdot, z) \, d\ell, \quad \psi \in H^1_\diamond(D),$$

because $\partial w/\partial \nu = \partial G(\cdot, z)/\partial \nu$ on ∂D. Again, the solution exists and is unique. Application of Green's first theorem in $B \setminus D$ yields

$$\int_{\partial D} \frac{\partial}{\partial \nu} G(\cdot, z) \, d\ell = \int_{\partial B} \frac{\partial}{\partial \nu} G(\cdot, z) \, d\ell = 0.$$

Therefore, the previous variational equation holds also for constants and thus for all $\psi \in H^1(D)$. Now let $\psi \in H^1_\diamond(B)$ be a test function on B. Then

$$\int_D \nabla u \cdot \nabla \psi \, dx = \int_D \nabla w \cdot \nabla \psi \, dx - \int_{\partial D} \psi \frac{\partial}{\partial \nu} G(\cdot, z) \, d\ell$$

$$= \int_D \nabla w \cdot \nabla \psi \, dx + \int_{B \setminus D} \nabla G(\cdot, z) \cdot \nabla \psi \, dx = \int_B \nabla w \cdot \nabla \psi \, dx.$$

Therefore, the definition $h = \nabla u$ in D yields $A^* h = w|_{\partial B} = \phi_z$ and thus $\phi_z \in \mathcal{R}(A^*)$.

Now we prove the opposite direction. Let $z \notin D$. We have to show that ϕ_z is not contained in the range of A^* and assume, on the contrary, that $\phi_z = A^* h$ for some $h \in L^2(D)^2$. Let $v \in H^1_\diamond(B)$ be the corresponding solution of (6.18). Therefore, the function $w = v - G(\cdot, z)$ vanishes on ∂B and solves the following equations in the weak form

$$\Delta w = 0 \text{ in } B \setminus D_\varepsilon(z), \quad \frac{\partial w}{\partial \nu} = 0 \text{ on } \partial B,$$

for every $\varepsilon > 0$ such that $D_\varepsilon(z) := D \cup B(z, \varepsilon) \subset B$; that is,

$$\int_{B \setminus D_\varepsilon(z)} \nabla w \cdot \nabla \psi \, dx = 0$$

for all $\psi \in H^1\big(B \setminus D_\varepsilon(z)\big)$ with $\psi = 0$ on $\partial D_\varepsilon(z)$. We extend w by zero into the exterior of B. Then $w \in H^1\big(\mathbb{R}^2 \setminus D_\varepsilon(z)\big)$ because $w = 0$ on ∂B^6 and

$$\int_{\mathbb{R}^2 \setminus D_\varepsilon(z)} \nabla w \cdot \nabla \psi \, dx = 0$$

for all $\psi \in H^1\big(\mathbb{R}^2 \setminus D_\varepsilon(z)\big)$ which vanish on $\partial D_\varepsilon(z)$. This is the variational form of $\Delta w = 0$ in $\mathbb{R}^2 \setminus D_\varepsilon(z)$. Since this holds for all sufficiently small $\varepsilon > 0$ we conclude that $\Delta w = 0$ in the exterior $\Omega := \mathbb{R}^2 \setminus \big(\overline{D} \cup \{z\}\big)$ of $\overline{D} \cup \{z\}$. Now we use without proof[7] that w is analytic in this set Ω and thus satisfies the unique continuation principle, see, e.g., Theorem 4.39 of [161]. Therefore, because it vanishes in the exterior of B it has to vanish in all of the connected set Ω. (Here we make use of the assumption that $B \setminus \overline{D}$ is connected.) Therefore, $v = G(\cdot, z)$ in $B \setminus \big(\overline{D} \cup \{z\}\big)$.

The point z can either be on the boundary ∂D or in the exterior of $\overline{D}$. In either case there is a cone C_0 of the form $C_0 = \big\{z + r\big(\begin{smallmatrix}\cos t \\ \sin t\end{smallmatrix}\big) : 0 < r < \varepsilon_0, \ |\theta - t| < \delta \big\}$ with $C_0 \subset B \setminus \overline{D}$. (Here we use the fact that every component of D satisfies the exterior cone condition.) It is $v|_{C_0} \in L^2(C_0)$ because even $v \in H^1_\diamond(B)$. However, Lemma 6.13 yields that $\|G(\cdot, z)\|_{L^2(C_\varepsilon)} \to \infty$ for $\varepsilon \to 0$ where $C_\varepsilon = \big\{z + r\big(\begin{smallmatrix}\cos t \\ \sin t\end{smallmatrix}\big) : \varepsilon < r < \varepsilon_0, \ |\theta - t| < \delta \big\}$. This is a contradiction because $v = G(\cdot, z)$ in C_0 and ends the proof. $\qquad\square$

Therefore, we have shown an explicit characterization of the unknown domain D by the range of the operator A^*. This operator, however, is also unknown: only $\Lambda_1 - \Lambda$ is known! The operators A^* and $\Lambda_1 - \Lambda$ are connected by the factorization $\Lambda_1 - \Lambda = A^* T A$. We can easily derive a second factorization of $\Lambda_1 - \Lambda$. The operator $\Lambda_1 - \Lambda$ is self-adjoint and compact as an operator from $L^2_\diamond(\partial B)$ into itself. Therefore, there exists a spectral decomposition of the form

$$(\Lambda_1 - \Lambda) f = \sum_{n=1}^\infty \lambda_j \, (f, \psi_j)_{L^2(\partial B)} \, \psi_j \,,$$

where $\lambda_j \in \mathbb{R}$ denote the eigenvalues and $\psi_j \in L^2_\diamond(\partial B)$ the corresponding orthonormal eigenfunctions of $\Lambda_1 - \Lambda$ (see Theorem A.53 of Appendix A.6). Furthermore, from the factorization and the coercivity of T it follows that

---

[6]It is not quite obvious that the extension is in $H^1\big(\mathbb{R}^2 \setminus D_\varepsilon(z)\big)$, see, e.g., Corollary 5.13 in [161]

[7]see again [161], Theorem 4.38 and Corollary 3.4

$\left((\Lambda_1 - \Lambda)f, f\right)_{L^2_\diamond(\partial B)} \geq 0$ for all $f \in L^2_\diamond(\partial B)$. This implies $\lambda_j \geq 0$ for all j. Therefore, we can define

$$Wf = \sum_{n=1}^{\infty} \sqrt{\lambda_j}\, (f, \psi_j)_{L^2(\partial B)}\, \psi_j,$$

and have a second factorization in the form $WW = \Lambda_1 - \Lambda$. We write $(\Lambda_1 - \Lambda)^{1/2}$ for W. The operator $(\Lambda_1 - \Lambda)^{1/2}$ is also self-adjoint, and we have

$$(\Lambda_1 - \Lambda)^{1/2}(\Lambda_1 - \Lambda)^{1/2} \;=\; \Lambda_1 - \Lambda \;=\; A^* T A. \tag{6.25}$$

We show that the ranges of $(\Lambda_1 - \Lambda)^{1/2}$ and A^* coincide.[8] This follows directly from the following functional analytic result:

Lemma 6.15 *Let X and Y be Hilbert spaces, $B : X \to X$, $A : X \to Y$, and $T : Y \to Y$ linear and bounded such that $B = A^* T A$. Furthermore, let T be self-adjoint and coercive; that is, there exists $c > 0$ such that $(Ty, y)_Y \geq c\|y\|_Y^2$ for all $y \in Y$. Then, for any $\phi \in X$, $\phi \neq 0$,*

$$\phi \in \mathcal{R}(A^*) \;\Longleftrightarrow\; \inf\{(Bx, x)_X : x \in X,\ (x, \phi)_X = 1\} \;>\; 0.$$

Proof: (i) First, let $\phi = A^* y \in \mathcal{R}(A^*)$ for some $y \in Y$. Then $y \neq 0$, and we estimate for arbitrary $x \in X$ with $(x, \phi)_X = 1$:

$$\begin{aligned}
(Bx, x)_X &= (A^* T A x, x)_X = (T A x, A x)_Y \geq c\|Ax\|_Y^2 \\
&= \frac{c}{\|y\|_Y^2}\|Ax\|_Y^2\|y\|_Y^2 \geq \frac{c}{\|y\|_Y^2}\left|(Ax, y)_Y\right|^2 \\
&= \frac{c}{\|y\|_Y^2}\left|(x, A^* y)_X\right|^2 = \frac{c}{\|y\|_Y^2}\left|(x, \phi)_X\right|^2 = \frac{c}{\|y\|_Y^2}.
\end{aligned}$$

Therefore, we have found a positive lower bound for the infimum.

(ii) Second, let $\phi \notin \mathcal{R}(A^*)$. Define the closed subspace

$$V \;=\; \{x \in X : (\phi, x)_X = 0\} \;=\; \{\phi\}^\perp.$$

We show that the image $A(V)$ is dense in the closure of the range of A. Indeed, let $y \in \text{closure}(\mathcal{R}(A))$ such that $y \perp Ax$ for all $x \in V$; that is, $0 = (Ax, y)_Y = (x, A^* y)$ for all $x \in V$; that is, $A^* y \in V^\perp = \text{span}\{\phi\}$. Because $\phi \notin \mathcal{R}(A^*)$ we conclude that $A^* y = 0$. Therefore, $y \in \text{closure}(\mathcal{R}(A)) \cap \mathcal{N}(A^*)$. This yields $y = 0$.[9] Therefore, $A(V)$ is dense in $\text{closure}(\mathcal{R}(A))$. Because $A\phi/\|\phi\|_X^2$ is in the range of A there exists a sequence $\tilde{x}_n \in V$ such that $A\tilde{x}_n \to -A\phi/\|\phi\|_X^2$. We define $x_n := \tilde{x}_n + \phi/\|\phi\|_X^2$. Then $(x_n, \phi)_X = 1$ and $Ax_n \to 0$ for $n \to \infty$, and we estimate

$$(Bx_n, x_n)_X \;=\; (T A x_n, A x_n)_Y \;\leq\; \|T\|_{\mathcal{L}(Y)}\|Ax_n\|_Y^2 \;\longrightarrow\; 0, \quad n \to \infty,$$

---

[8] This is also known as Douglas' Lemma, see [75].
[9] Take a sequence (x_j) in X such that $Ax_j \to y$. Then $0 = (A^* y, x_j)_X = (y, Ax_j)_Y \to (y, y)_Y$; that is, $y = 0$.

and thus $\inf\{(Bx,x)_X : x \in X,\ (x,\phi)_X = 1\} = 0$. $\square$

We apply this result to both of the factorizations of (6.25). In both cases, $B = \Lambda_1 - \Lambda$ and $X = L_\diamond^2(\partial B)$. First, we set $Y = L^2(D)^2$ and $A : L_\diamond^2(\partial B) \to L^2(D)^2$ and $T : L^2(D)^2 \to L^2(D)^2$ as in the second factorization of (6.25). Because T is self-adjoint and coercive we conclude for any $\phi \in L_\diamond^2(\partial B)$, $\phi \neq 0$, that

$$\phi \in \mathcal{R}(A^*) \iff \inf\{((\Lambda_1 - \Lambda)f, f)_{L^2(\partial B)} : f \in L_\diamond^2(\partial B),\ (f, \phi)_{L^2(\partial B)} = 1\} > 0.$$

Second, we consider the first factorization of (6.25) with T being the identity. For $\phi \in L_\diamond^2(\partial B)$, $\phi \neq 0$, we conclude that

$$\phi \in \mathcal{R}\big((\Lambda_1 - \Lambda)^{1/2}\big) \iff \inf\{((\Lambda_1 - \Lambda)f, f)_{L^2(\partial B)} : (f, \phi)_{L^2(\partial B)} = 1\} > 0.$$

The right-hand sides of the characterizations only depend on $\Lambda_1 - \Lambda$, therefore, we conclude that

$$\mathcal{R}\big((\Lambda_1 - \Lambda)^{1/2}\big) = \mathcal{R}(A^*). \tag{6.26}$$

Application of Theorem 6.14 yields the main result of the factorization method:

Theorem 6.16 *Let Assumptions 6.10 be satisfied. For fixed $a \in \mathbb{R}^2$ with $a \neq 0$ and every $z \in B$ let $\phi_z \in L_\diamond^2(\partial B)$ be defined by (6.23); that is, $\phi_z(x) = a \cdot \nabla_z N(x, z)$, $x \in \partial B$, where N denotes the Green's function for Δ with respect to the Neumann boundary conditions. Then*

$$z \in D \iff \phi_z \in \mathcal{R}\big((\Lambda_1 - \Lambda)^{1/2}\big). \tag{6.27}$$

We now rewrite the right-hand side with Picard's Theorem A.58 of Appendix A.6. First, we show injectivity of the operator $\Lambda_1 - \Lambda$.

Theorem 6.17 *The operator $\Lambda_1 - \Lambda$ is one-to-one.*

Proof: From

$$((\Lambda_1 - \Lambda)f, f)_{L^2(\partial B)} = (A^* T A f, f)_{L^2(\partial B)} = (T A f, A f)_{L^2(D)^2}$$
$$\geq c\|Af\|_{L^2(D)^2}^2 \quad \text{for } f \in L_\diamond^2(\partial B)$$

it suffices to prove injectivity of A. Let $Af = \nabla u_1|_D = 0$ where $u_1 \in H_\diamond^1(B)$ denotes the weak solution of $\Delta u_1 = 0$ in B and $\partial u_1/\partial \nu = f$ on ∂B. Therefore, ∇u_1 is constant in every component of D. Without proof, we use again the regularity result that u_1 is analytic in B. The derivatives $v_j = \partial u_1/\partial x_j$ are solutions of $\Delta v_j = 0$ in B and $v_j = 0$ in D. The unique continuation property yields $v_j = \partial u_1/\partial x_j = 0$ in all of B and thus $f = 0$. $\square$

Therefore, the operator $\Lambda_1 - \Lambda$ is self-adjoint, compact, one-to-one, and all eigenvalues are positive. Let $\{\lambda_j, \psi_j\}$ be an eigensystem of $\Lambda_1 - \Lambda$; that is, $\lambda_j > 0$ are the eigenvalues of $\Lambda_1 - \Lambda$ and $\psi_j \in L^2_\diamond(\partial B)$ are the corresponding orthonormal eigenfunctions (see Theorem A.53 of Appendix A.6). The set $\{\psi_j : j = 1, 2, \ldots\}$ is complete by the spectral theorem. Therefore, $\{\sqrt{\lambda_j}, \psi_j, \psi_j\}$ is a singular system of the operator $(\Lambda_1 - \Lambda)^{1/2}$. Application of Picard's Theorem A.58 of Appendix A.6 yields

Theorem 6.18 *Let Assumptions 6.10 be satisfied. For fixed $a \in \mathbb{R}^2$ with $a \neq 0$ and for every $z \in B$ let $\phi_z \in L^2_\diamond(\partial B)$ be defined by (6.23); that is, $\phi_z(x) = a \cdot \nabla_z N(x, z)$, $x \in \partial B$. Then*

$$z \in D \quad \Longleftrightarrow \quad \sum_{j=1}^{\infty} \frac{(\phi_z, \psi_j)^2_{L^2(\partial B)}}{\lambda_j} < \infty \tag{6.28}$$

or, equivalently,

$$z \in D \quad \Longleftrightarrow \quad \chi(z) := \left[\sum_{j=1}^{\infty} \frac{(\phi_z, \psi_j)^2_{L^2(\partial B)}}{\lambda_j} \right]^{-1} > 0. \tag{6.29}$$

Here we agreed on the setting that the inverse of the series is zero in the case of divergence. Therefore, χ vanishes outside of D and is positive in the interior of D. The function

$$\text{sign } \chi(z) = \begin{cases} 1, & \chi(z) > 0, \\ 0, & \chi(z) = 0, \end{cases}$$

is thus the characteristic function of D.

We finish this section with some further remarks.

We leave it to the reader to show (see Problems 6.2–6.4) that in the case of $B = B(0, 1)$ being the unit disk and $D = B(0, R)$ the disk of radius $R < 1$ the ratios $(\phi_z, \psi_j)^2_{L^2(\partial B)}/\lambda_j$ behave as $(|z|/R)^{2j}$. Therefore, convergence holds if and only if $|z| < R$ which confirms the assertion of the last theorem.

In practice, only finitely many measurements are available; that is, the data operator $\Lambda_1 - \Lambda$ is replaced by a matrix $M \in \mathbb{R}^{m \times m}$. The question of convergence of the series is obsolete because the sum consists of only finitely many terms. However, in practice, it is observed that the value of this sum is much smaller for points z inside of D than for points z outside of D. Some authors (see [123]) suggest to test the "convergence" by determining the slope of the straight line that best fits the curve $j \mapsto \ln[(\phi_z, \psi_j)^2_{L^2(\partial B)}/\lambda_j]$ (for some j only). The points z for which the slope is negative provide a good picture of D.

A rigorous justification of a projection method to approximate the (infinite) series by a (finite) sum has been given in [177].

In the implementation of the factorization method, only the relative data operator $\Lambda_1 - \Lambda$ has to be known and no other information on D. For example, it is allowed (see Assumption 6.10) that D consist of several components. Furthermore, the fact that the medium D is penetrable is not used. If one imposes some boundary condition on ∂D, the same characterization as in Theorem 6.18

holds. For example, in [123], the factorization method has been justified for an insulating object D. In particular, the factorization method provides a proof of uniqueness of D independent of the nature of D; that is, whether it is finitely conducting, a perfect conductor (Dirichlet boundary conditions on ∂D), a perfect insulator (Neumann boundary conditions on ∂D), or a boundary condition of Robin-type.

6.5   Problems

6.1 Let D be a domain with $\overline{D} \subset B$ and $\gamma \in L^\infty(B)$ piecewise constant with $\gamma = \gamma_0$ in D for some $\gamma_0 \in \mathbb{R}$ and $\gamma = 1$ in $B \setminus D$. Let $u \in H^1_\diamond(B) \cap C^2(B \setminus \partial D)$ be a solution of the variational equation (6.4) and assume that $u|_D$ and $u|_{B \setminus \overline{D}}$ have differentiable extensions to $\overline{D}$ and $\overline{B} \setminus D$, respectively.

Show that u solves $\Delta u = 0$ in $B \setminus \partial D$ and $\partial u / \partial \nu = f$ on ∂B and $u|_+ = u|_-$ on ∂D and $\gamma_0 \partial u|_- / \partial \nu = \partial u|_+ / \partial \nu$ on ∂D.

Hint: Use Green's first theorem.

For the following problems let B be the unit disk in $\mathbb{R}^2$ with center at the origin.

6.2 Show that the fundamental solution Φ and the Green's function N are given in polar coordinates $(x = r(\cos t, \sin t)^\top$ and $z = \rho(\cos \tau, \sin \tau)^\top)$ as

$$\Phi(x, z) = -\frac{1}{2\pi} \ln r + \frac{1}{2\pi} \sum_{n=1}^\infty \frac{1}{n} \left(\frac{\rho}{r}\right)^n \cos n(t - \tau),$$

$$N(x, z) = \Phi(x, z) + \Phi\left(\frac{x}{|x|^2}, z\right)$$

$$= -\frac{1}{2\pi} \ln r + \frac{1}{2\pi} \sum_{n=1}^\infty \frac{1}{n} \rho^n \left(\frac{1}{r^n} + r^n\right) \cos n(t - \tau),$$

for $\rho = |z| < |x| = r$.

Hint: Write Φ in the form

$$\Phi(x, z) = -\frac{1}{2\pi} \ln r - \frac{1}{4\pi} \ln\left[1 + \left(\frac{\rho}{r}\right)^2 - 2\frac{\rho}{r} \cos(t - \tau)\right]$$

and show $\sum_{n=1}^\infty \frac{1}{n} \alpha^n \cos(ns) = -\frac{1}{2} \ln[1 + \alpha^2 - 2\alpha \cos s]$ by differentiation with respect to α and applying the geometric series formula for $\sum_{n=1}^\infty \alpha^{n-1} \exp(ins)$.

6.3 Show that ϕ_z from (6.23) is given by

$$\phi_z(x) = \frac{a \cdot (x - z)}{\pi |x - z|^2}, \qquad |x| = 1.$$

Also compute ϕ_z in polar coordinates for $a = (\cos \alpha, \sin \alpha)^\top$ by the formulas of Problem 6.2.

6.4 Compute the eigenvalues λ_n and the normalized eigenfunctions $\psi_n \in L^2_\diamond(\partial B)$ of $\Lambda_1 - \Lambda_{\gamma_R}$ and the coefficients $(\phi_z, \psi_n)_{L^2(\partial B)}$ for the case of Example 6.6. Compute the ratios $(\phi_z, \psi_n)^2_{L^2(\partial B)}/\lambda_n$ and validate the condition (6.29) of Theorem 6.18.

6.5 Consider the case of $D \subset B$ being the annulus $D = \{x \in B : R_1 < |x| < R_2\}$ for some $0 < R_1 < R_2 < 1$. Compute again the eigenvalues λ_n and the normalized eigenfunctions $\psi_n \in L^2_\diamond(\partial B)$ of $\Lambda_1 - \Lambda$ and the coefficients $(\phi_z, \psi_n)_{L^2(\partial B)}$. Verify that you can only determine the outer boundary $\{x : |x| = R_2\}$ by the factorization method.

6.6 Let $f : [a, b] \to \mathbb{R}_{>0}$ be a Lipschitz continuous function; that is, $f(x) > 0$ for all $x \in [a, b]$ and there exists $L > 0$ with $|f(x) - f(y)| \leq L|x - y|$ for all $x, y \in [a, b]$. Define $D := \{(x_1, x_2) \in [a, b] \times \mathbb{R} : 0 < x_2 < f(x_1) \text{ for } x_1 \in [a, b]\}$. Show that this Lipschitz domain $D \subset \mathbb{R}^2$ satisfies the exterior cone condition of Assumption 6.10.

Chapter 7

An Inverse Scattering Problem

7.1 Introduction

We consider acoustic waves that travel in a medium such as a fluid. Let $v(x,t)$ be the velocity vector of a particle at $x \in \mathbb{R}^3$ and time t. Let $p(x,t)$, $\rho(x,t)$, and $S(x,t)$ denote the pressure, density, and specific entropy, respectively, of the fluid. We assume that no exterior forces act on the fluid. Then the movement of the particle is described by the following equations.

$$\frac{\partial v}{\partial t} + (v \cdot \nabla)v + \gamma v + \frac{1}{\rho}\nabla p = 0 \quad \text{(Euler's equation)}, \tag{7.1a}$$

$$\frac{\partial \rho}{\partial t} + \operatorname{div}(\rho v) = 0 \quad \text{(continuity equation)}, \tag{7.1b}$$

$$f(\rho, S) = p \quad \text{(equation of state)}, \tag{7.1c}$$

$$\frac{\partial S}{\partial t} + v \cdot \nabla S = 0 \quad \text{(adiabatic hypothesis)}, \tag{7.1d}$$

where the function f depends on the fluid. γ is a damping coefficient, which we assume to be piecewise constant. This system is nonlinear in the unknown functions v, ρ, p, and S. Let the *stationary case* be described by $v_0 = 0$, time-independent distributions $\rho = \rho_0(x)$ and $S = S_0(x)$, and constant p_0 such that $p_0 = f\big(\rho_0(x), S_0(x)\big)$. The *linearization* of this nonlinear system is given by the (directional) derivative of this system at (v_0, p_0, ρ_0, S_0). For deriving the linearization, we set

$$v(x,t) = \varepsilon v_1(x,t) + \mathcal{O}(\varepsilon^2),$$
$$p(x,t) = p_0 + \varepsilon p_1(x,t) + \mathcal{O}(\varepsilon^2),$$
$$\rho(x,t) = \rho_0(x) + \varepsilon \rho_1(x,t) + \mathcal{O}(\varepsilon^2),$$
$$S(x,t) = S_0(x) + \varepsilon S_1(x,t) + \mathcal{O}(\varepsilon^2),$$

© Springer Nature Switzerland AG 2021
A. Kirsch, *An Introduction to the Mathematical Theory of Inverse Problems*,
Applied Mathematical Sciences 120,
https://doi.org/10.1007/978-3-030-63343-1_7

and we substitute this into (7.1a), (7.1b), (7.1c), and (7.1d). Ignoring terms with $\mathcal{O}(\varepsilon^2)$ leads to the linear system

$$\frac{\partial v_1}{\partial t} + \gamma v_1 + \frac{1}{\rho_0} \nabla p_1 = 0, \tag{7.2a}$$

$$\frac{\partial \rho_1}{\partial t} + \operatorname{div}(\rho_0 v_1) = 0, \tag{7.2b}$$

$$\frac{\partial f(\rho_0, S_0)}{\partial \rho} \rho_1 + \frac{\partial f(\rho_0, S_0)}{\partial S} S_1 = p_1, \tag{7.2c}$$

$$\frac{\partial S_1}{\partial t} + v_1 \cdot \nabla S_0 = 0. \tag{7.2d}$$

First, we eliminate S_1. Because

$$0 = \nabla f\big(\rho_0(x), S_0(x)\big) = \frac{\partial f(\rho_0, S_0)}{\partial \rho} \nabla \rho_0 + \frac{\partial f(\rho_0, S_0)}{\partial S} \nabla S_0,$$

we conclude by differentiating (7.2c) with respect to t and using (7.2d)

$$\frac{\partial p_1}{\partial t} = c(x)^2 \left[\frac{\partial \rho_1}{\partial t} + v_1 \cdot \nabla \rho_0 \right], \tag{7.2e}$$

where the *speed of sound* c is defined by

$$c(x)^2 := \frac{\partial}{\partial \rho} f\big(\rho_0(x), S_0(x)\big).$$

Now we eliminate v_1 and ρ_1 from the system. This can be achieved by differentiating (7.2e) with respect to time and using equations (7.2a) and (7.2b). This leads to the *wave equation* for p_1:

$$\frac{\partial^2 p_1(x,t)}{\partial t^2} + \gamma \frac{\partial p_1(x,t)}{\partial t} = c(x)^2 \rho_0(x) \operatorname{div}\left[\frac{1}{\rho_0(x)} \nabla p_1(x,t) \right]. \tag{7.3}$$

Now we assume that terms involving $\nabla \rho_0$ are negligible and that p_1 is time-periodic; that is, of the form

$$p_1(x,t) = \operatorname{Re}\left[u(x) e^{-i\omega t} \right]$$

with frequency $\omega > 0$ and a complex-valued function $u = u(x)$ depending only on the spatial variable. Substituting this into the wave equation (7.3) yields the three-dimensional *Helmholtz equation* for u:

$$\Delta u(x) + \frac{\omega^2}{c(x)^2}\left(1 + i\frac{\gamma}{\omega}\right) u = 0.$$

In free space, $c = c_0$ is constant and $\gamma = 0$. We define the *wave number* and the *index of refraction* by

$$k := \frac{\omega}{c_0} > 0 \quad \text{and} \quad n(x) := \frac{c_0^2}{c(x)^2}\left(1 + i\frac{\gamma}{\omega}\right). \tag{7.4}$$

The Helmholtz equation then takes the form

$$\Delta u + k^2 n\, u = 0 \tag{7.5}$$

where n is a complex-valued function with $\operatorname{Re} n(x) \geq 0$ and $\operatorname{Im} n(x) \geq 0$.

This equation holds in every source-free domain in $\mathbb{R}^3$. We assume in this chapter that there exists $a > 0$ such that $c(x) = c_0$ and $\gamma(x) = 0$ for all x with $|x| \geq a$; that is, $n(x) = 1$ for $|x| \geq a$. This means that the inhomogeneous medium $\{x \in \mathbb{R}^3 : n(x) \neq 1\}$ is bounded and contained in the ball $B(0, a) := \{y \in \mathbb{R}^3 : |y| < a\}$ of radius a. By $B[0, a] := \{y \in \mathbb{R}^3 : |y| \leq a\}$, we denote its closure. We further assume that the sources lie outside the ball $B[0, a]$.

These sources generate "incident" fields u^i, that satisfy the unperturbed Helmholtz equation $\Delta u^i + k^2 u^i = 0$ outside the sources. In this introduction, we assume that u^i is either a *point source* or a *plane wave*; that is, the time-dependent incident fields have the form

$$p_1^i(x, t) = \frac{1}{|x - z|} \operatorname{Re} e^{ik|x-z|-i\omega t}; \quad \text{that is,} \quad u^i(x) = \frac{e^{ik|x-z|}}{|x - z|},$$

for a source at $z \in \mathbb{R}^3$, or

$$p_1^i(x, t) = \operatorname{Re} e^{ik\hat{\theta}\cdot x - i\omega t}; \quad \text{that is,} \quad u^i(x) = e^{ik\hat{\theta}\cdot x},$$

for a unit vector $\hat{\theta} \in \mathbb{R}^3$.

In any case, u^i is a solution of the Helmholtz equation $\Delta u^i + k^2 u^i = 0$ in $\mathbb{R}^3 \setminus \{z\}$ or $\mathbb{R}^3$, respectively. In the first case, p_1^i describes a *spherical wave* that travels away from the source with velocity c_0. In the second case, p_1^i is a plane wave that travels in the direction $\hat{\theta}$ with velocity c_0.

The incident field is disturbed by the medium described by the index of refraction n and produces a "scattered wave" u^s. The total field $u = u^i + u^s$ satisfies the Helmholtz equation $\Delta u + k^2 n\, u = 0$ outside the sources; that is, the scattered field u^s satisfies the inhomogeneous equation

$$\Delta u^s + k^2 n\, u^s = k^2 (1 - n) u^i \tag{7.6}$$

where the right-hand side is a function of compact support in $B(0, a)$. Furthermore, we expect the scattered field u^s to behave as a spherical wave far away from the medium. This can be described by the following *radiation condition*

$$\frac{\partial u^s(x)}{\partial r} - ik\, u^s(x) = \mathcal{O}(1/r^2) \quad \text{as } r = |x| \longrightarrow \infty, \tag{7.7}$$

uniformly in $x/|x| \in S^2$. Here we denote by S^2 the unit sphere in $\mathbb{R}^3$. The smoothness of the solution u^s depends on the smoothness of the refractive index n. We refer to the beginning of Subsection 7.2 for more details. We have now derived a (almost) complete description of the direct scattering problem.

Let the wave number $k > 0$, the index of refraction $n \in L^\infty(\mathbb{R}^3)$ with $n(x) = 1$ for $|x| \geq a$, and the incident field u^i be given. Determine the scattered field u^s that satisfies the source equation (7.6) in some generalized sense (to make more precise later) and the radiation condition (7.7).

In the inverse problem, one tries to determine the index of refraction n from measurements of the field u outside of $B(0, a)$ for several different incident fields u^i and/or different wave numbers k. The following example shows that the radially symmetric case reduces to an ordinary differential equation.

Example 7.1

Let $n = n(r)$ be radially symmetric: n is independent of the spherical coordinates. Because in spherical polar coordinates (r, ϕ, θ),

$$\Delta = \frac{1}{r^2} \frac{\partial}{\partial r} \left(r^2 \frac{\partial}{\partial r} \right) + \frac{1}{r^2 \sin^2 \theta} \frac{\partial^2}{\partial \phi^2} + \frac{1}{r^2 \sin \theta} \frac{\partial}{\partial \theta} \left(\sin \theta \frac{\partial}{\partial \theta} \right),$$

the Helmholtz equation for radially symmetric $u = u(r)$ reduces to the following ordinary differential equation of second order,

$$\frac{1}{r^2} \left(r^2 u'(r) \right)' + k^2 n(r) u(r) = 0;$$

that is,

$$u''(r) + \frac{2}{r} u'(r) + k^2 n(r) u(r) = 0 \quad \text{for } r > 0. \tag{7.8a}$$

From the theory of linear ordinary differential equations of second order with singular coefficients, we know that in a neighborhood of $r = 0$ there exist two linearly independent solutions, a regular one and one with a singularity at $r = 0$. We construct them by making the substitution $u(r) = v(r)/r$ in (7.8a). This yields the equation

$$v''(r) + k^2 n(r) v(r) = 0 \quad \text{for } r > 0. \tag{7.8b}$$

For the simplest case, where $n(r) = 1$, we readily see that $u_1(r) = \alpha \sin(kr)/r$ and $u_2(r) = \beta \cos(kr)/r$ are two linearly independent solutions. u_1 is regular and u_2 is singular at the origin. Neither of them satisfies the radiation condition. However, the combination $u(r) = \gamma \exp(ikr)/r$ does satisfy the radiation condition because

$$u'(r) - iku(r) = -\gamma \frac{\exp(ikr)}{r^2} = \mathcal{O}\left(\frac{1}{r^2} \right)$$

as is readily seen. For the case of arbitrary n, we construct a fundamental system $\{v_1, v_2\}$ of (7.8b) (compare with Section 5.2); that is, v_1 and v_2 satisfy (7.8b) with $v_1(0) = 0$, $v_1'(0) = 1$, and $v_2(0) = 1$, $v_2'(0) = 0$. Then $u_1(r) = v_1(r)/r$ is the regular and $u_2(r) = v_2(r)/r$ is the singular solution.

In the next section, we rigorously formulate the direct scattering problem and prove the uniqueness and existence of a solution. The basic ingredients for the uniqueness proof are a result by Rellich (see [222]) and a unique continuation principle for solutions of the Helmholtz equation. We prove neither

Rellich's lemma nor the general continuation principle, but rather give a simple proof for a special case of a unique continuation principle that is sufficient for the uniqueness proof of the direct problem. This suggestion was made by Hähner (see [116]). We then show the equivalence of the scattering problem with an integral equation. Existence is then proven by an application of the Riesz theorem A.36 of Appendix A. Section 7.3 is devoted to the introduction of the far field patterns that describe the scattered fields "far away" from the medium. We collect some results on the far field operator, several of which are needed in Sections 7.5 and 7.7. The question of injectivity of the far field operator is closely related to an unusual eigenvalue problem which we call the interior transmission eigenvalue problem. We will investigate this eigenvalue problem in Section 7.6. In Section 7.4, we prove uniqueness of the inverse problem. Section 7.5 is devoted to the factorization method which corresponds to the method in Section 6.4 and provides a very simple characterization of the support of the contrast by the far field patterns. This method is rigorously justified under the assumption that, again, the wavenumber is not an interior transmission eigenvalue. Since the interior transmission eigenvalue problem is also an interesting problem in itself and widely studied during the past decade we include Section 7.6 for some aspects of this eigenvalue problem. Finally, in Section 7.7, we present three classical numerical algorithms for solving the inverse scattering problem.

7.2   The Direct Scattering Problem

In this section, we collect properties of solutions to the Helmholtz equation that are needed later. We prove uniqueness and existence of the direct scattering problem and introduce the far field pattern. In the remaining part of this chapter, we restrict ourselves to scattering problems for *plane incident fields*.

Throughout this chapter, we make the following assumptions. Let $n \in L^\infty(\mathbb{R}^3)$ and $a > 0$ with $n(x) = 1$ for almost all $|x| \geq a$. Assume that $\operatorname{Re} n(x) \geq 0$ and $\operatorname{Im} n(x) \geq 0$ for almost all $x \in \mathbb{R}^3$. Let $k \in \mathbb{R}$, $k > 0$, and $\hat\theta \in \mathbb{R}^3$ with $|\hat\theta| = 1$. We set $u^i(x) := \exp(ik\hat\theta \cdot x)$ for $x \in \mathbb{R}^3$. Then u^i solves the Helmholtz equation

$$\Delta u^i + k^2 u^i = 0 \quad \text{in } \mathbb{R}^3. \tag{7.9}$$

We again formulate the direct scattering problem. Given n, k, $\hat\theta$ satisfying the previous assumptions, determine $u \in H^2_{loc}(\mathbb{R}^3)$ such that

$$\Delta u + k^2 n\, u = 0 \quad \text{in } \mathbb{R}^3, \tag{7.10}$$

and $u^s := u - u^i$ satisfies the *Sommerfeld radiation condition*

$$\frac{\partial u^s}{\partial r} - iku^s = \mathcal{O}(1/r^2) \quad \text{for } r = |x| \to \infty, \tag{7.11}$$

uniformly in $x/|x| \in S^2$. Since the index function n is not smooth we cannot expect that the solution u is smooth either. Rather, it belongs to the (local)

Sobolev space $H^2_{loc}(\mathbb{R}^3)$. We recall that for any open set $\Omega \subset \mathbb{R}^3$ and $p \in \mathbb{N}$ the Sobolev space $H^p(\Omega)$ is defined as the completion of

$$\left\{ u \in C^p(\overline{\Omega}) : \frac{\partial^{|j|_1}}{\partial x_1^{j_1} \partial x_2^{j_2} \partial x_3^{j_3}} u \in L^2(\Omega) \text{ for all } j \in \mathbb{N}_0^3 \text{ with } |j|_1 \leq p \right\}$$

with respect to the norm

$$\|u\|_{H^p(\Omega)} = \sqrt{\sum_{\substack{j \in \mathbb{N}_0^3 \\ |j|_1 \leq p}} \int_\Omega \left| \frac{\partial^{|j|_1} u(x)}{\partial x_1^{j_1} \partial x_2^{j_2} \partial x_3^{j_3}} \right|^2 dx}.$$

Here we have set $|j|_1 = |j_1| + |j_2| + |j_3|$ for $j = (j_1, j_2, j_3) \in \mathbb{N}_0^3$. We refer to Chapter 6 where we already used Sobolev spaces of functions on two-dimensional domains B. The local spaces $H^p_{loc}(\Omega)$ are defined by

$$H^p_{loc}(\Omega) = \{ u : \Omega \to \mathbb{C} : u|_B \in H^p(B) \text{ for every bounded domain } B \text{ with } \overline{B} \subset \Omega \}.$$

For us, the spaces $H^1(\Omega)$ and $H^2(\Omega)$ (and their local analogies) are particularly important. We define the subspace $H_0^p(\Omega)$ of $H^p(\Omega)$ by the closure of the set $C_0^p(\Omega) = \{ \psi \in C^p(\overline{\Omega}) : \psi \text{ has compact support in } \Omega \}$ in $H^p(\Omega)$. By definition it is closed, and one can show that it is a strict subspace of $H^p(\Omega)$. The space $H_0^1(\Omega)$ models the class of functions which vanish on the boundary of Ω while for functions in $H_0^2(\Omega)$ also the normal derivatives $\partial \psi / \partial \nu$ vanish at $\partial \Omega$. The following versions of Green's theorem are not difficult to prove by approximating u and v by sequences of smooth functions.

Lemma 7.2 *Let $\Omega \subset \mathbb{R}^3$ be a bounded domain. Then*

$$\int_\Omega [u \, \Delta v + \nabla u \cdot \nabla v] \, dx = 0 \quad \text{for all } u \in H_0^1(\Omega), \ v \in H^2(\Omega), \qquad (7.12a)$$

$$\int_\Omega [u \, \Delta v - v \, \Delta u] \, dx = 0 \quad \text{for all } u \in H^2(\Omega), \ v \in H_0^2(\Omega). \qquad (7.12b)$$

The application of (7.12a) to solutions of the Helmholtz equation yields the following version.

Lemma 7.3 *Let $v, w \in H^2(B(0, b)))$ be solutions of the Helmholtz equation $\Delta u + k^2 u = 0$ in some annular region $A = \{ x \in \mathbb{R}^3 : a < |x| < b \}$. Then v and w are analytic in A, and for every $R \in (a, b)$ it holds that*

$$\int_{|x|=R} v \frac{\partial w}{\partial r} \, ds = \int_{|x|<R} [\nabla v \cdot \nabla w + v \, \Delta w] \, dx. \qquad (7.13)$$

In particular, the total field u and thus the scattered field $u^s = u - u^i$ are analytic for $|x| > a$, and the radiation condition (7.11) is well-defined.

Proof: The smoothness of v and w follow from general regularity results for solutions of the Helmholtz equation $\Delta u + k^2 u = 0$ and is not proven here, see, e.g., [161], Corollary 3.4, or [55], Theorem 2.2. To show (7.13) we choose $\rho \in C^\infty(\mathbb{R}^3)$ with compact support in $B(0, b)$ such that $\rho(x) = 1$ for $|x| \leq R$. Then $\rho v \in H_0^1(B(0, b))$ as easily seen and thus by (7.12a)

$$
\begin{aligned}
0 &= \int\limits_{|x|<b} \left[\rho v\, \Delta w + \nabla w \cdot \nabla(\rho v) \right] dx \\
&= \int\limits_{|x|<R} \left[v\, \Delta w + \nabla w \cdot \nabla v \right] dx \;+\; \int\limits_{R<|x|<b} \left[\rho v\, \Delta w + \nabla w \cdot \nabla(\rho v) \right] dx\,.
\end{aligned}
$$

Because in the annular region $\{x \in \mathbb{R}^3 : R < |x| < b\}$ the functions v and w are smooth we apply the classical Green's first formula which yields (note that ρ vanishes for $|x| = b$ and is equal to one for $|x| = R$)

$$
\int\limits_{R<|x|<b} \left[\rho v\, \Delta w + \nabla w \cdot \nabla(\rho v) \right] dx \;=\; - \int\limits_{|x|=R} v\, \frac{\partial w}{\partial r}\, ds\,.
$$

This proves the assertion. $\square$

We will also need the following characterization of $H_0^2(\Omega)$ which is not obvious at all and holds only for domains Ω with sufficiently regular boundaries.

Lemma 7.4 *Let $\Omega \subset \mathbb{R}^3$ be a bounded* Lipschitz domain.[1] *Then*

$$
H_0^2(\Omega) \;=\; \left\{ u|_\Omega : u \in H^2(\mathbb{R}^3),\; u = 0 \text{ in } \mathbb{R}^3 \setminus \Omega \right\}.
$$

We need some further results from the theory of the Helmholtz equation. We omit some of the proofs and refer to [161, 53, 55] for a detailed investigation of the direct scattering problems. The proof of uniqueness relies on the following very important theorem, which we state without proof.

Lemma 7.5 *(Rellich)*
Let u satisfy the Helmholtz equation $\Delta u + k^2 u = 0$ for $|x| > a$. Assume, furthermore, that

$$
\lim_{R \to \infty} \int\limits_{|x|=R} |u(x)|^2 ds(x) \;=\; 0\,. \tag{7.14}
$$

Then $u = 0$ for $|x| > a$.

---

[1]For a definition see, e.g., [191, 161].

For the proof, we refer to [161] (Lemma 3.21) or [55] (Lemma 2.12). In particular, the condition (7.14) of this lemma is satisfied if $u(x)$ decays faster that $1/|x|$. Note that the assertion of this lemma does not hold if the imaginary part of k is positive or if $k = 0$.

The second important tool for proving uniqueness is the unique continuation principle. For the uniqueness proof, only a special case is sufficient. We present a simple proof by Hähner (see [116]), which is an application of the following result on periodic differential equations with constant coefficients. This lemma is also needed in the uniqueness proof for the inverse problem (see Section 7.4). First, we define the cube $Q := (-\pi, \pi)^3 \in \mathbb{R}^3$. Then every element $g \in L^2(Q)$ can be expanded into a Fourier series in the form

$$g(x) = \sum_{j \in \mathbb{Z}^3} g_j\, e^{i\,j\cdot x}, \quad x \in \mathbb{R}^3, \tag{7.15a}$$

with Fourier coefficients

$$g_j = \frac{1}{(2\pi)^3} \int_Q g(y)\, e^{-i\,j\cdot y}\, dy, \quad j \in \mathbb{Z}^3. \tag{7.15b}$$

The convergence of the series is understood in the L^2-sense. (See Section A.2 of the Appendix.) Then Parseval's equation holds in the form

$$(2\pi)^3 \sum_{j \in \mathbb{Z}^3} |g_j|^2 = \int_Q |g(y)|^2\, dy. \tag{7.15c}$$

In particular, $L^2(Q)$ can be defined by those functions g such that $\sum_{j \in \mathbb{Z}^3} |g_j|^2$ converges. Analogously, as in the one dimensional case of Section A.4 of the Appendix, for $p \in \mathbb{N}$ one defines the Sobolev space $H_{per}^p(Q)$ of periodic functions by

$$H_{per}^p(Q) = \left\{ g \in L^2(Q) : \|g\|_{H_{per}^p(Q)}^2 := \sum_{j \in \mathbb{Z}^3} [1 + |j|^2]^p\, |g_j|^2 < \infty \right\}.$$

Here we have set $|j| = \sqrt{j \cdot j} = \sqrt{j_1^2 + j_2^2 + j_3^2}$ for $j = (j_1, j_2, j_3) \in \mathbb{Z}^3$. Note that $\|g\|_{L^2(Q)} = (2\pi)^{3/2} \|g\|_{H_{per}^0(Q)}$. Then it is not difficult to show (see Problem 7.1) that $H_{per}^p(Q) \subset H^p(Q)$ and $H_0^p(Q) \subset H_{per}^p(Q)$. Furthermore, we identify $L^2(Q)$ and $H_{per}^p(Q)$ with the spaces of 2π-periodic functions on $\mathbb{R}^3$ with respect to all variables: they satisfy $g(2\pi j + x) = g(x)$ for almost all $x \in \mathbb{R}^3$ and $j \in \mathbb{Z}^3$.

Lemma 7.6 *Let $p \in \mathbb{R}^3$, $\alpha \in \mathbb{R}$, and $\hat{e} = (1, i, 0)^\top \in \mathbb{C}^3$. Then, for every $t > 0$ and every $g \in L^2(Q)$, there exists a unique solution $w = w_t(g) \in H_{per}^2(Q)$ of the differential equation*

$$\Delta w + (2t\hat{e} - ip) \cdot \nabla w - (it + \alpha)\, w = g \quad \text{in } \mathbb{R}^3. \tag{7.16}$$

Furthermore, the following estimate holds

$$\|w\|_{L^2(Q)} \leq \frac{1}{t} \|g\|_{L^2(Q)} \quad \text{for all } g \in L^2(Q), \ t > 0. \tag{7.17}$$

In other words, there exists a linear and bounded solution operator

$$L_t : L^2(Q) \to L^2(Q), \quad g \mapsto w_t(g),$$

of (7.16) with the property $\|L_t\|_{\mathcal{L}(L^2(Q))} \le 1/t$ for all $t > 0$.

Proof: We expand g into the Fourier series (7.15a) with Fourier coefficients (7.15b). The representation $w(x) = \sum_{j \in \mathbb{Z}^3} w_j \exp(i\,j \cdot x)$ leads to the equation

$$w_j \left[-|j|^2 + i\,j \cdot (2t\hat{e} - ip) - (it + \alpha) \right] = g_j, \quad j \in \mathbb{Z}^3,$$

for the coefficients w_j. For fixed $t > 0$ we estimate

$$\left| -|j|^2 + i\,j \cdot (2t\hat{e} - ip) - (it + \alpha) \right|$$
$$\ge \left| \mathrm{Re}[\cdots] \right| = \left| |j|^2 + 2tj_2 - j \cdot p + \alpha \right| \ge \frac{1}{2}\left[1 + |j|^2 \right]$$

for all $j \in \mathbb{Z}^3$ with $|j| \ge j_0$ for some $j_0 \in \mathbb{N}$. Furthermore,

$$\left| -|j|^2 + i\,j \cdot (2t\hat{e} - ip) - (it + \alpha) \right| \ge \left| \mathrm{Im}[\cdots] \right| = t|2j_1 - 1| \ge t$$

for all $j \in \mathbb{Z}^3$ and $t > 0$. Therefore,

$$w_j = \frac{g_j}{-|j|^2 + i\,j \cdot (ip + 2t\hat{e}) - (it + \alpha)}$$

are well defined for all $j \in \mathbb{Z}^3$ and $w \in H^2_{per}(Q)$ because

$$\sum_{|j| \ge j_0} [1 + |j|^2]^2 \, |w_j|^2 \le 4 \sum_{|j| \ge j_0} |g_j|^2 .$$

Furthermore, the solution operator

$$(L_t g)(x) := \sum_{j \in \mathbb{Z}^3} \frac{g_j}{-|j|^2 + i\,j \cdot (ip + 2t\hat{e}) - (it + \alpha)} \, e^{i\,j \cdot x}, \quad g \in L^2(Q),$$

is bounded from $L^2(Q)$ into itself with $\|L_t\|_{\mathcal{L}(L^2(Q))} \le 1/t$ for every $t > 0$. $\square$

Now we can give a simple proof of the following version of a unique continuation principle.

Theorem 7.7 *Let* $n \in L^\infty(\mathbb{R}^3)$ *with* $n(x) = 1$ *for* $|x| \ge a$ *be given. Let* $u \in H^2(\mathbb{R}^3)$ *be a solution of the Helmholtz equation* $\Delta u + k^2 n\,u = 0$ *in* $\mathbb{R}^3$ *such that* $u(x) = 0$ *for all* $|x| \ge b$ *for some* $b \ge a$. *Then* u *has to vanish in all of* $\mathbb{R}^3$.

Proof: Define $\hat{e} = (1, i, 0)^\top \in \mathbb{C}^3$ as before, set $\rho = 2b/\pi$, and define the function

$$w(x) := e^{i/2\,x_1 - t\,\hat{e} \cdot x} u(\rho x), \quad x \in Q := (-\pi, \pi)^3,$$

for some $t > 0$. Then $w(x) = 0$ for all $|x| \geq \pi/2$, in particular near the boundary
of the cube Q. Extend w to a 2π-periodic function in $\mathbb{R}^3$ by $w(2\pi j + x) := w(x)$
for $x \in Q$ and all $j \in \mathbb{Z}^3$, $j \neq 0$. Then $w \in H^2_{per}(Q)$, and w satisfies the
differential equation

$$\Delta w + (2t\hat{e} - ip) \cdot \nabla w - (it + 1/4)\, w = -\rho^2 k^2 \tilde{n}\, w\,.$$

Here, we have set $p = (1, 0, 0)^\top$ and $\tilde{n}(2\pi j + x) := n(\rho x)$ for almost all $x \in$
$[-\pi, \pi]^3$ and $j \in \mathbb{Z}^3$. Application of the previous lemma to this differential
equation yields the existence of a linear bounded operator L_t from $L^2(Q)$ into
itself with $\|L_t\|_{\mathcal{L}(L^2(Q))} \leq 1/t$ such that the differential equation is equivalent
to

$$w = -\rho^2 k^2\, L_t\big(\tilde{n}w\big)\,.$$

Estimating

$$\|w\|_{L^2(Q)} \leq \frac{\rho^2 k^2}{t}\, \|\tilde{n}w\|_{L^2(Q)} \leq \frac{\rho^2 k^2 \|n\|_\infty}{t}\, \|w\|_{L^2(Q)}$$

yields $w = 0$ for sufficiently large $t > 0$. Thus, also u has to vanish.   □

The preceding theorem is a special case of a far more general unique continuation principle, which we formulate without proof here.

*Let $u \in H^2_{loc}(\Omega)$ be a solution of the Helmholtz equation $\Delta u + k^2 n u = 0$ in a
domain $\Omega \subset \mathbb{R}^3$ (i.e., Ω is open and connected). Furthermore, let $n \in L^\infty(\Omega)$
and $u(x) = 0$ on some open set. Then $u = 0$ in all of Ω.*

For a proof we refer to, for example, [55].

Now we can prove the following uniqueness result.

Theorem 7.8 *(Uniqueness)*
*The problem (7.10), (7.11) has at most one solution; that is, if u is a solution
corresponding to $u^i = 0$, then $u = 0$.*

Proof: Let $u^i = 0$. The radiation condition (7.11) yields

$$\mathcal{O}(1/R^2) = \int\limits_{|x|=R} \left|\frac{\partial u}{\partial r} - ik\, u\right|^2 ds \qquad\qquad (7.18)$$

$$= \int\limits_{|x|=R} \left(\left|\frac{\partial u}{\partial r}\right|^2 + k^2 |u|^2\right) ds + 2k\, \mathrm{Im} \int\limits_{|x|=R} u\, \frac{\partial \overline{u}}{\partial r}\, ds\,.$$

We transform the last integral using Green's formula (7.13) for $v = u$ and $w = \overline{u}$;
that is,

$$\int\limits_{|x|=R} u\, \frac{\partial \overline{u}}{\partial r}\, ds = \int\limits_{|x|<R} \left[|\nabla u|^2 - k^2 \overline{n}\, |u|^2\right] dx\,,$$

and thus

$$\text{Im} \int_{|x|=R} u \frac{\partial \overline{u}}{\partial r} \, ds \; = \; k^2 \int_{|x|<R} \text{Im}\, n \, |u|^2 dx \; \geq \; 0.$$

We substitute this into (7.18) and let R tend to infinity. This yields

$$0 \; \leq \; \limsup_{R\to\infty} \int_{|x|=R} \left(\left| \frac{\partial u}{\partial r} \right|^2 + k^2 |u|^2 \right) ds \; \leq \; 0,$$

and thus

$$\int_{|x|=R} |u|^2 ds \; \longrightarrow \; 0 \quad \text{as } R \to \infty.$$

Rellich's Lemma 7.5 implies $u = 0$ for $|x| > a$. Finally, the unique continuation principle of Theorem 7.7 yields $u = 0$ in $\mathbb{R}^3$. $\qquad\square$

Now let

$$\Phi(x,y) \; := \; \frac{e^{ik|x-y|}}{4\pi|x-y|} \quad \text{for } x, y \in \mathbb{R}^3 \, , \; x \neq y \, , \tag{7.19}$$

be the *fundamental solution* or *free space Green's function* of the Helmholtz equation. Properties of the fundamental solution are summarized in the following theorem.

Theorem 7.9 $\Phi(\cdot, y)$ *solves the Helmholtz equation* $\Delta u + k^2 u = 0$ *in* $\mathbb{R}^3 \setminus \{y\}$ *for every* $y \in \mathbb{R}^3$. *It satisfies the radiation condition*

$$\frac{x}{|x|} \cdot \nabla_x \Phi(x,y) - ik\, \Phi(x,y) \; = \; \mathcal{O}(1/|x|^2)$$

uniformly in $x/|x| \in S^2$ *and* $y \in Y$ *for every bounded subset* $Y \subset \mathbb{R}^3$. *In addition,*

$$\Phi(x,y) \; = \; \frac{e^{ik|x|}}{4\pi|x|} e^{-ik\hat{x}\cdot y} + \mathcal{O}(1/|x|^2) \tag{7.20}$$

uniformly in $\hat{x} = \frac{x}{|x|} \in S^2$ *and* $y \in Y$.

The proof is not difficult and is left to the reader.

Before we turn to the question of existence we prove a general regularity result.

Lemma 7.10 *Let* $D \subset \mathbb{R}^3$ *be some bounded domain,* $n \in L^\infty(D)$, *and* $f \in L^2(D)$.

(a) Let $w \in L^2(D)$ be an ultra-weak solution of $\Delta w + k^2 n w = f$ in D; that is, let w satisfy

$$\int_D w\,(\Delta \psi + k^2 n \psi)\,dx \;=\; \int_D f\,\psi\,dx \qquad (7.21)$$

for all $\psi \in C^2(D)$ with compact support in D. Then $w \in H^2_{loc}(D)$; that is, $w \in H^2(A)$ for all domains A with $\overline{A} \subset D$. Furthermore, $\Delta w + k^2 n w = f$ almost everywhere in D, and for every domain A with $\overline{A} \subset D$ there exists $c > 0$ (depending only on n, D, and A) such that $\|w\|_{H^2(A)} \leq c\,\big[\|f\|_{L^2(D)} + \|w\|_{L^2(D)}\big]$.

(b) Every solution $w \in H^2(D)$ of $\Delta w + k^2 n w = f$ in D is also an ultra-weak solution.

Proof: (a) Let A be any open bounded set such that $\overline{A} \subset D$. We choose $\rho \in C^\infty(D)$ with compact support in D such that $\rho = 1$ on A. Furthermore, let Q be a cube containing $\overline{D}$ in its interior. For any $\phi \in C^\infty(\mathbb{R}^3)$ we take $\psi = \rho\phi$ in (7.21). With $\Delta \psi = \rho \Delta \phi + 2\nabla \rho \cdot \nabla \phi + \phi \Delta \rho$ we have

$$\int_D \rho\,w\,[\phi - \Delta \phi]\,dx$$
$$= \int_D \big\{ w\,[(k^2 n + 1)\rho\phi + 2\nabla \rho \cdot \nabla \phi + \phi \Delta \rho] - f\rho\phi \big\}\,dx \qquad (7.22)$$
$$= \int_D \big[g\,\phi + h \cdot \nabla \phi\big]\,dx$$

with $g = w[(k^2 n + 1)\rho + \Delta \rho] - f\rho$ and $h = 2w\nabla \rho$. We can replace the region of integration by Q because ρ vanishes in $D \setminus Q$. Without loss of generality we assume that $Q = (-\pi, \pi)^3$. Since $g, h \in L^2(Q)$ we expand them into Fourier series $g(x) = \sum_{j \in \mathbb{Z}^3} g_j e^{ij \cdot x}$ and $h(x) = \sum_{j \in \mathbb{Z}^3} h_j e^{ij \cdot x}$ with $g_j \in \mathbb{C}$ and $h_j \in \mathbb{C}^3$ such that $\sum_{j \in \mathbb{Z}^3} |g_j|^2 < \infty$ and $\sum_{j \in \mathbb{Z}^3} |h_j|^2 < \infty$ and make the ansatz $(\rho w)(x) = \sum_{j \in \mathbb{Z}^3} w_j e^{ij \cdot x}$ in Q. We take $\phi(x) = e^{-i\ell \cdot x}$ for some $\ell \in \mathbb{Z}^3$ in (7.22) and have $(1 + |\ell|^2)\,w_\ell = g_\ell - i\ell \cdot h_\ell$ and thus

$$(1 + |\ell|^2)\,|w_\ell|^2 \;\leq\; \frac{2}{1 + |\ell|^2}\big[|g_\ell|^2 + |\ell|^2|h_\ell|^2\big] \;\leq\; 2\big[|g_\ell|^2 + |h_\ell|^2\big] \qquad (7.23)$$

Therefore, $\rho w \in H^1_{per}(Q)$ and

$$\|w\|_{H^1(A)} \;\leq\; c_1 \|\rho w\|_{H^1_{per}(Q)} \;\leq\; c_2\big[\|g\|_{L^2(Q)} + \|h\|_{L^2(Q)}\big]$$
$$\leq\; c_3\big[\|w\|_{L^2(D)} + \|f\|_{L^2(D)}\big].$$

Since A was arbitrary we have shown that $w \in H^1_{loc}(D)$. Now we repeat the first part of the proof but apply Green's first formula to the second term of the right hand side of (7.22). This yields

$$\int_Q \rho w \left[\phi - \Delta\phi \right] dx = \int_Q \phi \underbrace{\left[w(k^2 n + 1)\rho - 2 \operatorname{div}(w\nabla\rho) + w\,\Delta\rho - f\rho \right]}_{=:\, g} dx .$$

Now we argue in the same way but with $h = 0$. Estimate (7.23) yields $(1 + |\ell|^2)^2 |w_\ell|^2 \leq 2|g_\ell|^2$ and thus $\rho w \in H^2_{per}(Q)$ and

$$\begin{aligned}
\|w\|_{H^2(A)} &\leq c_1 \|\rho w\|_{H^2_{per}(Q)} \leq c_2 \|g\|_{L^2(Q)} \\
&\leq c_3 \left[\|w\|_{L^2(D)} + \|\nabla w\|_{L^2(B)} + \|f\|_{L^2(D)} \right]
\end{aligned}$$

where B is the support of ρ. Now we substitute the estimate for $\|w\|_{H^1(B)}$ which yields the desired estimate.

(b) This follows directly from Green's theorem in the form (7.12b).   □

Now we construct volume potentials with the fundamental solution (7.19).

Theorem 7.11 *Let $\Omega \subset \mathbb{R}^3$ be a bounded domain. For every $\phi \in L^2(\Omega)$ the volume potential*

$$v(x) := \int_\Omega \phi(y)\,\Phi(x,y)\,dy , \qquad x \in \mathbb{R}^3 , \tag{7.24}$$

yields a function $v \in H^2_{loc}(\mathbb{R}^3)$ that satisfies the radiation condition (7.11) and is the only radiating[2] solution of $\Delta v + k^2 v = -\phi$.

Furthermore, for every ball $B = B(0, R)$ containing Ω in its interior there exists $c > 0$ (only dependent on B, k, and Ω) such that

$$\|v\|_{H^2(B)} \leq c \|\phi\|_{L^2(\Omega)} . \tag{7.25}$$

Proof: First we state without proof (see, e.g., [161], Theorem 3.9) that for any $k \in \mathbb{C}$ and $\phi \in C^1_0(\Omega)$; that is, $\phi \in C^1(\overline{\Omega})$ with compact support in Ω, the potential v is in $C^2(\mathbb{R}^3)$ and solves $\Delta v + k^2 v = -\phi$ in Ω and $\Delta v + k^2 v = 0$ in the exterior of $\overline{\Omega}$.

Second, we fix $\phi \in L^2(\Omega)$ and choose a sequence $\phi_j \in C^1_0(\Omega)$ which converges to ϕ in $L^2(\Omega)$. Let v and v_j be the corresponding potentials, and let $\psi \in C^\infty(\mathbb{R}^3)$ some test function with compact support. Then $\Delta v_j + k^2 v_j = -\phi_j$ in $\mathbb{R}^3$ and thus by Green's second theorem

$$\int_{\mathbb{R}^3} v_j\,(\Delta\psi + k^2\psi)\,dx = -\int_\Omega \phi_j\,\psi\,dx .$$

Let $B(0, R)$ be a ball that contains the support of ψ. From the boundedness of the volume integral operator from $L^2(\Omega)$ into $L^2(B(0, R))$ we conclude that v_j converges to v in $L^2(B(0, R))$ as j tends to infinity. Therefore,

$$\int_{\mathbb{R}^3} v\,(\Delta\psi + k^2\psi)\,dx = -\int_\Omega \phi\,\psi\,dx$$

---

[2]that is, it satisfies the radiation condition (7.11)

for all $\psi \in C^\infty(\mathbb{R}^3)$ with compact support. Therefore, v is an ultra-weak solution of $\Delta v + k^2 v = -\phi$ in $\mathbb{R}^3$. The regularity result of Lemma 7.10 applied to $D = B(0, R+1)$ yields $v \in H^2_{loc}(B(0, R+1))$ and the estimate

$$\|v\|_{H^2(B(0,R))} \leq c_1 \left[\|v\|_{L^2(B(0,R+1))} + \|\phi\|_{L^2(\Omega)}\right] \leq c_2 \|\phi\|_{L^2(\Omega)}$$

where we used again the boundedness of the volume potential from $L^2(\Omega)$ to $L^2(B(0, R+1))$. $\qquad \square$

Now we can transform the scattering problem into a Fredholm integral equation of the second kind. The following theorem is needed quite often later on.

Theorem 7.12 (a) Let $u \in H^2_{loc}(\mathbb{R}^3)$ be a solution of the scattering problem (7.10), (7.11). Then $u|_{B(0,a)}$ belongs to $L^2(B(0,a))$ and solves the Lippmann–Schwinger integral equation

$$u(x) = u^i(x) - k^2 \int_{|y| < a} (1 - n(y)) \, \Phi(x,y) \, u(y) \, dy, \quad x \in B(0,a). \quad (7.26)$$

(b) If, on the other hand, $u \in L^2(B(0,a))$ is a solution of the integral equation (7.26), then u can be extended by the right-hand side of (7.26) to a solution $u \in H^2_{loc}(\mathbb{R}^3)$ of the scattering problem (7.10), (7.11).

Proof: (a) Let u be a solution of (7.10), (7.11) and v the volume potential with density $k^2 (1 - n) u \in L^2(B(0,a))$. By Theorem 7.11 we conclude that $v \in H^2_{loc}(\mathbb{R}^3)$ and $\Delta v + k^2 v = k^2(n-1) u$. From $\Delta u + k^2 u = k^2(1-n) u$ and $\Delta u^i + k^2 u^i = 0$ we conclude that $\Delta(v + u^s) + k^2(v + u^s) = 0$. Furthermore, v and u^s both satisfy the radiation condition (7.11). The uniqueness Theorem 7.8 yields that $v + u^s = 0$, thus $u = u^i + u^s = u^i - v$. This proves the first part.

(b) Let $u \in L^2(B(0,a))$ be a solution of (7.26). Again define v as the volume potential with density $k^2(1-n) u \in L^2(B(0,a))$. Then $u = u^i - v$ in $B(0,a)$. Extend u by the right-hand side of this formula to all of $\mathbb{R}^3$. Again, by Theorem 7.11, we conclude that $v \in H^2_{loc}(\mathbb{R}^3)$ and $\Delta v + k^2 v = k^2(n-1) u$. Therefore, also $u \in H^2_{loc}(\mathbb{R}^3)$ and $\Delta u + k^2 u = -(\Delta v + k^2 v) = k^2(1-n) u$; that is, $\Delta u + k^2 n u = 0$. Therefore, $u^s = -v$, which ends the proof. $\qquad \square$

As a corollary, we derive the following result on existence.

Theorem 7.13 Under the given assumptions on k, n, and $\hat{\theta}$, there exists a unique solution u of the scattering problem (7.10), (7.11) or, equivalently, the integral equation (7.26).

Proof: We apply the Riesz theory (Theorem A.36 of Appendix A) to the integral equation $u = u^i - Tu$, where the operator T from $L^2(B(0,a))$ into itself is defined by

$$(Tu)(x) := k^2 \int_{|y| < a} (1 - n(y)) \, \Phi(x,y) \, u(y) \, dy, \quad |x| < a. \quad (7.27)$$

This integral operator is compact. There are several ways to prove this. The simplest is perhaps the observation that this integral operator is bounded from $L^2(B(0,a))$ into the Sobolev space $H^2(B(0,a))$ by Theorem 7.11. Furthermore, by Rellich's embedding theorem (see [1, 191]) the Sobolev space $H^2(B(0,a))$ is compactly embedded in $L^2(B(0,a))$. One can also argue directly by observing that the kernel $\Phi(x,y)$ of this integral operator is weakly singular (see Theorem A.35 of Appendix A for the one-dimensional case). Therefore, it is sufficient to prove uniqueness of a solution to (7.26). This follows by Theorems 7.12 and 7.8. $\square$

Remark 7.14 *From the proof we observe directly that the operator $I + T$ is an isomorphism from $L^2(B(0,a))$ onto itself.*

As another application of the Lippmann–Schwinger integral equation, we derive the following asymptotic behavior of u.

Theorem 7.15 *Let u be the solution of the scattering problem (7.10), (7.11). Then*

$$u(x) \;=\; u^i(x) \;+\; \frac{e^{ik|x|}}{|x|}\, u_\infty(\hat{x}) \;+\; \mathcal{O}(1/|x|^2) \quad \text{as } |x| \to \infty \qquad (7.28)$$

uniformly in $\hat{x} = x/|x|$, where

$$u_\infty(\hat{x}) \;=\; \frac{k^2}{4\pi} \int\limits_{|y|<a} \big(n(y)-1\big)\, e^{-ik\hat{x}\cdot y}\, u(y)\, dy \quad \text{for } \hat{x} \in S^2 . \qquad (7.29)$$

The function $u_\infty : S^2 \to \mathbb{C}$ is called the far field pattern *or* scattering amplitude *of u. It is analytic on S^2 and determines u^s outside of $B(0,a)$ uniquely; that is, $u_\infty = 0$ on S^2 if and only if $u^s(x) = 0$ for $|x| > a$.*

Proof: Formulas (7.28) and (7.29) follow directly from the asymptotic behavior (7.20) of the fundamental solution Φ. The analyticity of u_∞ follows from (7.29). Finally, if $u_\infty = 0$, then an application of Rellich's lemma yields that $u^s = u - u^i = 0$ for all $|x| > a$. $\square$

The existence of a far field pattern; that is, a function u_∞ with

$$u^s(x) \;=\; \frac{e^{ik|x|}}{|x|}\, u_\infty(\hat{x}) \;+\; \mathcal{O}(1/|x|^2) \quad \text{as } |x| \to \infty , \qquad (7.30)$$

is not restricted to scattering problems. Indeed, Theorem 7.16 below assures the existence of the far field pattern for every radiating solution of the Helmholtz equation.

We now draw some further conclusions from the Lippmann–Schwinger integral equation $u + Tu = u^i$. First we note that we can also treat the integral equation in $L^\infty(B(0,a))$ or even in $C(B[0,a])$ because the volume potential maps L^∞-functions u into continuous functions. In the following we consider T as an operator from $C\big(B[0,a]\big)$ into itself. We estimate the norm $\|T\|_{\mathcal{L}(C(B(0,a)))}$ of the integral operator T of (7.27) with respect to the sup-norm:

$$|(Tu)(x)| \leq k^2 \|1-n\|_\infty \|u\|_\infty \max_{|x|\leq a} \int_{|y|<a} |\Phi(x,y)|\, dy \quad \text{for } x \in B[0,a]\,;$$

that is,

$$\|T\|_\infty \leq k^2 \|1-n\|_\infty \max_{|x|\leq a} \int_{|y|<a} \frac{1}{4\pi|x-y|}\, dy$$

$$= \frac{(ka)^2}{2} \|1-n\|_\infty\,; \tag{7.31}$$

see Problem 7.4. We conclude that $\|T\|_{\mathcal{L}(C(B(0,a)))} < 1$, provided $(ka)^2\|1-n\|_\infty < 2$. The contraction mapping Theorem A.31 yields uniqueness and existence of a solution of the integral equation (7.26) for $(ka)^2\|1-n\|_\infty < 2$. We know this already even for all values of $(ka)^2\|1-n\|_\infty$. But Theorem A.31 also tells us that for $(ka)^2\|1-n\|_\infty < 2$ the solution can be represented as a Neumann series in the form

$$u = \sum_{j=0}^\infty (-1)^j\, T^j u^i\,. \tag{7.32}$$

The first two terms of the series are

$$u^b(x) := u^i(x) - k^2 \int_{|y|<a} \big(1-n(y)\big)\, u^i(y)\, \Phi(x,y)\, dy\,, \quad x \in \mathbb{R}^3\,. \tag{7.33}$$

u^b is called the *Born approximation*. It provides a good approximation to u in $B[0,a]$ for small values of $(ka)^2\|1-n\|_\infty$ because

$$\|u-u^b\|_\infty \leq \sum_{j=2}^\infty \|T\|_\infty^j \|u^i\|_\infty = \|T\|_\infty^2 \frac{1}{1-\|T\|_\infty} \leq \frac{(ka)^4}{2}\|1-n\|_\infty^2$$

for $(ka)^2\|1-n\|_\infty < 1$.

The far field pattern depends on both, the direction $\hat{x} \in S^2$ of observation and the direction $\hat{\theta} \in S^2$ of the incident field u^i. Therefore, we often write $u_\infty(\hat{x};\hat{\theta})$ to indicate this dependence. For the Born approximation, we see from the asymptotic form (7.20) of $\Phi(x,y)$ that

$$u_\infty^b(\hat{x};\hat{\theta}) = \frac{k^2}{4\pi} \int_{\mathbb{R}^3} \big(n(y)-1\big)\, e^{ik\hat{\theta}\cdot y}\, e^{-ik\hat{x}\cdot y}\, dy$$

$$= \frac{k^2}{4\pi} \int_{\mathbb{R}^3} \big(n(y)-1\big)\, e^{ik(\hat{\theta}-\hat{x})\cdot y}\, dy\,, \tag{7.34}$$

and this is just the Fourier transform of $m := n - 1$:

$$u_\infty^b(\hat{x}; \hat{\theta}) = \frac{k^2}{4\pi} m^\sim(k\hat{x} - k\hat{\theta}), \quad \hat{x}, \hat{\theta} \in S^2, \tag{7.35}$$

where the Fourier transform is defined by

$$f^\sim(x) := \int_{\mathbb{R}^3} f(y) \, e^{-ix \cdot y} dy, \quad x \in \mathbb{R}^3.$$

From this, the *reciprocity principle* follows:

$$u_\infty^b(-\hat{\theta}; -\hat{x}) = u_\infty^b(\hat{x}; \hat{\theta}) \quad \text{for } \hat{x}, \hat{\theta} \in S^2. \tag{7.36}$$

We show that this relation holds for u_∞ itself. Before we can prove this principle for u_∞, we need the important Green's representation theorem which expresses radiating solutions of the Helmholtz equation in terms of the Dirichlet and Neumann boundary data.

Theorem 7.16 *(Green's representation theorem)*
Let $\Omega \subset \mathbb{R}^3$ be a bounded domain and $\Omega^c := \mathbb{R}^3 \setminus \overline{\Omega}$ its exterior. Let the boundary $\partial\Omega$ be sufficiently smooth so that Gauss' theorem holds. Let the unit normal vector $\nu(x)$ in $x \in \partial\Omega$ be directed into the exterior of Ω.
(a) Let $u \in C^2(\Omega) \cap C^1(\overline{\Omega})$. Then

$$u(x) = \int_{\partial\Omega} \left[\Phi(x, y) \frac{\partial}{\partial\nu} u(y) - u(y) \frac{\partial}{\partial\nu(y)} \Phi(x, y) \right] ds(y)$$

$$- \int_\Omega \Phi(x, y) \left[\Delta u(y) + k^2 u(y) \right] dy, \quad x \in \Omega. \tag{7.37a}$$

(b) Let $u^s \in C^2(\Omega^c) \cap C^1(\overline{\Omega^c})$ be a solution of the Helmholtz equation $\Delta u^s + k^2 u^s = 0$ in Ω^c, and let u^s satisfy the radiation condition (7.11). Then

$$\int_{\partial\Omega} \left[\Phi(x, \cdot) \frac{\partial}{\partial\nu} u^s - u^s \frac{\partial}{\partial\nu} \Phi(x, \cdot) \right] ds = \begin{cases} 0, & x \in \Omega, \\ -u^s(x), & x \notin \overline{\Omega}. \end{cases} \tag{7.37b}$$

The far field pattern of u^s has the representation

$$u_\infty(\hat{x}) = \frac{1}{4\pi} \int_{\partial\Omega} \left[u^s(y) \frac{\partial}{\partial\nu(y)} e^{-ik\hat{x} \cdot y} - e^{-ik\hat{x} \cdot y} \frac{\partial}{\partial\nu} u^s(y) \right] ds(y) \tag{7.38}$$

for $\hat{x} \in S^2$.

For a proof, we refer to [161], Theorems 3.3 and 3.6, or [55], Theorems 2.1 and 2.5. As a corollary, we prove the following useful lemma.

Lemma 7.17 *Let $\Omega \in \mathbb{R}^3$ be a domain that is decomposed into two disjoint subdomains: $\overline{\Omega} = \overline{\Omega}_1 \cup \overline{\Omega}_2$ such that $\Omega_1 \cap \Omega_2 = \emptyset$. Let the boundaries $\partial\Omega_1$ and $\partial\Omega_2$ be smooth (i.e., C^2). Let $u_j \in C^2(\Omega_j) \cap C^1(\overline{\Omega}_j)$ for $j = 1, 2$ be solutions of the Helmholtz equation $\Delta u_j + k^2 u_j = 0$ in Ω_j. Furthermore, let $u_1 = u_2$ on Γ and $\partial u_1/\partial\nu = \partial u_2/\partial\nu$ on Γ, where Γ denotes the common boundary $\Gamma := \partial\Omega_1 \cap \partial\Omega_2$. Then the function u, defined by*

$$u(x) = \begin{cases} u_1(x), & x \in \Omega_1, \\ u_2(x), & x \in \Omega_2, \end{cases}$$

can be extended to an analytic function in Ω that satisfies the Helmholtz equation $\Delta u + k^2 u = 0$ in Ω.

Proof: It follows from Green's representation theorem that u_1 and u_2 are analytic in Ω_1 and Ω_2, respectively. We fix $x_0 \in \Gamma \cap \Omega$ and choose a small ball $B(x_0, \varepsilon)$ that is entirely contained in Ω. Let $B_j := B(x_0, \varepsilon) \cap \Omega_j$, $j = 1, 2$, and $x \in B_1$. We apply Green's representation theorems to u_1 in B_1 and to u_2 in B_2 and arrive at

$$u_1(x) = \int_{\partial B_1} \left[\Phi(x, y) \frac{\partial}{\partial\nu} u_1(y) - u_1(y) \frac{\partial}{\partial\nu(y)} \Phi(x, y) \right] ds(y), \quad x \in B_1,$$

$$0 = \int_{\partial B_2} \left[\Phi(x, y) \frac{\partial}{\partial\nu} u_2(y) - u_2(y) \frac{\partial}{\partial\nu(y)} \Phi(x, y) \right] ds(y), \quad x \in B_1.$$

We add both equations and note that the contributions on $\Gamma \cap B(x_0, \varepsilon)$ cancel. This yields

$$u_1(x) = \int_{\partial B(x_0, \varepsilon)} \left[\Phi(x, y) \frac{\partial}{\partial\nu} u(y) - u(y) \frac{\partial}{\partial\nu(y)} \Phi(x, y) \right] ds(y), \quad x \in B_1.$$

Interchanging the roles of $j = 1$ and $j = 2$ yields

$$u_2(x) = \int_{\partial B(x_0, \varepsilon)} \left[\Phi(x, y) \frac{\partial}{\partial\nu} u(y) - u(y) \frac{\partial}{\partial\nu(y)} \Phi(x, y) \right] ds(y), \quad x \in B_2.$$

The right-hand side defines an analytic function in $B(x_0, \varepsilon)$. $\square$

7.3 Properties of the Far Field Patterns

First, we prove a reciprocity principle for u_∞. It states the (physically obvious) fact that it is the same if we illuminate an object from the direction $\hat{\theta}$ and observe it in the direction $-\hat{x}$ or the other way around: illumination from $\hat{x}$ and observation in $-\hat{\theta}$.

Theorem 7.18 *(Reciprocity principle)*
Let $u_\infty(\hat{x}; \hat{\theta})$ be the far field pattern corresponding to the direction $\hat{x}$ of observation and the direction $\hat{\theta}$ of the incident plane wave. Then

$$u_\infty(\hat{x}; \hat{\theta}) \; = \; u_\infty(-\hat{\theta}; -\hat{x}) \quad \text{for all } \hat{x}, \hat{\theta} \in S^2 . \tag{7.39}$$

Proof: First we observe that the solutions u and thus also u^s of the scattering problems are analytic outside of the ball $B(0, a)$. Therefore, the Green's theorems and also Theorem 7.16 are applicable. Application of Green's second formula to u^i and u^s in the interior and exterior of $\{x \in \mathbb{R}^3 : |x| = a\}$, respectively, yields

$$0 \; = \; \int\limits_{|y|=a} \left[u^i(y; \hat{\theta}) \frac{\partial}{\partial \nu} u^i(y; -\hat{x}) - u^i(y; -\hat{x}) \frac{\partial}{\partial \nu} u^i(y; \hat{\theta}) \right] ds(y) ,$$

$$0 \; = \; \int\limits_{|y|=a} \left[u^s(y; \hat{\theta}) \frac{\partial}{\partial \nu} u^s(y; -\hat{x}) - u^s(y; -\hat{x}) \frac{\partial}{\partial \nu} u^s(y; \hat{\theta}) \right] ds(y) .$$

(More precisely, to prove the second equation, one applies Green's second formula to u^s in the region $\{x \in \mathbb{R}^3 : a < |x| < R\}$ with $R > a$ and lets R tend to infinity.)

Now we use the representations (7.38) for the far field patterns $u_\infty(\hat{x}; \hat{\theta})$ and $u_\infty(-\hat{\theta}; -\hat{x})$:

$$4\pi\, u_\infty(\hat{x}; \hat{\theta}) \; = \; \int\limits_{|y|=a} \left[u^s(y; \hat{\theta}) \frac{\partial}{\partial \nu} u^i(y; -\hat{x}) - u^i(y; -\hat{x}) \frac{\partial}{\partial \nu} u^s(y; \hat{\theta}) \right] ds(y) ,$$

$$4\pi u_\infty(-\hat{\theta}; -\hat{x}) \; = \; \int\limits_{|y|=a} \left[u^s(y; -\hat{x}) \frac{\partial}{\partial \nu} u^i(y; \hat{\theta}) - u^i(y; \hat{\theta}) \frac{\partial}{\partial \nu} u^s(y; -\hat{x}) \right] ds(y) .$$

We subtract the last of these equations from the sum of the first three. This yields

$$4\pi \left[u_\infty(\hat{x}; \hat{\theta}) \; - \; u_\infty(-\hat{\theta}; -\hat{x}) \right]$$

$$= \; \int\limits_{|y|=a} \left[u(y; \hat{\theta}) \frac{\partial}{\partial \nu} u(y; -\hat{x}) - u(y; -\hat{x}) \frac{\partial}{\partial \nu} u(y; \hat{\theta}) \right] ds(y) .$$

We have to show that this expression vanishes. But this follows directly from subtracting Green's formula (7.13) applied to $(v, w) = (u(\cdot; \hat{\theta}), u(\cdot; -\hat{x}))$ from the one applied to $(v, w) = (u(\cdot; -\hat{x}), u(\cdot; \hat{\theta}))$.   □

The far field patterns $u_\infty(\hat{x}; \hat{\theta})$, $\hat{x}, \hat{\theta} \in S^2$, define the integral operator

$$(Fg)(\hat{x}) \; = \; \int\limits_{S^2} u_\infty(\hat{x}; \hat{\theta}) \, g(\hat{\theta}) \, ds(\hat{\theta}) \quad \text{for } \hat{x} \in S^2 , \tag{7.40}$$

which we call the *far field operator*. It is certainly compact in $L^2(S^2)$ and is related to the *scattering operator* $S : L^2(S^2) \to L^2(S^2)$ by

$$S = I + \frac{ik}{2\pi} F.$$

The next results prove some properties of these operators. Some of them are important in Sections 7.5 and 7.7. We begin with a technical lemma (see [54]).

Lemma 7.19 *For $g, h \in L^2(S^2)$, define the* Herglotz *wave functions v^i and w^i by*

$$v^i(x) = \int_{S^2} e^{ikx \cdot \hat{\theta}} g(\hat{\theta}) \, ds(\hat{\theta}), \quad x \in \mathbb{R}^3, \tag{7.41a}$$

$$w^i(x) = \int_{S^2} e^{ikx \cdot \hat{\theta}} h(\hat{\theta}) \, ds(\hat{\theta}), \quad x \in \mathbb{R}^3, \tag{7.41b}$$

respectively. Let v and w be the solutions of the scattering problem (7.10), (7.11) corresponding to incident fields v^i and w^i, respectively. Then

$$ik^2 \int_{B(0,a)} (\operatorname{Im} n) \, v \, \overline{w} \, dx$$

$$= 2\pi (Fg, h)_{L^2(S^2)} - 2\pi (g, Fh)_{L^2(S^2)} - ik(Fg, Fh)_{L^2(S^2)}. \tag{7.42}$$

Proof: Let $v^s = v - v^i$ and $w^s = w - w^i$ denote the scattered fields with far field patterns v_∞ and w_∞, respectively. Then, by linearity, $v_\infty = Fg$ and $w_\infty = Fh$. Green's formula in the form (7.13) yields

$$\int_{|x|=a} \overline{w} \frac{\partial v}{\partial \nu} \, ds = \int_{|x|<a} \left[\nabla v \cdot \nabla \overline{w} - k^2 n v \overline{w} \right] dx.$$

Now we interchange the roles of v and $\overline{w}$ which yields

$$\int_{|x|=a} v \frac{\partial \overline{w}}{\partial \nu} \, ds = \int_{|x|<a} \left[\nabla v \cdot \nabla \overline{w} - k^2 \overline{n} v \overline{w} \right] dx.$$

Subtracting the results yields

$$2ik^2 \int_{B(0,a)} (\operatorname{Im} n) \, v \, \overline{w} \, dx = \int_{|x|=a} \left[v \frac{\partial \overline{w}}{\partial \nu} - \overline{w} \frac{\partial v}{\partial \nu} \right] ds.$$

The integral on the right-hand side is split into four parts by decomposing $v = v^i + v^s$ and $w = w^i + w^s$. The integral

$$\int_{|x|=a} \left[v^i \frac{\partial \overline{w}^i}{\partial \nu} - \overline{w}^i \frac{\partial v^i}{\partial \nu} \right] ds$$

vanishes by Green's second formula because v^i and $\overline{w}^i$ are solutions of the Helmholtz equation $\Delta u + k^2 u = 0$. We write

$$\int_{|x|=a} \left[v^s \frac{\partial \overline{w}^s}{\partial \nu} - \overline{w}^s \frac{\partial v^s}{\partial \nu} \right] ds = \int_{|x|=R} \left[v^s \frac{\partial \overline{w}^s}{\partial \nu} - \overline{w}^s \frac{\partial v^s}{\partial \nu} \right] ds$$

and note that by the radiation condition (7.11) and the form (7.30)

$$v^s(x) \frac{\partial \overline{w^s}(x)}{\partial r} - \overline{w^s}(x) \frac{\partial v^s(x)}{\partial r} = -\frac{2ik}{r^2} v_\infty(\hat{x}) \overline{w_\infty}(\hat{x}) + \mathcal{O}(1/r^3).$$

From this

$$\int_{|x|=R} \left[v^s \frac{\partial \overline{w}^s}{\partial \nu} - \overline{w}^s \frac{\partial v^s}{\partial \nu} \right] ds \longrightarrow -2ik \int_{S^2} v_\infty \overline{w_\infty}\, ds = -2ik\, (Fg, Fh)_{L^2(S^2)}$$

follows as R tends to infinity. Finally, we use the definition of v^i and w^i and the representation (7.38) to compute

$$\int_{|x|=a} \left[v^i \frac{\partial \overline{w}^s}{\partial \nu} - \overline{w}^s \frac{\partial v^i}{\partial \nu} \right] ds$$

$$= \int_{S^2} g(\hat{\theta}) \int_{|x|=a} \left[e^{ikx\cdot\hat{\theta}} \frac{\partial \overline{w^s}(x)}{\partial \nu} - \overline{w^s}(x) \frac{\partial}{\partial \nu} e^{ikx\cdot\hat{\theta}} \right] ds(x)\, ds(\hat{\theta})$$

$$= -4\pi \int_{S^2} g(\hat{\theta}) \overline{w_\infty}(\hat{\theta})\, d(\hat{\theta}) = -4\pi(g, Fh)_{L^2(S^2)}.$$

Analogously, we have that

$$\int_{|x|=a} \left[v^s \frac{\partial \overline{w}^i}{\partial \nu} - \overline{w}^i \frac{\partial v^s}{\partial \nu} \right] ds = 4\pi(Fg, h)_{L^2(S^2)}.$$

This ends the proof. □

We can now give a simple proof of the unitarity of the scattering operator for real-valued n.

Theorem 7.20 *Let $n \in L^\infty(\mathbb{R}^3)$ be real-valued such that the support of $n-1$ is contained in $B(0,a)$. Then F is normal (i.e., $F^*F = FF^*$), and the scattering operator $S := I + (ik)/(2\pi)\, F$ is unitary (i.e., $S^*S = SS^* = I$).*

Proof: The preceding lemma implies that

$$ik(Fg, Fh)_{L^2(S^2)} = 2\pi(Fg, h)_{L^2(S^2)} - 2\pi(g, Fh)_{L^2(S^2)} \tag{7.43}$$

for all $g, h \in L^2(S^2)$. By reciprocity (Theorem 7.18), we conclude that

$$(F^*g)(\hat{x}) \;=\; \int_{S^2} \overline{u_\infty(\hat{\theta}; \hat{x})} \, g(\hat{\theta}) \, ds(\hat{\theta}) \;=\; \int_{S^2} \overline{u_\infty(-\hat{x}; -\hat{\theta})} \, g(\hat{\theta}) \, ds(\hat{\theta})$$

$$=\; \overline{\int_{S^2} u_\infty(-\hat{x}; \hat{\theta}) \, \overline{g(-\hat{\theta})} \, ds(\hat{\theta})}$$

and thus $F^*g = \overline{RFR\overline{g}}$, where $(Rh)(\hat{x}) := h(-\hat{x})$ for $\hat{x} \in S^2$. Noting that $(Rg, Rh)_{L^2(S^2)} = (g, h)_{L^2(S^2)} = (\overline{h}, \overline{g})_{L^2(S^2)}$ for all $g, h \in L^2(S^2)$ and using (7.43) twice, we conclude that

$$
\begin{aligned}
ik\,(F^*h, F^*g)_{L^2(S^2)} \;&=\; ik\,(RFR\overline{g}, RFR\overline{h})_{L^2(S^2)} \;=\; ik\,(FR\overline{g}, FR\overline{h})_{L^2(S^2)} \\
&=\; 2\pi\,(FR\overline{g}, R\overline{h})_{L^2(S^2)} \;-\; 2\pi\,(R\overline{g}, FR\overline{h})_{L^2(S^2)} \\
&=\; 2\pi\,(RFR\overline{g}, \overline{h})_{L^2(S^2)} \;-\; 2\pi\,(\overline{g}, RFR\overline{h})_{L^2(S^2)} \\
&=\; 2\pi\,(h, F^*g)_{L^2(S^2)} \;-\; 2\pi\,(F^*h, g)_{L^2(S^2)} \\
&=\; 2\pi\,(Fh, g)_{L^2(S^2)} \;-\; 2\pi\,(h, Fg)_{L^2(S^2)} \\
&=\; ik\,(Fh, Fg)_{L^2(S^2)}\,.
\end{aligned}
$$

This holds for all $g, h \in L^2(S^2)$; thus $F^*F = F F^*$.

Finally, from (7.43), we conclude that

$$-(g, ikF^*Fh)_{L^2(S^2)} \;=\; 2\pi\big(g, (F^* - F)h\big)_{L^2(S^2)} \quad \text{for all } g, h \in L^2(S^2)\,;$$

that is, $ikF^*F = 2\pi(F - F^*)$. This formula, together with the normality of F, yields $\mathcal{S}^*\mathcal{S} = \mathcal{S}\mathcal{S}^* = I$ by substituting the definition of $\mathcal{S}$ into $\mathcal{S}^*\mathcal{S}$ and $\mathcal{S}\mathcal{S}^*$. $\square$

It is well known that the eigenvalues of unitary operators all lie on the unit circle in $\mathbb{C}$. From the definition $\mathcal{S} = I + (ik)/(2\pi)F$, we conclude that the eigenvalues of F lie on the circle $|2\pi i/k - z| = 2\pi/k$ with center $2\pi i/k$ and radius $2\pi/k$. We later show (Lemma 7.36) that the eigenvalues tend to zero from the right half of this circle. These properties hold for real-valued indices of refraction n. For further results for absorbing media (i.e., for which n is complex-valued), we refer to the original literature [54].

A number of numerical methods for determining the shape D of the support of the contrast $n - 1$, for example, the dual space method by Colton and Monk (or "superposition of incident fields", see [59, 55]) or the linear sampling method (see [51, 160]) study the question of unique solvability of the far field equation $Fg = f$; that is,

$$\int_{S^2} u_\infty(\hat{x}; \hat{\theta}) \, g(\hat{\theta}) \, ds(\hat{\theta}) \;=\; f(\hat{x}), \quad \hat{x} \in S^2,$$

for different right-hand sides f. The question of injectivity of the far field
operator F is particularly important. We show that the null space of F is characterized by the following unusual eigenvalue problem, the *interior transmission
eigenvalue problem* which will be the subject of investigation in Section 7.6. Let
D be some bounded Lipschitz domain that contains the support of $m = n - 1$.

Interior Transmission Eigenvalue Problem: Determine $k > 0$ and $v, w \in L^2(D)$,
$(v, w) \neq (0, 0)$, such that

$$\Delta v + k^2 v = 0 \text{ in } D, \qquad \Delta w + k^2 n w = 0 \text{ in } D, \qquad (7.44a)$$

$$v = w \text{ on } \partial D, \qquad \frac{\partial v}{\partial \nu} = \frac{\partial w}{\partial \nu} \text{ on } \partial D. \qquad (7.44b)$$

We also consider an inhomogeneous version of this system:

Interior Transmission Problem: Given $f, g \in L^2(\partial D)$, determine $v, w \in L^2(D)$
such that

$$\Delta v + k^2 v = 0 \text{ in } D, \qquad \Delta w + k^2 n w = 0 \text{ in } D, \qquad (7.45a)$$

$$w - v = f \text{ on } \partial D, \qquad \frac{\partial w}{\partial \nu} - \frac{\partial v}{\partial \nu} = g \text{ on } \partial D. \qquad (7.45b)$$

The solutions of (7.44a), (7.44b) and (7.45a), (7.45b), respectively, have to be
understood in the ultra-weak sense. To motivate the formulation we multiply
the equation for v and w with $\phi \in C^2(\overline{D})$ and $\psi \in C^2(\overline{D})$, respectively, where
$\phi = \psi$ and $\partial\phi/\partial\nu = \partial\psi/\partial\nu$ on ∂D, and apply Green's second formula in D
formally.

$$\int_D (\Delta\psi + k^2 n\psi)\, w \, dx = \int_{\partial D} \left[w \frac{\partial\psi}{\partial\nu} - \psi \frac{\partial w}{\partial\nu} \right] ds,$$

$$\int_D (\Delta\phi + k^2 \phi)\, v \, dx = \int_{\partial D} \left[v \frac{\partial\phi}{\partial\nu} - \phi \frac{\partial v}{\partial\nu} \right] ds = \int_{\partial D} \left[v \frac{\partial\psi}{\partial\nu} - \psi \frac{\partial v}{\partial\nu} \right] ds.$$

Subtraction and insertion of the boundary conditions yields the following form.

$$\int_D (\Delta\psi + k^2 n\psi)\, w \, dx \; - \; \int_D (\Delta\phi + k^2 \phi)\, v \, dx = \int_{\partial D} \left[f \frac{\partial\psi}{\partial\nu} - g\,\psi \right] ds$$

for all $\phi, \psi \in C^2(\overline{D})$ with $\phi = \psi$ and $\partial\phi/\partial\nu = \partial\psi/\partial\nu$ on ∂D. We take this form
as the definition of an ultra-weak solution of (7.45a), (7.45b) (and, analogously,
of (7.44a), (7.44b) for $f = g = 0$).

Definition 7.21 *Let D be a bounded Lipschitz domain.*

(a) *The wave number k is an* interior transmission eigenvalue *if there exists
no nontrivial pair $(v, w) \in L^2(D) \times L^2(D)$ of fields that satisfies (7.44a)
and (7.44b) in the ultra-weak sense; that is,*

$$\int_D (\Delta\psi + k^2 n\psi)\, w \, dx \; - \; \int_D (\Delta\phi + k^2 \phi)\, v \, dx = 0 \qquad (7.46a)$$

for all $\phi, \psi \in H^2(D)$ with $\phi - \psi \in H_0^2(D)$.

(b) Let $f, g \in L^2(\partial D)$. The pair $(v, w) \in L^2(D) \times L^2(D)$ is called an ultra-weak solution of (7.45a), (7.45b) if

$$\int\limits_D (\Delta\psi + k^2 n\psi)\, w\, dx - \int\limits_D (\Delta\phi + k^2\phi)\, v\, dx = \int\limits_{\partial D} \left[f\, \frac{\partial\psi}{\partial\nu} - g\,\psi \right] ds \quad (7.46b)$$

for all $\phi, \psi \in C^2(\overline{D})$ such that $\phi - \psi$ has compact support in D.

Note that in part (b) we replaced the H^2–test functions with smooth test functions to avoid the notion of traces of H^2–functions.[3] However, if one likes to use the trace theorem then one can take the same test functions as in part (a) (density argument).

We show the following theorem (see [61, 62, 154]).

Theorem 7.22 Let $D \subset \mathbb{R}^3$ be a bounded Lipschitz domain such that the exterior of D is connected and $n = 1$ outside of D.

(a) $g \in L^2(S^2)$ is a solution of the homogeneous integral equation

$$\int\limits_{S^2} u_\infty(\hat{x}; \hat{\theta})\, g(\hat{\theta})\, ds(\hat{\theta}) = 0, \quad \hat{x} \in S^2, \quad (7.47)$$

if and only if there exist $v, w \in L^2(D)$ such that (v, w) solve (7.44a), (7.44b) in the ultra-weak sense of (7.46a), and v is the Herglotz wave function defined by

$$v(x) = \int\limits_{S^2} e^{ikx\cdot\hat{y}}\, g(\hat{y})\, ds(\hat{y}), \quad x \in \mathbb{R}^3. \quad (7.48)$$

In particular, F is one-to-one if the system (7.44a), (7.44b) is only solvable by the trivial solution $v = w = 0$ in D; that is, if k is not an interior transmission eigenvalue.

(b) Let $z \in D$ be fixed. The integral equation

$$\int\limits_{S^2} u_\infty(\hat{x}; \hat{\theta})\, g(\hat{\theta})\, ds(\hat{\theta}) = e^{-ikz\cdot\hat{x}}, \quad \hat{x} \in S^2, \quad (7.49)$$

of the first kind is solvable in $L^2(S^2)$ if and only if the interior transmission problem

---

[3]We mention that the unit normal vector $\nu(x)$ exists for almost all $x \in \partial D$ and defines a L^∞–vector field. Therefore, the integral is well defined.

$$\Delta v + k^2 v = 0 \text{ in } D, \qquad \Delta w + k^2 n w = 0 \text{ in } D, \qquad (7.50a)$$

$$w(x) - v(x) = \frac{\exp(ik|x - z|)}{|x - z|} \qquad \text{on } \partial D, \qquad (7.50b)$$

$$\frac{\partial w(x)}{\partial \nu} - \frac{\partial v(x)}{\partial \nu} = \frac{\partial}{\partial \nu} \frac{\exp(ik|x - z|)}{|x - z|} \qquad \text{on } \partial D, \qquad (7.50c)$$

has an ultra-weak solution $w, v \in L^2(D)$ *in the sense of (7.46b), and* v *is of the form (7.48).*

(c) *For* $z \notin D$ *the integral equation (7.49) is never solvable in* $L^2(S^2)$.

Proof: (a) Let $g \in L^2(S^2)$ be a solution of (7.47) and define v by (7.48). We observe that the left-hand side of (7.47) is a superposition of far field patterns. Therefore, the far field pattern w_∞ of the scattered field w^s that corresponds to the incident field $w^i = v$ vanishes. The corresponding total field $w = w^s + v \in H^2_{loc}(\mathbb{R}^3)$ satisfies the Helmholtz equation $\Delta w + k^2 n\, w = 0$ in $\mathbb{R}^3$. By Rellich's lemma (Lemma 7.5), the scattered field $w^s = w - v$ vanishes outside of D and thus $w - v \in H^2_0(D)$ by Lemma 7.4. Let now $\phi, \psi \in H^2(D)$ such that $\phi - \psi \in H^2_0(D)$. Then we have with Green's formula (7.12b) and the differential equations for w and v

$$\int_D (w - v)(\Delta \psi + k^2 n\psi)\, dx = \int_D \psi\left[\Delta(w - v) + k^2 n(w - v)\right] dx$$

$$= k^2 \int_D (1 - n)\, v\, \psi\, dx\,;$$

that is,

$$\int_D w\,(\Delta\psi + k^2 n\psi)\, dx = \int_D v\,(\Delta\psi + k^2\psi)\, dx\,.$$

Furthermore, again by (7.12b),

$$\int_D v\left[\Delta(\phi - \psi) + k^2(\phi - \psi)\right] dx = 0. \qquad (7.51)$$

Combining these equations yields

$$\int_D w\,(\Delta\psi + k^2 n\psi)\, dx = \int_D v\,(\Delta\phi + k^2\phi)\, dx$$

which proves the first direction.

Now let v be of the form (7.48) and let there exist $w \in L^2(D)$ such that (v, w) solves the eigenvalue problem (7.44a), (7.44b) in the sense of (7.46a). We extend

w to all of $\mathbb{R}^3$ by setting $w := v$ on $\mathbb{R}^3 \setminus D$. Let $\psi \in C^2(\mathbb{R}^3)$ with compact support. Then

$$
\begin{aligned}
\int_{\mathbb{R}^3} w\,(\Delta\psi + k^2 n\psi)\,dx &= \int_D w\,(\Delta\psi + k^2 n\psi)\,dx + \int_{\mathbb{R}^3 \setminus D} v\,(\Delta\psi + k^2\psi)\,dx \\
&= \int_D v\,(\Delta\psi + k^2\psi)\,dx + \int_{\mathbb{R}^3 \setminus D} v\,(\Delta\psi + k^2\psi)\,dx \\
&= \int_{\mathbb{R}^3} v\,(\Delta\psi + k^2\psi)\,dx = 0
\end{aligned}
$$

by Green's second formula. Here we have used (7.46a) for $\phi = \psi$ and the fact that v is a smooth solution of the Helmholtz equation. Therefore, $w \in L^2_{loc}(\mathbb{R}^3)$ is an ultra-weak solution of $\Delta w + k^2 n w = 0$ in $\mathbb{R}^3$, compare with Lemma 7.10. The regularity result of that lemma yields $w \in H^2_{loc}(\mathbb{R}^3)$.

The difference $w - v$ vanishes in the exterior of D and obviously satisfies the radiation condition. Therefore, $w \in H^2_{loc}(\mathbb{R}^3)$ is the unique total field corresponding to the incident field v. The far field pattern w_∞ of the corresponding scattered field $w^s = w - v$ vanishes. As in the previous part, we see that w is a superposition

$$
w(x) = \int_{S^2} u(x; \hat{\theta})\, g(\hat{\theta})\, ds(\hat{\theta})
$$

of total fields. For the corresponding far field patterns we conclude that

$$
0 = w_\infty(\hat{x}) = \int_{S^2} u_\infty(\hat{x}; \hat{\theta})\, g(\hat{\theta})\, ds(\hat{\theta})
$$

for all $\hat{x} \in S^2$. This proves part (a).

(b) The proof is very similar to the preceding one. Let $g \in L^2(S^2)$ be a solution of (7.49) and define v as in (7.48). As in part (a), the integral is the far field pattern w_∞ corresponding to the total field w that satisfies the Helmholtz equation. Now w_∞ does not vanish but is equal to the function $\exp(-ikz{\cdot}x)$. By Theorem 7.9, the only radiating solution of the Helmholtz equation with this far field pattern is the spherical wave $\exp(ik|x - z|)/|x - z|$. Because z is contained in D and the exterior of D is connected, the scattered waves $w(x) - v(x)$ and $\exp(ik|x-z|)/|x-z|$ have to coincide outside of D. Now we modify the function $x \mapsto \exp(ik|x - z|)/|x - z|$ in a small ball $B[z, \varepsilon] \subset D$ centered at z such that the modified function—which we call F—is smooth in $\mathbb{R}^3$ and coincides with $\exp(ik|\cdot{-}z|)/|\cdot{-}z|$ in $\mathbb{R}^3 \setminus D$. Now we argue as in part (a). For $\phi, \psi \in C^2(\overline{D})$ such that $\phi - \psi$ has compact support in D we apply (7.12b) where we use that $w - v - F \in H^2_0(D)$. This yields

$$
\int_D (w - v - F)\,(\Delta\psi + k^2 n\psi)\,dx = k^2 \int_D (1 - n)\,v\,\psi\,dx - \int_D \psi\,(\Delta F + k^2 n F)\,dx\,,
$$

thus, using also (7.51) which is unchanged,

$$\int_D w \, (\Delta \psi + k^2 n \psi) \, dx \;=\; \int_D v \, (\Delta \phi + k^2 \phi) \, dx \;+\; \int_D (F\Delta\psi - \psi\Delta F) \, dx \,.$$

Application of Green's second theorem (classical because ∂D, F, and ψ are sufficiently smooth) proves the first direction.

The second direction is proved in the same way as in part (a). If v is of the form (7.48) and $w \in L^2(D)$ such that (v,w) solves (7.46b) then we set $w(x) := v(x) + \exp(ik|x - z|)/|x - z|$ in the exterior, choose F as above, and show that w is an ultra-weak solution of $\Delta w + k^2 n w = 0$ in $\mathbb{R}^3$. Therefore, w is the total field corresponding to the incident field v with scattered field $\exp(ik|x - z|)/|x - z|$. Equation (7.49) follows since $\exp(-ikz \cdot \hat{x})$ is the far field pattern.

(c) Assume, on the contrary, that (7.49) is solvable for some $g \in L^2(S^2)$ and define v as in (7.48). Then, as in part (b), the spherical wave $\exp(ik|x-z|)/|x-z|$ coincides with v in the exterior of $D \cup \{z\}$. This leads to a contradiction as in the proof of Theorem 6.14 because v is bounded in z and the spherical wave is singular for $x = z$. Here we note that any Lipschitz domain satisfies the exterior cone condition (see Problem 6.6 for a two-dimensional example). □

As an application of Theorem 7.22, we give conditions under which the range of the far field operator F from (7.40) is dense in $L^2(S^2)$. From $\big(Fg, h\big)_{L^2(S^2)} = \big(g, F^*h\big)_{L^2(S^2)}$ for all $g, h \in L^2(S^2)$, it is seen that the orthogonal complement of the range of F is characterized by the null space of the adjoint F^* of F.

Theorem 7.23 *The null space $\{h \in L^2(S^2) : F^*h = 0\}$ consists exactly of those functions $h \in L^2(S^2)$ for which the corresponding Herglotz wave functions*

$$v(x) \;:=\; \int_{S^2} e^{ikx\cdot\hat{y}} \, \overline{h(-\hat{y})} \, ds(\hat{y}), \quad x \in \mathbb{R}^3,$$

satisfy the interior transmission eigenvalue problem (7.44a), (7.44b) for some $w \in L^2(D)$.

Proof: By using the reciprocity principle (Theorem 7.18), we conclude that

$$F^*h \;=\; 0 \quad \Longleftrightarrow \quad \int_{S^2} \overline{u_\infty(\hat{\theta}; \hat{x})} \, h(\hat{\theta}) \, ds(\hat{\theta}) \;=\; 0 \quad \text{for all } \hat{x} \in S^2$$

$$\Longleftrightarrow \quad \int_{S^2} \overline{u_\infty(-\hat{x}; -\hat{\theta})} \, h(\hat{\theta}) \, ds(\hat{\theta}) \;=\; 0 \quad \text{for all } \hat{x} \in S^2$$

$$\Longleftrightarrow \quad \int_{S^2} u_\infty(\hat{x}; \hat{\theta}) \, \overline{h(-\hat{\theta})} \, ds(\hat{\theta}) \;=\; 0 \quad \text{for all } \hat{x} \in S^2 \,.$$

Application of Theorem 7.22 yields the assertion.   □

By the previous theorem, F is one-to-one and the range of F is dense in $L^2(S^2)$ if k is not an interior transmission eigenvalue. We will investigate the interior transmission eigenvalue problem (7.44a), (7.44b) in more detail in Section 7.6 below. In particular, we will show that the set of eigenvalues is discrete and accumulates at infinity.

The results (b) and (c) of Theorem 7.22 indicate that it should be possible to characterize the unknown set D by a criterion that depends on the solvability of the integral equation (7.49) of the first kind. A mathematically rigorous formulation of this idea leads to the linear sampling method. We note, however, that even for $z \in D$ the integral equation (7.49) is not always (even very rarely) solvable because of the additional requirement that in the solution (v, w) of (7.50a)–(7.50c) the part v has to be a Herglotz wave function. This observation led to the development of the factorization method which we present in Section 7.5 below.

7.4 Uniqueness of the Inverse Problem

In this section, we want to determine if the knowledge of the far field pattern $u_\infty(\hat{x}; \hat{\theta})$ provides enough information to recover the index of refraction $n = n(x)$. Therefore, let two functions $n_1, n_2 \in L^\infty(\mathbb{R}^3)$ be given with $n_1(x) = n_2(x) = 1$ for $|x| \geq a$. We assume that the corresponding far field patterns $u_{1,\infty}$ and $u_{2,\infty}$ coincide, and we wish to show that n_1 and n_2 also coincide. As a first simple case, we consider the Born approximation again. Let

$$u_{1,\infty}^b(\hat{x}; \hat{\theta}) = u_{2,\infty}^b(\hat{x}; \hat{\theta}) \quad \text{for all } \hat{x} \in S^2 \text{ and some } \hat{\theta} \in S^2.$$

Formula (7.35) implies that $m_1^\sim(k\hat{x} - k\hat{\theta}) = m_2^\sim(k\hat{x} - k\hat{\theta})$ for all $\hat{x} \in S^2$. Here, $m_j := n_j - 1$ for $j = 1, 2$. Therefore, the Fourier transforms of m_1 and m_2 coincide on a sphere with center $k\hat{\theta}$ and radius $k > 0$. This, however, is not enough to conclude that m_1 and m_2 coincide.

Let us now assume that

$$u_{1,\infty}^b(\hat{x}; \hat{\theta}) = u_{2,\infty}^b(\hat{x}; \hat{\theta}) \quad \text{for } all \ \hat{x} \in S^2 \text{ and all } \hat{\theta} \in S^2.$$

Then $m_1^\sim(k\hat{x} - k\hat{\theta}) = m_2^\sim(k\hat{x} - k\hat{\theta})$ for all $\hat{x}, \hat{\theta} \in S^2$. Therefore, the Fourier transforms coincide on the set $\{k(\hat{x} - \hat{\theta}) : \hat{x}, \hat{\theta} \in S^2\}$, which describes a ball in $\mathbb{R}^3$ with center zero and radius $2k$. The Fourier transforms of m_1 and m_2 are analytic functions, therefore the unique continuation principle for analytic functions yields that $m_1^\sim$ and $m_2^\sim$ coincide on all of $\mathbb{R}^3$ and thus $m_1 = m_2$. Therefore, the knowledge of $\{u_\infty^b(\hat{x}; \hat{\theta}) : \hat{x}, \hat{\theta} \in S^2\}$ is (theoretically) sufficient to recover the refractive index.

The same arguments also show that the knowledge of $u_\infty^b(\hat{x}; \hat{\theta})$ for all $\hat{x} \in S^2$, some $\hat{\theta} \in S^2$, and all k from an interval of $\mathbb{R}_{>0}$ is sufficient to recover n. We refer to Problem 7.2 for an investigation of this case.

These arguments hold for the Born approximation to the far field pattern. We now prove an analogous uniqueness theorem for the actual far field pattern, which is due to A. Nachman [199], R. Novikov [211], and A. Ramm [220]. The proof consists of three steps, which we formulate as lemmata. For the first result, we consider a fixed refraction index $n \in L^\infty(\mathbb{R}^3)$ with $n(x) = 1$ for $|x| \geq a$ and show that the span of all total fields that correspond to scattering problems with plane incident fields is dense in the space of solutions of the Helmholtz equation in $B(0, a)$.

Lemma 7.24 *Let $n \in L^\infty(\mathbb{R}^3)$ with $n(x) = 1$ for $|x| \geq a$. Let $u(\cdot; \hat\theta)$ denote the total field corresponding to the incident field $e^{ik\hat\theta \cdot x}$. Define the space $H \subset L^2(B(0, a))$ by*

$$H := \text{closure}_{L^2(B(0,a))} \left\{ v \in H^2(B(0, a)) : \Delta v + k^2 n\, v = 0 \text{ in } B(0, a) \right\}. \tag{7.52}$$

Then $\text{span}\{u(\cdot; \hat\theta) : \hat\theta \in S^2\}$ is dense in H with respect to the L^2–norm.

Proof: By Lemma 7.10 the closed subspace H is contained in the closed space

$$\tilde{H} := \left\{ v \in L^2(B(0, a)) : \int\limits_{|x|<a} v\,[\Delta\psi + k^2 n\psi]\, dx = 0 \text{ for all } \psi \in H_0^2(B(0, a)) \right\}$$

of all ultra-weak solutions.[4] We show that $\text{span}\{u(\cdot; \hat\theta) : \hat\theta \in S^2\}$ is even dense in $\tilde{H}$. Let $v \in \tilde{H}$ such that

$$\big(v, u(\cdot; \hat\theta)\big)_{L^2} = \int\limits_{|x|<a} v(x)\,\overline{u(x; \hat\theta)}\, dx = 0 \quad \text{for all } \hat\theta \in S^2,$$

where we write $(\cdot, \cdot)_{L^2}$ instead of $(\cdot, \cdot)_{L^2(B(0,a))}$. The Lippmann–Schwinger equation (7.26) yields $u(\cdot; \hat\theta) = (I + T)^{-1} u^i(\cdot; \hat\theta)$ with $u^i(x; \hat\theta) = \exp(ikx \cdot \hat\theta)$; thus

$$\begin{aligned} 0 &= \big(v, (I + T)^{-1} u^i(\cdot; \hat\theta)\big)_{L^2} = \big((I + T^*)^{-1} v, u^i(\cdot; \hat\theta)\big)_{L^2} \\ &= \big(w, u^i(\cdot; \hat\theta)\big)_{L^2} \quad \text{for all } \hat\theta \in S^2 \end{aligned} \tag{7.53}$$

where $w := (I + T^*)^{-1} v$. Then $w \in L^2(B(0, a))$, and w satisfies the "adjoint equation"

$$v(x) = w(x) + k^2 \big(1 - \overline{n(x)}\big) \int\limits_{|x|<a} \overline{\Phi(x, y)}\, w(y)\, dy, \quad x \in B[0, a];$$

that is,

$$\overline{v(x)} = \overline{w(x)} + k^2\big(1 - n(x)\big) \int\limits_{|x|<a} \Phi(x, y)\,\overline{w(y)}\, dy, \quad x \in B[0, a]. \tag{7.54}$$

---

[4] H and $\tilde{H}$ even coincide but we do not need this.

Now set

$$\tilde{w}(x) := \int\limits_{|y|<a} \overline{w(y)}\, \Phi(x,y)\, dy \quad \text{for } x \in \mathbb{R}^3.$$

Then $\tilde{w}$ is a volume potential with L^2-density $\overline{w}$. We know from Theorem 7.11 that $\tilde{w} \in H^2_{loc}(\mathbb{R}^3)$ satisfies $\Delta\tilde{w} + k^2\tilde{w} = -\overline{w}$ in $\mathbb{R}^3$ almost everywhere. The far field pattern $\tilde{w}_\infty$ of $\tilde{w}$ vanishes by (7.53) because

$$\overline{\tilde{w}_\infty(\hat\theta)} = \frac{1}{4\pi}\int\limits_{|y|<a} w(y)\, e^{ik\hat\theta\cdot y}\, dy = \frac{1}{4\pi}\big(w, u^i(\cdot;-\hat\theta)\big)_{L^2} = 0$$

for all $\hat\theta \in S^2$. Rellich's lemma implies that $\tilde{w}(x) = 0$ for $|x| \geq a$; that is, $\tilde{w} \in H^2_0(B(0,a))$ by Lemma 7.4. From (7.54) we conclude that

$$\overline{v} = \overline{w} + k^2(1-n)\tilde{w} = -\Delta\tilde{w} - k^2 n\tilde{w} \quad \text{in } B(0,a),$$

and thus

$$\int\limits_{|x|<a} |v|^2\, dx = \int\limits_{|x|<a} v\overline{v}\, dx = -\int\limits_{|x|<a} v\left[\Delta\tilde{w} - k^2 n\tilde{w}\right] dx = 0$$

since $v \in \tilde{H}$ and $\tilde{w} \in H^2_0(B(0,a))$. Therefore, v vanishes. □

The second lemma proves a certain "orthogonality relation" between solutions of the Helmholtz equation with different indices of refraction n_1 and n_2.

Lemma 7.25 *Let $n_1, n_2 \in L^\infty(\mathbb{R}^3)$ be two indices of refraction with $n_1(x) = n_2(x) = 1$ for all $|x| \geq a$ and assume that $u_{1,\infty}(\hat x;\hat\theta) = u_{2,\infty}(\hat x;\hat\theta)$ for all $\hat x, \hat\theta \in S^2$. Then*

$$\int\limits_{|x|<a} v_1(x)\, v_2(x)\left[n_1(x) - n_2(x)\right] dx = 0 \tag{7.55}$$

for all solutions $v_j \in H^2(B(0,a))$ of the Helmholtz equation $\Delta v_j + k^2 n_j v_j = 0$, $j = 1,2$, in $B(0,a)$.

Proof: Let $v_1 \in H^2(B(0,a))$ be any fixed solution of $\Delta v_1 + k^2 n_1 v_1 = 0$ in $B(0,a)$. By the denseness result of Lemma 7.24 it is sufficient to prove the assertion for $v_2 := u_2(\cdot;\hat\theta)$ and arbitrary $\hat\theta \in S^2$. We set $u = u_1(\cdot,\hat\theta) - u_2(\cdot,\hat\theta)$ which is the same as the difference of the corresponding scattered fields. From $u_{1,\infty}(\cdot,\hat\theta) = u_{2,\infty}(\cdot,\hat\theta)$ and Rellich's Lemma 7.5, it follows that $u \in H^2(\mathbb{R}^3)$ vanishes outside of $B(0,a)$; that is, $u \in H^2_0(B(0,a))$ by Lemma 7.4. Furthermore, u satisfies the inhomogeneous Helmholtz equation

$$\Delta u + k^2 n_1 u = k^2(n_2 - n_1)\, u_2(\cdot,\hat\theta) = k^2(n_2 - n_1)\, v_2 \quad \text{in } \mathbb{R}^3.$$

We multiply this equation by v_1 and integrate over $B(0, a)$.

$$k^2 \int_{|x|<a} (n_2 - n_1)\, v_1\, v_2\, dx \;=\; \int_{|x|<a} v_1 \left[\Delta u + k^2 n_1\, u\right] dx \;=\; 0$$

by Green's formula (7.12b). $\square$

The original proof of the third important "ingredient" of the uniqueness proof was first given in [258]. It is of independent interest and states that the set of all products $v_1 v_2$ of functions v_j that satisfy the Helmholtz equations $\Delta v_j + k^2 n_j v_j = 0$ in some bounded region Ω is dense in $L^2(\Omega)$. This is exactly the kind of argument we have used already for the uniqueness proof in the linearized problem of impedance tomography (see Theorem 6.9). The situation in this chapter is more complicated because we have to consider products of solutions of different differential equations with nonconstant coefficients. The idea is to construct solutions u of the Helmholtz equation $\Delta u + k^2 n\, u = 0$ in $B(0, a)$ that behave asymptotically as $\exp(z \cdot x)$. Here we take $n = n_1$ or n_2. The following result is crucial.

Theorem 7.26 *Let again $n(x) = 1$ for $|x| \geq a$. Then there exist $T > 0$ and $C > 0$ such that for all $z \in \mathbb{C}^3$ with $z \cdot z = \sum_{j=1}^{3} z_j^2 = 0$ and $|z| \geq T$ there exists a solution $u_z \in H^2\big(B(0, a)\big)$ of the differential equation*

$$\Delta u_z + k^2 n\, u_z = 0 \quad in\ B(0, a) \tag{7.56}$$

of the form

$$u_z(x) = e^{z \cdot x}\big(1 + v_z(x)\big), \quad x \in B(0, a). \tag{7.57}$$

where $v_z \in H^2\big(B(0, a)\big)$ satisfies the estimate

$$\|v_z\|_{L^2(B(0,a))} \leq \frac{C}{|z|} \quad for\ all\ z \in \mathbb{C}^3\ with\ z \cdot z = 0\ and\ |z| \geq T. \tag{7.58}$$

Proof: The proof consists of two parts. First, we construct v_z for the special case $z = t\hat{e}$, where $\hat{e} = (1, i, 0)^\top \in \mathbb{C}^3$ and t being sufficiently large. In the second part, we consider the general case by rotating the geometry.

Let $z = t\hat{e}$ for some $t > 0$. By scaling the functions as in the proof of Theorem 7.7, we can assume without loss of generality that $B(0, a)$ is contained in the cube $Q = [-\pi, \pi]^3 \subset \mathbb{R}^3$. We substitute the ansatz

$$u(x) = e^{t\hat{e} \cdot x}\big[1 + \exp(-i/2\, x_1)\, w_t(x)\big]$$

into the Helmholtz equation (7.56). This yields the following differential equation for w_t:

$$\Delta w_t(x) + (2t\hat{e} - ip) \cdot \nabla w_t(x) - (it + 1/4)\, w_t(x)$$
$$= -k^2 n(x)\, w_t(x) - k^2 n(x) \exp(i/2\, x_1) \quad in\ Q,$$

where $p = (1,0,0)^\top \in \mathbb{R}^3$. We refer to the proof of the unique continuation principle (Theorem 7.7) for the same kind of transformation.

We determine a 2π-periodic solution of this equation. Because this equation has the form of (7.16) (for $\alpha = 1/4$), we use the solution operator L_t of Lemma 7.6 and write this equation in the form

$$w_t + k^2 L_t(n w_t) = L_t \tilde{n} \quad \text{in } Q, \tag{7.59}$$

where we have set $\tilde{n}(x) = -k^2 n(x) \exp(i/2\, x_1)$. For large values of t, the operator $K_t : w \mapsto k^2 L_t(nw)$ is a contraction mapping in $L^2(Q)$. This follows from the estimates

$$\|K_t w\|_{L^2(Q)} = k^2 \|L_t(nw)\|_{L^2(Q)} \leq \frac{k^2}{t} \|nw\|_{L^2(Q)}$$

$$\leq \frac{k^2 \|n\|_\infty}{t} \|w\|_{L^2(Q)},$$

which implies that $\|K_t\|_{\mathcal{L}(L^2(Q))} < 1$ for sufficiently large $t > 0$. For these values of t there exists a unique solution $w_t \in H^2_{per}(Q)$ of (7.59). The solution depends continuously on the right-hand side, therefore we conclude that there exists $c > 0$ with

$$\|w_t\|_{L^2(Q)} \leq c\|L_t \tilde{n}\|_{L^2(Q)} \leq \frac{ck^2}{t} \|n\|_\infty$$

for all $t \geq T$ and some $T > 0$. This proves the theorem for the special choice $z = t\hat{e}$.

Now let $z \in \mathbb{C}^3$ be arbitrary with $z \cdot z = 0$ and $|z| \geq T$. From this, we observe that $|\mathrm{Re}\, z| = |\mathrm{Im}\, z|$ and $(\mathrm{Re}\, z) \cdot (\mathrm{Im}\, z) = 0$. We decompose z in the unique form $z = t(\hat{a} + i\hat{b})$ with $\hat{a}, \hat{b} \in S^2$ and $t > 0$ and $\hat{a} \cdot \hat{b} = 0$. We define the cross-product $\hat{c} = \hat{a} \times \hat{b}$ and the orthogonal matrix $R = [\hat{a}\,\hat{b}\,\hat{c}] \in \mathbb{R}^{3\times 3}$. Then $t\,R\hat{e} = z$ and thus $R^\top z = t\hat{e}$. The substitution $x \mapsto Rx$ transforms the Helmholtz equation (7.56) into

$$\Delta w(x) + k^2 n(Rx)\, w(x) = 0, \quad x \in B(0,a),$$

for $w(x) = v(Rx)$, $x \in B(0,a)$. Application of the first part of this proof yields the existence of a solution w of this equation of the form

$$w(x) = e^{t\hat{e}\cdot x}\left[1 + \exp(-i/2\, x_1)\, w_t(x)\right],$$

where w_t satisfies $\|w_t\|_{L^2(Q)} \leq C/t$ for $t \geq T$. From $v(x) = w(R^\top x)$, we conclude that

$$v(x) = e^{t\hat{e}\cdot R^\top x}\left[1 + \exp(-i/2\,\hat{a}\cdot x)\, w_t(R^\top x)\right]$$

$$= e^{z\cdot x}\left[1 + \exp(-i/2\,\hat{a}\cdot x)\, w_t(R^\top x)\right],$$

which proves the theorem also for the general case. $\qquad \square$

Now we are able to prove the following analogy of Calderón's approach (compare with the proof of Theorem 6.9).

Theorem 7.27 *Let again* $n_1, n_2 \in L^\infty(\Omega)$ *such that* $n_1(x) = n_2(x) = 1$ *for* $|x| \geq a$. *Then the span of the set*

$$P := \left\{ u_1\, u_2 : u_j \in H^2\big(B(0,a)\big) \text{ solves } \Delta u_j + k^2 n_j u_j = 0 \right\}$$

of products is dense in $L^1\big(B(0,a)\big)$. *(We note that the product of two* L^2- *functions in in* L^1 *by the theorem of Cauchy–Schwarz.)*

Proof: Since $L^\infty\big(B(0,a)\big)$ is the dual space of $L^1\big(B(0,a)\big)$ we have to show that any $g \in L^\infty\big(B(0,a)\big)$ with $\int_{|x|<a} u\, g\, dx = 0$ for all $u \in P$ has to vanish. Therefore, let $g \in L^\infty\big(B(0,a)\big)$ such that

$$\int\limits_{|x|<a} g(x)\, u_1(x)\, u_2(x)\, dx \;=\; 0 \tag{7.60}$$

for all solutions $u_j \in H^2\big(B(0,a)\big)$ of the Helmholtz equation $\Delta u_j + k^2 n_j u_j = 0$ in $B(0,a)$, $j = 1, 2$.

Fix an arbitrary vector $y \in \mathbb{R}^3 \setminus \{0\}$ and a number $\rho > 0$. Choose a unit vector $\hat{a} \in \mathbb{R}^3$ and a vector $b \in \mathbb{R}^3$ with $|b|^2 = |y|^2 + \rho^2$ such that $\{y, \hat{a}, b\}$ forms an orthogonal system in $\mathbb{R}^3$. Set

$$z^1 := \frac{1}{2}b \;-\; \frac{i}{2}\,(y + \rho\,\hat{a}) \quad \text{and} \quad z^2 := -\frac{1}{2}b \;-\; \frac{i}{2}\,(y - \rho\,\hat{a}).$$

Then $z^j \cdot z^j = |\operatorname{Re} z^j|^2 - |\operatorname{Im} z^j|^2 + 2i \operatorname{Re} z^j \cdot \operatorname{Im} z^j = |b|^2/4 - (|y|^2 + \rho^2)/4 = 0$ and $|z^j|^2 = \big(|b|^2 + |y|^2 + \rho^2\big)/4 \geq \rho^2/4$. Furthermore, $z^1 + z^2 = -i\,y$.

Now we apply Theorem 7.26 with z^j to the Helmholtz equations $\Delta u_j + k^2 n_j u_j = 0$ in $B(0,a)$. We substitute the forms (7.57) of u_j into the orthogonality relation (7.60) and arrive at

$$\begin{aligned}
0 &= \int\limits_{|x|<a} e^{(z^1+z^2)\cdot x} \big[1 + v_1(x)\big]\big[1 + v_2(x)\big]\, g(x)\, dx \\
&= \int\limits_{|x|<a} e^{-i\,y\cdot x} \big[1 + v_1(x) + v_2(x) + v_1(x)\, v_2(x)\big]\, g(x)\, dx\,.
\end{aligned}$$

By Theorem 7.26, there exist constants $T > 0$ and $C > 0$ with

$$\|v_j\|_{L^2(B(0,a))} \;\leq\; \frac{C}{|z^j|} \;\leq\; \frac{2C}{\rho}$$

for all $\rho \geq T$. Now we use the Cauchy–Schwarz inequality and let ρ tend to infinity. This yields

$$\int\limits_{|x|<a} e^{-i\,y\cdot x}\, g(x)\, dx \;=\; 0\,.$$

Because the vector $y \in \mathbb{R}^3 \setminus \{0\}$ was arbitrary, we conclude that the Fourier transform of g (extended by zero into $\mathbb{R}^3$) vanishes. This yields $g = 0$.   □

As a corollary, we have the following uniqueness theorem.

Theorem 7.28 *Let* $n_1, n_2 \in L^\infty(\mathbb{R}^3)$ *be two indices of refraction with* $n_1(x) = n_2(x) = 1$ *for all* $|x| \geq a$. *Let* $u_{1,\infty}$ *and* $u_{2,\infty}$ *be the corresponding far field patterns, and assume that they coincide; that is,* $u_{1,\infty}(\hat{x}; \hat{\theta}) = u_{2,\infty}(\hat{x}; \hat{\theta})$ *for all* $\hat{x}, \hat{\theta} \in S^2$. *Then* $n_1 = n_2$.

Proof: We combine the orthogonality relation of Lemma 7.25 with the denseness result of Theorem 7.27. This yields that $n_1 - n_2 \in L^\infty(B(0, a))$ satisfies $\int_{|x|<a}(n_1 - n_2)\, h\, dx = 0$ for all $h \in L^1(B(0, a))$. Therefore, $n_1 - n_2$ has to vanish. $\square$

The proof of Theorem 7.27 does not work in $\mathbb{R}^2$ because in that case there is no corresponding decomposition of y. However, using more complicated families of solutions, uniqueness of the two-dimensional case has been shown by Bukhgeim in [28].

7.5   The Factorization Method

It is the aim of this section to transfer the factorization method of the previous chapter to the present scattering problem.[5] Therefore, in this section we are only interested in determining the support of $n - 1$. We make the same kind of assumptions on this support as in the previous chapter (compare to Assumptions 6.10).

Assumption 7.29 *Let there exist finitely many Lipschitz domains* D_j, $j = 1, \ldots, M$, *such that* $\overline{D_j} \cap \overline{D_k} = \emptyset$ *for* $j \neq k$ *and such that the complement* $\mathbb{R}^3 \setminus \overline{D}$ *of the closure of the union* $D = \bigcup_{j=1}^{M} D_j$ *is connected. Furthermore, let* $n \in L^\infty(\mathbb{R}^3)$ *be real-valued such that there exists* $c_0 > 0$ *with* $n = 1$ *on* $\mathbb{R}^3 \setminus D$ *and* $m = n - 1 \geq c_0$ *on* D.

The far field operator F from (7.40) plays the role of the difference $\Lambda - \Lambda_1$ of the Neumann-Dirichlet operators. The first ingredient of the factorization method is again the factorization of the data operator. To motivate the operators that appear in the factorization we write the Helmholtz equation (7.10) in terms of the scattered field as

$$\Delta u^s + k^2 n u^s = k^2(1 - n)\, u^i = -k^2 m\, u^i \quad \text{in } \mathbb{R}^3, \qquad (7.61)$$

where we have again defined the contrast m by $m = n - 1$. The source on the right-hand side is of a special form. We allow more general sources and consider radiating[6] solutions $v \in H^2_{loc}(\mathbb{R}^3)$ of equations of the form

$$\Delta v + k^2 n v = -m\, f \quad \text{in } \mathbb{R}^3 \qquad (7.62)$$

---

[5]Actually, it was a scattering problem for which the factorization method was first discovered ([157]) before it was applied to the problem of electrical impedance tomography in [23, 24].

[6]that is, satisfies the radiation condition (7.11)

for any $f \in L^2(D)$. Here we extended m and f by zero into $\mathbb{R}^3$. This radiation problem has a unique solution for every $f \in L^2(D)$. Indeed, we take again $a > 0$ such that $\overline{D} \subset B(0, a)$ and consider the integral equation $v + Tv = g$ with the integral operator T from (7.27) and $g \in L^2(B(0, a))$ given by

$$g(x) := \int_{|y|<a} (n(y) - 1) \, \Phi(x, y) \, f(y) \, dy, \quad |x| < a.$$

By Remark 7.14 this equation $v + Tv = g$ has a unique solution. We rewrite the equation as

$$v(x) = k^2 \int_{|y|<a} (n(y) - 1) \, \Phi(x, y) \, v(y) \, dy \; + \; \int_{|y|<a} (n(y) - 1) \, \Phi(x, y) \, f(y) \, dy,$$

which shows (Theorem 7.11) that $v \in H^2_{loc}(\mathbb{R}^3)$, and v is a radiating solution of $\Delta v + k^2 v = k^2 (1 - n)v + (1 - n)f$; that is, $\Delta v + k^2 n v = -mf$.

We define the operator G from $L^2(D)$ into $L^2(S^2)$ by $Gf = v_\infty$ where $v \in H^2_{loc}(\mathbb{R}^3)$ is the radiating solution of (7.62). Then we can prove the following factorization of the far field operator F.

Theorem 7.30 *Again let $G : L^2(D) \to L^2(S^2)$ be defined by $Gf = v_\infty$, where $v \in H^2_{loc}(\mathbb{R}^3)$ is the radiating solution of (7.62). Then*

$$F = 4\pi k^2 \, G \, S^* \, G^*, \tag{7.63}$$

where S^* is the L^2-adjoint of $S : L^2(D) \to L^2(D)$ defined by

$$(S\psi)(x) = \frac{1}{m(x)} \, \psi(x) - k^2 \int_D \psi(y) \, \Phi(x, y) \, dy, \quad x \in D. \tag{7.64}$$

We note that the integral in the definition of S is a volume potential with density ψ and can be extended to a function $w \in H^2_{loc}(\mathbb{R}^3)$ that radiates and is a solution of

$$\Delta w + k^2 w = -\psi \quad \text{in } \mathbb{R}^3. \tag{7.65}$$

Proof of Theorem 7.30:   From (7.61) and the definition of G we observe that $u_\infty = k^2 G u^i$. As an auxiliary operator we define $\mathcal{H} : L^2(S^2) \to L^2(D)$ by

$$(\mathcal{H}g)(x) = \int_{S^2} g(\hat{\theta}) \, e^{ik \, x \cdot \hat{\theta}} ds(\hat{\theta}) = \int_{S^2} g(\hat{\theta}) \, u^i(x; \hat{\theta}) \, ds(\hat{\theta}), \quad x \in D, \tag{7.66}$$

where $u^i(\cdot; \hat{\theta})$ denotes the incident field of direction $\hat{\theta}$. By the superposition principle, Fg is the far field pattern corresponding to the incident field $\mathcal{H}g$; that is by (7.61), $F = k^2 G \mathcal{H}$. Now we consider the adjoint $\mathcal{H}^*$ of $\mathcal{H}$ which is given by

$$(\mathcal{H}^* \psi)(\hat{x}) = \int_D \psi(y) \, e^{-ik \, y \cdot \hat{x}} dy, \quad \hat{x} \in S^2.$$

From the asymptotic behavior (7.20) of the fundamental solution Φ we observe that $\mathcal{H}^*\psi = 4\pi\, w_\infty$ where w_∞ is the far field pattern of the volume potential

$$w(x) \;=\; \int_D \psi(y)\,\Phi(x,y)\,dy\,, \quad x \in \mathbb{R}^3\,.$$

By Theorem 7.11 the potential $w \in H^2_{loc}(\mathbb{R}^3)$ satisfies (7.65); that is,

$$\Delta w + k^2 n w \;=\; -m\left(\frac{1}{m}\psi \,-\, k^2 w\right).$$

Using the definition of G this yields $\mathcal{H}^*\psi = 4\pi\, w_\infty = 4\pi G\big(\psi/m - k^2 w\big) = 4\pi G S\psi$; that is, $\mathcal{H}^* = 4\pi\,GS$ and thus $\mathcal{H} = 4\pi\,S^*G^*$. Substituting this into $F = k^2 G\mathcal{H}$ yields the assertion. $\square$

Therefore, we arrived at a factorization of the far field operator in the form $F = G\mathcal{T}G^*$ with $\mathcal{T} = 4\pi k^2 S^*$. It has the same form of (6.17) (with A^* replaced by G) but there is an essential difference: In contrast to the operator T in the factorization (6.17) the operator $\mathcal{T}$ (i.e., S) fails to be self-adjoint. Otherwise, the operator F would be self-adjoint which is not the case. F is only normal by Theorem 7.20. However, we can prove an analogous characterization of D by the range of G as in Theorem 6.14.

Theorem 7.31 *For any $z \in \mathbb{R}^3$ define the function $\phi_z \in L^2(S^2)$ by*

$$\phi_z(\hat{x}) \;=\; e^{-ik\,z\cdot\hat{x}}\,, \quad \hat{x} \in S^2\,. \tag{7.67}$$

Then z belongs to D if and only if ϕ_z belongs to the range $\mathcal{R}(G)$ of G.

Proof: It is very similar to the proof of Theorem 6.14.
    First let $z \in D$. Choose a ball $B(z,\varepsilon) = \{x \in \mathbb{R}^3 : |x - z| < \varepsilon\}$ with center z and radius $\varepsilon > 0$ such that its closure $B[z,\varepsilon] \subset D$. Furthermore, choose a function $\varphi \in C^\infty(\mathbb{R}^3)$ such that $\varphi(x) = 0$ for $|x - z| \leq \varepsilon/2$ and $\varphi(x) = 1$ for $|x - z| \geq \varepsilon$ and set $v(x) = 4\pi\varphi(x)\Phi(x,z)$ for $x \in \mathbb{R}^3$. Then v is a C^∞-function and coincides with $4\pi\Phi(\cdot,z)$ outside of D. By (7.20) the far field pattern of v is given by ϕ_z. Therefore, $\phi_z = Gf$ with $f = -(\Delta v + k^2 n v)/m$ in D which proves the first part.
    Now let $z \notin D$ and assume, on the contrary, that $\phi_z = Gf \in \mathcal{R}(G)$ for some $f \in L^2(D)$. Let $v \in H^2_{loc}(\mathbb{R}^3)$ be the corresponding radiating solution of (7.62). Because ϕ_z is the far field pattern of $4\pi\Phi(\cdot,z)$ and Gf is the far field pattern of v we conclude from Rellich's lemma that $4\pi\Phi(\cdot,z) = v$ in the exterior of $D \cup \{z\}$. Now one argues exactly as in the proof of Theorem 6.14 or Theorem 7.22, part (c). If $z \notin \overline{D}$ then v is smooth in z and $4\pi\Phi(\cdot,z)$ has a singularity in z that leads to a contradiction. For $z \in \partial D$ one again chooses a bounded piece $C_0 \subset \mathbb{R}^3$ of an open cone with vertex at z (see Assumption 6.10 for the definition in $\mathbb{R}^2$ with an obvious extension to $\mathbb{R}^3$) and $C_0 \cap D = \emptyset$ and shows that $\Phi(\cdot,z) \notin H^2(C_0)$. This contradicts $v \in H^2(C_0)$ and the fact that v and $4\pi\Phi(\cdot,z)$ coincide. $\square$

The third step in the factorization method expresses the range of the operator G by the known far field operator F. First we again collect properties of the middle operator S of the factorization (7.63).

Theorem 7.32 *Again let $S : L^2(D) \to L^2(D)$ defined by (7.64) and $m = n-1$.*

(a) *Let S_0 be given by $S_0\psi = \psi/m$ in D. Then S_0 is bounded, self-adjoint, and coercive:*

$$(S_0\psi, \psi)_{L^2(D)} \geq \frac{1}{\|m\|_\infty} \|\psi\|^2_{L^2(D)} \quad \text{for all } \psi \in L^2(D). \tag{7.68}$$

(b) *The difference $S - S_0$ is compact from $L^2(D)$ into itself.*

(c) *S is an isomorphism from $L^2(D)$ onto itself.*

(d) *$\text{Im}(S\psi, \psi)_{L^2(D)} \leq 0$ for all $\psi \in L^2(D)$. Also, if k^2 is not an interior transmission eigenvalue (see Definition 7.21) then $\text{Im}(S\psi, \psi)_{L^2(D)} < 0$ for all ψ in the L^2-closure of the range $\mathcal{R}(G^*)$ of G^* with $\psi \neq 0$.*

Proof: (a) This is obvious because $S_0\psi$ is just the multiplication of ψ by a function that is bounded below by $1/\|m\|_\infty$ and above by $1/c_0$.

(b) We have already used (see the proof of Theorem 7.13 where this operator appears in the Lippmann–Schwinger equation) that the volume potential $S\psi - S_0\psi$ defines a compact operator in $L^2(D)$.

(c) By parts (a) and (b) and the Theorem of Riesz (Theorem A.36 of the Appendix) it is sufficient to prove injectivity of S. Let $S\psi = 0$ in D. Setting $\varphi = \psi/m$ we conclude that

$$\varphi - k^2 \int_D m(y)\, \varphi(y)\, \Phi(\cdot, y)\, dy = 0 \quad \text{in } D.$$

Therefore, φ solves the homogeneous Lippmann–Schwinger integral equation (7.26) and has thus to vanish by the uniqueness of the scattering problem. Therefore, ψ also vanishes.

(d) Let $\psi \in L^2(D)$ be arbitrary. Extend ψ and m by zero into $\mathbb{R}^3$ and set $f = \psi - k^2 m w$ in $\mathbb{R}^3$ where $w \in H^2_{loc}(\mathbb{R}^3)$ is the volume potential with density ψ. Then $S\psi = f/m|_D$. Because w satisfies $\Delta w + k^2 w = -\psi$ we observe that w satisfies also $\Delta w + k^2 n w = -\psi + k^2 m w = -f$. Now we compute, by replacing ψ by $f + k^2 m w$,

$$(S\psi, \psi)_{L^2(D)} = \int_D \frac{1}{m} f\, [\overline{f} + k^2 m\, \overline{w}]\, dx = \int_D \frac{1}{m} |f|^2\, dx + k^2 \int_D f\, \overline{w}\, dx$$

$$= \int_D \frac{1}{m} |f|^2\, dx - k^2 \int_{|x|<R} [\Delta w + k^2 n w]\, \overline{w}\, dx \tag{7.69}$$

where $R > a$ and $a > 0$ is again chosen such that $\overline{D}$ is contained in $B(0, a)$. Application of Green's formula (7.13) yields

$$\int_{|x|<R} [\Delta w + k^2 n w] \, \overline{w} \, dx = \int_{|x|<R} [k^2 n |w|^2 - |\nabla w|^2] \, \overline{w} \, dx + \int_{|x|=R} \overline{w} \frac{\partial w}{\partial \nu} \, ds .$$

Substituting this into (7.69) and taking the imaginary part yields

$$\mathrm{Im}(S\psi, \psi)_{L^2(D)} = -k^2 \mathrm{Im} \int_{|x|=R} \overline{w} \frac{\partial w}{\partial \nu} \, ds .$$

Now we use the fact that $|w(x)|$ decays as $1/|x|$ and, by the radiation condition (7.11), $\partial w / \partial r - ikw$ decays as $1/|x|^2$. Therefore, we can replace $\partial w / \partial \nu$ by ikw in the last formula and have that

$$\mathrm{Im}(S\psi, \psi)_{L^2(D)} = -k^3 \int_{|x|=R} |w|^2 ds + \mathcal{O}(1/R) .$$

Letting R tend to infinity yields by the definition (7.30) of the far field pattern

$$\mathrm{Im}(S\psi, \psi)_{L^2(D)} = -k^3 \int_{S^2} |w_\infty|^2 ds$$

which is nonpositive.

Now let $\psi \in$ closure $\mathcal{R}(G^*) = \mathcal{N}(G)^\perp$ such that $\mathrm{Im}(S\psi, \psi)_{L^2(D)} = 0$. As we see from the previous equation, the far field pattern w_∞ of the corresponding volume potential w vanishes. Rellich's Lemma 7.5 and unique continuation yield that w vanishes outside of D, thus $w \in H_0^2(D)$ by Lemma 7.4. Let $\phi \in H_0^2(D)$. We extend ϕ by zero into the exterior of D. Then $\phi \in H^2(\mathbb{R}^3)$ and $\hat{\phi} := \frac{1}{m}[\Delta \phi + k^2 n \phi] \in \mathcal{N}(G)$ by the definition of G because $\Delta \overline{\phi} + k^2 n \overline{\phi} = m\hat{\phi}$ in $\mathbb{R}^3$ with vanishing far field pattern. Now we set $\tilde{w} := \frac{\psi}{m} \in L^2(D)$ and $\tilde{v} = \tilde{w} - k^2 w$. Then

$$0 = (\psi, \hat{\phi})_{L^2(D)} = \int_D \psi \overline{\hat{\phi}} \, dx = \int_D \tilde{w} \, [\Delta \phi + k^2 n \phi] \, dx$$

for all $\phi \in H_0^2(D)$. We show that $(\tilde{v}, \tilde{w}) \in L^2(D) \times L^2(D)$ satisfies the interior transmission eigenvalue problem (7.46a). Indeed, let $\phi_1, \phi_2 \in H^2(D)$ with $\phi_1 -$

$\phi_2 \in H_0^2(D)$. We take $\phi = \phi_1 - \phi_2$ in the previous equation for $\tilde{w}$; that is,

$$\int_D \tilde{w} \left[\Delta\phi_1 + k^2 n\phi_1\right] dx = \int_D \tilde{w} \left[\Delta\phi_2 + k^2 n\phi_2\right] dx$$

$$= \int_D \tilde{v} \left[\Delta\phi_2 + k^2 n\phi_2\right] dx + k^2 \int_D w \left[\Delta\phi_2 + k^2 n\phi_2\right] dx$$

$$= \int_D \tilde{v} \left[\Delta\phi_2 + k^2 \phi_2\right] dx + k^2 \int_D w \left[\Delta\phi_2 + k^2 \phi_2\right] dx$$

$$+ k^2 \int_D m\,\phi_2 \left[\tilde{v} + k^2 w\right] dx$$

$$= \int_D \tilde{v} \left[\Delta\phi_2 + k^2 \phi_2\right] dx + k^2 \int_D \left[\psi\,\phi_2 + w\left(\Delta\phi_2 + k^2\phi_2\right)\right] dx .$$

It remains to show that the last integral vanishes. Since $w \in H_0^2(D)$ we apply Green's formula (7.12b) and have that

$$\int_D w \left[\Delta\phi_2 + k^2\phi_2\right] dx = \int_D \phi_2 \left[\Delta w + k^2 w\right] dx = -\int_D \phi_2 \psi \, dx .$$

Therefore, $(\tilde{v}, \tilde{w}) \in L^2(D) \times L^2(D)$ satisfies the interior transmission eigenvalue problem (7.46a). By assumption $\tilde{v} = \tilde{w} = 0$ in D which implies that ψ vanishes in D and ends the proof.   □

Now we continue with the task of expressing the range of G by the known operator F. We make the assumption that k is not an interior transmission eigenvalue in the sense that (7.44a), (7.44b) is only solvable by the trivial solution. Then F is one-to-one by Theorem 7.22 and, furthermore, normal by Theorem 7.20 and certainly compact. Therefore, there exists a complete set of orthonormal eigenfunctions $\psi_j \in L^2(S^2)$ with corresponding eigenvalues $\lambda_j \in \mathbb{C}$, $j = 1, 2, 3, \ldots$ (see, e.g., [227]). Furthermore, because the operator $I + (ik)/(2\pi)\,F$ is unitary (see again Theorem 7.20), the eigenvalues λ_j of F lie on the circle of radius $1/r$ and center i/r where $r = k/(2\pi)$. We can now argue exactly as in the corresponding case of impedance tomography. The spectral theorem for normal operators yields that F has the form

$$F\psi = \sum_{j=1}^{\infty} \lambda_j (\psi, \psi_j)_{L^2(S^2)}\, \psi_j, \quad \psi \in L^2(S^2). \tag{7.70}$$

Therefore, F has a second factorization in the form

$$F = (F^*F)^{1/4}\, R\, (F^*F)^{1/4}, \tag{7.71}$$

where the self-adjoint operator $(F^*F)^{1/4} : L^2(S^2) \to L^2(S^2)$ and the signum $R : L^2(S^2) \to L^2(S^2)$ of F are given by

$$(F^*F)^{1/4}\psi = \sum_{j=1}^{\infty} \sqrt{|\lambda_j|}\, (\psi, \psi_j)_{L^2(S^2)}\, \psi_j\,, \quad \psi \in L^2(S^2)\,, \qquad (7.72)$$

$$R\psi = \sum_{j=1}^{\infty} \frac{\lambda_j}{|\lambda_j|}\, (\psi, \psi_j)_{L^2(S^2)}\, \psi_j\,, \quad \psi \in L^2(S^2)\,. \qquad (7.73)$$

Again, as in the case of impedance tomography (see (6.25)) we have thus derived two factorizations of F, namely

$$F = 4\pi k^2 G\, S^*\, G^* = (F^*F)^{1/4} R\, (F^*F)^{1/4}\,. \qquad (7.74)$$

We now show that these factorizations of F imply again that the ranges of G and $(F^*F)^{1/4}$ coincide. Application of Theorem 7.31 provides then the desired characterization of D by F.

The following functional analytic result is a slight extension of Lemma 6.15.

Lemma 7.33 *Let X and Y be Hilbert spaces and $F : X \to X$ and $G : Y \to X$ be linear bounded operators such that the factorization $F = GRG^*$ holds for some linear and bounded operator $R : Y \to Y$ that satisfies a coercivity condition of the form: there exists $c > 0$ with*

$$|(Ry, y)_Y| \geq c\|y\|_Y^2 \quad \text{for all } y \in \mathcal{R}(G^*) \subset Y\,. \qquad (7.75)$$

Then, for any $\phi \in X$, $\phi \neq 0$,

$$\phi \in \mathcal{R}(G) \iff \inf\{|(Fx, x)_X| : x \in X,\ (x, \phi)_X = 1\} > 0\,. \qquad (7.76)$$

We omit the proof because it follows exactly the same lines as the proof of Lemma 6.15 (see also [160]).

We note again that the inf-condition depends only on F and not on the factorization itself. Therefore, we have the following corollary.

Corollary 7.34 *Let X, Y_1, and Y_2 be Hilbert spaces. Furthermore, let $F : X \to X$ have two factorizations of the form $F = G_1 R_1 G_1^* = G_2 R_2 G_2^*$ with bounded operators $G_j : Y_j \to X$ and $R_j : Y_j \to Y_j$, which both satisfy the coercivity condition (7.75). Then the ranges of G_1 and G_2 coincide.*

In order to apply this corollary to the factorization (7.74) we have to prove that $S : L^2(D) \to L^2(D)$ and $R : L^2(S^2) \to L^2(S^2)$ from (7.64) and (7.73), respectively, satisfy the coercivity conditions (7.75). The coercivity condition for S follows from Theorem 7.32.

Lemma 7.35 *Let k^2 be no interior transmission eigenvalue in the sense of Definition 7.21. Then there exists $c_1 > 0$ such that*

$$|(S\varphi, \varphi)_{L^2(D)}| \geq c_1\|\varphi\|_{L^2(D)}^2 \quad \text{for all } \varphi \in \mathcal{R}(G^*) \subset L^2(D) \qquad (7.77)$$

where again $G^* : L^2(S^2) \to L^2(D)$ is the adjoint of G.

Proof: We assume, on the contrary, that there exists a sequence $\varphi_j \in \mathcal{R}(G^*)$ with $\|\varphi_j\|_{L^2(D)} = 1$ and $(S\varphi_j, \varphi_j)_{L^2(D)} \to 0$. The unit ball is weakly (sequentially) compact in $L^2(D)$. This is again a conclusion from the theorem of Alaoglu-Bourbaki (see Corollary A.78 of Appendix A.9). Therefore, there exists a weakly convergent subsequence of (φ_j). We denote this by $\varphi_j \rightharpoonup \varphi$ and note that φ belongs to the closure of the range of G^*. Let S_0 be the operator from Theorem 7.32. Then,

$$
\begin{aligned}
\big(\varphi - \varphi_j, S_0(\varphi - \varphi_j)\big)_{L^2(D)} & \hspace{3cm} (7.78) \\
= \ & \big(\varphi, S_0(\varphi - \varphi_j)\big)_{L^2(D)} - \big(\varphi_j, (S_0 - S)(\varphi - \varphi_j)\big)_{L^2(D)} \\
& + (\varphi_j, S\varphi_j)_{L^2(D)} - (\varphi_j, S\varphi)_{L^2(D)} \, .
\end{aligned}
$$

From the compactness of $S - S_0$ we note that $\|(S - S_0)(\varphi - \varphi_j)\|_{L^2(D)}$ tends to zero (see Theorem A.76, part (e)) and thus also $\big(\varphi_j, (S_0 - S)(\varphi - \varphi_j)\big)_{L^2(D)}$ by the Cauchy–Schwarz inequality. Therefore, the first three terms on the right-hand side of (7.78) converge to zero, the last one to $(\varphi, S\varphi)_{L^2(D)}$. Taking the imaginary part and noting that $\big(\varphi - \varphi_j, S_0(\varphi - \varphi_j)\big)_{L^2(D)}$ is real-valued yields $\varphi = 0$ by part (d) of Theorem 7.32. Now we write, using the coercivity of S_0 by part (a) of Theorem 7.32,

$$
\frac{1}{\|m\|_\infty} \leq (\varphi_j, S_0\varphi_j)_{L^2(D)} \leq \big|(\varphi_j, (S_0 - S)\varphi_j)_{L^2(D)}\big| + \big|(\varphi_j, S\varphi_j)_{L^2(D)}\big| \, ,
$$

and the right-hand side tends to zero which is certainly a contradiction.   □

Coercivity of the middle operator $R : L^2(S^2) \to L^2(S^2)$ in the second factorization of (7.74) can be proven by using the fact that the scattering operator is unitary. Before doing this we prove a result of independent interest.

Lemma 7.36 *Let k^2 be no interior transmission eigenvalue in the sense of Definition 7.21 and let $\lambda_j \in \mathbb{C}$, $j \in \mathbb{N}$, be the eigenvalues of the normal far field operator F. Then λ_j lie on the circle $|2\pi i/k - z| = 2\pi/k$ with center $2\pi i/k$ and radius $2\pi/k$ passing through the origin and converging to zero from the right; that is, $\mathrm{Re}\,\lambda_j > 0$ for sufficiently large j.*

Proof: The fact that λ_j lie on the circle with center $2\pi i/k$ passing through the origin follows from the unitarity of the scattering operator $S = I + (ik)/(2\pi)\,F$ (see Theorem 7.20). We have only to show that the eigenvalues tend to zero from the right. Let ψ_j again be the normalized and orthogonal eigenfunctions of F corresponding to the nonvanishing eigenvalues λ_j. From the factorization (7.63) it follows that

$$
4\pi\,k^2 (S^*\,G^*\psi_j\,,\,G^*\psi_\ell)_{L^2(D)} = (F\psi_j\,,\,\psi_\ell)_{L^2(S^2)} = \lambda_j\,\delta_{j,\ell}
$$

with the Kronecker symbol $\delta_{j,\ell} = 1$ for $j = \ell$ and 0 otherwise. We set

$$
\varphi_j = \frac{2k\sqrt{\pi}}{\sqrt{|\lambda_j|}}\,G^*\psi_j \quad \text{and} \quad s_j = \frac{\lambda_j}{|\lambda_j|} \, .
$$

Then $(\varphi_j, S\varphi_\ell)_{L^2(D)} = s_j \delta_{j,\ell}$. From the facts that λ_j lie on the circle with center $2\pi i/k$ passing through the origin and that λ_j tends to zero as j tends to infinity we conclude that the only accumulation points of the sequence (s_j) can be $+1$ or -1. The assertion of the theorem is proven once we have shown that $+1$ is the only possible accumulation point. Assume, on the contrary, that $s_j \to -1$ for a subsequence. From Lemma 7.35 we observe that the sequence (φ_j) is bounded in $L^2(D)$. Therefore, there exists a weakly convergent subsequence (see Corollary A.78 of Appendix A.9) that we denote by $\varphi_j \rightharpoonup \varphi$. Now we write exactly as in equation (7.78):

$$
\begin{aligned}
&\left(\varphi - \varphi_j, S_0(\varphi - \varphi_j)\right)_{L^2(D)} \\
&= \left(\varphi, S_0(\varphi - \varphi_j)\right)_{L^2(D)} - \left(\varphi_j, (S_0 - S)(\varphi - \varphi_j)\right)_{L^2(D)} \\
&\quad + \underbrace{(\varphi_j, S\varphi_j)_{L^2(D)}}_{= s_j} - (\varphi_j, S\varphi)_{L^2(D)}.
\end{aligned}
$$

The left-hand side is real-valued, the right-hand side tends to $-1 - (\varphi, S\varphi)_{L^2(D)}$. Taking the imaginary part again shows that φ has to vanish, thus as before

$$
0 \leq (\varphi_j, S_0\varphi_j)_{L^2(D)} = (\varphi_j, (S_0 - S)\varphi_j)_{L^2(D)} + (\varphi_j, S\varphi_j)_{L^2(D)}.
$$

The right-hand side converges to -1 which is impossible and ends the proof. $\square$

Now we can easily prove coercivity of the operator R in (7.74).

Lemma 7.37 *Assume that k is not an interior transmission eigenvalue. Then there exists $c_2 > 0$ with*

$$
\left|(R\psi, \psi)_{L^2(S^2)}\right| \geq c_2 \|\psi\|_{L^2(S^2)}^2 \quad \text{for all } \psi \in L^2(S^2). \tag{7.79}
$$

Proof: It is sufficient to prove (7.79) for $\psi \in L^2(S^2)$ of the form $\psi = \sum_j c_j \psi_j$ with $\|\psi\|_{L^2(S^2)}^2 = \sum_j |c_j|^2 = 1$. With the abbreviation $s_j = \lambda_j/|\lambda_j|$ it is

$$
\left|(R\psi, \psi)_{L^2(S^2)}\right| = \left|\left(\sum_{j=1}^\infty s_j c_j \psi_j, \sum_{j=1}^\infty c_j \psi_j\right)_{L^2(S^2)}\right| = \left|\sum_{j=1}^\infty s_j |c_j|^2\right|.
$$

The complex number $\sum_{j=1}^\infty s_j |c_j|^2$ belongs to the closure of the convex hull $\mathcal{C} = \text{conv}\{s_j : j \in \mathbb{N}\} \subset \mathbb{C}$ of the complex numbers s_j, see (A.53) of Appendix A.8). We conclude that

$$
\left|(R\psi, \psi)_{L^2(S^2)}\right| \geq \inf\{|z| : z \in \mathcal{C}\}
$$

for all $\psi \in L^2(S^2)$ with $\|\psi\|_{L^2(S^2)} = 1$. From the previous lemma we know that the set $\mathcal{C}$ is contained in the part of the upper half-disk that is above the line $\ell = \{t\hat{s} + (1-t)1 : t \in \mathbb{R}\}$ passing through $\hat{s}$ and 1. Here, $\hat{s}$ is the point in $\{s_j : j \in \mathbb{N}\}$ with the smallest real part. (The reader should draw a picture.)

Therefore, the distance of the origin to this convex hull $\mathcal{C}$ is positive; that is, there exists c_2 with (7.79).   □

By the range identity of Corollary 7.34 the ranges of G and $(F^*F)^{1/4}$ coincide. The combination of this result and Theorem 7.31 yields the main result of this section.

Theorem 7.38 *Assume that k^2 is not an interior transmission eigenvalue. For any $z \in \mathbb{R}^3$ again define $\phi_z \in L^2(S^2)$ by (7.67); that is,*

$$\phi_z(\hat{x}) := e^{-ik\,\hat{x}\cdot z}, \quad \hat{x} \in S^2.$$

Then

$$z \in D \iff \phi_z \in \mathcal{R}\big((F^*F)^{1/4}\big). \tag{7.80}$$

We want to rewrite this condition using Picard's Theorem A.58 of Appendix A.6. Again let $\lambda_j \in \mathbb{C}$ be the eigenvalues of the normal operator F with corresponding normalized eigenfunctions $\psi_j \in L^2(S^2)$. Then we note that $(\sqrt{|\lambda_j|}, \psi_j, \psi_j)$ is a singular system of $(F^*F)^{1/4}$. Therefore, Picard's theorem A.58 converts the condition $\phi_z \in \mathcal{R}\big((F^*F)^{1/4}\big)$ into a decay behavior of the expansion coefficients.

Theorem 7.39 *Under the assumptions of the previous theorem a point $z \in \mathbb{R}^3$ belongs to D if and only if the series*

$$\sum_j \frac{\big|(\phi_z, \psi_j)_{L^2(S^2)}\big|^2}{|\lambda_j|} \tag{7.81}$$

converges.

If we agree on the notation $1/\infty = 0$ and $\text{sign}(t) = 1$ for $t > 0$ and $\text{sign}(t) = 0$ for $t = 0$ then

$$\chi(z) = \text{sign}\left[\sum_j \frac{\big|(\phi_z, \psi_j)_{L^2(S^2)}\big|^2}{|\lambda_j|}\right]^{-1}, \quad z \in \mathbb{R}^3, \tag{7.82}$$

is just the characteristic function of D. Formula (7.82) provides a simple and fast technique to visualize the object D. One simply plots the inverse of the series (7.81). In practice, this is a finite sum instead of a series, but the value of the finite sum is much larger for points z outside than for points inside D. We refer to the original paper [157] and to [160] for some typical plots.

We conclude this section with some further remarks on the factorization method. The characteristic function χ derived in the previous theorem depends only on the operator F. Nothing else about the scattering medium has to be known for plotting this function. In particular, it is not assumed that the support D of $n - 1$ is connected; it can very well consist of several components.

Also, the function χ can be plotted in every case where the scattering operator S is unitary (and thus F is normal). This is the case, for example, if the medium is perfectly soft or hard; that is, if a Dirichlet or Neumann boundary condition, respectively, on ∂D is imposed. The theoretical justification of the factorization method however (i.e., the proof that χ is indeed the characteristic function of D), has to be given in every single case. For the Dirichlet and Neumann boundary condition and also for the impedance boundary condition $\partial u/\partial \nu + \lambda u = 0$ on ∂D with a real-valued function λ this can be shown (see [160]). This implies, in particular, a general uniqueness result. It is not possible that different "scattering supports" D give rise to the same far field operator F. There are, however, cases of unitary scattering operators for which the factorization method has not yet been justified. For example, if D consists of two components D_1 and D_2 (separated from each other) and $n \geq 1 + c_0$ on D_1 but $n \leq 1 - c_0$ on D_2 it is not known whether the factorization method is valid. The same open question arises for the case where the Dirichlet boundary condition is imposed on ∂D_1 and the Neumann boundary condition on ∂D_2. The main problem is the range identity; that is, the characterization of the range of G by the known operator F.

There also exist extensions of the factorization method for absorbing media. In these cases, the far field operator fails to be normal. Although some results exist on the existence of eigenvalues (see, e.g., [54]) the methods to construct the second factorization as in (7.74) fail. Instead, one considers factorizations of the self-adjoint operator $F_\# = |\operatorname{Re} F| + |\operatorname{Im} F|$ where $\operatorname{Re} F = (F + F^*)/2$ and $\operatorname{Im} F = (F - F^*)/(2i)$ are the self-adjoint parts of F, and $|A|$ of a self-adjoint operator A is defined by its spectral system. We refer to [160] for a comprehensive study of these cases.

7.6   The Interior Transmission Eigenvalue Problem

As we have just seen, it is an important assumption for the factorization method to work that the wavenumber k is not an eigenvalue of the interior transmission eigenvalue problem (7.44a), (7.44b). This is one of the motivations to study this eigenvalue problem in more detail. We mention the monographs [34] and [55] with chapters on transmission eigenvalues and also the special issue [37] of the journal *Inverse Problems* which indicates the importance of this topic. First, we recall the Definition 7.21 for the convenience of the reader.

The wave number k is an *interior transmission eigenvalue* if there exists no nontrivial pair $(v, w) \in L^2(D) \times L^2(D)$ of fields that satisfies

$$\Delta v + k^2 v = 0 \text{ in } D, \qquad \Delta w + k^2 n w = 0 \text{ in } D, \qquad (7.83a)$$

$$v = w \text{ on } \partial D, \qquad \frac{\partial v}{\partial \nu} = \frac{\partial w}{\partial \nu} \text{ on } \partial D. \qquad (7.83b)$$

in the ultra-weak sense; that is,

$$\int_D (\Delta\psi + k^2 n\psi)\, w\, dx \;-\; \int_D (\Delta\phi + k^2\phi)\, v\, dx \;=\; 0 \qquad (7.84)$$

for all $\phi, \psi \in H^2(D)$ with $\phi - \psi \in H_0^2(D)$.

In the case when the index of refraction has a nonvanishing imaginary part (i.e., the medium is *absorbing*), there exist no eigenvalues.

Theorem 7.40 *If* $\operatorname{Im} n(x) \geq 0$ *on* D *and* $\operatorname{Im} n(x) > 0$ *on some open set* $A \subset D$ *then the eigenvalue problem (7.44a), (7.44b) has no real eigenvalues* $k > 0$.

Proof: Let $(v, w) \in L^2(D) \times L^2(D)$ be a solution of (7.44a) and (7.44b) corresponding to some eigenvalue $k > 0$. We choose any $\psi \in C^\infty(\mathbb{R}^3)$ with compact support and set $\phi = \psi$ in (7.84). Then

$$\int_D v\, [\Delta\psi + k^2\psi]\, dx \;=\; \int_D w\, [\Delta\psi + k^2 n\psi]\, dx\,;$$

that is,

$$\int_D (v - w)\, [\Delta\psi + k^2\psi]\, dx \;=\; k^2 \int_D (n - 1)\, w\,\psi\, dx\,.$$

Set

$$u \;=\; \begin{cases} v - w & \text{in } D\,, \\ 0 & \text{in } \mathbb{R}^3 \setminus D\,. \end{cases}$$

Then

$$\int_{\mathbb{R}^3} u\, [\Delta\psi + k^2\psi]\, dx \;=\; k^2 \int_D (n - 1)\, w\,\psi\, dx$$

for all $\psi \in C^\infty(\mathbb{R}^3)$ with compact support. The regularity result of Lemma 7.10 implies $u \in H^2(\mathbb{R}^3)$ and

$$\Delta u + k^2 u \;=\; k^2 (n - 1)\, w \quad \text{almost everywhere in } \mathbb{R}^3 \qquad (7.85)$$

where we have set the right hand side to zero outside of D. In particular, $u \in H_0^2(D)$ by Lemma 7.4. Setting $\psi = \bar{u}$ and $\phi = 0$ in (7.84) we conclude that

$$\int_D w\, [\Delta\bar{u} + k^2 n\bar{u}]\, dx \;=\; 0\,.$$

Multiplication of (7.85) by $\bar{u}$ and integration yields, using the previous formula,

$$\int_D \bar{u}\, [\Delta u + k^2 u]\, dx \;=\; \int_D [k^2 n\, w\,\bar{u} - k^2 w\,\bar{u}]\, dx$$

$$=\; -\int_D w\, [\Delta\bar{u} + k^2\bar{u}]\, dx$$

$$=\; -k^2 \int_D (\bar{n} - 1)\, |w|^2 dx\,.$$

Since $u \in H_0^2(D)$ we can apply Green's first formula (7.12a) which shows that the left hand side is real valued. Now we take the imaginary part and arrive at

$$\int_D \operatorname{Im} n \, |w|^2 \, dx \;=\; 0 \,.$$

Because $\operatorname{Im} n(x) \geq 0$ for all x and $\operatorname{Im} n(x) > 0$ on the open set A we conclude that w has to vanish in A. The general unique continuation principle (see the remark following Theorem 7.7) implies that w vanishes in all of D; that is, $\Delta u + k^2 u = 0$ in $\mathbb{R}^3$ and thus $u = 0$ in D by the unique continuation principle. Therefore, also v vanishes in D, and k cannot be an eigenvalue.   □

Therefore, in the following, we always assume that the refractive index n is real valued. We note that in the case where D is not penetrable but acoustically soft (i.e., $u = 0$ on ∂D) the corresponding eigenvalue problem (with respect to the justification of the factorization method) is just the classical eigenvalue problem for $-\Delta$ in D with respect to the Dirichlet boundary condition $u = 0$ on ∂D. Compared to this classical case the interior transmission eigenvalue problem is much less understood. Under certain assumptions on n we will show below in Subsection 7.7.3 that the spectrum is discrete and accumulates at most at infinity (see [52]), if eigenvalues exist at all. It took almost 20 years for the proof of *existence* of real eigenvalues (see [213, 35, 36, 33]). The reason for this gap is partially because the interior transmission eigenvalue problem is not self-adjoint. Looking back, it is surprising that it took so long to prove existence of real eigenvalues because the proof is rather elementary, and we will present it in Subsection 7.7.3.

The fact that the transmission eigenvalue problem fails to be self-adjoint raises the question whether or not complex eigenvalues exist. In the general case the answer is totally open. For the case of D being a ball and n being radially symmetric a huge amount of work related to the existence and location of complex eigenvalues has appeared in the past decade. We will answer this question only for the special case of constant n in the following subsection (see Theorem 7.48).

We refer to [64] for a survey prior to 2007 and to [158, 134] for interior transmission eigenvalue problems for other types of elliptic operators. Also we refer again to the forth edition of the monograph [55] where a chapter on transmission eigenvalues has been added.

As just announced, we consider the special case of D being the unit ball and n being radially symmetric.

7.6.1   The Radially Symmetric Case

Let D be the unit ball and let n depend only on $r = |x|$. We assume that $n \in C^2[0,1]$ is positive on $[0,1]$ and search for eigenfunctions v and w which are also radially symmetric; that is, depend only on r. The Helmholtz equations for v and w reduce to ordinary differential equations, see (7.8a). In Example 7.1

we proved the representation $v(r) = \alpha \sin(kr)/r$, $0 < r < 1$, for some $\alpha \in \mathbb{R}$ and $w(r) = \tilde{w}(r)/r$ where $\tilde{w}(0) = 0$ and $\tilde{w}$ satisfies $\tilde{w}''(r) + k^2 n(r)\tilde{w}(r) = 0$ in $(0,1)$. Therefore, $\tilde{w}$ is a multiple of the function y_k; that is, $w(r) = \beta\, y_k(r)/r$ for some $\beta \in \mathbb{C}$, where y_k satisfies the initial value problem

$$y_k''(r) + k^2 n(r)\, y_k(r) = 0, \quad r \in [0,1], \quad y_k(0) = 0, \ y_k'(0) = 1. \qquad (7.86)$$

Then w is regular at 0 and solves the equation $\Delta w + k^2 n w = 0$ in D. Furthermore, (v,w) solves the eigenvalue problem (7.44a), (7.44b) if and only if α and β satisfy the system

$$\begin{aligned}
\alpha \sin k \ - \ \beta\, y_k(1) &= 0, \\
\alpha k \cos k \ - \ \beta\, y_k'(1) &= 0.
\end{aligned}$$

This system has a non trivial solution if its determinant vanishes; that is, for those values of $k \in \mathbb{C}$ which are zeros of

$$d(k) := y_k'(1)\,\frac{\sin k}{k} - y_k(1)\cos k, \quad k \in \mathbb{C}. \qquad (7.87)$$

We call d the *characteristic function* of the radially symmetric transmission eigenvalue problem because it corresponds exactly to the characteristic function $f = f(\lambda)$ of the spectral problem studied in Chapter 5.

In the following it will be important to study the asymptotic behavior of $y_k(1)$ and $y_k'(1)$ when $|k|$ tends to infinity. We first use the *Liouville transformation* (see Section 5.1)

$$s = s(r) := \int_0^r \sqrt{n(t)}\, dt, \quad z_k(s) := n\big(r(s)\big)^{1/4} y_k\big(r(s)\big) \qquad (7.88a)$$

again. Here, $s \mapsto r(s)$ denotes the inverse function of the monotonic function $r \mapsto s(r)$. It transforms the differential equation (7.86) into the following form for z_k;

$$z_k''(s) + \big(k^2 - q(s)\big) z_k(s) = 0 \quad \text{for } 0 < s < \eta, \qquad (7.88b)$$

$$q(s) := \frac{1}{4}\left[n(r)^{-3/4}\left(n(r)^{-5/4}\, n'(r) \right)' \right]_{r=r(s)} \qquad (7.88c)$$

and

$$\eta = s(1) = \int_0^1 \sqrt{n(t)}\, dt. \qquad (7.88d)$$

The initial conditions $y_k(0) = 0$ and $y_k'(0) = 1$ transform to $z_k(0) = 0$ and $z_k'(0) = n(0)^{-1/4}$, respectively. The quantity k^2 plays the role of λ of Section 5.1. Then $z_k(s) = n(0)^{-1/4} u_2(s, k^2, q)$ where again $u_2 = u_2(s, k^2, q)$ denotes the function of the fundamental system corresponding to (5.7b) with $u_2(0) = 0$ and $u_2'(0) = 1$. With the asymptotic form of $u_2(\eta)$ and $u_2'(\eta)$ for $|k| \to \infty$ (see

Theorem 5.5), we have after the back transform $y_k(r) = n(r)^{-1/4} z_k(s(r)) = [n(r)n(0)]^{-1/4} u_2(s(r), k^2, q)$

$$y_k(1) = \frac{1}{[n(0)n(1)]^{1/4}} \frac{\sin(k\eta)}{k} + \mathcal{O}\big(\exp(|\operatorname{Im} k|\eta)/|k|^2\big), \qquad (7.89a)$$

$$y'_k(1) = \left[\frac{n(1)}{n(0)}\right]^{1/4} \cos(k\eta) + \mathcal{O}\big(\exp(|\operatorname{Im} k|\eta)/|k|\big), \qquad (7.89b)$$

where again $\eta = \int_0^1 \sqrt{n(s)}\,ds$. We substitute this into the definition (7.87) of $d(k)$ which yields

$$d(k) = \frac{1}{[n(0)n(1)]^{1/4}} \frac{f(k)}{k} + \mathcal{O}\left(\frac{\exp(|\operatorname{Im} k|(\eta+1))}{|k|^2}\right) \qquad (7.90)$$

as $|k| \to \infty$ where

$$\begin{aligned} f(k) &= \sqrt{n(1)} \sin k \cos(k\eta) - \sin(k\eta) \cos k \\ &= A \sin(k(1+\eta)) + B \sin(k(1-\eta)) \end{aligned}$$

with

$$A = \frac{1}{2}\big(\sqrt{n(1)} - 1\big) \quad \text{and} \quad B = \frac{1}{2}\big(\sqrt{n(1)} + 1\big). \qquad (7.91)$$

We note that in the case $n(1) = 1$; that is, $A = 0$, the estimate (7.90) is useful only for real values of k because for $|\operatorname{Im} k| \to \infty$ the term $f(k)$ is absorbed in $\mathcal{O}\big(\exp(|\operatorname{Im} k|(\eta+1))/|k|^2\big)$.

If $\eta \neq 1$ we observe that the amplitude B of the second term of $f(k)$ dominates the first term. Therefore, f has a zero in every interval $\big(\frac{2\pi m + \pi/2}{|1-\eta|}, \frac{2\pi m + 3\pi/2}{|1-\eta|}\big)$ for $m \in \mathbb{N}$ and therefore also d for sufficiently large m. From this we conclude that the determinant vanishes at infinitely many discrete real values of k. If $\eta = 1$ and $n(1) \neq 1$ there is only the first term with $A \neq 0$, and we can argue analogously with the extreme values of $k \mapsto \sin(k(1+\eta))$.

Therefore, we have shown:

Theorem 7.41 *Let $n \in C^2[0,1]$ be positive and $\int_0^1 \sqrt{n(t)}\,dt \neq 1$ or $n(1) \neq 1$. Then there exists an infinite number of real eigenvalues of (7.83a) and (7.83b), and these eigenvalues tend to infinity.*

We note that the corresponding eigenfunctions v are Herglotz functions of the form (7.48) because, by the following lemma, we have that

$$v(x) = \alpha \frac{\sin(kr)}{r} = \frac{\alpha k}{4\pi} \int_{S^2} e^{ikx \cdot \hat{y}}\, ds(\hat{y}), \quad x \in \mathbb{R}^3.$$

Lemma 7.42 *For $x \in \mathbb{R}^3$ we have that*

$$\int_{S^2} e^{ix \cdot \hat{y}}\, ds(\hat{y}) = 4\pi \frac{\sin|x|}{|x|}.$$

Proof: Because the integrand is spherically symmetric, we can assume without loss of generality that $x = \rho \hat{x}$ for some $\rho \geq 0$ and $\hat{x}$ is the "north pole", that is, $\hat{x} = (0, 0, 1)^\top$. Then

$$\int_{S^2} e^{i\rho\hat{x}\cdot\hat{y}}\, ds(\hat{y}) \;=\; \int_0^{2\pi}\int_0^{\pi} e^{i\rho\cos\theta}\, \sin\theta\, d\theta\, d\phi$$

$$\qquad\qquad =\; 2\pi \int_{-1}^{1} e^{i\rho s}\, ds \;=\; 4\pi\,\frac{\sin\rho}{\rho}\,. \qquad\qquad \square$$

Before we consider the question of existence of *complex* eigenvalues k; that is, of complex zeros of the determinant d from (7.87), for constant values of n, we have to recall some results from the theory of entire functions. We follow the presentation in [55] (see also [56]) but simplify some of the proofs.

Definition 7.43 *Let $f : \mathbb{C} \to \mathbb{C}$ be an entire function; that is, holomorphic in all of $\mathbb{C}$. The order ρ of f is defined as*

$$\rho \;=\; \limsup_{r\to\infty} \frac{\ln\big(\ln \|f\|_r\big)}{\ln r}\,,$$

where $\|f\|_r = \max\big\{|f(z)| : |z| = r\big\}$ for $r > 0$.

Lemma 7.44 *The characteristic function d from (7.87) is an even entire function of order one provided $n(1) \neq 1$ or $\eta := \int_0^1 \sqrt{n(s)}ds \neq 1$.*

Proof: d is an even function because y_k and also $k \mapsto \sin k/k$ and $\cos$ are all functions of k^2. Furthermore, from the Volterra integral equation (5.12b) for the part $u_2(\cdot, \lambda, q)$ of the fundamental system $\{u_1, u_2\}$ of (5.7a), (5.7b) we derive easily that $\lambda \mapsto u_2(\eta, \lambda, q)$ is holomorphic in all of $\mathbb{C}$ and thus also $k \mapsto y_k(1) = n(1)^{-1/4}z_k(\eta) = [n(0)n(1)]^{-1/4}u_2(\eta, k^2, q)$. The same holds for the derivative. Therefore, d is an entire function.

It remains to compute the order of d. From (7.90) and the estimate $|\sin z| \leq e^{|z|}$ we observe that $|d(k)| \leq c\,e^{|k|(\eta+1)}$ for some $c > 0$. Therefore, for $|k| = r$,

$$\frac{\ln(\ln|d(k)|)}{\ln r} \;\leq\; \frac{\ln[\ln c + (\eta+1)r]}{\ln r} \;\leq\; \frac{\ln[(\eta + 2)r]}{\ln r} \;=\; \frac{\ln(\eta + 2)}{\ln r} + 1$$

for large values of r which proves that $\rho \leq 1$. Now we set $k = ti$ for $t \in \mathbb{R}_{>0}$. If $n(1) \neq 1$ then it is easy to see that $|f(it)| \geq \frac{|A|}{2}\, e^{t(\eta+1)}$ for large values of t (where f if given below (7.90)) and thus, again from (7.90), $|d(it)| \geq c\,e^{t(\eta+1)}$ for some $c > 0$. This yields that $\rho \geq 1$ and ends the proof if $n(1) \neq 1$. If $n(1) = 1$ then $\eta \neq 1$ and $A = 0$ and one gets $|d(it)| \geq c\,e^{t|\eta-1|}$ and again $\rho \geq 1$.
$\square$

The following theorem is a special case of the more general factorization theorem of Hadamard which we cite without proof (see, e.g., [19], Chapter 2).

Theorem 7.45 *(Hadamard)*
Let f be an entire function of order one, let $m \geq 0$ be the order[7] of the root $z = 0$ of f (in particular, $m = 0$ if $f(0) \neq 0$), and let $\{a_j : j \in \mathbb{N}\}$ be the nonzero roots of f repeated according to multiplicity. Then there exists a polynomial p of degree at most one such that

$$f(z) = e^{p(z)} z^m \prod_{j=1}^{\infty} \left(1 - \frac{z}{a_j} \right) e^{z/a_j}, \quad z \in \mathbb{C}. \tag{7.92}$$

With this theorem we can prove a slight generalization of a theorem of Laguerre.

Theorem 7.46 *Let f be an entire function of order one which is real for real values of z. Suppose that f has infinitely many real zeros and only a finite number of complex ones. Then f has a single critical point[8] on each interval formed by two consecutive real zeros of f provided this interval is sufficiently far away from the origin.*

Proof: Since f has only finitely many complex zeros (which will occur in conjugate pairs) there exists a polynomial q whose roots are exactly all those complex ones of f as well as a possible root at $z = 0$ which has the same order as the possible zero of f at $z = 0$. We factorize $f(z) = q(z)g(z)$ where all the roots of g are given by $\{a_j : j \in \mathbb{N}\}$ in nondecreasing magnitude (where multiple roots are repeated according to their multiplicities). We apply Hadamard's theorem to the function g and have for real values of $x \notin \{a_j : j \in \mathbb{N}\}$

$$\frac{g'(x)}{g(x)} = \frac{d}{dx} \ln g(x) = p'(x) + \sum_{j=1}^{\infty} \frac{d}{dx} \left[\ln \left(1 - \frac{x}{a_j} \right) + \frac{x}{a_j} \right]$$

$$= \alpha + \sum_{j=1}^{\infty} \left[\frac{1}{x - a_j} + \frac{1}{a_j} \right]$$

where $\alpha = p'(x)$ is a real constant (since p is a polynomial of order at most one). Differentiating this expression yields

$$\frac{d}{dx} \left(\frac{f'(x)}{f(x)} \right) = \frac{d}{dx} \left(\frac{q'(x)}{q(x)} \right) + \frac{d}{dx} \left(\frac{g'(x)}{g(x)} \right)$$

$$= \frac{d}{dx} \left(\frac{q'(x)}{q(x)} \right) - \sum_{j=1}^{\infty} \frac{1}{(x - a_j)^2}, \quad x \notin \{a_j : j \in \mathbb{N}\}.$$

Since q is a polynomial with no real zeros (except possibly $z = 0$) there exists $c > 0$ with $\left| \frac{d}{dx} \left(\frac{q'(x)}{q(x)} \right) \right| \leq \frac{c}{|x|^2}$ for all $|x| \geq 1$. Choose $N \in \mathbb{N}$ with $N > 2c$ and

---

[7]Note that the order of a root of a holomorphic function is always finite.
[8]that is, a zero of f'

then $R > 0$ with $|a_j - x|^2 \leq 2|x|^2$ for all $|x| \geq R$ and $j = 1, \ldots, N$. Then

$$\frac{d}{dx}\left(\frac{f'(x)}{f(x)}\right) \leq \frac{c}{|x|^2} - \sum_{j=1}^{N} \frac{1}{(x-a_j)^2} \leq \frac{c}{|x|^2} - \frac{N}{2|x|^2} = \frac{2c-N}{2|x|^2} < 0$$

for $|x| \geq R$, $x \notin \{a_j : j \in \mathbb{N}\}$. Therefore, f'/f is strictly decreasing in every interval $(a_\ell, a_{\ell+1})$ with $a_\ell < a_{\ell+1}$ and ℓ large enough. Furthermore, from

$$\frac{f'(x)}{f(x)} = \frac{q'(x)}{q(x)} + \frac{g'(x)}{g(x)} = \frac{q'(x)}{q(x)} + \alpha + \sum_{j=1}^{\infty}\left[\frac{1}{x-a_j} + \frac{1}{a_j}\right] \qquad (7.93)$$

$$= \frac{q'(x)}{q(x)} + \alpha + \sum_{j\neq\ell,\ell+1}\left[\frac{1}{x-a_j} + \frac{1}{a_j}\right]$$

$$+ \left[\frac{1}{x-a_\ell} + \frac{1}{a_\ell}\right] + \left[\frac{1}{x-a_{\ell+1}} + \frac{1}{a_{\ell+1}}\right]$$

we conclude that $\lim\limits_{x\to a_\ell+} \frac{f'(x)}{f(x)} = +\infty$ and $\lim\limits_{x\to a_{\ell+1}-} \frac{f'(x)}{f(x)} = -\infty$. Therefore, f'/f has exactly one zero in the interval $(a_\ell, a_{\ell+1})$ which ends the proof. $\square$

As a corollary we obtain Laguerre's theorem (see [19], Chapter 2).

Corollary 7.47 *(Laguerre)*
Let f be an entire function of order one which is real for real values of z and all of its zeros are real. Then all of the zeros of f' are real as well and interlace those of f.

Proof: In this case we set $q(z) = z^m$ where m is the order of the zero $z = 0$. Then, by analytic extension, (7.93) holds also for complex z instead of $x \in \mathbb{R}$. If $z \in \mathbb{C}$ is a critical point then from (7.93)

$$0 = \frac{m}{z} + \alpha + \sum_{j=1}^{\infty}\left[\frac{1}{z-a_j} + \frac{1}{a_j}\right] = \frac{m\bar{z}}{|z|^2} + \alpha + \sum_{j=1}^{\infty}\left[\frac{\bar{z}-a_j}{|z-a_j|^2} + \frac{1}{a_j}\right].$$

Taking the imaginary part yields

$$0 = -\operatorname{Im} z \frac{m}{|z|^2} - \operatorname{Im} z \sum_{j=1}^{\infty} \frac{1}{|z-a_j|^2}$$

which implies that z is real. Now we follow exactly the proof of the preceding theorem. $\square$

Now we are able to answer the question of existence of complex eigenvalues for the special case of n being constant. For constant n the function y_k from

(7.86) is given by $y_k(r) = \frac{\sin(k\eta r)}{k\eta}$ for $r \geq 0$ where $\eta = \sqrt{n}$. Therefore, $d(k) = \frac{1}{k\eta} f_\eta(k)$ where

$$f_\eta(k) = \eta \cos(k\eta) \sin k - \cos k \sin(k\eta), \quad k \in \mathbb{C}. \qquad (7.94)$$

Here we indicated the dependence on $\eta > 0$. For the author the following result is rather surprising.

Theorem 7.48 *If $\eta = \sqrt{n} \neq 1$ is an integer or the reciprocal of an integer then no complex eigenvalues with eigenfunctions depending only on r exist. Otherwise, there exist infinitely many complex eigenvalues.*

Proof: We consider the function f_η from (7.94) and note first that $-\eta f_{1/\eta}(\eta k) = f_\eta(k)$ for all $k \in \mathbb{C}$. Therefore, a complex zero of f_η exists if, and only if, a complex zero of $f_{1/\eta}$ exists. It is thus sufficient to study the case $\eta > 1$.

Let first $\eta > 1$ be an integer. Then $f_\eta(k) = 0$ if, and only if, k is a critical point of the entire function $g(k) = \frac{\sin(\eta k)}{\sin k}$ because $g'(k) = f_\eta(k)/\sin^2 k$. Since all of the zeros of g are real, by Corollary 7.47 also its critical points are real; that is, all of the zeros of f_η are real.

Let now $\eta > 1$ not be an integer. We will construct a sequence I_ℓ of intervals which tend to infinity such that f_η does not change its sign on I_ℓ and each I_ℓ contains two consecutive real critical points of f_η. By Theorem 7.46 this is only possible if there exist infinitely many complex zeros of f_η.

From

$$f_\eta'(k) = (1 - \eta^2) \sin(k\eta) \sin k$$

we observe that $\{\frac{j\pi}{\eta} : j \in \mathbb{N}\}$ and $\{j\pi : j \in \mathbb{N}\}$ are the critical points of f_η. Choose a sequence $m_\ell \in \mathbb{N}$ converging to infinity such that $\eta\, m_\ell \notin \mathbb{N}$ for all $\ell \in \mathbb{N}$. Fix $\ell \in \mathbb{N}$ in the following and set $m = m_\ell$ for abbreviation. The interval $(m\eta - 1, m\eta)$ contains an integer j. Set $\varepsilon = m\eta - j \in (0, 1)$. The two points $\frac{j\pi}{\eta}$ and $m\pi$ are consecutive zeros of f_η' because their distance is $m\pi - \frac{j\pi}{\eta} = \varepsilon \frac{\pi}{\eta} < \frac{\pi}{\eta}$. Furthermore,

$$
\begin{aligned}
f_\eta\left(\frac{j\pi}{\eta}\right) &= \eta \cos(j\pi) \sin\frac{j\pi}{\eta} - \cos\frac{j\pi}{\eta} \sin(j\pi) \\
&= \eta(-1)^j \sin\left(m\pi - \frac{\varepsilon\pi}{\eta}\right) = \eta(-1)^{j+1+m} \sin\frac{\varepsilon\pi}{\eta} \quad \text{and} \\
f_\eta(m\pi) &= \eta \cos(m\pi\eta) \sin(m\pi) - \cos(m\pi) \sin(m\pi\eta) \\
&= (-1)^{m+1} \sin(j\pi + \varepsilon\pi) = (-1)^{m+1+j} \sin(\varepsilon\pi).
\end{aligned}
$$

We observe that the signs of $f_\eta\left(\frac{j\pi}{\eta}\right)$ and $f_\eta(m\pi)$ coincide because $\varepsilon, \varepsilon/\eta \in (0,1)$. Furthermore, f_η has no zero between $\frac{j\pi}{\eta}$ and $m\pi$ because otherwise there would be another critical point between them. Therefore, the interval $I = \left[\frac{j\pi}{\eta}, m\pi\right]$ has the desired properties. $\quad \square$

We refer to Problem 7.5 for two explicit examples where the assertions of this theorem are illustrated.

There exist also results on the existence of complex transmission eigenvalues for non-constant refractive indices under certain conditions. For example, as shown in [55] (see also [58]), if either $1 < \sqrt{n(1)} < \eta$ or $\eta < \sqrt{n(1)} < 1$ (where η is again given by (7.88d)) there exist infinitely many real and infinitely many complex eigenvalues. Also, all complex eigenvalues lie in a strip around the real axis if $n(1) \neq 1$. For a more detailed investigation of the location for the constant case we refer to [257] and to [55] and the references therein.

7.6.2   Discreteness And Existence in the General Case

We continue now with the general case; that is, D is a bounded Lipschitz domain and n is real-valued with $n(x) \geq 1 + q_0$ on D for some $q_0 > 0$. For the definition of a Lipschitz domain we refer again to [191] or [161]. For Lipschitz domains we can give an alternative characterization of an interior transmission eigenvalue. We allow $k \neq 0$ to be complex.

Theorem 7.49 *Let D be a Lipschitz domain. $k \in \mathbb{C} \setminus \{0\}$ is an interior transmission eigenvalue if and only if there exist $u \in H_0^2(D)$ and $v \in L^2(D)$, not vanishing simultaneously, such that*

$$\Delta u + k^2 n u = k^2(n-1)v \quad \text{a.e. in } D \quad \text{and} \qquad (7.95a)$$

$\Delta v + k^2 v = 0$ *in D in the ultra weak sense; that is,*

$$\int_D v\left[\Delta\varphi + k^2\varphi\right] dx = 0 \quad \text{for all } \varphi \in H_0^2(D). \qquad (7.95b)$$

Proof: Let first $u \in H_0^2(D)$ and $v \in L^2(D)$ with (7.95a), (7.95b). Set $w = v - u$ and let $\phi, \psi \in H^2(D)$ with $\phi - \psi \in H_0^2(D)$. Then

$$\int_D v\left[\Delta\phi + k^2\phi\right] - w\left[\Delta\psi + k^2 n\psi\right] dx$$

$$= \int_D v\left[\Delta(\phi - \psi) + k^2(\phi - \psi)\right] dx + \int_D k^2 v(1 - n)\psi + u\left[\Delta\psi + k^2 n\psi\right] dx$$

The first integral on the right hand side vanishes because v is an ultra weak solution of $\Delta v + k^2 v = 0$. For the second integral we use Green's second formula (7.12b) which yields

$$\int_D u\left[\Delta\psi + k^2 n\psi\right] dx = \int_D \psi\left[\Delta u + k^2 n u\right] dx = k^2 \int_D \psi(n-1)v\, dx.$$

Therefore, the pair $(v, w) \in L^2(D) \times L^2(D)$ satisfies (7.84). This proves the first direction.

For the reverse direction let $v, w \in L^2(D)$ with (7.84). Define $u = v - w$ in D and $u = 0$ in $\mathbb{R}^3 \setminus D$. Furthermore, set $\varphi = \psi$ for some $\psi \in C^\infty(\mathbb{R}^3)$ with compact support in (7.84). Then

$$\int_{\mathbb{R}^3} u\left[\Delta\psi + k^2 n\psi\right] dx = k^2 \int_D (n-1)\, v\,\psi\, dx\,;$$

that is, u is an ultra weak solution of $\Delta u + k^2 n u = k^2(n-1)v$ in $\mathbb{R}^3$. The regularity result of Lemma 7.10 yields $u \in H^2(\mathbb{R}^2)$ with $u = 0$ in $\mathbb{R}^3 \setminus D$; that is, $u \in H_0^2(D)$ by Lemma 7.4 and $\Delta u + k^2 n u = k^2(n-1)v$ almost everywhere in D. Finally, let $\varphi \in H_0^2(D)$ and set $\psi = 0$ in (7.84). Then $\int_D v\left[\Delta\varphi + k^2\varphi\right] dx = 0$. This ends the proof.   □

This theorem makes it possible to eliminate the function v from the system. Indeed, let $u \in H_0^2(D)$ and $v \in L^2(D)$ satisfy (7.95a) and (7.95b). We devide (7.95a) by $n - 1$ (note that $n(x) - 1 \geq q_0$ on D by assumption), multiply the resulting equation by $\Delta\psi + k^2\psi$ for some $\psi \in H_0^2(D)$ and integrate. This gives

$$\int_D \left[\Delta u + k^2 n u\right]\left[\Delta\psi + k^2\psi\right] \frac{dx}{n-1} = k^2 \int_D v\left[\Delta\psi + k^2\psi\right] dx = 0$$

by (7.95b); that is, replacing ψ by its complex conjugate,

$$\int_D \left[\Delta u + k^2 n u\right]\left[\Delta\overline{\psi} + k^2\overline{\psi}\right] \frac{dx}{n-1} = 0 \quad \text{for all } \psi \in H_0^2(D). \tag{7.96}$$

On the other side, if $u \in H_0^2(D)$ satisfies (7.96) then we set $v = \frac{1}{k^2(n-1)}[\Delta u + k^2 n u] \in L^2(D)$ which satisfies (7.95b). In this sense the system (7.95a), (7.95b) is equivalent to (7.96). Equation (7.96) is the weak form of the fourth order equation $[\Delta + k^2]\frac{1}{n-1}[\Delta u + k^2 n u] = 0$.

With respect to (7.96) it is convenient to introduce a new inner product $(\cdot, \cdot)_*$ in $H_0^2(D)$ by

$$(u, \psi)_* := \int_D \Delta u\, \overline{\Delta\psi}\, \frac{dx}{n-1}, \quad u, \psi \in H_0^2(D). \tag{7.97}$$

Lemma 7.50 $(\cdot, \cdot)_*$ *defines an inner product in $H_0^2(D)$ with corresponding norm $\|\cdot\|_*$ which is equivalent to the ordinary norm $\|\cdot\|_{H^2(D)}$ in $H_0^2(D)$.*

Proof: By the definition of $H_0^2(D)$ and a denseness argument it is sufficient to prove the existence of $c_1, c_2 > 0$ with $c_1\|\psi\|_* \leq \|\psi\|_{H^2(D)} \leq c_2\|\psi\|_*$ for all $\psi \in C^\infty(D)$. The left estimate is obvious because $n - 1 \geq q_0 > 0$ on D. To prove the right estimate we choose a cube $Q \subset \mathbb{R}^3$ which contains $\overline{D}$ in its interior, extend any $\psi \in C_0^\infty(D)$ by zero into Q, and then periodically (with respect to Q) into $\mathbb{R}^3$. Without loss of generality we assume that $Q = (-\pi, \pi)^3$.

Then $\psi \in H^2_{per}(Q)$ and $\|\psi\|_{H^2(D)} \leq c\|\psi\|_{H^2_{per}(Q)}$ where $c > 0$ is independent of ψ. With the Fourier coefficients ψ_j of ψ (see (7.15b)) we have

$$
\begin{aligned}
\|\psi\|^2_{H^2_{per}(Q)} &= \sum_{j \in \mathbb{Z}^3} [1 + |j|^2]^2 |\psi_j|^2 \\
&= |\psi_0|^2 + \sum_{0 \neq j \in \mathbb{Z}^3} [1 + 2|j|^2 + |j|^4]|\psi_j|^2 \\
&\leq |\psi_0|^2 + 4 \sum_{0 \neq j \in \mathbb{Z}^3} |j|^4 |\psi_j|^2 = |\psi_0|^2 + \frac{4}{(2\pi)^3}\|\Delta\psi\|^2_{L^2(Q)} \\
&\leq |\psi_0|^2 + 4 \frac{\|n - 1\|_\infty}{(2\pi)^3}\|\psi\|^2_* .
\end{aligned}
$$

It remains to estimate $|\psi_0|$. For $x = (\pi, 0, 0)^\top \in \partial Q$ we conclude from the boundary condition that $\sum_j \psi_j e^{ij_1 \pi} = 0$ and thus

$$
|\psi_0| \leq \sum_{0 \neq j \in \mathbb{Z}^3} |\psi_j| = \sum_{j \neq 0} \frac{1}{|j|^2}|j|^2|\psi_j| \leq \sqrt{\sum_{j \neq 0} \frac{1}{|j|^4}}\sqrt{\sum_{j \neq 0}|j|^4|\psi_j|^2} \leq c\|\psi\|_*
$$

as before.   □

Now we rewrite (7.96) in the form

$$
(u, \psi)_* + k^2 a(u, \psi) + k^4 b(u, \psi) = 0 \quad \text{for all } \psi \in H^2_0(D), \tag{7.98}
$$

where

$$
\begin{aligned}
a(u, \psi) &= \int_D [n\, u\, \Delta\overline{\psi} + \overline{\psi}\, \Delta u] \frac{dx}{n - 1} \\
&= \int_D [u\, \Delta\overline{\psi} + \overline{\psi}\, \Delta u] \frac{dx}{n - 1} + \int_D u\, \Delta\overline{\psi}\, dx, \\
b(u, \psi) &= \int_D n\, u\, \overline{\psi} \frac{dx}{n - 1}
\end{aligned}
$$

for $u, \psi \in H^2_0(D)$. The sesqui-linear forms a and b are hermitian. For a this is seen from the second form and Green's second identity (7.12b). By the representation theorem of Riesz (Theorem A.23 of the Appendix) in the Hilbert space $H^2_0(D)$ for every $u \in H^2_0(D)$ there exists a unique $u' \in H^2_0(D)$ with $a(u, \psi) = (u', \psi)_*$ for all $\psi \in H^2_0(D)$. We define the operator A from $H^2_0(D)$ into itself by $Au = u'$. Therefore, $a(u, \psi) = (Au, \psi)_*$ for all $u, \psi \in H^2_0(D)$. Then it is not difficult to prove that A is linear, self-adjoint, and compact.[9]

---

[9] For the proof of compactness one needs, however, the fact that $H^1_0(D)$ is compactly imbedded in $L^2(D)$, see Problem 7.1.

Analogously, there exists a linear, self-adjoint, and compact operator B from $H_0^2(D)$ into itself with $b(u, \psi) = (Bu, \psi)_*$ for all $u, \psi \in H_0^2(D)$. Then (7.96) can be written as $(u, \psi)_* + k^2(Au, \psi)_* + k^4(Bu, \psi)_* = 0$ for all $\psi \in H_0^2(D)$; that is,

$$u + k^2 Au + k^4 Bu = 0 \quad \text{in } H_0^2(D). \tag{7.99}$$

This is a quadratic eigenvalue problem in the parameter $\tau = k^2$. We can reduce it to a linear eigenvalue problem for a compact operator. Indeed, since B is positive (that is, $(Bu, u)_* > 0$ for all $u \neq 0$) there exists a positive and compact operator $B^{1/2}$ from $H_0^2(D)$ into itself with $B^{1/2}B^{1/2} = B$ (see formula (A.47) of the Appendix). If u satisfies (7.99) for some $k \in \mathbb{C} \setminus \{0\}$ then set $u_1 = u$ and $u_2 = k^2 B^{1/2}u$. Then the pair $\binom{u_1}{u_2} \in H_0^2(D) \times H_0^2(D)$ satisfies

$$\begin{pmatrix} u_1 \\ u_2 \end{pmatrix} + k^2 \begin{pmatrix} A & B^{1/2} \\ -B^{1/2} & 0 \end{pmatrix} \begin{pmatrix} u_1 \\ u_2 \end{pmatrix} = \begin{pmatrix} 0 \\ 0 \end{pmatrix}. \tag{7.100}$$

Therefore $1/k^2$ is an eigenvalue of the compact (but not self-adjoint) operator $\begin{pmatrix} -A & -B^{1/2} \\ B^{1/2} & 0 \end{pmatrix}$. Conversely, if $\binom{u_1}{u_2} \in H_0^2(D) \times H_0^2(D)$ satisfies (7.100) then $u = u_1$ satisfies (7.99). Therefore, well-known results on the spectrum of compact operators (see, e.g., [151]) imply the following theorem.

Theorem 7.51 *There exists at most a countable number of eigenvalues $k \in \mathbb{C}$ with no accumulation point in $\mathbb{C}$.*

By different methods the discreteness can be shown under the weaker assumption that $n > 1$ only in a neighborhood of the boundary ∂D (see, e.g., [256, 159] or [34]).

The question of the existence of real eigenvalue was open for about 20 years.[10]. The idea of the proof of the following result goes back to Päivärinta and Sylvester ([213]) and was generalized to general refractive indices by Cakoni, Gintides, and Haddar ([36]).

Theorem 7.52 *There exists a countable number of real eigenvalues $k > 0$ which converge to infinity.*

Proof: With the Riesz representations A and B of the hermitian forms a and b, respectively, from above we define the family of operators $\Phi(\kappa) = I + \kappa A + \kappa^2 B$ for $\kappa \geq 0$ with corresponding sesqui-linear forms

$$
\begin{aligned}
\phi(\kappa; u, \psi) &= \int_D [\Delta u + \kappa n u][\Delta \overline{\psi} + \kappa \overline{\psi}] \frac{dx}{n-1} \\
&= \int_D [\Delta u + \kappa u][\Delta \overline{\psi} + \kappa \overline{\psi}] \frac{dx}{n-1} - \kappa \int_D u[\Delta \overline{\psi} + \kappa \overline{\psi}] dx; \tag{7.101}
\end{aligned}
$$

---

[10]Note that the matrix operator in (7.100) fails to be self-adjoint!

that is, $\phi(\kappa; u, \psi) = \big(\Phi(\kappa)u, \psi\big)_*$ for all $u, \psi \in H_0^2(D)$. Here, κ plays the role of k^2 and is considered as a parameter. We search for parameters $\kappa > 0$ such that 0 is an eigenvalue of $\Phi(\kappa)$; that is, -1 is an eigenvalue of the compact and self-adjoint operator $\kappa A + \kappa^2 B$.

For $\kappa \geq 0$ let $\lambda_j(\kappa) \in \mathbb{R}$, $j \in \mathbb{N}$, be the nonzero eigenvalues of the compact and self-adjoint operators $\kappa A + \kappa^2 B$. They converge to zero as j tends to infinity (if there exist infinitely many). The corresponding eigenspaces are finite-dimensional. Let the negative eigenvalues be sorted as $\lambda_1^-(\kappa) \leq \lambda_2^-(\kappa) \leq \cdots < 0$ where the entries are repeated with its multiplicity. In general, they could be none or finitely many or infinitely many of them.

Let $m \in \mathbb{N}$ be any natural number. First we construct $\hat{\kappa} > 0$ and a subspace V_m of $H_0^2(D)$ of dimension m such that $\phi(\hat{\kappa}; u, u) \leq 0$ for all $u \in V_m$: We choose $\varepsilon > 0$ and m pairwise disjoint balls $B_j = B(z_j, \varepsilon)$, $j = 1, \ldots, m$, of radius ε with $\bigcup_{j=1}^m \overline{B_j} \subset D$. Setting $n_0 := 1 + q_0$ we note that $n(x) \geq n_0$ on D. In every ball B_j we consider the interior transmission eigenvalue problem with constant refractive index n_0. By Theorem 7.41 infinitely many real and positive transmission eigenvalues exist. Note that these eigenvalues do not depend on j because a translation of a domain results in the same interior transmission eigenvalues. Let $\hat{k} > 0$ be the smallest one with corresponding eigenfunctions $u_j \in H_0^2(B_j)$. Set $\hat{\kappa} = \hat{k}^2$ and let $\phi_j(\kappa; u, \psi)$ be the sesqui-linear form corresponding to n_0 in B_j. Then $\phi_j(\hat{\kappa}; u_j, \psi) = 0$ for all $\psi \in H_0^2(B_j)$. We extend each u_j by zero to a function in $H_0^2(D)$. (We use Lemma 7.4 again.) Then $\{u_j : j = 1, \ldots, m\}$ are certainly linearly independent because their supports are pairwise disjoint. We define $V_m = \text{span}\{u_j : j = 1, \ldots, m\}$ as a m−dimensional subspace of $H_0^2(D)$ and compute for $u = \sum_{j=1}^m \alpha_j u_j \in V_m$, using (7.101),

$$
\begin{aligned}
\phi(\hat{\kappa}; u, u) &= \sum_{j=1}^m |\alpha_j|^2 \phi(\hat{\kappa}; u_j, u_j) \\
&= \sum_{j=1}^m |\alpha_j|^2 \left[\int_{B_j} |\Delta u_j + \hat{\kappa} u_j|^2 \frac{dx}{n-1} - \hat{\kappa} \int_{B_j} u_j \big[\Delta \overline{u_j} + \hat{\kappa}\overline{u_j}\big] dx \right] \\
&\leq \sum_{j=1}^m |\alpha_j|^2 \left[\int_{B_j} |\Delta u_j + \hat{\kappa} u_j|^2 \frac{dx}{n_0-1} - \hat{\kappa} \int_{B_j} u_j \big[\Delta \overline{u_j} + \hat{\kappa}\overline{u_j}\big] dx \right] \\
&= \sum_{j=1}^m |\alpha_j|^2 \phi_j(\hat{\kappa}; u_j, u_j) = 0.
\end{aligned}
$$

This shows that $\big(\Phi(\hat{\kappa})u, u\big)_* \leq 0$ for all $u \in V_m$. By Corollary A.55 we conclude that there exist eigenvalues $\lambda_\ell(\hat{\kappa}) \leq -1$ of $\hat{\kappa} A + \hat{\kappa}^2 B$ for $\ell = 1, \ldots, m$. Furthermore, again by Corollary A.55 of Appendix A.6 the eigenvalues $\lambda_\ell(\kappa)$ depend continuously on κ and $\lambda_\ell(0) = 0$ for all ℓ. Therefore, for every $\ell \in \{1, \ldots, m\}$ there exists $\kappa_\ell \in (0, \hat{\kappa}]$ with $\lambda_\ell(\kappa_\ell) = -1$. The corresponding eigenfunctions u_ℓ, $\ell = 1, \ldots, m$, satisfy $\Phi(\kappa_\ell)u_\ell = 0$ which implies that $\{\sqrt{\kappa_\ell} : \ell = 1, \ldots, m\}$ are interior transmission eigenvalues. Since m was arbitrary the existence of infinitely many real eigenvalues is shown.   □

The—very natural—question of existence of complex transmission eigenvalues in this general situation is totally open. Since even in the case of a constant refractive index in a ball both, existence and nonexistence, of complex eigenvalues can occur (see, for example Theorem 7.48) the problem is certainly hard to solve. Aspects concerning the asymptotic distribution of the eigenvalues (Weyl's formula), location and properties of the eigenfunctions have attracted a lot of attention, also for much more general types of elliptic problems. We only mention the special issue ([37]) of *Inverse Problems* in 2013, the already mentioned monographs [34, 55], and refer to the references therein.

7.6.3   The Inverse Spectral Problem for the Radially Symmetric Case

After the study of the existence and discreteness of the interior transmission eigenvalues the natural question arises where these eigenvalues determine the refractive index $n(x)$ uniquely. For spherically stratified media studied in Subsection 7.6.1, this question is the analogue of the inverse Sturm-Liouville problem of Chapter 5 and has been subject of intensive research. It started with the work by J. McLaughlin and P. Polyakov ([190]) and was picked up in, e.g., [56–58, 46, 45]. We will follow the presentations of [34, 55].

We assume as at the beginning of Subsection 7.6.1 that $n \in C^2[0,1]$ is positive on $[0,1]$. This smoothness assumption is mainly necessary to apply the Liouville transform. We have seen in Subsection 7.6.1 that the interior transmission eigenvalue eigenvalues of (7.44a), (7.44b) are—for radially symmetric eigenfunctions—just the zeros of the characteristic function d, given by (7.87); that is,

$$d(k) \;=\; y_k'(1)\,\frac{\sin k}{k} \;-\; y_k(1)\cos k\,, \quad k \in \mathbb{C}\,, \tag{7.102}$$

where y_k solves the initial value problem

$$y_k''(r) + k^2 n(r) y_k(r) = 0 \text{ in } (0,1)\,, \quad y_k(0) = 0\,,\ y_k'(0) = 1\,. \tag{7.103}$$

Therefore, the inverse problem is to recover the refractive index $n = n(r)$ from the zeros of the characteristic function $d = d(k)$. This corresponds exactly to the situation of Chapter 5 where the inverse spectral problem was to recover $q = q(x)$ from the zeros of the characteristic function $f = f(\lambda)$. Therefore, it is not surprising that we use similar arguments.

We saw already (in Lemma 7.44) that the characteristic function d is an even entire function of order one provided $n(1) \neq 1$ or $\eta := \int_0^1 \sqrt{n(s)}\,ds \neq 1$. We prove a further property of d.

Lemma 7.53 *Let d be the characteristic function from (7.102). Then*

$$\lim_{k \to 0} \frac{d(k)}{k^2} \;=\; \frac{1}{3} \;-\; \int_0^1 n(s)\,s^2 ds\,.$$

which is not zero if, for example, $n(r) \leq 1$ for all $r \in [0,1]$ and $n \not\equiv 1$. Therefore, under this condition $k = 0$ is a root of d of order 2.

Proof: For low values of $|k|$ it is convenient to use an integral equation argument directly for y_k. Indeed, it is not difficult to show that $y_k \in C^2[0,1]$ solves (7.103) if, and only if $y_k \in C[0,1]$ satisfies the Volterra equation

$$y_k(r) \ = \ r \ - \ k^2 \int_0^r (r-s)\, n(s)\, y_k(s)\, ds \,, \quad 0 \leq r \leq 1. \tag{7.104}$$

For sufficiently small $|k|$ (actually, for all values of k) this fixed point equation is solved by the Neumann series (see Theorem A.31 of Appendix A). The first two terms yield

$$y_k(r) \ = \ r \ - \ k^2 \int_0^r (r-s)\, n(s)\, s\, ds \ + \ \mathcal{O}(|k|^4)\,.$$

Also, for the derivative we get

$$y_k'(r) \ = \ 1 \ - \ k^2 \int_0^r n(s)\, y_k(s)\, ds \ = \ 1 \ - \ k^2 \int_0^r n(s)\, s\, ds \ + \ \mathcal{O}(|k|^4)\,.$$

Substituting these expansions for $r = 1$ and the power series of $\sin k / k$ and $\cos k$ into the definition of d yields after collecting the terms with k^2

$$d(k) \ = \ k^2 \left[\frac{1}{3} \ - \ \int_0^1 n(s)\, s^2 ds \right] \ + \ \mathcal{O}(|k|^4)\,.$$

This ends the proof. $\square$

From the properties derived in Lemmas 7.44 and 7.53 we have the following form of the characteristic function, provided $\int_0^1 n(s) s^2 ds \neq 1/3$ holds and $n(1) \neq 1$ or $\eta \neq 1$.

$$d(k) \ = \ \gamma k^2 \prod_{j=1}^{\infty} \left(1 - \frac{k^2}{k_j^2} \right), \quad k \in \mathbb{C}, \tag{7.105}$$

for some $\gamma \in \mathbb{C}$ where $k_j \in \mathbb{C}$ are all nonzero roots of d repeated according to multiplicity.

This follows directly from Theorem 7.45 and the fact that with k also $-k$ is a zero. Indeed, let $\{k_j : j \in J\}$ be the set of zeros in the half plane $\{z \in \mathbb{C} : \mathrm{Re}\, z > 0 \text{ or } z = it,\ t > 0\}$. Then the disjoint union $\{k_j : j \in J\} \cup \{-k_j : j \in J\}$ cover all of the roots. We group the factors of Hadamard's formula (7.5) into pairs $\left(1 - \frac{k}{k_j}\right) e^{k/k_j} \left(1 + \frac{k}{k_j}\right) e^{-k/k_j} = 1 - \frac{k^2}{k_j^2}$ for $j \in J$. Furthermore, the polynomial $p(k)$ must be constant because d is even. This shows that

$$d(k) \ = \ e^p\, k^2 \prod_{j \in J} \left(1 - \frac{k^2}{k_j^2} \right) \ = \ \frac{1}{2}\, e^p\, k^2 \prod_{j=1}^{\infty} \left(1 - \frac{k^2}{k_j^2} \right).$$

After these preparations we turn to the **inverse problem** to determine $n(r)$ from (all of) the roots $k_j \in \mathbb{C}$ of the characteristic function d. Our main concern is again the question of uniqueness, namely, do the refractive indices n_1 and n_2 have to coincide if the roots of the corresponding characteristic functions coincide?

Therefore, we assume that we have two positive refractive indices $n_1, n_2 \in C^2[0,1]$ such that the zeros k_j of their characteristic functions d_1 and d_2 coincide. Furthermore, we assume that $\int_0^1 n_\ell(s) s^2 ds \neq 1/3$ for $\ell = 1, 2$ and also $n_\ell(1) \neq 1$ for $\ell = 1, 2$. Then the characteristic functions $d_\ell(k)$ have the representations (7.105) with constants γ_ℓ for $\ell = 1, 2$. Then we conclude that $d_1(k)/\gamma_1 = d_2(k)/\gamma_2$ for all $k \in \mathbb{C}$; that is, *we can determine $d(k)/\gamma$ from the data $\{k_j : j \in \mathbb{N}\}$*.

In the following we proceed in several steps.

Step A: We determine η from the data; that is, we show that $\eta_1 = \eta_2$ where $\eta_\ell = \int_0^1 \sqrt{n_\ell(s)}\, ds$ for $\ell = 1, 2$ if the transmission eigenvalues corresponding to n_1 and n_2 coincide. Fix $a \in \mathbb{R}$ and set $k = a + it$ for $t > 0$. From the asymptotic behavior (7.90) we conclude that

$$
\begin{aligned}
k\frac{d(k)}{\gamma} &= \frac{1}{\gamma\left[n(0)n(1)\right]^{1/4}}\left[A\sin\big(k(1+\eta)\big) + B\sin\big(k(1-\eta)\big)\right] \\
&\quad + \mathcal{O}\left(\frac{\exp(t(\eta+1))}{t}\right) \\
&= -\frac{A}{\gamma\left[n(0)n(1)\right]^{1/4}2i}e^{-ia(\eta+1)}e^{t(\eta+1)}\left[1 + \mathcal{O}(1/t)\right], \quad t \to \infty,
\end{aligned}
$$

because $1 + \eta > |1 - \eta|$. Here, A and B are given by (7.91) and $A \neq 0$ because of $n(1) \neq 1$. Analogously, we have for the complex conjugate $\overline{k} = a - it$

$$
\overline{k}\frac{d(\overline{k})}{\gamma} = \frac{A}{\gamma\left[n(0)n(1)\right]^{1/4}2i}e^{ia(\eta+1)}e^{t(\eta+1)}\left[1 + \mathcal{O}(1/t)\right], \quad t \to \infty.
$$

Therefore, also the ratio is known and also

$$
\psi(a) := \lim_{t\to\infty}\frac{k\,d(k)}{\overline{k}\,d(\overline{k})} = -e^{2ia(1+\eta)} \quad \text{for all } a \in \mathbb{R}.
$$

This determines η through $2i(1 + \eta) = \psi'(0)$.

Step B: Under the assumption $\eta \neq 1$ we determine $n(1)$ and $\gamma n(0)^{1/4}$ from the data. Let now $k > 0$ be real valued. Estimate (7.90) has the form

$$
k\frac{d(k)}{\gamma} = \frac{1}{\gamma\left[n(0)n(1)\right]^{1/4}}\left[A\sin\big(k(1+\eta)\big) + B\sin\big(k(1-\eta)\big)\right] + R(k)
$$

with $|R(k)| \leq c_1/k$ for $k \geq 1$ and some $c_1 > 0$. The left hand side is known and also η from Step A. Therefore, for fixed $a > 0$, also

$$\psi_1(T) = \frac{1}{T} \int_a^T k \frac{d(k)}{\gamma} \sin(k(1+\eta)) \, dk$$

$$= \frac{1}{\gamma \left[n(0)n(1)\right]^{1/4}} \frac{1}{T} \int_a^T \left[A \sin^2(k(1+\eta)) \right. +$$

$$\left. +B \sin(k(1-\eta)) \sin(k(1+\eta))\right] dk + \frac{1}{T} \int_a^T R(k) \sin(k(1+\eta)) \, dk$$

and

$$\psi_2(T) = \frac{1}{T} \int_a^T k \frac{d(k)}{\gamma} \sin(k(1-\eta)) \, dk$$

$$= \frac{1}{\gamma \left[n(0)n(1)\right]^{1/4}} \frac{1}{T} \int_a^T \left[A \sin(k(1+\eta)) \sin(k(1-\eta)) \right. +$$

$$\left. +B \sin^2(k(1-\eta))\right] dk + \frac{1}{T} \int_a^T R(k) \sin(k(1-\eta)) \, dk$$

are known for $T \geq a$. The terms involving $R(k)$ tend to zero as T tends to infinity because $\frac{1}{T} \int_a^T |R(k)| \, dk \leq \frac{c_1}{T} \int_a^T \frac{dk}{k} = \frac{c_1 \ln(T/a)}{T}$. The other elementary integrals can be computed explicitly which yields that $\lim_{T\to\infty} \frac{1}{T} \int_a^T \sin(k(1+\eta)) \sin(k(1-\eta)) \, dk = 0$ and $\lim_{T\to\infty} \frac{1}{T} \int_a^T \sin^2(k(1\pm\eta)) \, dk = \frac{1}{2}$ (note that $\eta \neq 1$). Therefore, the limits $\lim_{T\to\infty} \psi_1(T) = \frac{A}{2\gamma[n(0)n(1)]^{1/4}}$ and $\lim_{T\to\infty} \psi_2(T) = \frac{B}{2\gamma[n(0)n(1)]^{1/4}}$ are known and thus also

$$\lim_{T\to\infty} \frac{\psi_1(T)}{\psi_2(T)} = \frac{A}{B} = \frac{\sqrt{n(1)}-1}{\sqrt{n(1)}+1}.$$

This determines $n(1)$, thus also A and B and therefore also $\gamma \, n(0)^{1/4}$.

From now on we assume that $n(r) \leq 1$ for all $r \in [0,1]$ and $n(1) < 1$. Then $\eta < 1$ and $\int_0^1 n(s) \, s^2 ds < \frac{1}{3}$. Therefore, all of the assumptions are satisfied for the determination of η, $n(1)$, and $\gamma \, n(0)^{1/4}$.

Step C: Now we use the Liouville transform $y_k(r) = n(s(r))^{-1/4} z_k(s(r))$ from (7.88a)–(7.88d) again where $s(r) = \int_0^r \sqrt{n(s)} ds$ and $z_k(s)$ solves (7.88b) in $(0,\eta)$ with $z_k(0) = 0$ and $z_k'(0) = n(0)^{-1/4}$. The application of Theorem 5.19 in Example 5.20 yields an explicit form of $n(0)^{1/4} z_k(r)$ in terms of the solution

$K \in C^2(\overline{\Delta_0})$ of the Gousat problem[11]

$$K_{ss}(s,t) \; - \; K_{tt}(s,t) \; - \; q(s)\,K(s,t) \; = \; 0 \quad \text{in } \Delta_0, \tag{7.106a}$$

$$K(s,0) \;\; = \;\; 0, \quad 0 \le s \le \eta, \tag{7.106b}$$

$$K(s,s) \;\; = \;\; \frac{1}{2} \int_0^s q(\sigma)\,d\sigma, \quad 0 \le s \le \eta, \tag{7.106c}$$

where $q(s)$ is related to $n(r)$ by (7.88c) and $\Delta_0 = \{(s,t) \in \mathbb{R}^2 : 0 < t < s < \eta\}$. Indeed, equation (5.52) yields

$$y_k(r) \;\; = \;\; \frac{1}{[n(0)n(r)]^{1/4}} \left[\frac{\sin(ks(r))}{k} \; + \; \int_0^{s(r)} K(s(r),t)\, \frac{\sin(kt)}{k}\, dt \right] \tag{7.107}$$

for $0 \le r \le 1$ and thus

$$y_k(1) \;\; = \;\; \frac{1}{[n(0)n(1)]^{1/4}} \left[\frac{\sin(k\eta)}{k} \; + \; \int_0^{\eta} K(\eta,t)\, \frac{\sin(kt)}{k}\, dt \right].$$

Differentiation of (7.107) and setting $r = 1$ yields

$$y_k'(1) \;\; = \;\; \left[\frac{n(1)}{n(0)}\right]^{1/4} \left[\cos(k\eta) + \frac{\sin(k\eta)}{2k} \int_0^{\eta} q(s)\,ds + \int_0^{\eta} \frac{\partial K(\eta,t)}{\partial s}\, \frac{\sin(kt)}{k}\, dt \right]$$

$$- \frac{n'(1)}{4\,n(0)^{1/4}\,n(1)^{3/4}} \left[\frac{\sin(k\eta)}{k} \; + \; \int_0^{\eta} K(\eta,t)\, \frac{\sin(kt)}{k}\, dt \right]$$

Step D: We determine $K(\eta,t)$ for $t \in [0,\eta]$. With the constant γ from Hadamard's formula (7.105) we have for $k = \ell\pi$, $\ell \in \mathbb{N}$,

$$\ell\pi \frac{d(\ell\pi)}{\gamma} \;\; = \;\; \ell\pi \frac{y_{\ell\pi}(1)}{\gamma} (-1)^{\ell+1}$$

$$= \;\; \frac{(-1)^{\ell+1}}{[\gamma\, n(0)^{1/4}]\, n(1)^{1/4}} \left[\sin(\ell\pi\eta) \; + \; \int_0^{\eta} K(\eta,t)\, \sin(\ell\pi t)\, dt \right].$$

The left hand side and the factor in front of the bracket are known which implies that $\int_0^{\eta} K(\eta,t)\, \sin(\ell\pi t)\, dt$ is determined from the data for all $\ell \in \mathbb{N}$. This determines $K(\eta,\cdot)$ because $\{\sin(\ell\pi t) : \ell \in \mathbb{N}\}$ is complete in $L^2(0,\eta)$ since $\eta < 1$. Indeed, extend $K(\eta,\cdot)$ by zero into $(0,1)$ and then to an odd function

---

[11]The interval $[0,1]$ is now replaced by $[0,\eta]$ which does not affect the result.

into $(-1, 1)$. Then $\int_0^\eta K(\eta, t) \sin(\ell\pi t)\, dt$ are the Fourier coefficients of this odd extension which determine $K(\eta, \cdot)$ uniquely.

Step E: Determination of $n'(1)$. From the previous arguments $y_k(1)/\gamma$ is known and thus also $\psi(k) := k \frac{d(k)}{\gamma} + k \frac{y_k(1)}{\gamma} \cos k = \frac{y_k'(1)}{\gamma} \sin k$. We determine the asymptotic behavior of this expression as $k \in \mathbb{R}$ tends to infinity. The integrals $\int_0^\eta K(\eta, t) \sin(kt)\, dt$ and $\int_0^\eta \partial K(\eta, t)/\partial s \sin(kt)\, dt$ tend to zero as $\mathcal{O}(1/k)$ as seen from partial integration. Therefore,

$$
\begin{aligned}
\psi(k) &= k \frac{d(k)}{\gamma} + k \frac{y_k(1)}{\gamma} \cos k = \frac{y_k'(1)}{\gamma} \sin k \\
&= \frac{n(1)^{1/4}}{\gamma\, n(0)^{1/4}} \left[\cos(k\eta) + \frac{\sin(k\eta)}{2k} \int\limits_0^\eta q(s)\, ds \right] \sin k \\
&\quad - \frac{n'(1)}{4\,\gamma\, n(0)^{1/4}\, n(1)^{3/4}} \frac{\sin(k\eta)}{k} \sin k + \mathcal{O}(1/k^2).
\end{aligned}
$$

The left hand side is known and also the first term on the right hand side because $\int_0^\eta q(s)\, ds = 2K(\eta, \eta)$. This determines $n'(1)$ from the data.

Step F: Determination of $\partial K(\eta, t)/\partial s$ for $t \in [0, \eta]$. We compute $\psi'(\ell\pi)$ from the previously defined function ψ as

$$
\begin{aligned}
\psi'(\ell\pi) &= \frac{y_{\ell\pi}'(1)}{\gamma} \cos(\ell\pi) = \frac{n(1)^{1/4}}{\gamma\, n(0)^{1/4}} (-1)^\ell \left[\cos(\ell\pi\eta) + \frac{\sin(\ell\pi\eta)}{\ell\pi} K(\eta, \eta) \right. \\
&\quad \left. + \int\limits_0^\eta \frac{\partial K(\eta, t)}{\partial s} \frac{\sin(\ell\pi t)}{\ell\pi}\, dt \right] \\
&\quad - \frac{n'(1)}{4\,\gamma\, n(0)^{1/4}\, n(1)^{3/4}} (-1)^\ell \left[\frac{\sin(\ell\pi\eta)}{\ell\pi} + \int\limits_0^\eta K(\eta, t) \frac{\sin(\ell\pi t)}{\ell\pi}\, dt \right].
\end{aligned}
$$

From this we conclude that also $\int_0^\eta \partial K(\eta, t)/\partial s \sin(\ell\pi t)\, dt$ is known for all $\ell \in \mathbb{N}$ and thus also $\partial K(\eta, t)/\partial s$ by the same arguments as in Step D.

Step G: Determination of $q = q(s)$ for $0 \leq s \leq \eta$. We recall from Steps D and F, that $K(\eta, t)$ and $\partial K(\eta, t)/\partial s$ are determined from the data for all $t \in [0, \eta]$. More precisely, if K_ℓ denote the solutions of (7.106a)–(7.106c) for $q = q_\ell$, $\ell = 1, 2$, then $\eta_1 = \eta_2 =: \eta$ and $K_1(\eta, \cdot) = K_2(\eta, \cdot)$ and $\partial K_1(\eta, \cdot)/\partial s = \partial K_2(\eta, \cdot)/\partial s$. The difference $K := K_1 - K_2$ satisfies

$$
K_{ss}(s, t) - K_{tt}(s, t) - q_1(s) K(s, t) = [q_1(s) - q_2(s)] K_2(s, t) \quad \text{in } \Delta_0,
$$

and $K(\cdot, 0) = 0$ on $[0, \eta]$ and $K(s, s) = \frac{1}{2} \int\limits_0^s [q_1(\sigma) - q_2(\sigma)]\, d\sigma$ for $0 \leq s \leq \eta$. Furthermore, $K(\eta, \cdot) = \partial K(\eta, \cdot)/\partial s = 0$ on $[0, \eta]$. Now we apply Theorem 5.18 from Chapter 5 with $q = q_1$, $F = K_2$, and $f = g = 0$. We observe that the pair

$(K, q_1 - q_2)$ solves the homogeneous system (5.43a)–(5.43d). The uniqueness result of this theorem yields that $K = 0$ and $q_1 = q_2$. This proves Part G.

Step H: In this final part we have to determine n from q where their relationship is given by (7.88c). Let again n_1 and n_2 be two indices with the same transmission eigenvalues. Define u_ℓ by

$$u_\ell(s) = \left[n_\ell\big(r_\ell(s)\big) \right]^{1/4}, \quad s \in [0, \eta_\ell] = [0, \eta],$$

where again $r_\ell = r_\ell(s)$ is the inverse of $s_\ell = s_\ell(r) = \int_0^r \sqrt{n_\ell(\sigma)}\, d\sigma$ for $\ell = 1, 2$. An elementary computation (using the chain rule and the derivative of the inverse function and (7.88c)) yields that u_ℓ satisfies the ordinary linear differential equation

$$u_\ell''(s) = q_\ell(s)\, u_\ell(s), \quad 0 \le s \le \eta,$$

with end conditions $u_\ell(\eta) = n_\ell(1)^{1/4}$ and $u_\ell'(\eta) = \frac{n_\ell'(1)}{4\, n_\ell(1)^{5/4}}$. From $q_1 = q_2$ and $n_1(1) = n_2(1)$ and $n_1'(1) = n_2'(1)$ and the uniqueness of this initial value problem we conclude that $u_1(s) = u_2(s)$ for all s; that is,

$$s_1'\big(r_1(s)\big) = \sqrt{n_1\big(r_1(s)\big)} = \sqrt{n_2\big(r_2(s)\big)} = s_2'\big(r_2(s)\big).$$

On the other hand, differentiating $s_\ell\big(r_\ell(s)\big) = s$ yields $s_\ell'\big(r_\ell(s)\big) r_\ell'(s) = 1$ which implies that $r_1' = r_2'$, thus $r_1 = r_2$ and, finally, $n_1 = n_2$.

We summarize the result in the following theorem.

Theorem 7.54 *Let $n_j \in C^2[0, 1]$, $j = 1, 2$, be positive with $n_j(r) \le 1$ on $[0, 1]$ and $n_j(1) < 1$ such that all of the corresponding transmission eigenvalues with radially symmetric eigenfunctions coincide. Then n_1 and n_2 have to coincide.*

7.7   Numerical Methods

In this section, we describe three types of numerical algorithms for the approximate solution of the inverse scattering problem for the determination of n and not only of the support D of $n - 1$. We assume—unless stated otherwise—that $n \in L^\infty(\mathbb{R}^3)$ with $n(x) = 1$ outside some ball $B = B(0, a)$ of radius $a > 0$.

The numerical methods we describe now are all based on the Lippmann–Schwinger integral equation. We define the volume potential $V\phi$ with density ϕ by

$$(V\phi)(x) := \int\limits_{|y| < a} \frac{e^{ik|x-y|}}{4\pi|x-y|}\, \phi(y)\, dy, \quad x \in B. \tag{7.108}$$

Then the Lippmann–Schwinger equation (7.26) takes the form

$$u - k^2 V(mu) = u^i \quad \text{in } B, \tag{7.109}$$

where again $m = n - 1$ and $u^i(x, \hat{\theta}) = \exp(ik\hat{\theta} \cdot x)$. The far field pattern of $u^s = k^2 V(mu)$ is given by

$$u_\infty(\hat{x}) = \frac{k^2}{4\pi} \int_B m(y)\, u(y)\, e^{-ik\hat{x}\cdot y}\, dy, \quad \hat{x} \in S^2. \tag{7.110}$$

Defining the integral operator $W : L^\infty(B) \to L^\infty(S^2)$ by

$$(W\psi)(\hat{x}) := \frac{k^2}{4\pi} \int_B \psi(y)\, e^{-ik\hat{x}\cdot y}\, dy, \quad \hat{x} \in S^2, \tag{7.111}$$

we note that the inverse scattering problem is to solve the system of equations

$$u - k^2 V(mu) = u^i \quad \text{in } B, \tag{7.112a}$$

$$W(mu) = u^\infty \quad \text{on } S^2, \tag{7.112b}$$

for m and u. Here, u^∞ denotes the measured far field pattern (in contrast to u_∞ which is the true far field pattern). From the uniqueness results of Section 7.4, we expect that the far field patterns of more than one incident field have to be known. Therefore, from now on, we consider $u^i = u^i(x, \hat{\theta}) = \exp(ik\hat{\theta} \cdot x)$, $u = u(x, \hat{\theta})$, and $u^\infty = u^\infty(\hat{x}, \hat{\theta})$ to be functions of two variables. The operators V and W from equations (7.108) and (7.111) can be considered as linear and bounded operators

$$V: \quad L^\infty(B \times S^2) \longrightarrow L^\infty(B \times S^2), \tag{7.113a}$$

$$W: \quad L^\infty(B \times S^2) \longrightarrow L^\infty(S^2 \times S^2). \tag{7.113b}$$

In the next sections, we discuss three methods for solving the inverse scattering problem, the first two of which are based on the system (7.112a), (7.112b). We formulate the algorithms and prove convergence results only for the setting in function spaces, although for the practical implementations these algorithms have to be discretized. The methods suggested by Gutman and Klibanov [112, 113] and Kleinman and van den Berg [164] are iteration methods based on the system (7.112a), (7.112b). The first one is a regularized simplified Newton method, the second a modified gradient method. In Section 7.7.3, we describe a different method, which has been proposed by Colton and Monk in several papers (see [59]–[63]) and can be considered as an intermediate step towards the development of the linear sampling method (see [160]).

The system (7.112a), (7.112b) describes a nonlinear equation for the pair (m, u). It can be shown that this equation is locally improperly posed in the sense of Definition 4.1. In principle, all of the methods of Chapter 4 as Tikhonov's regularization or Landweber's method can be applied. Convergence results are available once the assumptions such as the source condition or the tangential cone condition can be verified. For the inverse scattering problem discussed in this chapter T. Hohage and F. Weidling (see [138]) were able to verify the variational source condition for an index function of logarithmic type.

To the author's knowledge, the validation of the tangential cone condition is still open. Therefore, the results of Section 4.2 on the Tikhonov regularization technique are applicable.

7.7.1   A Simplified Newton Method

For simplicity of the presentation we assume for this section that n is continuous; that is, $n \in C(\overline{D})$ for some bounded domain D and $n = 1$ outside of $\overline{D}$. By scaling the problem, we assume throughout this section that D is contained in the cube $Q = [-\pi, \pi]^3 \subset \mathbb{R}^3$. We define the nonlinear mapping

$$T : C(Q) \times C(Q \times S^2) \longrightarrow C(Q \times S^2) \times C(S^2 \times S^2) \qquad (7.114a)$$

by

$$T(m, u) := \left(u - k^2 V(mu), \, W(mu) \right) \qquad (7.114b)$$

for $m \in C(Q)$ and $u \in C(Q \times S^2)$. Then the inverse problem can be written in the form

$$T(m, u) = (u^i, u^\infty).$$

The Newton method is to compute iterations (m_ℓ, u_ℓ), $\ell = 0, 1, 2, \dots$ by

$$(m_{\ell+1}, u_{\ell+1}) = (m_\ell, u_\ell) - T'(m_\ell, u_\ell)^{-1} \left[T(m_\ell, u_\ell) - (u^i, u^\infty) \right] \qquad (7.115)$$

for $\ell = 0, 1, 2, \dots$. The components of the mapping T are bilinear, thus it is not difficult to see that the Fréchet derivative $T'(m, u)$ of T at (m, u) is given by

$$T'(m, u)(\mu, v) = \left(k^2 V(\mu u) + v - k^2 V(mv), \, W(\mu u) + W(mv) \right) \qquad (7.116)$$

for $\mu \in C(Q)$ and $v \in C(Q \times S^2)$.

The simplified Newton method is to replace $T'(m_\ell, u_\ell)$ by some fixed $T'(\hat{m}, \hat{u})$ (see Theorem A.65 of Appendix A). Then it is known that under certain assumptions linear convergence can be expected. We choose $\hat{m} = 0$ and $\hat{u} = u^i$. Then the simplified Newton method sets $m_{\ell+1} = m_\ell + \mu$ and $u_{\ell+1} = u_\ell + v$, where (μ, v) solves $T'(0, u^i)(\mu, v) = (u^i, u^\infty) - T(m_\ell, u_\ell)$. Using the characterization of T', we are led to the following algorithm.

(A) Set $m_0 = 0$, $u_0 = u^i$, and $\ell = 0$.

(B) Determine $(\mu, v) \in C(Q) \times C(Q \times S^2)$ from the system of equations

$$k^2 V(\mu u^i) - v = u^i - u_\ell + k^2 V(m_\ell u_\ell), \qquad (7.117a)$$
$$W(\mu u^i) = u^\infty - W(m_\ell u_\ell). \qquad (7.117b)$$

(C) Set $m_{\ell+1} = m_\ell + \mu$ and $u_{\ell+1} = u_\ell + v$, replace ℓ by $\ell + 1$, and continue with step (B).

We assume in the following that the given far field pattern u^∞ is continuous (that is, $u^\infty \in C(S^2 \times S^2)$). Solving an equation of the form $W(\mu u^i) = \rho$ means solving the integral equation of the first kind,

$$\int_Q \mu(y) \, e^{iky \cdot (\hat\theta - \hat x)} \, dy = -\frac{4\pi}{k^2} \rho(\hat x, \hat\theta), \quad \hat x, \hat\theta \in S^2. \tag{7.118}$$

We approximately solve this equation by a special collocation method. We observe that the left-hand side is essentially the Fourier transform $\tilde\mu$ of μ evaluated at $\xi = k(\hat x - \hat\theta)$. As in Gutman and Klibanov [113], we define $N \in \mathbb{N}$ to be the largest integer not exceeding $2k/\sqrt 3$, the set

$$\mathcal{Z}_N := \{ j \in \mathbb{Z}^3 : |j_s| \leq N, \; s = 1, 2, 3 \}$$

of grid points, and the finite-dimensional space

$$X_N := \left\{ \sum_{j \in \mathcal{Z}_N} a_j \, e^{i j \cdot x} : a_j \in \mathbb{C} \right\}. \tag{7.119}$$

Then, for every $j \in \mathcal{Z}_N$, there exist unit vectors $\hat x_j, \hat\theta_j \in S^2$ with $j = k(\hat x_j - \hat\theta_j)$ (note that $|j|/k \leq 2$). This is easily seen from the fact that the intersection of S^2 with the sphere of radius 1 and center j/k is not empty. For every $j \in \mathcal{Z}_N$, we fix the unit vectors $\hat x_j$ and $\hat\theta_j$ such that $j = k(\hat x_j - \hat\theta_j)$.

We solve (7.118) approximately by substituting $\hat x_j$ and $\hat\theta_j$ into this equation. This yields

$$\int_Q \mu(y) \, e^{-i j \cdot y} \, dy = -\frac{4\pi}{k^2} \rho(\hat x_j, \hat\theta_j), \quad j \in \mathcal{Z}_N. \tag{7.120}$$

The left-hand sides are just the first Fourier coefficients of μ, therefore the unique solution of (7.120) in X_N is given by $\mu = L_1\rho$, where the operator $L_1 : C(S^2 \times S^2) \to X_N$ is defined by

$$(L_1\rho)(x) = -\frac{1}{2\pi^2 k^2} \sum_{j \in \mathcal{Z}_N} \rho(\hat x_j, \hat\theta_j) \, e^{i j \cdot x}. \tag{7.121}$$

The regularized algorithm now takes the form

(A$_r$) Set $m_0 = 0$, $u_0 = u^i$, and $\ell = 0$.

(B$_r$) Set

$$\mu := L_1 \big[u^\infty - W(m_\ell u_\ell) \big] \quad \text{and}$$

$$v := u^i - u_\ell - k^2 V(m_\ell u_\ell) - k^2 V(\mu u^i).$$

(C$_r$) Set $m_{\ell+1} = m_\ell + \mu$ and $u_{\ell+1} = u_\ell + v$, replace ℓ by $\ell + 1$, and continue with step (B$_r$).

Then we can prove the following (see [113]).

Theorem 7.55 *There exists $\varepsilon > 0$ such that, if $m \in C(Q)$ with $\|m\|_\infty \leq \varepsilon$ and $u = u(x, \hat{\theta})$ is the corresponding total field with exact far field pattern $u^\infty(\hat{x}, \hat{\theta}) = u_\infty(\hat{x}, \hat{\theta})$, then the sequence (m_ℓ, u_ℓ) constructed by the regularized algorithm (A_r), (B_r), (C_r) converges to some $(\tilde{m}, \tilde{u}) \in X_N \times C(Q \times S^2)$ that satisfies the scattering problem with refraction contrast $\tilde{m}$. Its far field pattern $\tilde{u}_\infty$ coincides with u_∞ at the points $(\hat{x}_j, \hat{\theta}_j) \in S^2 \times S^2$, $j \in \mathbb{Z}_N$. If, in addition, the exact solution m satisfies $m \in X_N$, then the sequence (m_ℓ, u_ℓ) converges to (m, u).*

Proof: We define the operator

$$L : C(Q \times S^2) \times C(S^2 \times S^2) \longrightarrow X_N \times C(Q \times S^2)$$

by

$$L(w, \rho) := \left(L_1 \rho, \ w - k^2 V(u^i L_1 \rho) \right).$$

Then L is a left inverse of $T'(0, u^i)$ on $X_N \times C(Q \times S^2)$; that is,

$$L T'(0, u^i)(\mu, v) = (\mu, v) \quad \text{for all } (\mu, v) \in X_N \times C(Q \times S^2).$$

Indeed, let $(\mu, v) \in X_N \times C(Q \times S^2)$ and set $(w, \rho) = T'(0, u^i)(\mu, v)$, i.e., $w = v + k^2 V(\mu u^i)$ and $\rho = W(\mu u^i)$. The latter equation implies that

$$\int_Q \mu(y) e^{-ij \cdot y} \, dy = -\frac{4\pi}{k^2} \rho(\hat{x}_j, \hat{\theta}_j), \quad j \in \mathbb{Z}_N.$$

Because $\mu \in X_N$, this yields $\mu = L_1 \rho$ and thus $L(w, \rho) = (\mu, v)$.

With the abbreviations $z_\ell = (m_\ell, u_\ell)$ and $R = (u^i, u^\infty)$, we can write the regularized algorithm in the form

$$z_{\ell+1} = z_\ell - L \left[T(z_\ell) - R \right], \quad \ell = 0, 1, 2, \ldots$$

in the space $X_N \times C(Q \times S^2)$. We can now apply a general result about local convergence of the simplified Newton method (see Appendix A, Theorem A.65). This yields the existence of a unique solution $(\tilde{m}, \tilde{u}) \in X_N \times C(Q \times S^2)$ of $L \left[T(\tilde{m}, \tilde{u}) - R \right] = 0$ and linear convergence of the sequence (m_ℓ, u_ℓ) to $(\tilde{m}, \tilde{u})$. The equation $\tilde{u} + k^2 V(\tilde{m}\tilde{u}) = u^i$ is equivalent to the scattering problem by Theorem 7.12. The equation $L_1 W(\tilde{m}\tilde{u}) = L_1 f$ is equivalent to $\tilde{u}_\infty(\hat{x}_j, \hat{\theta}_j) = u^\infty(\hat{x}_j, \hat{\theta}_j)$ for all $j \in \mathbb{Z}_N$. Finally, if $m \in X_N$, then (m, u) satisfies $L T(m, u) = L R$ and thus $(\tilde{m}, \tilde{u}) = (m, u)$. This proves the assertion.   □

We have formulated the algorithm with respect to the Lippmann–Schwinger integral equation because our analysis of existence and continuous dependence is based on this setting. There is an alternative way to formulate the simplified Newton method in terms of the original scattering problems; see [113]. We note also that our analysis can easily be modified to treat the case where only $n \in L^\infty(B)$. For numerical examples, we refer to [113].

7.7.2 A Modified Gradient Method

The idea of the numerical method proposed and numerically tested by Kleinman and van den Berg (see [164]) is to solve (7.112a), (7.112b) by a gradient-type method. For simplicity, we describe the method again in the function space setting and refer for discretization aspects to the original literature [164]. Again let $B = B(0, a)$ contain the support of $m = 1 - n$.

(A) Choose $m_0 \in L^\infty(B)$, $u_0 \in L^2(B \times S^2)$, and set $\ell = 0$.

(B) Choose directions $e_\ell \in L^2(B \times S^2)$ and $d_\ell \in L^\infty(B)$, and set

$$u_{\ell+1} = u_\ell + \alpha_\ell \, e_\ell, \quad m_{\ell+1} = m_\ell + \beta_\ell \, d_\ell. \tag{7.122}$$

The stepsizes $\alpha_\ell, \beta_\ell > 0$ are chosen in such a way that they minimize the functional

$$\Psi_\ell(\alpha, \beta) := \frac{\|r_{\ell+1}\|^2_{L^2(B \times S^2)}}{\|u^i\|^2_{L^2(B \times S^2)}} + \frac{\|s_{\ell+1}\|^2_{L^2(S^2 \times S^2)}}{\|f\|^2_{L^2(S^2 \times S^2)}}, \tag{7.123a}$$

where the defects $r_{\ell+1}$ and $s_{\ell+1}$ are defined by

$$r_{\ell+1} := u^i - u_{\ell+1} - k^2 V(m_{\ell+1} u_{\ell+1}), \tag{7.123b}$$

$$s_{\ell+1} := u^\infty - W(m_{\ell+1} u_{\ell+1}). \tag{7.123c}$$

(C) Replace ℓ by $\ell + 1$ and continue with step (B).

There are different choices for the directions d_ℓ and e_ℓ. In [164],

$$d_\ell(x) = -\int_{S^2} \tilde{d}_\ell(x, \hat{\theta}) \, \overline{u_\ell(x, \hat{\theta})} \, ds(\hat{\theta}), \quad x \in B, \quad \text{and} \quad e_\ell := r_\ell \tag{7.124}$$

have been chosen where

$$\tilde{d}_\ell = -W^*\big(W(m_\ell u_\ell) - u^\infty\big) \in L^\infty(\overline{B} \times S^2).$$

In this case, d_ℓ is the steepest descent direction of $\mu \mapsto \|W(\mu u_\ell) - u^\infty\|^2_{L^2(S^2 \times S^2)}$. In [266], for d_ℓ and e_ℓ Polak–Ribière conjugate gradient directions are chosen. A rigorous convergence analysis of either method has not been carried out.

A severe drawback of the methods discussed in Sections 7.7.1 and 7.7.2 is that they iterate on functions $m_\ell = m_\ell(x)$ and $u_\ell = u_\ell(x, \hat{\theta})$. To estimate the storage requirements, we choose a grid of order $N \cdot N \cdot N$ grid points in B and M directions $\theta_1, \ldots, \theta_M \in S^2$. Then both methods iterate on vectors of dimension $N^6 \cdot M$. From the uniqueness results, M is expected to be large, say, of order N^2. For large values of M, the method described next has proven to be more efficient.

7.7.3   The Dual Space Method

The method described here is due to Colton and Monk [61, 62] based on their earlier work for inverse obstacle scattering problems (see [59, 60]). There exist various modifications of this method, but we restrict ourselves to the simplest case.

This method consists of two steps. In the first step, one tries to determine a superposition of the incident fields $u^i = u^i(\cdot, \hat{\theta})$ such that the corresponding far field pattern $u_\infty(\cdot, \hat{\theta})$ is (close to) the far field pattern of radiating multipoles. In the second step, the function $m = n - 1$ is determined from an interior transmission problem.

We describe both steps separately. Assume for the following that the origin is contained in $B = B(0, a)$. By u^∞ we denote again the measured far field pattern.

Step 1: Determine $g \in L^2(S^2)$ with

$$\int_{S^2} u^\infty(\hat{x}, \hat{\theta}) \, g(\hat{\theta}) \, ds(\hat{\theta}) = 1, \quad \hat{x} \in S^2. \tag{7.125}$$

In Theorem 7.22, we have proven that for the exact far field pattern $u_\infty(\hat{x}, \hat{\theta})$ this integral equation of the first kind is solvable in $L^2(S^2)$ if and only if the interior transmission problem

$$\Delta v + k^2 v = 0 \text{ in } B, \qquad \Delta w + k^2 n w = 0 \text{ in } B, \tag{7.126a}$$

$$w(x) - v(x) = \frac{e^{ik|x|}}{|x|} \quad \text{on } \partial B, \tag{7.126b}$$

$$\frac{\partial w(x)}{\partial \nu} - \frac{\partial v(x)}{\partial \nu} = \frac{\partial}{\partial \nu} \frac{e^{ik|x|}}{|x|} \quad \text{on } \partial B, \tag{7.126c}$$

has a solution $v, w \in L^2(B)$ in the ultra weak sense of Definition 7.21 such that

$$v(x) = \int_{S^2} e^{ikx \cdot \hat{y}} \, g(\hat{y}) \, ds(\hat{y}), \quad x \in \mathbb{R}^3. \tag{7.127}$$

The kernel of the integral operator in (7.125) is (for the exact far field pattern) analytic with respect to both variables, thus (7.125) represents a severely ill-posed—but linear—equation and can be treated by Tikhonov's regularization method as described in Chapter 2 in detail. (In this connection, see the remark following Theorem 7.23.)

We formulate the interior transmission problem (7.126a)–(7.126c) as an integral equation.

Lemma 7.56   *(a)Let $v, w \in L^2(B)$ solve the boundary value problem (7.126a)–(7.126c) in the ultra weak sense of (7.46b). Define $u = w - v$ in B and*

$u(x) = \exp(ik|x|)/|x|$ in $\mathbb{R}^3 \setminus B$. Then $u \in H^2_{loc}(\mathbb{R}^3)$ and u and $w \in L^2(B)$ solve

$$u(x) = k^2 \int\limits_{|y|<a} m(y)\, w(y)\, \Phi(x,y)\, dy\,, \quad x \in \mathbb{R}^3\,, \tag{7.128a}$$

where again $m = n - 1$. Furthermore, v satisfies $\Delta v + k^2 v = 0$ in B in the ultra weak sense; that is,

$$\int\limits_B (\Delta\phi + k^2\phi)\, v\, dx = 0 \quad \text{for all } \phi \in H^2_0(B)\,. \tag{7.128b}$$

(b) Let $u \in H^2_{loc}(\mathbb{R}^3)$ and $w \in L^2(B)$ solve (7.128a) and $u(x) = \exp(ik|x|)/|x|$ in $\mathbb{R}^3 \setminus B$. Furthermore, let $v := w - u \in L^2(B)$ be an ultra weak solution of $\Delta v + k^2 v = 0$ in B in the sense of (7.128b). Then $v, w \in L^2(B)$ solve the boundary value problem (7.126a)–(7.126c) in the ultra weak sense of (7.46b).

Proof: (a) Let $v, w \in L^2(B)$ solve the boundary value problem (7.126a)–(7.126c) in the ultra weak sense. Equation (7.128b) follows immediately from (7.46b) by choosing $\phi \in H^2_0(B)$ and $\psi = 0$ in B.

Let now $\psi \in C^2(\mathbb{R}^3)$ with compact support and set $\phi = \psi$ in $\mathbb{R}^3$. Substituting this into (7.46b) yields

$$\int\limits_B (\Delta\psi + k^2\psi)\,(w - v)\, dx \;+\; k^2 \int\limits_B m\,\psi\, w\, dx$$

$$= \; 4\pi \int\limits_{\partial B} \left[\Phi(\cdot,0) \frac{\partial\psi}{\partial\nu} - \psi\, \frac{\partial\Phi(\cdot,0)}{\partial\nu} \right] ds$$

$$= \; -4\pi \int\limits_{\mathbb{R}^3\setminus B} \Phi(\cdot,0)\,(\Delta\psi + k^2\psi)\, dx$$

for all $\psi \in C^2(\mathbb{R}^3)$ with compact support. Here we have used Green's theorem in the exterior of B (note that ψ has compact support). Using the definition of u in B and in the exterior of B yields

$$\int\limits_{\mathbb{R}^3} (\Delta\psi + k^2\psi)u\, dx \; = \; -k^2 \int\limits_B m\psi w\, dx$$

for all $\psi \in C^2(\mathbb{R}^3)$ with compact support. The regularity result of Lemma 7.10 yields $u \in H^2_{loc}(\mathbb{R}^3)$ and $\Delta u + k^2 u = -k^2 mw$ in $\mathbb{R}^3$. This equation is uniquely solved by the volume potential with density $k^2 mw$ (see Theorem 7.11). This proves the first part.

(b) Since u is the volume potential we conclude from Theorem 7.11 that u is the radiating solution of $\Delta u + k^2 u = -k^2 mw$ in $\mathbb{R}^3$. Let $\phi, \psi \in C^2(\mathbb{R})$ with compact support such that $\phi = \psi$ in the exterior of B. Then, since $w = v + u$ in B,

$$\int_B (\Delta\psi + k^2 n\psi)\, w\, dx \ - \ \int_B (\Delta\phi + k^2\phi)\, v\, dx$$

$$= \ \int_B (\Delta\psi + k^2\psi)\,(v + u)\, dx \ + \ k^2 \int_B m\psi w\, dx \ - \ \int_B (\Delta\phi + k^2\phi)\, v\, dx$$

$$= \ \int_B \left[\Delta(\psi - \phi) + k^2(\psi - \phi)\right] v\, dx \ + \ k^2 \int_B m\psi w\, dx$$

$$+ \ \int_{\mathbb{R}^3} (\Delta\psi + k^2\psi)u\, dx \ - \ \int_{\mathbb{R}^3 \setminus B} (\Delta\psi + k^2\psi)u\, dx \, .$$

The first integral on the right hand side vanishes because v is an ultra weak solution of $\Delta v + k^2 v = 0$ in B and $\psi - \phi \in H_0^2(B)$. The sum of the second and the third integral vanishes as well after application of Green's second formula (7.12b) and $\Delta u + k^2 u = -k^2 m w$ in $\mathbb{R}^3$. For the last integral we use that $u(x) = \exp(ik|x|)/|x|$ in $\mathbb{R}^3 \setminus B$ and apply Green's theorem in the exterior of B which yields

$$\int_{\mathbb{R}^3 \setminus B} (\Delta\psi + k^2\psi)u\, dx \ = \ -\int_{\partial B} \left[\frac{\partial\psi(x)}{\partial\nu}\frac{e^{ik|x|}}{|x|} - \psi(x)\frac{\partial}{\partial\nu}\frac{e^{ik|x|}}{|x|}\right] ds$$

which ends the proof.   □

Motivated by this characterization, we describe the second step.

Step 2: With the (approximate) solution $g \in L^2(S^2)$ of (7.125), define the function $v = v_g$ by (7.127). Determine m and w such that m, v_g, and w solve the interior boundary value problem (7.126a)–(7.126c) or, equivalently, the system

$$w - v_g - k^2 V(mw) \ = \ 0 \quad \text{in } B\,, \tag{7.129a}$$

$$k^2 V(mw) - 4\pi\, \Phi(\cdot, 0) \ = \ 0 \quad \text{on } \partial B\,, \tag{7.129b}$$

where V again denotes the volume potential operator (7.108) and Φ the fundamental solution (7.19). Here we used the trace theorem for H^2–functions and the fact that $k^2 V(mw) = 4\pi\, \Phi(\cdot, 0)$ on ∂B is equivalent to $k^2 V(mw) = 4\pi\, \Phi(\cdot, 0)$ in the exterior of B by the uniqueness of the exterior Dirichlet problem.

Instead of solving both steps separately, we can combine them and solve the following optimization problem. Given a compact subset $\mathcal{C} \subset L^\infty(B)$, some $\varepsilon > 0$ and $\lambda_1, \lambda_2 > 0$,

$$\text{minimize } J(g, w, m) \text{ on } L^2(S^2) \times L^2(B) \times \mathcal{C}, \tag{7.130a}$$

where

$$J(g, w, m) \quad := \quad \|Fg - 1\|^2_{L^2(S^2)} + \varepsilon \|g\|^2_{L^2(S^2)} \tag{7.130b}$$
$$+ \lambda_1 \|w - v_g - k^2 V(mw)\|^2_{L^2(B)}$$
$$+ \lambda_2 \|k^2 V(mw) - 4\pi \, \Phi(\cdot, 0)\|^2_{L^2(\partial B)},$$

and the far field operator $F : L^2(S^2) \to L^2(S^2)$ is defined by (see (7.40))

$$(Fg)(\hat{x}) \quad := \quad \int_{S^2} u^\infty(\hat{x}, \hat{\theta}) \, g(\hat{\theta}) \, ds(\hat{\theta}), \quad \hat{x} \in S^2.$$

Theorem 7.57 *This optimization problem (7.130a), (7.130b) has an optimal solution (g, w, m) for every choice of $\varepsilon, \lambda_1, \lambda_2 > 0$ and every compact subset $\mathcal{C} \subset L^\infty(B)$.*

Proof: Let $(g_j, w_j, m_j) \in L^2(S^2) \times L^2(B) \times \mathcal{C}$ be a minimizing sequence; that is, $J(g_j, w_j, m_j) \to J^*$ where the optimal value J^* is defined by

$$J^* \quad := \quad \inf\{J(g, w, m) : (g, w, m) \in L^2(S^2) \times L^2(B) \times \mathcal{C}\}.$$

We can assume that (m_j) converges to some $m \in \mathcal{C}$ because $\mathcal{C}$ is compact. Several tedious applications of the parallelogram equality

$$\|a + b\|^2 \quad = \quad -\|a - b\|^2 + 2\|a\|^2 + 2\|b\|^2$$

and the binomial formula

$$\|b\|^2 \quad = \quad \|a\|^2 + 2 \, \mathrm{Re}\,(a, b - a) + \|a - b\|^2$$

yield

$$-J^* \quad \geq \quad -J\left(\frac{1}{2}(g_j + g_\ell), \frac{1}{2}(w_j + w_\ell), m_j\right)$$
$$= \quad -\frac{1}{2} J(g_j, w_j, m_j) - \frac{1}{2} J(g_\ell, w_\ell, m_j)$$
$$+ \frac{1}{4} \|F(g_j - g_\ell)\|^2_{L^2(S^2)} + \frac{\varepsilon}{4} \|g_j - g_\ell\|^2_{L^2(S^2)}$$
$$+ \frac{\lambda_1}{4} \|(w_j - w_\ell) - v_{g_j - g_\ell} - k^2 V(m_j(w_j - w_\ell))\|^2_{L^2(B)}$$
$$+ \frac{\lambda_2 k^4}{4} \|V(m_j(w_j - w_\ell))\|^2_{L^2(\partial B)}.$$

From this we conclude that

$$-J^* + \frac{1}{2} J(g_j, w_j, m_j) + \frac{1}{2} J(g_\ell, w_\ell, m_j)$$
$$\geq \quad \frac{\varepsilon}{4} \|g_j - g_\ell\|^2_{L^2(S^2)} + \frac{\lambda_1}{4} \|(w_j - w_\ell) - v_{g_j - g_\ell} - k^2 V(m_j(w_j - w_\ell))\|^2_{L^2(B)}.$$

The left-hand side tends to zero as j and ℓ tend to infinity, therefore we conclude that (g_j) is a Cauchy sequence, thus converging $g_j \to g$ in $L^2(S^2)$. Furthermore, from

$$\|(w_j - w_\ell) - v_{g_j - g_\ell} - k^2 V\big(m_j(w_j - w_\ell)\big)\|_{L^2(B)} \longrightarrow 0$$

as $\ell, j \to \infty$ and the convergence $g_j \to g$ we conclude that

$$\|(w_j - w_\ell) - k^2 V\big(m_j(w_j - w_\ell)\big)\|_{L^2(B)} \longrightarrow 0$$

as $\ell, j \to \infty$. The operators $I - k^2 V(m_j \cdot)$ converge to the isomorphism $I - k^2 V(m \cdot)$ in the operator norm of $L^2(B)$. Therefore, by Theorem A.37 of Appendix A, we conclude that (w_j) is a Cauchy sequence and thus is convergent in $L^2(B)$ to some w. The continuity of J implies that $J(g_j, w_j, m_j) \to J(g, w, m)$. Therefore, (g, w, m) is optimal.   □

7.8   Problems

7.1 Let $Q = (-\pi, \pi)^3 \subset \mathbb{R}^3$ be the cube and $H^p(Q)$, $H_0^p(Q)$, and $H_{per}^p(Q)$ be the Sobolev spaces defined at the beginning of this chapter. Show that $H_{per}^p(Q) \subset H^p(Q)$ and $H_0^p(Q) \subset H_{per}^p(Q)$ with bounded inclusions. Use this result to show that $H_0^p(Q)$ is compactly imbedded in $L^2(Q)$.

7.2 Let $u_{1,\infty}^b(\hat{x}, \hat{\theta}, k)$ and $u_{2,\infty}^b(\hat{x}, \hat{\theta}, k)$ be the far field patterns of the Born approximations corresponding to observation $\hat{x}$, angle of incidence $\hat{\theta}$, wave number k, and indices of refraction n_1 and n_2, respectively. Assume that

$$u_{1,\infty}^b(\hat{x}, \hat{\theta}, k) = u_{2,\infty}^b(\hat{x}, \hat{\theta}, k)$$

for all $\hat{x} \in S^2$ and $k \in [k_1, k_2] \subset \mathbb{R}^+$ and some $\hat{\theta} \in S^2$. Prove that $n_1 = n_2$.

7.3 Prove the following result, sometimes called *Karp's theorem*. Let $u_\infty(\hat{x}; \hat{\theta})$, $\hat{x}, \hat{\theta} \in S^2$, be the far field pattern and assume that there exists a function $f : [-1, 1] \to \mathbb{C}$ with

$$u_\infty(\hat{x}; \hat{\theta}) = f(\hat{x} \cdot \hat{\theta}) \quad \text{for all } \hat{x}, \hat{\theta} \in S^2.$$

Prove that the index of refraction n has to be radially symmetric: $n = n(r)$.

Hint: Rotate the geometry and use the uniqueness result.

7.4 Show that for any $a > 0$

$$\max_{|x| \le a} \int_{|y| < a} \frac{1}{|x - y|} \, dy = 2\pi \, a^2.$$

Hint: Define $u(x)$ as the volume integral. Apply Theorem 7.11 to show that $u \in C^1(\mathbb{R}^3)$ satisfies $\Delta u = -4\pi$ for $|x| < a$ and $\Delta u = 0$ for $|x| > a$, and solve this elliptic equation explicitly by separation of variables.

7.5 Show that the characteristic functions f_η from (7.94) for $\eta = \sqrt{n} = 1/2$ and $\eta = 2/3$ have the forms

$$
\begin{aligned}
f_{1/2}(k) &= \sin^3 \frac{k}{3}, \quad k \in \mathbb{C}, \\
f_{2/3}(k) &= \frac{2}{3} \sin^3 \frac{k}{3} \left[3 + 2 \cos \frac{2k}{3} \right], \quad k \in \mathbb{C},
\end{aligned}
$$

respectively. Discuss the existence of zeros of these functions and justify for these examples the assertions of Theorem 7.48.

Hint: Use the addition formulas for the trigonometric functions to express f_η in terms of $\sin(k/3)$ and $\cos(k/3)$.

Appendix A

Basic Facts from Functional Analysis

In this appendix, we collect some of the basic definitions and theorems from functional analysis. We prove only those theorems whose proofs are not easily accessible. We recommend the monographs [151, 168, 230, 271] for a comprehensive treatment of linear and nonlinear functional analysis.

A.1 Normed Spaces and Hilbert Spaces

First, we recall two basic definitions.

Definition A.1 *(Scalar Product, Pre-Hilbert Space)*
Let X be a vector space over the field $\mathbb{K} = \mathbb{R}$ or $\mathbb{K} = \mathbb{C}$. A scalar product *or* inner product *is a mapping*

$$(\cdot, \cdot)_X : X \times X \longrightarrow \mathbb{K}$$

with the following properties:

(i) $(x + y, z)_X = (x, z)_X + (y, z)_X$ for all $x, y, z \in X$,

(ii) $(\alpha x, y)_X = \alpha\,(x, y)_X$ for all $x, y \in X$ and $\alpha \in \mathbb{K}$,

(iii) $(x, y)_X = \overline{(y, x)_X}$ for all $x, y \in X$,

(iv) $(x, x)_X \in \mathbb{R}$ and $(x, x)_X \geq 0$, for all $x \in X$,

(v) $(x, x)_X > 0$ if $x \neq 0$.

A vector space X over $\mathbb{K}$ with inner product $(\cdot, \cdot)_X$ is called a pre-Hilbert space *over $\mathbb{K}$.*

© Springer Nature Switzerland AG 2021
A. Kirsch, *An Introduction to the Mathematical Theory of Inverse Problems*,
Applied Mathematical Sciences 120,
https://doi.org/10.1007/978-3-030-63343-1

The following properties are easily derived from the definition:

(vi) $(x, y + z)_X = (x, y)_X + (x, z)_X$ for all $x, y, z \in X$,

(vii) $(x, \alpha y)_X = \overline{\alpha}(x, y)_X$ for all $x, y \in X$ and $\alpha \in \mathbb{K}$.

Definition A.2 *(Norm)*
Let X be a vector space over the field $\mathbb{K} = \mathbb{R}$ or $\mathbb{K} = \mathbb{C}$. A norm on X is a mapping
$$\| \cdot \|_X : X \longrightarrow \mathbb{R}$$
with the following properties:

(i) $\|x\|_X > 0$ for all $x \in X$ with $x \neq 0$,

(ii) $\|\alpha x\|_X = |\alpha| \, \|x\|_X$ for all $x \in X$ and $\alpha \in \mathbb{K}$,

(iii) $\|x + y\|_X \leq \|x\|_X + \|y\|_X$ for all $x, y \in X$.

A vector space X over $\mathbb{K}$ with norm $\| \cdot \|_X$ is called normed space *over $\mathbb{K}$.*

Property (iii) is called *triangle inequality*. Applying it to the identities $x = (x - y) + y$ and $y = (y - x) + x$ yields the second triangle inequality $\|x - y\|_X \geq \big| \|x\|_X - \|y\|_X \big|$ for all $x, y \in X$.

Theorem A.3 *Let X be a pre-Hilbert space. The mapping $\| \cdot \|_X : X \longrightarrow \mathbb{R}$ defined by*
$$\|x\|_X := \sqrt{(x, x)_X}, \quad x \in X,$$
is a norm; that is, it has properties (i), (ii), and (iii) of Definition A.2. Furthermore,

(iv) $|(x, y)_X| \leq \|x\|_X \|y\|_X$ for all $x, y \in X$ (Cauchy–Schwarz inequality),

(v) $\|x \pm y\|_X^2 = \|x\|_X^2 + \|y\|_X^2 \pm 2 \operatorname{Re}(x, y)_X$ for all $x, y \in X$

   (binomial formula),

(vi) $\|x + y\|_X^2 + \|x - y\|_X^2 = 2\|x\|_X^2 + 2\|y\|_X^2$ for all $x, y \in X$.

In the following example, we list some of the most important pre-Hilbert and normed spaces.

Example A.4

(a) $\mathbb{C}^n$ is a pre-Hilbert space of dimension n over $\mathbb{C}$ with inner product $(x, y)_2 := \sum_{k=1}^{n} x_k \overline{y}_k$.

(b) $\mathbb{C}^n$ is a pre-Hilbert space of dimension $2n$ over $\mathbb{R}$ with inner product $(x, y)_2 := \operatorname{Re} \sum_{k=1}^{n} x_k \overline{y}_k$.

(c) $\mathbb{R}^n$ is a pre-Hilbert space of dimension n over $\mathbb{R}$ with inner product $(x, y)_2 := \sum_{k=1}^{n} x_k y_k$.

(d) For $p \geq 1$ define the set ℓ^p of complex-valued sequences by

$$\ell^p := \left\{ (x_k) : \sum_{k=1}^{\infty} |x_k|^p < \infty \right\}. \tag{A.1}$$

Then ℓ^p is a linear space over $\mathbb{C}$ because if $(x_k), (y_k) \in \ell^p$, then (λx_k) and $(x_k + y_k)$ are also in ℓ^p. The latter follows from the inequality $|x_k + y_k|^p \leq (2 \max\{|x_k|, |y_k|\})^p \leq 2^p(|x_k|^p + |y_k|^p)$.

$$\|x\|_{\ell^p} := \left(\sum_{k=1}^{\infty} |x_k|^p \right)^{1/p}, \quad x = (x_k) \in \ell^p,$$

defines a norm in ℓ^p. The triangle inequality in the case $p > 1$ is known as the *Minkowski inequality*. In the case $p = 2$, the sesquilinear form

$$(x, y)_{\ell^2} := \sum_{k=1}^{\infty} x_k \overline{y_k}, \quad x = (x_k), \ y = (y_k) \in \ell^2,$$

defines an inner product on ℓ^2. It is well-defined by the Cauchy–Schwarz inequality.

(e) The space $C[a, b]$ of (real- or complex-valued) continuous functions on $[a, b]$ is a pre-Hilbert space over $\mathbb{R}$ or $\mathbb{C}$ with inner product

$$(x, y)_{L^2} := \int_a^b x(t) \overline{y(t)} \, dt, \quad x, y \in C[a, b]. \tag{A.2a}$$

The corresponding norm is called the *Euclidean norm* and is denoted by

$$\|x\|_{L^2} := \sqrt{(x, x)_{L^2}} = \sqrt{\int_a^b |x(t)|^2 dt}, \quad x \in C[a, b]. \tag{A.2b}$$

(f) On the same vector space $C[a, b]$ as in example (e), we introduce a norm by

$$\|x\|_\infty := \max_{a \leq t \leq b} |x(t)|, \quad x \in C[a, b], \tag{A.3}$$

that we call the *supremum norm*.

(g) Let $m \in \mathbb{N}$ and $\alpha \in (0, 1]$. We define the spaces $C^m[a, b]$ and $C^{m,\alpha}[a, b]$ by

$$C^m[a, b] := \left\{ x \in C[a, b] : \begin{array}{l} x \text{ is } m \text{ times continuously} \\ \text{differentiable on } [a, b] \end{array} \right\},$$

$$C^{m,\alpha}[a, b] := \left\{ x \in C^m[a, b] : \sup_{t \neq s} \frac{|x^{(m)}(t) - x^{(m)}(s)|}{|t - s|^\alpha} < \infty \right\},$$

and we equip them with norms

$$\|x\|_{C^m} \quad := \quad \max_{0 \le k \le m} \|x^{(k)}\|_\infty , \tag{A.4a}$$

$$\|x\|_{C^{m,\alpha}} \quad := \quad \|x\|_{C^m} + \sup_{s \ne t} \frac{|x^{(m)}(t) - x^{(m)}(s)|}{|t - s|^\alpha} . \tag{A.4b}$$

Every normed space carries a topology introduced by the norm; that is, we can define open, closed, and compact sets; convergent sequences; continuous functions; etc. We introduce balls of radius r and center $x \in X$ by

$$B(x,r) := \{y \in X : \|y - x\|_X < r\}, \quad B[x,r] := \{y \in X : \|y - x\|_X \le r\}.$$

Definition A.5 *Let X be a normed space over the field $\mathbb{K} = \mathbb{R}$ or $\mathbb{C}$.*

(a) *A subset $M \subset X$ is called* bounded *if there exists $r > 0$ with $M \subset B(0,r)$. The set $M \subset X$ is called* open *if for every $x \in M$ there exists $\varepsilon > 0$ such that $B(x,\varepsilon) \subset M$. The set $M \subset X$ is called* closed *if the complement $X \setminus M$ is open.*

(b) *A sequence $(x_k)_k$ in X is called* bounded *if there exists $c > 0$ such that $\|x_k\|_X \le c$ for all k. The sequence $(x_k)_k$ in X is called* convergent *if there exists $x \in X$ such that $\|x - x_k\|_X$ converges to zero in $\mathbb{R}$. We denote the limit by $x = \lim_{k \to \infty} x_k$, or we write $x_k \to x$ as $k \to \infty$. The sequence $(x_k)_k$ in X is called a* Cauchy *sequence if for every $\epsilon > 0$ there exists $N \in \mathbb{N}$ with $\|x_m - x_k\|_X < \epsilon$ for all $m, k \ge N$.*

(c) *Let $(x_k)_k$ be a sequence in X. A point $x \in X$ is called an* accumulation point *if there exists a subsequence $(a_{k_n})_n$ that converges to x.*

(d) *A set $M \subset X$ is called* compact *if every sequence in M has an accumulation point in M.*

Example A.6
Let $X = C[0,1]$ over $\mathbb{R}$ and $x_k(t) = t^k$, $t \in [0,1]$, $k \in \mathbb{N}$. The sequence $(x_k)_k$ converges to zero with respect to the Euclidean norm $\|\cdot\|_{L^2}$ introduced in (A.2b). With respect to the supremum norm $\|\cdot\|_\infty$ of (A.3), however, the sequence does not converge to zero.

It is easy to prove (see Problem A.1) that a set M is closed if and only if the limit of every convergent sequence $(x_k)_k$ in M also belongs to M. The sets

$$\text{int}(M) := \{x \in M : \text{there exists } \varepsilon > 0 \text{ with } B(x,\varepsilon) \subset M\}$$

and

$$\text{closure}(M) := \{x \in X : \text{there exists } (x_k)_k \text{ in } M \text{ with } x = \lim_{k \to \infty} x_k\}$$

are called the *interior* and *closure*, respectively, of M. The set $M \subset X$ is called *dense* in X if closure$(M) = X$.

In general, the topological properties depend on the norm in X as we have seen already in Example A.6. For finite-dimensional spaces, however, these properties are *independent* of the norm. This is seen from the following theorem.

Theorem A.7 *Let X be a finite-dimensional space with norms $\|\cdot\|_1$ and $\|\cdot\|_2$. Then both norms are equivalent; that is, there exist constants $c_2 \geq c_1 > 0$ with*

$$c_1 \|x\|_1 \leq \|x\|_2 \leq c_2 \|x\|_1 \quad \text{for all } x \in X.$$

In other words, every ball with respect to $\|\cdot\|_1$ contains a ball with respect to $\|\cdot\|_2$ and vice versa.

Further properties are collected in the following theorem.

Theorem A.8 *Let X be a normed space over $\mathbb{K}$ and $M \subset X$ be a subset.*

(a) *M is closed if and only if $M = \text{closure}(M)$, and M is open if and only if $M = \text{int}(M)$.*

(b) *If $M \neq X$ is a linear subspace, then $\text{int}(M) = \emptyset$, and $\text{closure}(M)$ is also a linear subspace.*

(c) *In finite-dimensional spaces, every subspace is closed.*

(d) *Every compact set is closed and bounded. In finite-dimensional spaces, the reverse is also true (Theorem of Bolzano–Weierstrass): In a finite-dimensional normed space, every closed and bounded set is compact.*

A crucial property of the set of real numbers is its *completeness*. It is also a necessary assumption for many results in functional analysis.

Definition A.9 *(Banach Space, Hilbert Space)*
A normed space X over $\mathbb{K}$ is called complete *or a* Banach space *if every Cauchy sequence converges in X. A complete pre-Hilbert space is called a* Hilbert space.

The spaces $\mathbb{C}^n$ and $\mathbb{R}^n$ are Hilbert spaces with respect to their canonical inner products. The space $C[a, b]$ is not complete with respect to the inner product $(\cdot, \cdot)_{L^2}$ of (A.2a)! As an example, we consider the sequence $x_k(t) = t^k$ for $0 \leq t \leq 1$ and $x_k(t) = 1$ for $1 \leq t \leq 2$. Then $(x_k)_k$ is a Cauchy sequence in $C[0, 2]$ but does not converge in $C[0, 2]$ with respect to $(\cdot, \cdot)_{L^2}$ because it converges to the function

$$x(t) = \begin{cases} 0, & t < 1, \\ 1, & t \geq 1, \end{cases}$$

that is not continuous. The space $\left(C[a, b], \|\cdot\|_\infty\right)$, however, is a Banach space.

Every normed space or pre-Hilbert space X can be "completed"; that is, there exists a "smallest" Banach or Hilbert space $\tilde{X}$, respectively, that extends X (that is, $\|x\|_X = \|x\|_{\tilde{X}}$ or $(x, y)_X = (x, y)_{\tilde{X}}$, respectively, for all $x, y \in X$). More precisely, we have the following (formulated only for normed spaces).

Theorem A.10 *Let X be a normed space with norm $\|\cdot\|_X$. There exist a Banach space $(\tilde{X}, \|\cdot\|_{\tilde{X}})$ and an injective linear operator $J : X \to \tilde{X}$ such that*

(i) The range $J(X) \subset \tilde{X}$ is dense in $\tilde{X}$, and

(ii) $\|Jx\|_{\tilde{X}} = \|x\|_X$ for all $x \in X$; that is, J preserves the norm.

Furthermore, $\tilde{X}$ is uniquely determined in the sense that if $\hat{X}$ is a second space with properties (i) and (ii) with respect to a linear injective operator $\hat{J}$, then the operator $\hat{J} J^{-1} : J(X) \to \hat{J}(X)$ has an extension to a norm-preserving isomorphism from $\tilde{X}$ onto $\hat{X}$. In other words, $\tilde{X}$ and $\hat{X}$ can be identified.

We denote the completion of the pre-Hilbert space $(C[a,b], (\cdot,\cdot)_{L^2})$ by $L^2(a,b)$. Using Lebesgue integration theory, it can be shown that the space $L^2(a,b)$ is characterized as follows. (The notions "measurable," "almost everywhere" (a.e.), and "integrable" are understood with respect to the Lebesgue measure.) First, we define the vector space

$$\mathcal{L}^2(a,b) := \{x : (a,b) \to \mathbb{C} : x \text{ is measurable and } |x|^2 \text{ integrable}\},$$

where addition and scalar multiplication are defined pointwise almost everywhere. Then $\mathcal{L}^2(a,b)$ is a vector space because, for $x, y \in \mathcal{L}^2(a,b)$ and $\alpha \in \mathbb{C}$, $x + y$ and αx are also measurable and $\alpha x, x + y \in \mathcal{L}^2(a,b)$, the latter by the binomial theorem $|x(t) + y(t)|^2 \leq 2|x(t)|^2 + 2|y(t)|^2$. We define a sesquilinear form on $\mathcal{L}^2(a,b)$ by

$$\langle x,y \rangle := \int_a^b x(t)\,\overline{y(t)}\,dt, \quad x,y \in \mathcal{L}^2(a,b).$$

$\langle \cdot,\cdot \rangle$ is not an inner product on $\mathcal{L}^2(a,b)$ because $\langle x,x \rangle = 0$ only implies that x vanishes almost everywhere; that is, that $x \in \mathcal{N}$, where $\mathcal{N}$ is defined by

$$\mathcal{N} := \{x \in \mathcal{L}^2(a,b) : x(t) = 0 \text{ a.e. on } (a,b)\}.$$

Now we define $L^2(a,b)$ as the factor space

$$L^2(a,b) := \mathcal{L}^2(a,b)/\mathcal{N}$$

and equip $L^2(a,b)$ with the inner product

$$\big([x],[y]\big)_{L^2} := \int_a^b x(t)\,\overline{y(t)}\,dt, \quad x \in [x],\ y \in [y].$$

Here, $[x], [y] \in L^2(a,b)$ are equivalence classes of functions in $\mathcal{L}^2(a,b)$. Then it can be shown that this definition is well defined and yields an inner product on $L^2(a,b)$. From now on, we write $x \in L^2(a,b)$ instead of $x \in [x] \in L^2(a,b)$. Furthermore, it can be shown by fundamental results of Lebesgue integration theory that $L^2(a,b)$ is complete; that is, a Hilbert space and contains $C[a,b]$ as a dense subspace.

Definition A.11 *(Separable Space)*
The normed space X is called separable *if there exists a countable dense subset $M \subset X$; that is, if there exist M and a bijective mapping $j : \mathbb{N} \to M$ with* closure$(M) = X$.

The spaces $\mathbb{C}^n$, $\mathbb{R}^n$, $L^2(a,b)$, and $C[a,b]$ are all separable. For the first two examples, let M consist of all vectors with rational coefficients; for the latter examples, take polynomials with rational coefficients.

Definition A.12 *(Orthogonal Complement)*
Let X be a pre-Hilbert space (over $\mathbb{K} = \mathbb{R}$ or $\mathbb{C}$).

(a) Two elements x and y are called orthogonal *if $(x,y)_X = 0$.*

(b) Let $M \subset X$ be a subset. The set

$$M^\perp := \{x \in X : (x,y)_X = 0 \text{ for all } y \in M\}$$

is called the orthogonal complement *of M.*

$M^\perp$ is always a closed subspace and $M \subset (M^\perp)^\perp$. Furthermore, $A \subset B$ implies that $B^\perp \subset A^\perp$.

The following theorem is a fundamental result in Hilbert space theory and relies heavily on the completeness property.

Theorem A.13 *(Projection Theorem)*
Let X be a pre-Hilbert space and $V \subset X$ be a complete subspace. Then $V = (V^\perp)^\perp$. Every $x \in X$ possesses a unique decomposition of the form $x = v + w$, where $v \in V$ and $w \in V^\perp$. The operator $P : X \to V$, $x \mapsto v$, is called the orthogonal projection operator *onto V and has the properties*

(a) $Pv = v$ for $v \in V$; that is, $P^2 = P$;

(b) $\|x - Px\|_X \le \|x - v'\|_X$ for all $v' \in V$.

This means that $Px \in V$ is the best approximation of $x \in X$ in the subspace V.

A.2   Orthonormal Systems

In this section, let X always be a *separable* Hilbert space over the field $\mathbb{K} = \mathbb{R}$ or $\mathbb{C}$.

Definition A.14 *(Orthonormal System)*
A countable set of elements $A = \{x_k : k = 1, 2, 3, \ldots\}$ is called an orthonormal system *(ONS) if*

(i) $(x_k, x_j)_X = 0$ for all $k \ne j$ and

(ii) $\|x_k\|_X = 1$ *for all* $k \in \mathbb{N}$.

A is called a complete *or a* maximal *orthonormal system if, in addition, there is no ONS B with* $A \subset B$ *and* $A \neq B$.

One can show using Zorn's Lemma that every separable Hilbert possesses a maximal ONS. Furthermore, it is well known from linear algebra that every countable set of linearly independent elements of X can be orthonormalized. For any set $A \subset X$, let

$$\operatorname{span} A := \left\{ \sum_{k=1}^{n} \alpha_k\, x_k : \alpha_k \in \mathbb{K},\ x_k \in A,\ n \in \mathbb{N} \right\} \qquad (A.5)$$

be the subspace of X spanned by A.

Theorem A.15 *Let* $A = \{x_k : k = 1, 2, 3, \ldots\}$ *be an orthonormal system. Then*

(a) Every finite subset of A is linearly independent.

(b) If A is finite; that is, $A = \{x_k : k = 1, 2, \ldots, n\}$, then for every $x \in X$ there exist uniquely determined coefficients $\alpha_k \in \mathbb{K}$, $k = 1, \ldots, n$, such that

$$\left\| x - \sum_{k=1}^{n} \alpha_k x_k \right\|_X \leq \|x - a\|_X \quad \text{for all } a \in \operatorname{span} A. \qquad (A.6)$$

The coefficients α_k are given by $\alpha_k = (x, x_k)_X$ for $k = 1, \ldots, n$.

(c) For every $x \in X$, the following Bessel inequality *holds:*

$$\sum_{k=1}^{\infty} \left| (x, x_k)_X \right|^2 \leq \|x\|_X^2, \qquad (A.7)$$

and the series $\sum_{k=1}^{\infty} (x, x_k)_X x_k$ converges in X.

(d) A is complete if and only if $\operatorname{span} A$ *is dense in X.*

(e) A is complete if and only if for all $x \in X$ the following Parseval equation *holds:*

$$\sum_{k=1}^{\infty} \left| (x, x_k)_X \right|^2 = \|x\|_X^2. \qquad (A.8)$$

(f) A is complete if and only if every $x \in X$ has a (generalized) Fourier expansion of the form

$$x = \sum_{k=1}^{\infty} (x, x_k)_X\, x_k, \qquad (A.9)$$

where the convergence is understood in the norm of X. In this case, the Parseval equation holds in the following more general form:

$$(x, y)_X = \sum_{k=1}^{\infty} (x, x_k)_X \overline{(y, x_k)_X}. \tag{A.10}$$

This important theorem includes, as special examples, the classical Fourier expansion of periodic functions and the expansion with respect to orthogonal polynomials. We recall two examples.

Example A.16 *(Fourier Expansion)*
(a) The functions $x_k(t) := \exp(ikt)/\sqrt{2\pi}$, $k \in \mathbb{Z}$, form a complete system of orthonormal functions in $L^2(0, 2\pi)$. By part (f) of the previous theorem, every function $x \in L^2(0, 2\pi)$ has an expansion of the form

$$x(t) = \frac{1}{2\pi} \sum_{k=-\infty}^{\infty} e^{ikt} \int_0^{2\pi} x(s) e^{-iks} ds,$$

where the convergence is understood in the sense of L^2; that is,

$$\int_0^{2\pi} \left| x(t) - \frac{1}{2\pi} \sum_{k=-M}^{N} e^{ikt} \int_0^{2\pi} x(s) e^{-iks} ds \right|^2 dt \longrightarrow 0$$

as M, N tend to infinity. Parseval's identity holds the form

$$\sum_{k \in \mathbb{Z}} |a_k|^2 = \frac{1}{2\pi} \|x\|_{L^2}^2, \quad a_k = \frac{1}{2\pi} \int_0^{2\pi} x(s) e^{-iks} ds. \tag{A.11}$$

For smooth periodic functions, one can even show uniform convergence (see Section A.4).

(b) The *Legendre polynomials* P_k, $k = 0, 1, \ldots$, form a maximal orthonormal system in $L^2(-1, 1)$. They are defined by

$$P_k(t) = \gamma_k \frac{d^k}{dt^k} (1 - t^2)^k, \quad t \in (-1, 1), \ k \in \mathbb{N}_0,$$

with normalizing constants

$$\gamma_k = \sqrt{\frac{2k+1}{2}} \frac{1}{k! \, 2^k}.$$

We refer to [135] for details.

Other important examples will be given later.

A.3 Linear Bounded and Compact Operators

For this section, let X and Y always be normed spaces and $A : X \to Y$ be a linear operator.

Definition A.17 *(Boundedness, Norm of A)*
The linear operator A is called bounded *if there exists $c > 0$ such that*

$$\|Ax\|_Y \leq c\|x\|_X \quad \text{for all } x \in X.$$

The smallest of these constants is called the norm *of A; that is,*

$$\|A\|_{\mathcal{L}(X,Y)} := \sup_{x \neq 0} \frac{\|Ax\|_Y}{\|x\|_X}. \tag{A.12}$$

Theorem A.18 *The following assertions are equivalent:*

(a) A is bounded.

(b) A is continuous at $x = 0$; that is, $x_j \to 0$ implies that $Ax_j \to 0$.

(c) A is continuous for every $x \in X$.

The space $\mathcal{L}(X, Y)$ of all linear bounded mappings from X to Y with the operator norm is a normed space; that is, the operator norm has properties (i), (ii), and (iii) of Definition A.2 and the following: Let $B \in \mathcal{L}(X, Y)$ and $A \in \mathcal{L}(Y, Z)$; then $AB \in \mathcal{L}(X, Z)$ and $\|AB\|_{\mathcal{L}(X,Z)} \leq \|A\|_{\mathcal{L}(Y,Z)}\|B\|_{\mathcal{L}(X,Y)}$.

Integral operators are the most important examples for our purposes.

Theorem A.19 *(a) Let $k \in L^2\big((c,d) \times (a,b)\big)$. The operator*

$$(Ax)(t) := \int_a^b k(t,s)\,x(s)\,ds\,, \quad t \in (c,d)\,, \quad x \in L^2(a,b)\,, \tag{A.13}$$

is well-defined, linear, and bounded from $L^2(a,b)$ into $L^2(c,d)$. Furthermore,

$$\|A\|_{\mathcal{L}(L^2(a,b),L^2(c,d))} \leq \sqrt{\int_c^d \int_a^b |k(t,s)|^2\,ds\,dt}\,.$$

(b) Let k be continuous on $[c,d] \times [a,b]$. Then A is also well-defined, linear, and bounded from $C[a,b]$ into $C[c,d]$ and

$$\|A\|_\infty := \|A\|_{\mathcal{L}(C[a,b],C[c,d])} = \max_{t \in [c,d]} \int_a^b |k(t,s)|\,ds\,.$$

We can extend this theorem to integral operators with weakly singular kernels. We recall that a kernel k is called *weakly singular* on $[a, b] \times [a, b]$ if k is defined and continuous for all $t, s \in [a, b]$, $t \neq s$, and there exist constants $c > 0$ and $\alpha \in [0, 1)$ such that

$$|k(t, s)| \leq c |t - s|^{-\alpha} \quad \text{for all } t, s \in [a, b], \ t \neq s.$$

Theorem A.20 *Let k be weakly singular on $[a, b]$. Then the integral operator A, defined by (A.13) for $[c, d] = [a, b]$, is well-defined and bounded as an operator in $L^2(a, b)$ as well as in $C[a, b]$.*

For the special case $Y = \mathbb{K}$, we denote by $X^* := \mathcal{L}(X, \mathbb{K})$ the *dual space* of X. Often we write $\langle \ell, x \rangle_{X^*, X}$ instead of $\ell(x)$ for $\ell \in X^*$ and $x \in X$ and call $\langle \cdot, \cdot \rangle_{X^*, X}$ the dual pairing. The dual pairing is a bilinear form from $X^* \times X$ into $\mathbb{K}$. The space $X^{**} = (X^*)^*$ is called the *bidual* of X. The canonical embedding $J : X \hookrightarrow X^{**}$, defined by

$$(Jx)\ell := \langle \ell, x \rangle_{X^*, X}, \quad x \in X, \ \ell \in X^*, \tag{A.14}$$

is linear, bounded, one-to-one, and satisfies $\|Jx\|_{X^{**}} = \|x\|_X$ for all $x \in X$.

We recall some important examples of dual spaces (where we write $\langle \cdot, \cdot \rangle$ for the dual pairing).

Example A.21
Let again $\mathbb{K} = \mathbb{R}$ or $\mathbb{K} = \mathbb{C}$.

(a) The dual of $\mathbb{K}^n$ can be identified with $\mathbb{K}^n$ itself. The identification $I : \mathbb{K}^n \to (\mathbb{K}^n)^*$ is given by $\langle Ix, y \rangle = \sum_{j=1}^n x_j y_j$ for $x, y \in \mathbb{K}^n$.

(b) Let $p > 1$ and $q > 1$ with $\frac{1}{p} + \frac{1}{q} = 1$. The dual of ℓ^p (see Example A.4) can be identified with ℓ^q. The identification $I : \ell^q \to (\ell^p)^*$ is given by

$$\langle Ix, y \rangle = \sum_{j=1}^{\infty} x_j y_j \quad \text{for } x = (x_j) \in \ell^q \text{ and } y = (y_j) \in \ell^p.$$

(c) The dual $(\ell^1)^*$ of the space ℓ^1 can be identified with the space ℓ^∞ of bounded sequences (equipped with the supremum norm). The identification $I : \ell^\infty \hookrightarrow (\ell^1)^*$ is given by the form as in (b) for $x = (x_j) \in \ell^\infty$ and $y = (y_j) \in \ell^1$.

(d) Let $c_0 \subset \ell^\infty$ be the space of sequences in $\mathbb{K}$ which converge to zero, equipped with the supremum norm. Then c_0^* can be identified with ℓ^1. The identification $I : \ell^1 \hookrightarrow c_0^*$ is given by the form as in (b) for $x \in \ell^1$ and $y \in c_0$.

Definition A.22 *(Reflexive Space)*
The normed space X is called reflexive *if the canonical embedding of X into X^{**} is surjective; that is, a norm-preserving isomorphism from X onto the bidual space X^{**}.*

The spaces ℓ^p for $p > 1$ of Example A.21 are typical examples of reflexive spaces. The spaces ℓ^1, ℓ^∞, and c_0 fail to be reflexive.

The following important result gives a characterization of X^* in Hilbert spaces.

Theorem A.23 *(Riesz)*
Let X be a Hilbert space. For every $x \in X$, the functional $\ell_x(y) := (y, x)_X$, $y \in X$, defines a linear bounded mapping from X to $\mathbb{K}$; that is, $\ell_x \in X^*$. Furthermore, for every $\ell \in X^*$ there exists one and only one $x \in X$ with $\ell(y) = (y, x)_X$ for all $y \in X$ and

$$\|\ell\|_{X^*} := \sup_{y \neq 0} \frac{|\langle \ell, y \rangle|}{\|y\|_X} = \|x\|_X .$$

This theorem implies that every Hilbert space is reflexive. It also yields the existence of a unique adjoint operator for every linear bounded operator $A : X \longrightarrow Y$. We recall that for any linear and bounded operator $A : X \to Y$ between normed spaces X and Y the *dual operator* $A^* : Y^* \to X^*$ is defined as $T^* \ell = \ell \circ A$ for all $\ell \in Y^*$. Here $\ell \circ A$ is the composition of A and ℓ; that is, $(\ell \circ A)x = \ell(Ax)$ for $x \in X$.

Theorem A.24 *(Adjoint Operator)*
Let $A : X \longrightarrow Y$ be a linear and bounded operator between Hilbert spaces. Then there exists one and only one linear bounded operator $A^* : Y \longrightarrow X$ with the property

$$(Ax, y)_Y = (x, A^*y)_X \quad \text{for all } x \in X, \ y \in Y .$$

This operator $A^* : Y \longrightarrow X$ is called the adjoint operator *to A. For $X = Y$, the operator A is called* self-adjoint *if $A^* = A$.*

Example A.25
(a) Let $X = L^2(a, b)$, $Y = L^2(c, d)$, and $k \in L^2\big((c, d) \times (a, b)\big)$. The adjoint A^* of the integral operator

$$(Ax)(t) = \int_a^b k(t, s)\, x(s)\, ds , \quad t \in (c, d), \quad x \in L^2(a, b) ,$$

is given by

$$(A^*y)(t) = \int_c^d \overline{k(s, t)}\, y(s)\, ds , \quad t \in (a, b), \quad y \in L^2(c, d) .$$

(b) Let the space $X = C[a, b]$ of continuous function over $\mathbb{C}$ be supplied with the L^2-inner product. Define $f, g : C[a, b] \to \mathbb{C}$ by

$$f(x) := \int_a^b x(t)\, dt \quad \text{and} \quad g(x) := x(a) \quad \text{for } x \in C[a, b] .$$

Both f and g are linear. f is bounded but g is unbounded. There is an extension of f to a linear bounded functional (also denoted by f) on $L^2(a,b)$; that is, $f \in L^2(a,b)^*$. By Theorem A.23, we can identify $L^2(a,b)^*$ with $L^2(a,b)$ itself. For the given f, the representation function is just the constant function 1 because $f(x) = (x,1)_{L^2}$ for $x \in L^2(a,b)$. The adjoint of f is calculated by

$$f(x) \cdot \overline{y} \;=\; \int_a^b x(t)\,\overline{y}\,dt \;=\; (x,y)_{L^2} \;=\; \left(x, f^*(y)\right)_{L^2}$$

for all $x \in L^2(a,b)$ and $y \in \mathbb{C}$. Therefore, $f^*(y) \in L^2(a,b)$ is the constant function with value y.

(c) Let X be the *Sobolev space* $H^1(a,b)$; that is, the space of L^2-functions that possess generalized L^2-derivatives:

$$H^1(a,b) \;:=\; \left\{ x \in L^2(a,b) : \begin{array}{l} \text{there exists } \alpha \in \mathbb{K} \text{ and } y \in L^2(a,b) \text{ with} \\ x(t) = \alpha + \int_a^t y(s)\,ds \text{ for } t \in (a,b) \end{array} \right\}.$$

We denote the generalized derivative $y \in L^2(a,b)$ by x'. We observe that $H^1(a,b) \subset C[a,b]$ with bounded embedding. As an inner product in $H^1(a,b)$, we define

$$(x,y)_{H^1} \;:=\; x(a)\,\overline{y(a)} \;+\; (x',y')_{L^2}, \quad x,y \in H^1(a,b).$$

Now let $Y = L^2(a,b)$ and $A : H^1(a,b) \longrightarrow L^2(a,b)$ be the operator $x \mapsto x'$ for $x \in H^1(a,b)$. Then A is well-defined, linear, and bounded. It is easily seen that the adjoint of A is given by

$$(A^*y)(t) \;=\; \int_a^t y(s)\,ds, \quad t \in (a,b), \quad y \in L^2(a,b).$$

In the following situation, we consider the case that a Banach space V is contained in a Hilbert space X with bounded imbedding $j : V \hookrightarrow X$ such that also $j(V)$ is dense in X. We have in mind examples such as $H^1(0,1) \subset L^2(0,1)$. Then the dual operator j^* is a linear bounded operator from X^* into V^* with dense range (the latter follows from the injectivity of j). Also, j^* is one-to-one because $j(V)$ is dense in X, see Problem A.3. Now we use the fact that X and X^* are anti-isomorphic by the Theorem A.23 of Riesz; that is, the operator $j_R : X \ni x \mapsto \ell_x \in X^*$ where $\ell_x(z) = (z,x)_X$ for $z \in X$ is bijective and anti-linear; that is, satisfies $j_R(\lambda x + \mu y) = \overline{\lambda} j_R x + \overline{\mu} j_R y$ for all $x,y \in X$ and $\lambda, \mu \in \mathbb{K}$. Therefore, also the composition $j^* \circ j_R : X \to V^*$ is anti-linear. For this reason we define the *anti-dual space* V' of V by

$$V' \;=\; \overline{V^*} \;=\; \{\ell : V \to \mathbb{K} : \overline{\ell} \in V^*\} \tag{A.15}$$

where $\overline{\ell}(v) = \overline{\ell(v)}$ for all $v \in V$. Then the operator $j' := \overline{j^* \circ j_R} : X \to V'$ is linear and one-to-one with dense range, thus an imbedding. In this sense, X is densely imbedded in V'. We denote the application of $\ell \in V'$ to $v \in V$ by $\langle \ell, v \rangle$ and note that $(\ell, v) \mapsto \langle \ell, v \rangle$ is a sesquilinear form on $V' \times V$. (It should not be mixed up with the *dual pairing* $\langle \cdot, \cdot \rangle_{V^*,V} : V^* \times V \to \mathbb{K}$ which is bilinear.) From this analysis, we conclude that $(x, v)_X = \overline{\ell_x(v)} = \langle j'x, v \rangle$ for all $x \in X$ and $v \in V$ and thus

$$\left| (x, v)_X \right| \leq \| j'x \|_{V'} \| v \|_V \quad \text{for all } x \in X, \ v \in V. \tag{A.16}$$

Definition A.26 *(a) A* Gelfand triple *(or rigged Hilbert space, see [100])* $V \subset X \subset V'$ *consists of a reflexive Banach space* V, *a separable Hilbert space* X, *and the anti-dual space* V' *of* V *(all over the same field* $\mathbb{K} = \mathbb{R}$ *or* $\mathbb{K} = \mathbb{C}$*) such that* V *is a dense subspace of* X, *and the imbedding* $j : V \hookrightarrow X$ *is bounded. Furthermore, the sesquilinear form* $\langle \cdot, \cdot \rangle : V' \times V \to \mathbb{K}$ *is an extension of the inner product in* X; *that is,* $\langle x, v \rangle = (x, v)_X$ *for all* $v \in V$ *and* $x \in X$.

(b) A linear bounded operator $K : V' \to V$ *is called* coercive *if there exists* $\gamma > 0$ *with*

$$\left| \langle x, Kx \rangle \right| \geq \gamma \| x \|_{V'}^2 \quad \text{for all } x \in V'. \tag{A.17}$$

The operator K *satisfies* Gårding's inequality *if there exists a linear compact operator* $C : V' \to V$ *such that* $K + C$ *is coercive; that is,*

$$\left| \langle x, (K + C)x \rangle \right| \geq \gamma \| x \|_{V'}^2 \quad \text{for all } x \in V'.$$

By the same arguments as in the proof of the Lax–Milgram theorem (see [129]), it can be shown that every coercive operator is an isomorphism from V' onto V. Coercive operators play an important role in the study of partial differential equations and integral equations by variational methods. Often, the roles of V and V' are interchanged. For integral operators that are "smoothing", our definition seems more appropriate. However, both definitions are equivalent in the sense that the inverse operator $K^{-1} : V \to V'$ is coercive in the usual sense with γ replaced by $\gamma / \| K \|_{\mathcal{L}(V',V)}^2$.

The following theorems are two of the most important results of linear functional analysis.

Theorem A.27 *(Open Mapping Theorem)*
Let X, Y *be Banach spaces and* $A : X \to Y$ *a linear bounded operator from* X *onto* Y. *Then* A *is open; that is, the images* $A(U) \subset Y$ *are open in* Y *for all open sets* $U \subset X$. *In particular, if* A *is a bounded isomorphism from* X *onto* Y, *then the inverse* $A^{-1} : Y \to X$ *is bounded. This result is sometimes called the* Banach–Schauder theorem.

Theorem A.28 *(Banach–Steinhaus, Principle of Uniform Boundedness)*
Let X *be a Banach space,* Y *be a normed space,* I *be an index set, and* $A_\alpha \in \mathcal{L}(X, Y)$, $\alpha \in I$, *be a collection of linear bounded operators such that*

$$\sup_{\alpha \in I} \| A_\alpha x \|_Y < \infty \quad \text{for every } x \in X.$$

Then $\sup_{\alpha \in I} \| A_\alpha \|_{\mathcal{L}(X,Y)} < \infty$.

As an immediate consequence, we have the following.

Theorem A.29 *Let X be a Banach space, Y be a normed space, $D \subset X$ be a dense subspace, and $A_n \in \mathcal{L}(X, Y)$ for $n \in \mathbb{N}$. Then the following two assertions are equivalent:*

(i) $A_n x \to 0$ as $n \to \infty$ for all $x \in X$.

(ii) $\sup_{n \in \mathbb{N}} \|A_n\|_{\mathcal{L}(X,Y)} < \infty$ and $A_n x \to 0$ as $n \to \infty$ for all $x \in D$.

We saw in Theorem A.10 that every normed space X possesses a unique completion $\tilde{X}$. Every linear bounded operator defined on X can also be extended to $\tilde{X}$.

Theorem A.30 *Let $\tilde{X}, \tilde{Y}$ be Banach spaces, $X \subset \tilde{X}$ a dense subspace, and $A : X \to \tilde{Y}$ be linear and bounded. Then there exists a linear bounded operator $\tilde{A} : \tilde{X} \to \tilde{Y}$ with*

(i) $\tilde{A}x = Ax$ for all $x \in X$; that is, $\tilde{A}$ is an extension of A, and

(ii) $\|\tilde{A}\|_{\mathcal{L}(\tilde{X},\tilde{Y})} = \|A\|_{\mathcal{L}(X,Y)}$.

Furthermore, the operator $\tilde{A}$ is uniquely determined.

We now study equations of the form

$$x - Kx = y, \tag{A.18}$$

where the operator norm of the linear bounded operator $K : X \to X$ is small. The following theorem plays an essential role in the study of Volterra integral equations.

Theorem A.31 *(Contraction Theorem, Neumann Series)*
Let X be a Banach space over $\mathbb{R}$ or $\mathbb{C}$ and $K : X \to X$ be a linear bounded operator with

$$\limsup_{n \to \infty} \|K^n\|_{\mathcal{L}(X)}^{1/n} < 1. \tag{A.19}$$

Then $I - K$ is invertible, the Neumann series $\sum_{n=0}^{\infty} K^n$ converges in the operator norm, and

$$\sum_{n=0}^{\infty} K^n = (I - K)^{-1}.$$

Condition (A.19) is satisfied if, for example, $\|K^m\|_{\mathcal{L}(X)} < 1$ for some $m \in \mathbb{N}$.

Example A.32
Let $\Delta := \{(t, s) \in \mathbb{R}^2 : a < s < t < b\}$.

(a) Let $k \in L^2(\Delta)$. Then the Volterra operator

$$(Kx)(t) := \int_a^t k(t,s)\,x(s)\,ds, \quad a < t < b, \ x \in L^2(a,b), \qquad \text{(A.20)}$$

is bounded in $L^2(a,b)$. There exists $m \in \mathbb{N}$ with $\|K^m\|_{\mathcal{L}(L^2(a,b))} < 1$. The Volterra equation of the second kind

$$x(t) - \int_a^t k(t,s)\,x(s)\,ds = y(t), \quad a < t < b, \qquad \text{(A.21)}$$

is uniquely solvable in $L^2(a,b)$ for every $y \in L^2(a,b)$, and the solution x depends continuously on y. The solution $x \in L^2(a,b)$ has the form

$$x(t) = y(t) + \int_a^t r(t,s)\,y(s)\,ds, \quad t \in (a,b),$$

with some kernel $r \in L^2(\Delta)$.

(b) Let $k \in C(\overline{\Delta})$. Then the operator K defined by (A.20) is bounded in $C[a,b]$, and there exists $m \in \mathbb{N}$ with $\|K^m\|_\infty < 1$. Equation (A.21) is also uniquely solvable in $C[a,b]$ for every $y \in C[a,b]$, and the solution x depends continuously on y.

For the remaining part of this section, we assume that X and Y are normed spaces and $K : X \to Y$ a linear and bounded operator.

Definition A.33 *(Compact Operator)*
The operator $K : X \to Y$ is called compact *if it maps every bounded set S into a relatively compact set $K(S)$.*

We recall that a set $M \subset Y$ is called *relatively compact* if every *bounded* sequence $(y_j)_j$ in M has an accumulation point in closure(M); that is, if the closure closure(M) is compact. The set of all compact operators from X into Y is a closed subspace of $\mathcal{L}(X,Y)$ and even a two-sided ideal by part (c) of the following theorem.

Theorem A.34 *(a) If K_1 and K_2 are compact from X into Y, then so are $K_1 + K_2$ and λK_1 for every $\lambda \in \mathbb{K}$.*

(b) Let $K_n : X \longrightarrow Y$ be a sequence of compact operators between Banach spaces X and Y. Let $K : X \longrightarrow Y$ be bounded, and let K_n converge to K in the operator norm; that is,

$$\|K_n - K\|_{\mathcal{L}(X,Y)} := \sup_{x \neq 0} \frac{\|K_n x - K x\|_Y}{\|x\|_X} \longrightarrow 0 \ (n \longrightarrow \infty).$$

Then K is also compact.

(c) If $L \in \mathcal{L}(X, Y)$ and $K \in \mathcal{L}(Y, Z)$, and L or K is compact, then KL is also compact.

(d) Let $A_n \in \mathcal{L}(X, Y)$ be pointwise convergent to some $A \in \mathcal{L}(X, Y)$; that is, $A_n x \to Ax$ for all $x \in X$. If $K : Z \to X$ is compact, then $\|A_n K - AK\|_{\mathcal{L}(Z,Y)} \to 0$; that is, the operators $A_n K$ converge to AK in the operator norm.

(e) The identity operator $x \mapsto x$ is compact as an operator from X into itself if, and only if, X is finite dimensional.

(f) Every bounded operator K from X into Y with finite-dimensional range is compact.

Theorem A.35 *(a) Let $k \in L^2\big((c, d) \times (a, b)\big)$. The operator $K : L^2(a, b) \to L^2(c, d)$, defined by*

$$(Kx)(t) := \int_a^b k(t, s)\, x(s)\, ds\,, \quad t \in (c, d)\,, \quad x \in L^2(a, b)\,, \tag{A.22}$$

is compact from $L^2(a, b)$ into $L^2(c, d)$.

(b) Let k be continuous on $[c, d] \times [a, b]$ or weakly singular on $[a, b] \times [a, b]$ (in this case $[c, d] = [a, b]$). Then K defined by (A.22) is also compact as an operator from $C[a, b]$ into $C[c, d]$.

We now study equations of the form

$$x \, - \, Kx \, = \, y\,, \tag{A.23}$$

where the linear operator $K : X \to X$ is compact. The following theorem extends the well-known existence results for finite linear systems of n equations and n variables to compact perturbations of the identity.

Theorem A.36 *(Riesz)*
Let X be a normed space and $K : X \to X$ be a linear compact operator.

(a) The null space $\mathcal{N}(I - K) = \{x \in X : x = Kx\}$ is finite-dimensional and the range $\mathcal{R}(I - K) = (I - K)(X)$ is closed in X.

(b) If $I - K$ is one-to-one, then $I - K$ is also surjective, and the inverse $(I - K)^{-1}$ is bounded. In other words, if the homogeneous equation $x - Kx = 0$ admits only the trivial solution $x = 0$, then the inhomogeneous equation $x - Kx = y$ is uniquely solvable for every $y \in X$ and the solution x depends continuously on y.

The next theorem studies approximations of equations of the form $Ax = y$. Again, we have in mind that $A = I - K$.

Theorem A.37 *Assume that the operator $A : X \to Y$ between Banach spaces X and Y has a bounded inverse A^{-1}. Let $A_n \in \mathcal{L}(X,Y)$ be a sequence of bounded operators that converge in norm to A; that is, $\|A_n - A\|_{\mathcal{L}(X,Y)} \to 0$ as $n \to \infty$. Then, for sufficiently large n, more precisely for all n with*

$$\|A^{-1}(A_n - A)\|_{\mathcal{L}(X)} < 1, \qquad (A.24)$$

the inverse operators $A_n^{-1} : Y \to X$ exist and are uniformly bounded by

$$\|A_n^{-1}\|_{\mathcal{L}(Y,X)} \leq \frac{\|A^{-1}\|_{\mathcal{L}(Y,X)}}{1 - \|A^{-1}(A_n - A)\|_{\mathcal{L}(X)}} \leq c. \qquad (A.25)$$

For the solutions of the equations

$$Ax = y \quad and \quad A_n x_n = y_n,$$

the error estimate

$$\|x_n - x\|_X \leq c\{\|A_n x - Ax\|_Y + \|y_n - y\|_Y\} \qquad (A.26)$$

holds with the constant c from (A.25).

A.4   Sobolev Spaces of Periodic Functions

In this section, we recall definitions and properties of Sobolev (Hilbert) spaces of periodic functions. A complete discussion including proofs can be found in the monograph [168].

From Parseval's identity (A.11), we note that $x \in L^2(0, 2\pi)$ if and only if the Fourier coefficients

$$a_k = \frac{1}{2\pi} \int_0^{2\pi} x(s) e^{-iks} \, ds, \quad k \in \mathbb{Z}, \qquad (A.27)$$

are square summable. In this case

$$\sum_{k \in \mathbb{Z}} |a_k|^2 = \frac{1}{2\pi} \|x\|_{L^2}^2.$$

If x is periodic and continuously differentiable on $[0, 2\pi]$, partial integration of (A.27) yields the formula

$$a_k = \frac{-i}{2\pi k} \int_0^{2\pi} x'(s) e^{-iks} \, ds;$$

that is, ika_k are the Fourier coefficients of x' and are thus square summable. This motivates the introduction of subspaces $H_{per}^r(0, 2\pi)$ of $L^2(0, 2\pi)$ by requiring for their elements a certain decay of the Fourier coefficients a_k. In the following we set

$$\hat{\psi}_k(t) = e^{ikt} \quad \text{for } k \in \mathbb{Z} \text{ and } t \in (0, 2\pi).$$

Definition A.38 *(Sobolev space of periodic functions)*
For $r \geq 0$, the Sobolev space $H^r_{per}(0, 2\pi)$ of order r is defined by

$$H^r_{per}(0, 2\pi) := \left\{ x = \sum_{k \in \mathbb{Z}} a_k \,\hat{\psi}_k \in L^2(0, 2\pi) : \sum_{k \in \mathbb{Z}} (1 + k^2)^r \, |a_k|^2 < \infty \right\}.$$

We note that $H^0_{per}(0, 2\pi)$ coincides with $L^2(0, 2\pi)$.

Theorem A.39 *The Sobolev space $H^r_{per}(0, 2\pi)$ is a Hilbert space with the inner product defined by*

$$(x, y)_{H^r_{per}} := \sum_{k \in \mathbb{Z}} (1 + k^2)^r \, a_k \, \overline{b_k}, \tag{A.28}$$

where $x = \sum_{k \in \mathbb{Z}} a_k \, \hat{\psi}_k$ and $y = \sum_{k \in \mathbb{Z}} b_k \, \hat{\psi}_k$. The norm in $H^r_{per}(0, 2\pi)$ is given by

$$\|x\|_{H^r_{per}} = \left(\sum_{k \in \mathbb{Z}} (1 + k^2)^r \, |a_k|^2 \right)^{1/2}.$$

The Sobolev space $H^r_{per}(0, 2\pi)$ is dense in $L^2(0, 2\pi)$.

We note that $\|x\|_{L^2} = \sqrt{2\pi} \, \|x\|_{H^0_{per}}$, that is, the norms $\|x\|_{L^2}$ and $\|x\|_{H^0_{per}}$ are equivalent on $L^2(0, 2\pi)$.

Theorem A.40 *(a) For $r \in \mathbb{N}_0 := \mathbb{N} \cup \{0\}$, the space $C^r_{per}[0, 2\pi] = \{x \in C^r[0, 2\pi] : x \text{ is } 2\pi - \text{ periodic}\}$ is boundedly embedded in $H^r_{per}(0, 2\pi)$.*
(b) The space $\mathcal{T}$ of all trigonometric polynomials

$$\mathcal{T} := \left\{ \sum_{k=-n}^{n} a_k \, \hat{\psi}_k : a_k \in \mathbb{C}, \, n \in \mathbb{N} \right\}$$

is dense in $H^r_{per}(0, 2\pi)$ for every $r \geq 0$.

We consider $H^r_{per}(0, 2\pi) \subset L^2(0, 2\pi)$ as a Banach space (that is, forget the inner product for a moment) with bounded imbedding $j : H^r_{per}(0, 2\pi) \hookrightarrow L^2(0, 2\pi)$ which has a dense range. Therefore, we can consider the corresponding Gelfand triple $H^r_{per}(0, 2\pi) \subset L^2(0, 2\pi) \subset H^r_{per}(0, 2\pi)'$ where $H^r_{per}(0, 2\pi)'$ denotes the space of all anti-linear functionals on $H^r_{per}(0, 2\pi)$, see Definition A.26. We make the following definition.

Definition A.41 *For $r \geq 0$, we denote by $H^{-r}_{per}(0, 2\pi) = H^r_{per}(0, 2\pi)'$ the anti-dual space of $H^r_{per}(0, 2\pi)$; that is, the space of all anti-linear bounded functionals on $H^r_{per}(0, 2\pi)$. Then $H^r_{per}(0, 2\pi) \subset L^2(0, 2\pi) \subset H^{-r}_{per}(0, 2\pi)$ with bounded and dense imbeddings. The corresponding sesquilinear form $\langle \cdot, \cdot \rangle : H^{-r}_{per}(0, 2\pi) \times H^r_{per}(0, 2\pi) \to \mathbb{C}$ extends the inner product in $L^2(0, 2\pi)$; that is,*

$$\langle \psi, \phi \rangle = (\psi, \phi)_{L^2} = \int_0^{2\pi} \psi(t) \, \overline{\phi(t)} \, dt$$

for all $\psi \in L^2(0, 2\pi)$ and $\phi \in H^r_{per}(0, 2\pi)$.

The following theorems give characterizations in terms of the Fourier coefficients.

Theorem A.42 *Let again $\hat{\psi}_k(t) = e^{ikt}$ for $k \in \mathbb{Z}$ and $t \in (0, 2\pi)$.*

(a) *Let $\ell \in H_{per}^{-r}(0, 2\pi) = H_{per}^r(0, 2\pi)'$ and define $c_k := \langle \ell, \hat{\psi}_k \rangle$ for $k \in \mathbb{Z}$. Then*

$$\|\ell\|_{H_{per}^{-r}} = \left(\sum_{k \in \mathbb{Z}} (1 + k^2)^{-r} |c_k|^2 \right)^{1/2}$$

and

$$\langle \ell, x \rangle = \sum_{k \in \mathbb{Z}} c_k \overline{a_k} \quad \text{for all } x = \sum_{k \in \mathbb{Z}} a_k \hat{\psi}_k \in H_{per}^r(0, 2\pi). \tag{A.29}$$

(b) *Conversely, let $c_k \in \mathbb{C}$ satisfy*

$$\sum_{k \in \mathbb{Z}} (1 + k^2)^{-r} |c_k|^2 < \infty.$$

Then ℓ, defined by (A.29), is in $H_{per}^{-r}(0, 2\pi)$ with $\sum_{k \in \mathbb{Z}} (1 + k^2)^{-r} |c_k|^2 = \|\ell\|_{H_{per}^{-r}}^2$.

Proof: (a) Set $z^N = \sum_{|k| \leq N} c_k (1 + k^2)^{-r} \hat{\psi}_k$ for $N \in \mathbb{N}$. Then (note that ℓ is anti-linear)

$$\sum_{|k| \leq N} (1 + k^2)^{-r} |c_k|^2 = \langle \ell, z^N \rangle \leq \|\ell\|_{H_{per}^{-r}} \|z^N\|_{H_{per}^r}$$

$$= \|\ell\|_{H_{per}^{-r}} \sqrt{\sum_{|k| \leq N} (1 + k^2)^{-2r} |c_k|^2 (1 + k^2)^r}$$

$$= \|\ell\|_{H_{per}^{-r}} \sqrt{\sum_{|k| \leq N} (1 + k^2)^{-r} |c_k|^2}$$

which proves $\sum_{k \in \mathbb{Z}} (1 + k^2)^{-r} |c_k|^2 \leq \|\ell\|_{H_{per}^{-r}}^2$ by letting N tend to infinity. In particular, the series converges. Furthermore, for $x = \sum_{k \in \mathbb{Z}} a_k \hat{\psi}_k \in H_{per}^r(0, 2\pi)$ we set $x^N = \sum_{|k| \leq N} a_k \hat{\psi}_k$, and have that $\langle \ell, x^N \rangle = \sum_{|k| \leq N} \overline{a_k} c_k$ and thus

$$|\langle \ell, x^N \rangle| = \left| \sum_{|k| \leq N} \overline{a_k} c_k \right| \leq \sum_{|k| \leq N} |a_k| (1 + k^2)^{r/2} |c_k| (1 + k^2)^{-r/2}$$

$$\leq \sqrt{\sum_{|k| \leq N} |a_k|^2 (1 + k^2)^r} \sqrt{\sum_{|k| \leq N} |c_k|^2 (1 + k^2)^{-r}}$$

$$= \|x^N\|_{H_{per}^r} \sqrt{\sum_{|k| \leq N} |c_k|^2 (1 + k^2)^{-r}}$$

and thus also $\langle \ell, x \rangle = \sum_{k \in \mathbb{Z}} \overline{a_k}\, c_k$ and $\|\ell\|^2_{H^{-r}_{per}} \leq \sum_{k \in \mathbb{Z}} (1+k^2)^{-r} |c_k|^2$ by letting N tend to infinity.

(b) This is shown in the same way.   □

This theorem makes it possible to equip the Banach space $H^{-r}_{per}(0, 2\pi) = H^r_{per}(0, 2\pi)'$ with an inner product and make it a Hilbert space.

Theorem A.43 *Let $r > 0$. On $L^2(0, 2\pi)$, we define the inner product and norm by*

$$(x, y)_{-r} := \sum_{k \in \mathbb{Z}} (1+k^2)^{-r} a_k \overline{b_k}\,, \tag{A.30a}$$

$$\|x\|_{-r} := \sqrt{\sum_{k \in \mathbb{Z}} (1+k^2)^{-r} |a_k|^2}\,, \tag{A.30b}$$

respectively, where $x = \sum_{k \in \mathbb{Z}} a_k \hat{\psi}_k$ and $y = \sum_{k \in \mathbb{Z}} b_k \hat{\psi}_k$. Then the completion $\tilde{H}^{-r}_{per}(0, 2\pi)$ of $L^2(0, 2\pi)$ with respect to $\|\cdot\|_{-r}$ can be identified with $H^{-r}_{per}(0, 2\pi)$. The isomorphism is given by the extension of $J : L^2(0, 2\pi) \to H^{-r}_{per}(0, 2\pi)$, where

$$\langle Jx, y \rangle := \sum_{k \in \mathbb{Z}} a_k \overline{b_k} \quad for \; x = \sum_{k \in \mathbb{Z}} a_k \hat{\psi}_k \in H^r_{per}(0, 2\pi)$$

and $y = \sum_{k \in \mathbb{Z}} b_k \hat{\psi}_k \in L^2(0, 2\pi)$. Therefore, we identify $\|p\|_{-r}$ with $\|Jp\|_{H^{-r}_{per}}$ and simply write $\|p\|_{H^{-r}_{per}}$.

Proof: First we show that $Jx \in H^{-r}_{per}(0, 2\pi)$. Indeed, by the Cauchy–Schwarz inequality, we have for $y = \sum_{k \in \mathbb{Z}} b_k \hat{\psi}_k \in H^r_{per}(0, 2\pi)$

$$|\langle Jx, y \rangle| \leq \sum_{k \in \mathbb{Z}} \{(1+k^2)^{-r/2}|a_k|\} \{(1+k^2)^{r/2}|b_k|\}$$

$$\leq \left(\sum_{k \in \mathbb{Z}} (1+k^2)^{-r}|a_k|^2 \right)^{1/2} \left(\sum_{|k| \leq N} (1+k^2)^r |b_k|^2 \right)^{1/2}$$

$$\leq \|x\|_{H^{-r}_{per}} \|y\|_r\,,$$

and thus $Jx \in H^{-r}_{per}(0, 2\pi)$ with $\|Jx\|_{H^{-r}_{per}} \leq \|x\|_{-r}$ for all $x \in L^2(0, 2\pi)$. By the previous theorem, applied to $\ell = Jx$, we have $\|Jx\|^2_{H^{-r}_{per}} = \sum_{k \in \mathbb{Z}} (1+k^2)^{-r} |c_k|^2$ with $c_k = \langle Jx, \hat{\psi}_k \rangle = a_k$. Therefore, $\|Jx\|_{H^{-r}_{per}} = \|x\|_{-r}$, and J can be extended to a bounded operator from $\tilde{H}^{-r}_{per}(0, 2\pi)$ into $H^{-r}_{per}(0, 2\pi)$ (by Theorem A.30). It remains to show that J is surjective. Let $\ell \in H^{-r}_{per}(0, 2\pi) = H^r_{per}(0, 2\pi)'$ and define $c_k = \langle \ell, \hat{\psi}_k \rangle$ and $x^N = \sum_{|k| \leq N} c_k \hat{\psi}_k$ for $N \in \mathbb{N}$. For $y = \sum_{k \in \mathbb{Z}} b_k \hat{\psi}_k \in H^r_{per}(0, 2\pi)$ we have $Jx^N = \sum_{|k| \leq N} \overline{b_k}\, c_k = \langle \ell, \sum_{|k| \leq N} b_k \hat{\psi}_k \rangle$ which converges to $\langle \ell, y \rangle$. Furthermore, (x^N) is a Cauchy sequence with respect to $\|\cdot\|_{-r}$ because of the convergence of $\sum_{k \in \mathbb{Z}}(1+k^2)^{-r}|c_k|^2$ by the previous theorem, part (a). □

Theorem A.44 *(a) For $r > s$, the Sobolev space $H^r_{per}(0, 2\pi)$ is a dense subspace of $H^s_{per}(0, 2\pi)$. The inclusion operator from $H^r_{per}(0, 2\pi)$ into $H^s_{per}(0, 2\pi)$ is compact.*

(b) For all $r \geq 0$ and $x \in L^2(0, 2\pi)$ and $y \in H^r_{per}(0, 2\pi)$, there holds

$$\left|(x, y)_{L^2}\right| = 2\pi \left|(x, y)_{H^0_{per}}\right| \leq 2\pi \|x\|_{H^{-r}_{per}} \|y\|_{H^r_{per}}. \tag{A.31}$$

We note that the estimate (A.31) is in accordance with (A.16) because, with the imbedding j' from $L^2(0, 2\pi)$ into $H^{-r}_{per}(0, 2\pi)$ and the identification J from $\tilde{H}^{-r}_{per}(0, 2\pi)$ onto $H^{-r}_{per}(0, 2\pi)$ of the previous theorem we show easily that the imbedding of $L^2(0, 2\pi)$ into $\tilde{H}^{-r}_{per}(0, 2\pi)$ is given by $J^{-1} \circ j'$ where $(J^{-1} \circ j')x = 2\pi x$ for $x \in L^2(0, 2\pi)$.

Proof: (a) Denseness is easily seen by truncating the Fourier series.

Compactness is shown by defining the finite-dimensional operators J_N from $H^r_{per}(0, 2\pi)$ into $H^s_{per}(0, 2\pi)$ by $J_N x = \sum_{|k| \leq N} a_k \hat{\psi}_k$ where $x = \sum_{k \in \mathbb{Z}} a_k \hat{\psi}_k$. Then J_N is compact by part (f) of Theorem A.34 and

$$\|J_N x - x\|^2_{H^s_{per}} = \sum_{|k| > N}(1+k^2)^s|a_k|^2 \leq \frac{1}{(1+N^2)^{r-s}}\sum_{|k|>N}(1+k^2)^r|a_k|^2$$

$$\leq \frac{1}{N^{2(r-s)}}\|x\|^2_{H^r_{per}}.$$

Therefore, J_N converges in the operator norm to the imbedding J which implies, again by Theorem A.34, that also J is compact.

(b) This is seen as in the proof of Theorem A.43 with the Cauchy–Schwarz inequality. □

Theorems A.40 and A.43 imply that the space $\mathcal{T}$ of all trigonometric polynomials is dense in $H^r_{per}(0, 2\pi)$ for every $r \in \mathbb{R}$. Now we study the orthogonal projection and the interpolation operators with respect to equidistant knots and the $2n$-dimensional space

$$\mathcal{T}_n := \left\{\sum_{k=-n}^{n-1} a_k \hat{\psi}_k : a_k \in \mathbb{C}\right\} \tag{A.32}$$

where again $\hat{\psi}_k(t) = e^{ikt}$ for $t \in [0, 2\pi]$ and $k \in \mathbb{Z}$.

Lemma A.45 *Let $r, s \in \mathbb{R}$ with $r \geq s$.*

(a) The following stability estimate holds

$$\|z_n\|_{H^r_{per}} \leq c n^{r-s} \|z_n\|_{H^s_{per}} \quad \text{for all } z_n \in \mathcal{T}_n.$$

(b) Let $P_n : L^2(0, 2\pi) \to \mathcal{T}_n \subset L^2(0, 2\pi)$ be the orthogonal projection operator. Then P_n is given by

$$P_n x = \sum_{k=-n}^{n-1} a_k \hat{\psi}_k, \quad x \in L^2(0, 2\pi), \tag{A.33}$$

where

$$a_k = \frac{1}{2\pi} \int_0^{2\pi} x(s) e^{-iks} \, ds, \quad k \in \mathbb{Z},$$

are the Fourier coefficients of x. Furthermore, the following estimate holds:

$$\|x - P_n x\|_{H_{per}^s} \leq \frac{1}{n^{r-s}} \|x\|_{H_{per}^r} \quad \text{for all } x \in H_{per}^r(0, 2\pi), \tag{A.34}$$

Proof: (a) With $z_n = \sum_{k=-n}^{n-1} a_k \hat{\psi}_k$ this follows simply by

$$\|z_n\|_{H_{per}^r}^2 \leq \sum_{|k| \leq n} (1 + k^2)^r |a_k|^2 = \sum_{|k| \leq n} (1 + k^2)^{r-s} (1 + k^2)^s |a_k|^2$$

$$\leq (1 + n^2)^{r-s} \|z_n\|_{H_{per}^s}^2 \leq (2n^2)^{r-s} \|z_n\|_{H_{per}^s}^2.$$

(b) Let $x = \sum_{k \in \mathbb{Z}} a_k \hat{\psi}_k \in L^2(0, 2\pi)$ and define the right-hand side of (A.33) by z; that is, $z = \sum_{k=-n}^{n-1} a_k \hat{\psi}_k \in \mathcal{T}_n$. The orthogonality of $\hat{\psi}_k$ implies that $x - z$ is orthogonal to $\mathcal{T}_n$. This proves that z coincides with $P_n x$. Now let $x \in H_{per}^r(0, 2\pi)$. Then

$$\|x - P_n x\|_{H_{per}^s}^2 \leq \sum_{|k| \geq n} (1 + k^2)^s |a_k|^2$$

$$= \sum_{|k| \geq n} (1 + k^2)^{-(r-s)} \left[(1 + k^2)^r |a_k|^2 \right]$$

$$\leq (1 + n^2)^{s-r} \|x\|_{H_{per}^r}^2 \leq n^{2(s-r)} \|x\|_{H_{per}^r}^2. \qquad \square$$

Now let $t_j := j \frac{\pi}{n}, j = 0, \ldots, 2n - 1$, be equidistantly chosen points in $[0, 2\pi]$. Interpolation of smooth periodic functions by trigonometric polynomials can be found in numerous books as, for example, in [72]. Interpolation in Sobolev spaces of integer orders can be found in [42]. We give a different and much simpler proof of the error estimates that are optimal and hold in Sobolev spaces of fractional order.

Theorem A.46 For every $n \in \mathbb{N}$ and every 2π-periodic function $x \in C[0, 2\pi]$, there exists a unique $p_n \in \mathcal{T}_n$ with $x(t_j) = p_n(t_j)$ for all $j = 0, \ldots, 2n - 1$. The *trigonometric interpolation operator* $Q_n : C_{per}[0, 2\pi] = \{x \in C[0, 2\pi] : x(0) = x(2\pi)\} \to \mathcal{T}_n$ *has the form*

$$Q_n x = \sum_{k=0}^{2n-1} x(t_k) L_k$$

with Lagrange interpolation basis functions

$$L_k(t) = \frac{1}{2n} \sum_{m=-n}^{n-1} e^{im(t-t_k)}, \quad k = 0, \ldots, 2n-1. \qquad (A.35)$$

The interpolation operator Q_n *has an extension to a bounded operator from* $H^r_{per}(0, 2\pi)$ *into* $\mathcal{T}_n \subset H^r_{per}(0, 2\pi)$ *for all* $r > \frac{1}{2}$. *Furthermore,* Q_n *obeys estimates of the form*

$$\|x - Q_n x\|_{H^s_{per}} \leq \frac{c}{n^{r-s}} \|x\|_{H^r_{per}} \quad \text{for all } x \in H^r_{per}(0, 2\pi), \qquad (A.36)$$

where $0 \leq s \leq r$ *and* $r > \frac{1}{2}$. *The constant* c *depends only on* s *and* r. *In particular,* $\|Q_n\|_{\mathcal{L}(H^r_{per}(0, 2\pi))}$ *is uniformly bounded with respect to* n.

Proof: The proof of the first part can be found in, for example, [168]. Let $x(t) = \sum_{m \in \mathbb{Z}} a_m \exp(imt)$. Direct calculation shows that for smooth functions x the interpolation is given by

$$(Q_n x)(t) = \sum_{j=-n}^{n-1} \hat{a}_j e^{ijt} \quad \text{with}$$

$$\hat{a}_j = \frac{1}{2n} \sum_{k=0}^{2n-1} x(t_k) e^{-ijk\pi/n}, \quad j = -n, \ldots, n-1.$$

The connection between the continuous and discrete Fourier coefficients is simply

$$\hat{a}_j = \frac{1}{2n} \sum_{k=0}^{2n-1} \sum_{m \in \mathbb{Z}} a_m e^{imk\pi/n - ijk\pi/n}$$

$$= \frac{1}{2n} \sum_{m \in \mathbb{Z}} a_m \sum_{k=0}^{2n-1} \left[e^{i(m-j)\pi/n} \right]^k = \sum_{\ell \in \mathbb{Z}} a_{j+2n\ell}$$

where $x = \sum_{m \in \mathbb{Z}} a_m \hat{\psi}_m$. It is sufficient to estimate $P_n x - Q_n x$ because the required estimate holds for $x - P_n x$ by formula (A.34). We have

$$P_n x - Q_n x = \sum_{m=-n}^{n-1} [a_m - \hat{a}_m] \psi_m$$

and thus by the Cauchy–Schwarz inequality

$$\|P_n x - Q_n x\|_{H^s_{per}}^2$$

$$= \sum_{m=-n}^{n-1} |a_m - \hat{a}_m|^2 (1 + m^2)^s \le cn^{2s} \sum_{m=-n}^{n-1} \left| \sum_{\ell \neq 0} a_{m+2n\ell} \right|^2$$

$$\le cn^{2s} \sum_{m=-n}^{n-1} \left| \sum_{\ell \neq 0} [(1 + (m+2n\ell)^2)^{r/2} a_{m+2n\ell}] \frac{1}{(1 + (m+2n\ell)^2)^{r/2}} \right|^2$$

$$\le cn^{2s} \sum_{m=-n}^{n-1} \left[\sum_{\ell \neq 0} (1 + (m+2n\ell)^2)^r |a_{m+2n\ell}|^2 \sum_{\ell \neq 0} \frac{1}{(1 + (m+2n\ell)^2)^r} \right].$$

From the obvious estimate

$$\sum_{\ell \neq 0} (1 + (m+2n\ell)^2)^{-r} \le (2n)^{-2r} \sum_{\ell \neq 0} \left(\frac{m}{2n} + \ell \right)^{-2r} \le cn^{-2r}$$

for all $|m| \le n$ and $n \in \mathbb{N}$, we conclude that

$$\|P_n x - Q_n x\|_{H^s_{per}}^2 \le cn^{2(s-r)} \sum_{m=-n}^{n-1} \sum_{\ell \neq 0} (1 + (m+2n\ell)^2)^r |a_{m+2n\ell}|^2$$

$$\le cn^{2(s-r)} \|x\|_{H^r_{per}}^2. \qquad \square$$

For real-valued functions, it is more convenient to study the orthogonal projection and interpolation in the $2n$-dimensional space

$$\left\{ \sum_{j=0}^{n} a_j \cos(jt) + \sum_{j=1}^{n-1} b_j \sin(jt) : a_j, b_j \in \mathbb{R} \right\}.$$

In this case, the Lagrange interpolation basis functions L_k are given by (see [168])

$$L_k(t) = \frac{1}{2n} \left\{ 1 + 2 \sum_{m=1}^{n-1} \cos m(t - t_k) + \cos n(t - t_k) \right\}, \qquad (A.37)$$

$k = 0, \ldots, 2n - 1$, and the estimates (A.34) and (A.36) are proven by the same arguments.

Theorem A.47 Let $r \in \mathbb{N}$ and $k \in C^r([0, 2\pi] \times [0, 2\pi])$ be 2π-periodic with respect to both variables. Then the integral operator K, defined by

$$(Kx)(t) := \int_0^{2\pi} k(t, s) x(s) \, ds, \quad t \in (0, 2\pi), \qquad (A.38)$$

can be extended to a bounded operator from $H^p_{per}(0, 2\pi)$ into $H^r_{per}(0, 2\pi)$ for every $-r \le p \le r$.

Proof: Let $x \in L^2(0, 2\pi)$. From

$$\frac{d^j}{dt^j}(Kx)(t) \;=\; \int\limits_0^{2\pi} \frac{\partial^j k(t, s)}{\partial t^j}\, x(s)\, ds\,, \quad j = 0, \ldots, r\,,$$

we conclude from Theorem A.44 that for $x \in L^2(0, 2\pi)$

$$\left| \frac{d^j}{dt^j}(Kx)(t) \right| \;\leq\; 2\pi \left\| \frac{\partial^j k(t, \cdot)}{\partial t^j} \right\|_{H^r_{per}} \|x\|_{H^{-r}_{per}}$$

and thus

$$\|Kx\|_{H^r_{per}} \;\leq\; c_1 \|Kx\|_{C^r} \;\leq\; c_2 \|x\|_{H^{-r}_{per}}$$

for all $x \in L^2(0, 2\pi)$. Application of Theorem A.30 yields the assertion because $L^2(0, 2\pi)$ is dense in $H^{-r}_{per}(0, 2\pi)$. $\square$

A.5   Sobolev Spaces on the Unit Disc

Let $B = \{x \in \mathbb{R}^2 : |x| < 1\}$ be the open unit disc with boundary ∂B. In this section, we consider functions from B into $\mathbb{C}$ that we describe by Cartesian coordinates $x = (x_1, x_2)$ or by polar coordinates (r, φ). Functions on the boundary are identified with 2π-periodic functions on $\mathbb{R}$. As in the case of the Sobolev spaces $H^s_{per}(0, 2\pi)$ we define the Sobolev space $H^1(B)$ by completion.

Definition A.48 *The Sobolev space $H^1(B)$ is defined as the completion of $C^\infty(\overline{B})$ with respect to the norm*

$$\|f\|_{H^1(B)} \;=\; \sqrt{ \int\limits_B \left[|f(x)|^2 + |\nabla f(x)|^2 \right] dx }\,. \tag{A.39}$$

We express the norm (A.39) in polar coordinates (r, φ). The gradient is given in polar coordinates as

$$\nabla f(r, \varphi) \;=\; \frac{\partial f(r, \varphi)}{\partial r}\, \hat{r} \;+\; \frac{1}{r} \frac{\partial f(r, \varphi)}{\partial \varphi}\, \hat{\varphi}\,,$$

where $\hat{r} = \binom{\cos\varphi}{\sin\varphi}$ and $\hat{\varphi} = \binom{-\sin\varphi}{\cos\varphi}$ denote the unit vectors. We fix $r > 0$ and expand the function $f(r, \cdot)$ (formally) into a Fourier series with respect to φ:

$$f(r, \varphi) \;=\; \sum_{m \in \mathbb{Z}} f_m(r)\, e^{im\varphi}$$

with Fourier coefficients

$$f_m(r) \;=\; \frac{1}{2\pi} \int\limits_0^{2\pi} f(r, t)\, e^{-imt}\, dt\,, \quad m \in \mathbb{Z}\,,$$

that depend on r. Therefore,

$$\frac{\partial f(r,\varphi)}{\partial r} = \sum_{m \in \mathbb{Z}} f'_m(r) e^{im\varphi}, \qquad \frac{\partial f(r,\varphi)}{\partial \varphi} = i \sum_{m \in \mathbb{Z}} f_m(r) m e^{im\varphi}.$$

The norm in $H^1(B)$ is given by

$$\|f\|^2_{H^1(B)} = 2\pi \sum_{m \in \mathbb{Z}} \int_0^1 \left[\left(1 + \frac{m^2}{r^2}\right) |f_m(r)|^2 + |f'_m(r)|^2 \right] r\, dr, \qquad (A.40)$$

because

$$|\nabla f(r,\varphi)|^2 = \left| \frac{\partial f(r,\varphi)}{\partial r} \right|^2 + \frac{1}{r^2} \left| \frac{\partial f(r,\varphi)}{\partial \varphi} \right|^2$$

and

$$\int_0^{2\pi} \left| \sum_{m \in \mathbb{Z}} f_m(r) e^{im\varphi} \right|^2 d\varphi = 2\pi \sum_{m \in \mathbb{Z}} |f_m(r)|^2.$$

To every function $f \in C^\infty(\overline{B})$, one can assign the trace $f|_{\partial B}$ on ∂B. We denote this mapping by τ, thus $\tau : C^\infty(\overline{B}) \to C^\infty(\partial B)$ is defined as $\tau f = f|_{\partial B}$. The following result is central.

Theorem A.49 *(Trace Theorem)*

The trace operator τ has an extension to a bounded operator from $H^1(B)$ to $H^{1/2}(\partial B)$, where again $H^{1/2}(\partial B)$ is identified with the Sobolev space $H^{1/2}_{per}(0, 2\pi)$ of periodic functions (see Definition A.38). Furthermore, $\tau : H^1(B) \to H^{1/2}(\partial B)$ is surjective. More precisely, there exists a bounded linear operator $E : H^{1/2}(\partial B) \to H^1(B)$ with $\tau \circ E = I$ on $H^{1/2}(\partial B)$ (that is, E is a right inverse of τ).

Proof: Let $f \in C^\infty(\overline{B})$. Then (see (A.40) and (A.28))

$$\|f\|^2_{H^1(B)} = 2\pi \sum_{m \in \mathbb{Z}} \int_0^1 \left[\left(1 + \frac{m^2}{r^2}\right) |f_m(r)|^2 + |f'_m(r)|^2 \right] r\, dr,$$

$$\|\tau f\|^2_{H^{1/2}(\partial B)} = \sum_{m \in \mathbb{Z}} \sqrt{1 + m^2}\, |f_m(1)|^2.$$

We estimate, using the fundamental theorem of calculus and the inequality of Cauchy–Schwarz,

$$|f_m(1)|^2 = \int_0^1 \frac{d}{dr}\left(r^2|f_m(r)|^2\right)dr$$

$$= 2\int_0^1 |f_m(r)|^2 r\,dr + 2\,\mathrm{Re}\int_0^1 f_m(r)\overline{f'_m(r)}\,r^2\,dr$$

$$\leq 2\int_0^1 |f_m(r)|^2 r\,dr + 2\sqrt{\int_0^1 |f_m(r)|^2 r^2\,dr}\sqrt{\int_0^1 |f'_m(r)|^2 r^2\,dr}\,.$$

Using the inequality $2ab \leq \sqrt{1+m^2}\,a^2 + \dfrac{b^2}{\sqrt{1+m^2}}$ yields

$$\sqrt{1+m^2}\,|f_m(1)|^2 \leq 2\sqrt{1+m^2}\int_0^1 |f_m(r)|^2 r\,dr$$

$$+ (1+m^2)\int_0^1 |f_m(r)|^2 r^2\,dr + \int_0^1 |f'_m(r)|^2 r^2\,dr$$

$$\leq 3\,(1+m^2)\int_0^1 |f_m(r)|^2 r\,dr + \int_0^1 |f'_m(r)|^2 r\,dr$$

$$\leq 3\int_0^1 \left(1 + \frac{m^2}{r^2}\right)|f_m(r)|^2 r\,dr + \int_0^1 |f'_m(r)|^2 r\,dr\,,$$

where we have also used the estimates $r^2 \leq r$ for $r \in [0,1]$ and $\sqrt{1+m^2} \leq 1+m^2$. By summation, we conclude that

$$\|\tau f\|^2_{H^{1/2}(\partial B)} \leq \frac{3}{2\pi}\|f\|^2_{H^1(B)} \quad \text{for all } f \in C^\infty(\overline{B}). \tag{A.41}$$

Therefore, the trace operator is bounded with respect to the norms of $H^1(B)$ and $H^{1/2}(\partial B)$. By the general functional analytic Theorem A.30, the operator has an extension to a bounded operator from $H^1(B)$ into $H^{1/2}(\partial B)$.

We define the operator $E : C^\infty(\partial B) \to C^\infty(\overline{B})$ by

$$(Ef)(r,\varphi) = \sum_{m\in\mathbb{Z}} f_m\, r^{|m|}\, e^{im\varphi}, \quad r \in [0,1],\ \varphi \in [0,2\pi].$$

Here, again, f_m are the Fourier coefficients of $f \in C^\infty(\partial B)$.

Obviously, $(\tau E f)(\varphi) = \sum_{m\in\mathbb{Z}} f_m\, e^{im\varphi} = f(\varphi)$; that is, E is a right inverse of τ.

It remains to show the boundedness of E.

$$
\begin{aligned}
\|Ef\|^2_{H^1(B)} &= 2\pi \sum_{m\in\mathbb{Z}} \int_0^1 \left[\left(1 + \frac{m^2}{r^2}\right) |f_m|^2 r^{2|m|} + |f_m|^2 m^2 r^{2|m|-2} \right] r\,dr, \\
&= 2\pi \sum_{m\in\mathbb{Z}} |f_m|^2 \left(\frac{1}{2|m|+2} + |m| \right) \leq 2\pi \sum_{m\in\mathbb{Z}} |f_m|^2 (1+|m|) \\
&\leq \sqrt{2}\, 2\pi \sum_{m\in\mathbb{Z}} |f_m|^2 \sqrt{1+m^2} = \sqrt{2}\, 2\pi \|f\|^2_{H^{1/2}(\partial B)}
\end{aligned}
$$

where we used the inequality $1 + |m| \leq \sqrt{2}\sqrt{1+m^2}$. Therefore, E also possesses an extension to a bounded operator from $H^{1/2}(\partial B)$ to $H^1(B)$. $\quad\square$

Remark: The trace operator is compact when considered as an operator from $H^1(B)$ to $L^2(\partial B)$ because it is the composition of the bounded operator $\tau : H^1(B) \to H^{1/2}(\partial B)$ and the compact embedding $j : H^{1/2}(\partial B) \to L^2(\partial B)$.

We now consider the subspaces

$$
\begin{aligned}
L^2_\diamond(\partial B) &= \left\{ f \in L^2(\partial B) : \int_{\partial B} f\,d\ell = 0 \right\}, \\
H^{1/2}_\diamond(\partial B) &= \left\{ f \in H^{1/2}(\partial B) : \int_{\partial B} f\,d\ell = 0 \right\}, \\
H^1_\diamond(B) &= \left\{ f \in H^1(B) : \int_{\partial B} \tau f\,d\ell = 0 \right\}.
\end{aligned}
$$

Because $\int_0^{2\pi} \exp(im\varphi)\,d\varphi = 0$ for $m \neq 0$, the spaces $H^{1/2}_\diamond(\partial B)$ and $H^1_\diamond(B)$ consist exactly of the functions with the representations

$$
f(\varphi) = \sum_{\substack{m\in\mathbb{Z}\\ m\neq 0}} f_m e^{im\varphi} \quad \text{and}
$$

$$
f(r,\varphi) = \sum_{m\in\mathbb{Z}} f_m(r) e^{im\varphi},
$$

that satisfy the summation conditions

$$
\sum_{\substack{m\in\mathbb{Z}\\ m\neq 0}} \sqrt{1+m^2}\, |f_m|^2 < \infty \quad \text{and}
$$

$$
\sum_{m\in\mathbb{Z}} \int_0^1 \left[\left(1 + \frac{m^2}{r^2}\right) |f_m(r)|^2 + |f'_m(r)|^2 \right] r\,dr < \infty \quad \text{and} \quad f_0(1) = 0,
$$

respectively.

We can define an equivalent norm in the subspace $H^1_\diamond(B)$. This is a consequence of the following result.

Theorem A.50 (*Friedrich's Inequality*)

For all $f \in H^1_\diamond(B)$, we have

$$\|f\|_{L^2(B)} \leq \sqrt{2}\,\|\nabla f\|_{L^2(B)}. \qquad (A.42)$$

Proof: Again, we use the representation of the norm in polar coordinates:

$$r\,|f_m(r)|^2 = \int_0^r \frac{d}{ds}\left(s\,|f_m(s)|^2\right) ds$$

$$= \int_0^r |f_m(s)|^2\,ds \;+\; 2\,\mathrm{Re}\int_0^r f_m(s)\,\overline{f'_m(s)}\,s\,ds$$

$$\leq \int_0^1 |f_m(s)|^2\,ds \;+\; 2\sqrt{\int_0^1 |f_m(s)|^2\,s\,ds}\sqrt{\int_0^1 |f'_m(s)|^2\,s\,ds}$$

$$\leq \int_0^1 |f_m(s)|^2\,(1+s)\,ds \;+\; \int_0^1 |f'_m(s)|^2\,s\,ds,$$

where we again used $2ab \leq a^2 + b^2$ in the last step.

First let $|m| \geq 1$. By $1 + s \leq 2 \leq 2m^2/s$, it is

$$r\,|f_m(r)|^2 \leq 2\int_0^1 \frac{m^2}{s^2}\,|f_m(s)|^2\,s\,ds \;+\; \int_0^1 |f'_m(s)|^2\,s\,ds,$$

and thus by integration

$$\int_0^1 r\,|f_m(r)|^2\,dr \leq 2\int_0^1 \left(\frac{m^2}{s^2}\,|f_m(s)|^2 \;+\; |f'_m(s)|^2\right) s\,ds. \qquad (A.43)$$

We finally consider f_0. It is $f_0(r) = -\int_r^1 f'_0(s)\,ds$ because $f_0(1) = 0$, thus

$$r\,|f_0(r)|^2 \leq (1-r)\int_r^1 |f'_0(s)|^2 r\,ds \leq \int_r^1 |f'_0(s)|^2 s\,ds \leq \int_0^1 |f'_0(s)|^2 s\,ds.$$

Therefore, (A.43) also holds for $m = 0$. Summation with respect to m yields the assertion. $\square$

Remark: Therefore, $f \mapsto \|\nabla f\|_{L^2(B)}$ defines an equivalent norm to $\|\cdot\|_{H^1(B)}$ in $H^1_\diamond(B)$. Indeed, for $f \in H^1_\diamond(B)$ it holds by Friedrich's inequality:

$$\|f\|^2_{H^1(B)} = \|f\|^2_{L^2(B)} + \|\nabla f\|^2_{L^2(B)} \leq 3\,\|\nabla f\|^2_{L^2(B)},$$

thus

$$\frac{1}{\sqrt{3}} \, \|f\|_{H^1(B)} \; \leq \; \|\nabla f\|_{L^2(B)} \; \leq \; \|f\|_{H^1(B)} \quad \text{for all } f \in H^1_\diamond(B) . \qquad (A.44)$$

So far, we considered spaces of *complex-valued* functions. The spaces of *real-valued* functions are closed subspaces. In the Fourier representation, one has

$$\overline{f(r, \varphi)} \; = \; \sum_{m \in \mathbb{Z}} \overline{f_m(r)} \, e^{-im\varphi} \; = \; \sum_{m \in \mathbb{Z}} \overline{f_{-m}(r)} \, e^{im\varphi} \; = \; f(r, \varphi) \; = \; \sum_{m \in \mathbb{Z}} f_m(r) \, e^{im\varphi}$$

because $\overline{f(r, \varphi)} = f(r, \varphi)$. Therefore, $\overline{f_{-m}} = f_m$ for all m. All of the theorems remain valid also for Sobolev spaces of real-valued functions.

A.6   Spectral Theory for Compact Operators in Hilbert Spaces

Definition A.51 *(Spectrum)*
Let X be a normed space and $A : X \longrightarrow X$ be a linear operator. The spectrum *$\sigma(A)$ is defined as the set of (complex) numbers λ such that the operator $A - \lambda I$ does not have a bounded inverse on X. Here, I denotes the identity on X. $\lambda \in \sigma(A)$ is called an* eigenvalue *of A if $A - \lambda I$ is not one-to-one. If λ is an eigenvalue, then the nontrivial elements x of the kernel $\mathcal{N}(A - \lambda I) = \{x \in X : Ax - \lambda x = 0\}$ are called* eigenvectors *of A.*

This definition makes sense for arbitrary linear operators in normed spaces. For noncompact operators A, it is possible that even for $\lambda \neq 0$ the operator $A - \lambda I$ is one-to-one but fails to be bijective. As an example, we consider $X = \ell^2$ and define A by

$$(Ax)_k \; := \; \begin{cases} 0, & \text{if} \quad k = 1, \\ x_{k-1}, & \text{if} \quad k \geq 2, \end{cases}$$

for $x = (x_k) \in \ell^2$. Then $\lambda = 1$ belongs to the spectrum of A but is not an eigenvalue of A, see Problem A.4.

Theorem A.52 *Let $A : X \to X$ be a linear bounded operator.*

(a) Let $x_j \in X$, $j = 1, \ldots, n$, be a finite set of eigenvectors corresponding to pairwise different eigenvalues $\lambda_j \in \mathbb{C}$. Then $\{x_1, \ldots, x_n\}$ are linearly independent. If X is a Hilbert space and A is self-adjoint (that is, $A^* = A$), then all eigenvalues λ_j are real-valued and the corresponding eigenvectors $x_1, \ldots, x_n$ are pairwise orthogonal.

(b) Let X be a Hilbert space and $A : X \to X$ be self-adjoint. Then

$$\|A\|_{\mathcal{L}(X)} \; = \; \sup_{\|x\|_X = 1} |(Ax, x)_X| \; = \; r(A) ,$$

where $r(A) = \sup\{|\lambda| : \lambda \in \sigma(A)\}$ is called the spectral radius *of A.*

The situation is simpler for compact operators. We collect the most important results in the following fundamental theorem.

Theorem A.53 *(Spectral Theorem for Compact Self-Adjoint Operators)*
Let $K : X \to X$ be compact and self-adjoint (and $K \neq 0$). Then the following holds:

(a) *The spectrum consists only of eigenvalues and possibly 0. Every eigenvalue of K is real-valued. K has at least one but at most a countable number of eigenvalues with 0 as the only possible accumulation point.*

(b) *For every eigenvalue $\lambda \neq 0$, there exist only finitely many linearly independent eigenvectors; that is, the eigenspaces are finite-dimensional. Eigenvectors corresponding to different eigenvalues are orthogonal.*

(c) *We order the eigenvalues in the form*

$$|\lambda_1| \geq |\lambda_2| \geq |\lambda_3| \geq \cdots$$

and denote by $P_j : X \to \mathcal{N}(K - \lambda_j I)$ the orthogonal projection onto the eigenspace corresponding to λ_j. If there exist only a finite number $\lambda_1, \ldots, \lambda_m$ of eigenvalues, then

$$K = \sum_{j=1}^{m} \lambda_j P_j \,.$$

If there exists an infinite sequence (λ_j) of eigenvalues, then

$$K = \sum_{j=1}^{\infty} \lambda_j P_j \,,$$

where the series converges in the operator norm. Furthermore,

$$\left\| K - \sum_{j=1}^{m} \lambda_j P_j \right\|_{\mathcal{L}(X)} = |\lambda_{m+1}| \,.$$

(d) *Let H be the linear span of all of the eigenvectors corresponding to the eigenvalues $\lambda_j \neq 0$ of K. Then*

$$X = \mathrm{closure}(H) \oplus \mathcal{N}(K) \,.$$

Sometimes, part (d) is formulated differently. For a common treatment of the cases of finitely and infinitely many eigenvalues, we introduce the index set $J \subset \mathbb{N}$, where J is finite in the first case and $J = \mathbb{N}$ in the second case. For every eigenvalue λ_j, $j \in J$, we choose an orthonormal basis of the corresponding eigenspace $\mathcal{N}(K - \lambda_j I)$. Again, let the eigenvalues $\lambda_j \neq 0$ be ordered in the form

$$|\lambda_1| \geq |\lambda_2| \geq |\lambda_3| \geq \cdots > 0 \,.$$

By counting every $\lambda_j \neq 0$ relative to its multiplicity, we can assign an eigenvector x_j to every eigenvalue λ_j. Then every $x \in X$ possesses an abstract Fourier expansion of the form

$$x = x_0 + \sum_{j \in J}(x, x_j)_X \, x_j$$

for some $x_0 \in \mathcal{N}(K)$ and

$$Kx = \sum_{j \in J}\lambda_j \, (x, x_j)_X \, x_j.$$

As a corollary, we observe that the set $\{x_j : j \in J\}$ of all eigenvectors forms a complete system in X if K is one-to-one.

The eigenvalues can be expressed by Courant's max-inf and min-sup principle. We need it in the following form.

Theorem A.54 *Let $K : X \to X$ be compact and self-adjoint (and $K \neq 0$) and let $\{\lambda_j^- : j = 1, \ldots, n_-\}$ and $\{\lambda_j^+ : j = 1, \ldots, n_+\}$ be its negative and positive eigenvalues, respectively, ordered as*

$$\lambda_1^- \leq \lambda_2^- \leq \cdots \leq \lambda_j^- \leq \cdots < 0 < \cdots \leq \lambda_j^+ \leq \cdots \leq \lambda_2^+ \leq \lambda_1^+$$

and counted according to multiplicity. Here $n_\pm$ can be zero (if no positive or negative, respectively, eigenvalues occur), finite, or infinity.

(a) If there exist positive eigenvalues λ_m^+ then

$$\lambda_m^+ = \min_{\substack{V \subset X \\ \dim V = m-1}} \sup_{x \in V^\perp} \frac{(Kx, x)_X}{\|x\|_X^2}.$$

(b) If there exist negative eigenvalues λ_m^- then

$$\lambda_m^- = \max_{\substack{V \subset X \\ \dim V = m-1}} \inf_{x \in V^\perp} \frac{(Kx, x)_X}{\|x\|_X^2}.$$

Proof: We only prove part (b) because this is needed in the proof of Theorem 7.52. Set

$$\mu_m = \sup_{\substack{V \subset X \\ \dim V = m-1}} \inf_{x \in V^\perp} \frac{(Kx, x)_X}{\|x\|_X^2}$$

for abbreviation. For any $x \in X$ we use the representation as $x = x_0 + \sum_j c_j^- x_j^- + \sum_j c_j^+ x_j^+$ with $Kx_0 = 0$. Here, $x_j^\pm$ are the eigenvectors corresponding to $\lambda_j^\pm$. Then $(Kx, x)_X = \sum_j \lambda_j^- |c_j^-|^2 + \sum_j \lambda_j^+ |c_j^+|^2 \geq \sum_j \lambda_j^- |c_j^-|^2$.

For $V = \text{span}\{x_j^- : j = 1, \ldots, m-1\}$ and $x \in V^\perp$ we have that $c_j^- = (x, x_j^-)_X = 0$ for $j = 1, \ldots, m-1$ and thus

$$\inf_{x \in V^\perp} \frac{(Kx, x)_X}{\|x\|_X^2} \geq \inf_{x \in V^\perp} \frac{\sum_{j \geq m} \lambda_j^- |c_j^-|^2}{\|x_0\|_X^2 + \sum_j |c_j^-|^2 + \sum_j |c_j^+|^2} \geq \lambda_m^-$$

because $\lambda_j^- < 0$. Therefore, $\mu_m \geq \lambda_m^-$. We note that for this choice of V the infimum is attained for $x = x_m^-$.

Let, on the other hand, $V \subset X$ be an arbitrary subspace of dimension $m-1$. Choose a basis $\{v_j : j = 1, \ldots, m-1\}$ of V. Construct $\hat{x} \in \text{span}\{x_j^- : j = 1, \ldots, m\}$ such that $\hat{x} \perp V$ and $\|\hat{x}\|_X = 1$. This is possible. Indeed, the ansatz $\hat{x} = \sum_{j=1}^m c_j x_j^-$ leads to the system $\sum_{j=1}^m c_j (x_j^-, v_\ell)_X = 0$ for $\ell = 1, \ldots, m-1$. This system of $m-1$ equations and m variables has a nontrivial solution which can be normalized such that $\|\hat{x}\|_X^2 = \sum_{j=1}^m |c_j|^2 = 1$. Therefore,

$$\inf_{x \in V^\perp} \frac{(Kx, x)_X}{\|x\|_X^2} \leq (K\hat{x}, \hat{x})_X = \sum_{j=1}^m \lambda_j^- |c_j|^2 \leq \lambda_m^- \sum_{j=1}^m |c_j|^2 = \lambda_m^-.$$

This shows $\mu_m \leq \lambda_m^-$ and ends the proof. $\qquad\square$

The following corollary is helpful.

Corollary A.55 *Let $K : X \to X$ and $\lambda_j^\pm$ as in the previous theorem.*

(a) *If there exists a subspace $W \subset X$ of dimension m with $(Kx, x)_X + \|x\|_X^2 \leq 0$ for all $x \in W$ then (at least) m negative eigenvalues exist and $\lambda_1^- \leq \cdots \leq \lambda_m^- \leq -1$.*

(b) *The eigenvalues $\lambda_j^\pm$ depend continuously on K in the operator norm. More precisely[1], if eigenvalues $\lambda_1^- \leq \cdots \lambda_m^- < 0$ of K exist and if $S : X \to X$ is also compact and self-adjoint with sufficiently small $\|S - K\|_{\mathcal{L}(X)}$ then also eigenvalues $\mu_1^- \leq \cdots \mu_m^- < 0$ of S exist and $|\mu_j^- - \lambda_j^-| \leq \|S - K\|_{\mathcal{L}(X)}$ for all $j = 1, \ldots, m$.*

Proof: (a) Let $V \subset X$ be an arbitrary subspace of dimension $m - 1$. We construct $w \in W$ with $\|w\|_X = 1$ such that $w \perp V$ as in the proof of the previous theorem. Then

$$\inf_{x \in V^\perp} \frac{(Kx, x)_X}{\|x\|_X^2} \leq (Kw, w)_X \leq -\|w\|_X^2 = -1.$$

Since this holds for every such subspace V, we conclude from the previous theorem that $\lambda_m^- \leq -1$.

---

[1] formulated for the negative eigenvalues

(b) Assume that some negative eigenvalue $\lambda_1^- \leq \cdots \leq \lambda_m^- < 0$ for K exist. Then, for any $j \in \{1, \ldots, m\}$ and any subspace $V \subset X$ of dimension $j - 1$,

$$
\begin{aligned}
\inf_{x \in V^\perp} \frac{(Sx, x)_X}{\|x\|_X^2} &= \inf_{x \in V^\perp} \left[\frac{((S-K)x, x)_X}{\|x\|_X^2} + \frac{(Kx, x)_X}{\|x\|_X^2} \right] \\
&\leq \|S - K\|_{\mathcal{L}(X)} + \inf_{x \in V^\perp} \frac{(Kx, x)_X}{\|x\|_X^2} \\
&\leq \|S - K\|_{\mathcal{L}(X)} + \lambda_j^-
\end{aligned}
$$

which is negative for sufficiently small $\|S - K\|_{\mathcal{L}(X)}$. Therefore, S has negative eigenvalues as well and, by taking the supremum, $\mu_j^- \leq \lambda_j^- + \|S - K\|_{\mathcal{L}(X)}$. Now we can interchange the roles of S and K which yields $|\mu_j^- - \lambda_j^-| \leq \|S - K\|_{\mathcal{L}(X)}$.
□

The spectral theorem for compact self-adjoint operators has an extension to non-self-adjoint operators $K : X \to Y$. First, we have the following definition.

Definition A.56 *(Singular Values)*
Let X and Y be Hilbert spaces and $K : X \to Y$ be a compact operator with adjoint operator $K^* : Y \to X$. The square roots $\mu_j = \sqrt{\lambda_j}$, $j \in J$, of the eigenvalues λ_j of the self-adjoint operator $K^*K : X \to X$ are called singular values *of K. Here again, $J \subset \mathbb{N}$ could be either finite or $J = \mathbb{N}$.*

Note that every eigenvalue λ of K^*K is non-negative because $K^*Kx = \lambda x$ implies that $\lambda (x, x)_X = (K^*Kx, x)_X = (Kx, Kx)_Y \geq 0$; that is, $\lambda \geq 0$.

Theorem A.57 *(Singular Value Decomposition)*
Let $K : X \longrightarrow Y$ be a linear compact operator, $K^* : Y \longrightarrow X$ its adjoint operator, and $\mu_1 \geq \mu_2 \geq \mu_3 \ldots > 0$ the ordered sequence of the positive singular values of K, counted relative to its multiplicity. Then there exist orthonormal systems $\{x_j : j \in J\} \subset X$ and $\{y_j : j \in J\} \subset Y$ with the following properties:

$$
Kx_j = \mu_j y_j \quad \text{and} \quad K^*y_j = \mu_j x_j \quad \text{for all } j \in J.
$$

The system $\{\mu_j, x_j, y_j : j \in J\}$ is called a singular system *for K. Every $x \in X$ possesses the* singular value decomposition

$$
x = x_0 + \sum_{j \in J} (x, x_j)_X x_j
$$

for some $x_0 \in \mathcal{N}(K)$ and

$$
Kx = \sum_{j \in J} \mu_j (x, x_j)_X y_j .
$$

Note that $J \subset \mathbb{N}$ can be finite or infinite; that is, $J = \mathbb{N}$.

The following theorem characterizes the range of a compact operator with the help of a singular system.

Theorem A.58 *(Picard)*
Let $K : X \longrightarrow Y$ be a linear compact operator with singular system $\{\mu_j, x_j, y_j : j \in J\}$. The equation

$$Kx = y \tag{A.45}$$

is solvable if and only if

$$y \in \mathcal{N}(K^*)^{\perp} \quad and \quad \sum_{j \in J} \frac{1}{\mu_j^2} |(y, y_j)_Y|^2 < \infty. \tag{A.46}$$

In this case

$$x = \sum_{j \in J} \frac{1}{\mu_j} (y, y_j)_Y \, x_j$$

is a solution of (A.45).

We note that the solvability conditions (A.46) require a fast decay of the Fourier coefficients of y with respect to the orthonormal system $\{y_j : j \in J\}$ in order for the series

$$\sum_{j=1}^{\infty} \frac{1}{\mu_j^2} |(y, y_j)_Y|^2$$

to converge. Of course, this condition is only necessary for the important case where there exist infinitely many singular values. As a simple example, we study the following integral operator.

Example A.59
Let $K : L^2(0, 1) \longrightarrow L^2(0, 1)$ be defined by

$$(Kx)(t) := \int_0^t x(s) \, ds, \quad t \in (0, 1), \; x \in L^2(0, 1).$$

Then

$$(K^*y)(t) = \int_t^1 y(s) \, ds \quad and \quad (K^*Kx)(t) = \int_t^1 \left(\int_0^s x(\tau) \, d\tau \right) ds.$$

The eigenvalue problem $K^*Kx = \lambda x$ is equivalent to

$$\lambda x(t) = \int_t^1 \left(\int_0^s x(\tau) \, d\tau \right) ds, \quad t \in [0, 1].$$

Differentiating twice, we observe that for $\lambda \neq 0$ this is equivalent to the eigenvalue problem

$$\lambda x'' + x = 0 \text{ in } (0,1), \quad x(1) = x'(0) = 0.$$

Solving this yields

$$x_j(t) = \sqrt{\frac{2}{\pi}} \cos \frac{2j-1}{2} \pi t, \ t \in [0,1], \quad \text{and} \quad \lambda_j = \frac{4}{(2j-1)^2 \pi^2}$$

for $j \in \mathbb{N}$. The singular values μ_j and the ONS $\{y_j : j \in \mathbb{N}\}$ are given by

$$\mu_j = \frac{2}{(2j-1)\pi}, \ j \in \mathbb{N}, \quad \text{and}$$

$$y_j(t) = \sqrt{\frac{2}{\pi}} \sin \frac{2j-1}{2} \pi t, \ j \in \mathbb{N}.$$

The singular value decomposition of Theorem A.57 makes it possible to define, for every continuous function $f : \left[0, \|K\|^2_{\mathcal{L}(X,Y)}\right] \to \mathbb{R}$, the operator $f(K^*K)$ from X into itself by

$$f(K^*K)x = \sum_{j \in J} f(\mu_j^2)\, (x, x_j)_X\, x_j, \quad x \in X. \tag{A.47}$$

This operator is always well-defined, linear, and bounded. It is compact if, and only if, $f(0) = 0$ (see Problem A.5). The special cases $f(t) = t$ and $f(t) = \sqrt{t}$ are of particular importance. From this definition we note that $\mathcal{R}\big(\sqrt{K^*K}\big) = \mathcal{R}(K^*)$.

For the operator K of the previous Example A.59 we note that $x \in \mathcal{R}\big((K^*K)^{\sigma/2}\big)$ if, and only if,

$$\sum_{j=1}^{\infty} \frac{1}{(2j-1)^{2\sigma}} |c_j|^2 < \infty \quad \text{where} \quad c_j = \sqrt{\frac{2}{\pi}} \int_0^1 x(t) \cos \frac{2j-1}{2} \pi t\, dt$$

are the Fourier coefficients of x. Therefore, $\mathcal{R}\big((K^*K)^{\sigma/2}\big)$ plays the role of the periodic Sobolev spaces $H^{\sigma}_{per}(0, 2\pi)$ of Section A.4.

A.7 The Fréchet Derivative

In this section, we briefly recall some of the most important results for nonlinear mappings between normed spaces. The notions of continuity and differentiability carry over in a very natural way.

Definition A.60 *Let X and Y be normed spaces over the field $\mathbb{K} = \mathbb{R}$ or $\mathbb{C}$, $U \subset X$ an open subset, $\hat{x} \in U$, and $T : X \supset U \to Y$ be a (possibly nonlinear) mapping.*

(a) T is called continuous *at $\hat{x}$ if for every $\varepsilon > 0$ there exists $\delta > 0$ such that $\|T(x) - T(\hat{x})\|_Y \leq \varepsilon$ for all $x \in U$ with $\|x - \hat{x}\|_X \leq \delta$.*

(b) T is called Fréchet differentiable *at $\hat{x} \in U$ if there exists a linear bounded operator $A : X \rightarrow Y$ (depending on $\hat{x}$) such that*

$$\lim_{h \to 0} \frac{1}{\|h\|_X} \|T(\hat{x} + h) - T(\hat{x}) - Ah\|_Y = 0. \tag{A.48}$$

We write $T'(\hat{x}) := A$. In particular, $T'(\hat{x}) \in \mathcal{L}(X, Y)$.

(c) The mapping T is called continuously Fréchet differentiable *at $\hat{x} \in U$ if T is Fréchet differentiable in a neighborhood V of $\hat{x}$ and the mapping $T' : V \rightarrow \mathcal{L}(X, Y)$ is continuous in $\hat{x}$.*

Continuity and differentiability of a mapping depend on the norms in X and Y, in contrast to the finite-dimensional case. If T is differentiable in $\hat{x}$, then the linear bounded mapping A in part (b) of Definition A.60 is unique. Therefore, $T'(\hat{x}) := A$ is well-defined. If T is differentiable in x, then T is also continuous in x. In the finite-dimensional case $X = \mathbb{K}^n$ and $Y = \mathbb{K}^m$, the linear bounded mapping $T'(x)$ is given by the Jacobian (with respect to the Cartesian coordinates).

Example A.61 (Integral Operator)
Let $f : [c, d] \times [a, b] \times \mathbb{K} \rightarrow \mathbb{K}$, $f = f(t, s, r)$, $\mathbb{K} = \mathbb{R}$ or $\mathbb{K} = \mathbb{C}$, be continuous with respect to all arguments and also $\partial f / \partial r \in C([c, d] \times [a, b] \times \mathbb{R})$.

(a) Let the mapping $T : C[a, b] \rightarrow C[c, d]$ be defined by

$$T(x)(t) := \int_a^b f(t, s, x(s)) \, ds, \quad t \in [c, d], \; x \in C[a, b]. \tag{A.49}$$

We equip the normed spaces $C[c, d]$ and $C[a, b]$ with the maximum norm. Then T is continuously Fréchet differentiable with derivative

$$(T'(x)z)(t) = \int_a^b \frac{\partial f}{\partial r}(t, s, x(s)) \, z(s) \, ds, \quad t \in [c, d], \; x, z \in C[a, b].$$

Indeed, let $(Az)(t)$ be the term on the right-hand side. By assumption, $\partial f / \partial r$ is continuous on $[a, b] \times [c, d] \times \mathbb{R}$, thus uniformly continuous on the compact set $M = \{(t, s, r) \in [a, b] \times [c, d] \times \mathbb{K} : |r| \leq \|x\|_\infty + 1\}$. Let $\varepsilon > 0$. Choose $\delta \in (0, 1)$ with $\left|\frac{\partial f}{\partial r}(t, s, r) - \frac{\partial f}{\partial r}(t, s, \tilde{r})\right| \leq \frac{\varepsilon}{b-a}$ for all $(t, s, r), (t, s, \tilde{r}) \in M$ with

$|r - \tilde{r}| \leq \delta$. We estimate for $z \in C[a, b]$ with $\|z\|_\infty \leq \delta$:

$$|T(x + z)(t) - T(x)(t) - (Az)(t)|$$

$$= \left| \int_a^b \left[f\big(t, s, x(s) + z(s)\big) - f\big(t, s, x(s)\big) - \frac{\partial f}{\partial r}\big(t, s, x(s)\big) z(s) \right] ds \right|$$

$$= \left| \int_a^b \int_0^1 \left(\frac{d}{dr} f\big(t, s, x(s) + rz(s)\big) - \frac{\partial f}{\partial r}\big(t, s, x(s)\big) z(s) \right) dr \, ds \right|$$

$$\leq \int_a^b \int_0^1 \left| \frac{\partial f}{\partial r}\big(t, s, x(s) + rz(s)\big) - \frac{\partial f}{\partial r}\big(t, s, x(s)\big) \right| |z(s)| \, dr \, ds$$

$$\leq \int_a^b \|z\|_\infty \frac{\varepsilon}{b - a} \, ds \; = \; \varepsilon \|z\|_\infty .$$

This holds for all $t \in [c, d]$, thus also for the maximum. Therefore,

$$\frac{\|T(x + z) - T(x) - Az\|_\infty}{\|z\|_\infty} \leq \varepsilon \quad \text{for all } z \in C[a, b] \text{ with } \|z\|_\infty \leq \delta .$$

This proves the differentiability of T from $C[a, b]$ into $C[c, d]$.

(b) Let now in addition $\partial f / \partial r \in C\big([a, b] \times [c, d] \times \mathbb{K}\big)$ be Lipschitz continuous with respect to r; that is, there exists $\kappa > 0$ with $|\partial f(t, s, r)/\partial r - \partial f(t, s, \tilde{r})| \leq \kappa |r - \tilde{r}|$ for all $(t, s, r), (t, s, \tilde{r}) \in [a, b] \times [c, d] \times \mathbb{K}$. Then the operator T of (A.49) is also Fréchet-differentiable as an operator from $L^2(a, b)$ into $C[c, d]$ (and thus into $L^2(c, d)$) with the same representation of the derivative.

Indeed, define again the operator A as in part (a). Then A is bounded because

$$|(Az)(t)| \;\leq\; \int_a^b \left| \frac{\partial f}{\partial r}\big(t, s, x(s)\big) \right| |z(s)| \, ds$$

$$\leq\; \int_a^b \left[\left| \frac{\partial f}{\partial r}(t, s, 0) \right| + \kappa |x(s)| \right] |z(s)| \, ds$$

$$\leq\; \left[\max_{c \leq \tau \leq d} \|\partial f(\tau, \cdot, 0)/\partial r\|_{L^2(a,b)} + \kappa \|x\|_{L^2(a,b)} \right] \|z\|_{L^2(a,b)}$$

for all $t \in [c, d]$.

Concerning the computation of the derivative we can proceed in the same way as in part (a):

$$|T(x+z)(t) - T(x)(t) - (Az)(t)|$$

$$\leq \int_a^b \int_0^1 \left| \frac{\partial f}{\partial r}(t,s,x(s)+rz(s)) - \frac{\partial f}{\partial r}(t,s,x(s)) \right| |z(s)| \, dr \, ds$$

$$\leq \kappa \int_a^b \int_0^1 r \, |z(s)|^2 \, dr \, ds \;=\; \frac{\kappa}{2} \int_a^b |z(s)|^2 \, ds \;=\; \frac{\kappa}{2} \|z\|_{L^2(a,b)}^2 \,.$$

This holds for all $t \in [c,d]$. Therefore, $\|T(x+z) - T(x) - Az\|_\infty \leq \frac{\kappa}{2} \|z\|_{L^2(a,b)}^2$ which proves the differentiability.

The following theorem collects further properties of the Fréchet derivative.

Theorem A.62   *(a) Let $T, S : X \supset U \to Y$ be Fréchet differentiable at $x \in U$. Then $T + S$ and λT are also Fréchet differentiable for all $\lambda \in \mathbb{K}$ and*

$$(T+S)'(x) \;=\; T'(x) + S'(x) \,, \qquad (\lambda T)'(x) \;=\; \lambda T'(x) \,.$$

(b) Chain rule: Let $T : X \supset U \to V \subset Y$ and $S : Y \supset V \to Z$ be Fréchet differentiable at $x \in U$ and $T(x) \in V$, respectively. Then ST is also Fréchet differentiable at x and

$$(ST)'(x) \;=\; \underbrace{S'\big(T(x)\big)}_{\in \mathcal{L}(Y,Z)} \; \underbrace{T'(x)}_{\in \mathcal{L}(X,Y)} \;\in\; \mathcal{L}(X,Z) \,.$$

(c) Special case: If $\hat{x}, h \in X$ are fixed and $T : X \to Y$ is Fréchet differentiable on X, then $\psi : \mathbb{K} \to Y$, defined by $\psi(t) := T(\hat{x} + th)$, $t \in \mathbb{K}$, is differentiable on $\mathbb{K}$ and $\psi'(t) = T'(\hat{x} + th)h \in Y$. Note that originally $\psi'(t) \in \mathcal{L}(\mathbb{K}, Y)$. In this case, one identifies the linear mapping $\psi'(t) : \mathbb{K} \to Y$ with its generating element $\psi'(t) \in Y$.

If T' is Lipschitz continuous then the following estimate holds.

Lemma A.63 *Let T be differentiable in the ball $B(\overline{x}, \rho)$ centered at $\overline{x} \in U$ with radius $\rho > 0$, and let there exists $\gamma > 0$ with $\|T'(x) - T'(\overline{x})\|_{\mathcal{L}(X,Y)} \leq \gamma \|x - \overline{x}\|_X$ for all $x \in B(\overline{x}, \rho)$. Then*

$$\big\| T(x) - T(\overline{x}) - T'(\overline{x})(x - \overline{x}) \big\|_Y \;\leq\; \frac{\gamma}{2} \|x - \overline{x}\|_X^2 \quad \text{for all } x \in B(\overline{x}, \rho) \,.$$

Proof: Let $\ell \in Y^*$ and $x \in B(\overline{x}, \rho)$ kept fixed. Set $h = x - \overline{x}$ and define the scalar function $f(t) = \langle \ell, T(\overline{x} + th) \rangle$ for $|t| < \rho/\|h\|_X$ where $\langle \cdot, \cdot \rangle = \langle \cdot, \cdot \rangle_{Y^*, Y}$ denotes the dual pairing in $\langle Y^*, Y \rangle$. Then, by the chain rule of the previous Theorem, $f'(t) = \langle \ell, T'(\overline{x} + th)h \rangle$ and thus

$$
\begin{aligned}
& \left| \langle \ell, T(x) - T(\overline{x}) - T'(\overline{x})h \rangle \right| \\
&= \left| \langle \ell, T(x) \rangle - \langle \ell, T(\overline{x}) \rangle - \langle \ell, T'(\overline{x})h \rangle \right| \\
&= \left| f(1) - f(0) - \langle \ell, T'(\overline{x})h \rangle \right| = \left| \int_0^1 [f'(t) - \langle \ell, T'(\overline{x})h \rangle] \, dt \right| \\
&= \left| \int_0^1 [\langle \ell, T'(\overline{x} + th)h \rangle - \langle \ell, T'(\overline{x})h \rangle] \, dt \right| \\
&= \left| \int_0^1 \langle \ell, [T'(\overline{x} + th) - T'(\overline{x})]h \rangle \, dt \right| \\
&\leq \|\ell\|_{Y^*} \|h\|_X \int_0^1 \|T'(\overline{x} + th) - T'(\overline{x})\|_{\mathcal{L}(X,Y)} \, dt \\
&\leq \gamma \|\ell\|_{Y^*} \|h\|_X^2 \int_0^1 t \, dt = \frac{\gamma}{2} \|\ell\|_{Y^*} \|h\|_X^2 .
\end{aligned}
$$

We set $y := T(x) - T(\overline{x}) - T'(\overline{x})h$ and choose $\ell \in Y^*$ with $\|\ell\|_{Y^*} = 1$ and $\langle \ell, y \rangle = \|y\|_Y$. This is possible by a well known consequence of the Hahn-Banach theorem (see [151], Chap.V, §7, Theorem 2). Then the assertion follows.   □

We recall Banach's contraction mapping principle (compare with Theorem A.31 for the linear case).

Theorem A.64 *(Contraction Mapping Principle)*
Let $C \subset X$ be a closed subset of the Banach space X and $T : X \supset C \to X$ a (nonlinear) mapping with the properties

(a) T maps C into itself; that is, $T(x) \in C$ for all $x \in C$, and

(b) T is a contraction on C; that is, there exists $c < 1$ with

$$\|T(x) - T(y)\|_X \leq c \, \|x - y\|_X \quad \text{for all } x, y \in C. \tag{A.50}$$

Then there exists a unique $\tilde{x} \in C$ with $T(\tilde{x}) = \tilde{x}$. The sequence (x_ℓ) in C, defined by $x_{\ell+1} := T(x_\ell)$, $\ell = 0, 1, \ldots$ converges to $\tilde{x}$ for every $x_0 \in C$. Furthermore, the following error estimates hold:

$$\|x_{\ell+1} - \tilde{x}\|_X \leq c \, \|x_\ell - \tilde{x}\|_X, \quad \ell = 0, 1, \ldots ; \tag{A.51a}$$

that is, the sequence converges linearly to $\tilde{x}$,

$$\|x_\ell - \tilde{x}\|_X \leq \frac{c^\ell}{1 - c} \|x_1 - x_0\|_X, \quad \text{(a priori estimate)} \tag{A.51b}$$

$$\|x_\ell - \tilde{x}\|_X \leq \frac{1}{1 - c} \|x_{\ell+1} - x_\ell\|_X, \quad \text{(a posteriori estimate)} \tag{A.51c}$$

for $\ell = 1, 2, \ldots$

The Newton method for systems of nonlinear equations has a direct analogy for equations of the form $T(x) = y$, where $T : X \to Y$ is a continuously Fréchet differentiable mapping between Banach spaces X and Y. We formulate a simplified Newton method and prove local linear convergence. It differs from the ordinary Newton method not only by replacing the derivative $T'(x_\ell)$ by $T'(\hat{x})$ but also by requiring only the existence of a left inverse.

Theorem A.65 *(Simplified Newton Method)*
Let $T : X \to Y$ be continuously Fréchet differentiable between Banach spaces X and Y. Let $V \subset X$ be a closed subspace, $\hat{x} \in V$ and $\hat{y} := T(\hat{x}) \in Y$. Let $L : Y \to V$ be linear and bounded such that L is a left inverse of $T'(\hat{x}) : X \to Y$ on V; that is, $L\,T'(\hat{x})v = v$ for all $v \in V$.

Then there exists $\varepsilon > 0$ such that for any $\bar{y} = T(\bar{x})$ with $\bar{x} \in X$ and $\|\bar{x} - \hat{x}\|_X \leq \varepsilon$ the following algorithm converges linearly to some $\tilde{x} \in V$:

$$x_0 = \hat{x}, \quad x_{\ell+1} = x_\ell - L\big[T(x_\ell) - \bar{y}\big], \quad \ell = 0, 1, 2, \ldots. \qquad (A.52)$$

The limit $\tilde{x} \in V$ satisfies $L[T(\tilde{x}) - \bar{y}] = 0$.

Proof:　　We apply the contraction mapping principle of the preceding theorem to the mapping

$$S(x) := x - L[T(x) - \bar{y}] = L\big[T'(\hat{x})x - T(x) + T(\bar{x})\big]$$

from V into itself on some closed ball $B[\hat{x}, \rho] \subset V$. We estimate

$$
\begin{aligned}
\|S(x) - S(z)\|_X &\leq \|L\|_{\mathcal{L}(Y,X)}\|T'(\hat{x})(x - z) + T(z) - T(x)\|_Y \\
&\leq \|L\|_{\mathcal{L}(Y,X)}\|x - z\|_X \Big\{ \|T'(\hat{x}) - T'(z)\|_{\mathcal{L}(X,Y)} \\
&\quad + \frac{\|T(z) - T(x) + T'(z)(x - z)\|_Y}{\|x - z\|_X} \Big\}
\end{aligned}
$$

and

$$
\begin{aligned}
\|S(x) - \hat{x}\|_X &\leq \|L\|_{\mathcal{L}(Y,X)}\|T'(\hat{x})(x - \hat{x}) - T(x) + T(\bar{x})\|_Y \\
&\leq \|L\|_{\mathcal{L}(Y,X)}\|T'(\hat{x})(x - \hat{x}) + T(\hat{x}) - T(x)\|_Y \\
&\quad + \|L\|_{\mathcal{L}(Y,X)}\|T(\hat{x}) - T(\bar{x})\|_Y
\end{aligned}
$$

First, we choose $\rho > 0$ such that

$$\|L\|_{\mathcal{L}(Y,X)} \left[\|T'(\hat{x}) - T'(z)\|_{\mathcal{L}(X,Y)} + \frac{\|T(z) - T(x) + T'(z)(x - z)\|_Y}{\|x - z\|_X} \right] \leq \frac{1}{2}$$

for all $x, z \in B[\hat{x}, \rho]$. This is possible because T is continuously differentiable. Next, we choose $\varepsilon > 0$ such that

$$\|L\|_{\mathcal{L}(Y,X)}\|T(\hat{x}) - T(\bar{x})\|_Y \leq \frac{\rho}{2}$$

for $\|\bar{x} - \hat{x}\|_X \leq \varepsilon$. Then we conclude that

$$\|S(x) - S(z)\|_X \leq \frac{1}{2}\|x - z\|_X \quad \text{for all } x, z \in B[\hat{x}, \rho],$$

$$\|S(x) - \hat{x}\|_X \leq \frac{1}{2}\|x - \hat{x}\|_X + \frac{1}{2}\rho \leq \rho \quad \text{for all } x \in B[\hat{x}, \rho].$$

Application of the contraction mapping principle ends the proof. □

The notion of *partial derivatives* of mappings $T : X \times Z \to Y$ is introduced just as for functions of two scalar variables as the Fréchet derivative of the mappings $T(\cdot, z) : X \to Y$ for $z \in Z$ and $T(x, \cdot) : Z \to Y$ for $x \in X$. We denote the partial derivatives in $(x, z) \in X \times Z$ by

$$\frac{\partial}{\partial x}T(x, z) \in \mathcal{L}(X, Y) \quad \text{and} \quad \frac{\partial}{\partial z}T(x, z) \in \mathcal{L}(Z, Y).$$

Theorem A.66 *(Implicit Function Theorem)*
Let $T : X \times Z \to Y$ be continuously Fréchet differentiable with partial derivatives $\frac{\partial}{\partial x}T(x, z) \in \mathcal{L}(X, Y)$ and $\frac{\partial}{\partial z}T(x, z) \in \mathcal{L}(Z, Y)$. Furthermore, let $T(\hat{x}, \hat{z}) = 0$ and $\frac{\partial}{\partial z}T(\hat{x}, \hat{z}) : Z \to Y$ be a norm-isomorphism from Z onto Y. Then there exists a neighborhood U of $\hat{x}$ and a Fréchet differentiable function $\psi : U \to Z$ such that $\psi(\hat{x}) = \hat{z}$ and $T(x, \psi(x)) = 0$ for all $x \in U$. The Fréchet derivative $\psi' \in \mathcal{L}(X, Z)$ is given by

$$\psi'(x) = -\left[\frac{\partial}{\partial z}T(x, \psi(x))\right]^{-1}\frac{\partial}{\partial x}T(x, \psi(x)), \quad x \in U.$$

The following special case is particularly important.
Let $Z = Y = \mathbb{K}$; thus $T : X \times \mathbb{K} \to \mathbb{K}$ and $T(\hat{x}, \hat{\lambda}) = 0$ and $\frac{\partial}{\partial \lambda}T(\hat{x}, \hat{\lambda}) \neq 0$. Then there exists a neighborhood U of $\hat{x}$ and a Fréchet differentiable function $\psi : U \to \mathbb{K}$ such that $\psi(\hat{x}) = \hat{\lambda}$ and $T(x, \psi(x)) = 0$ for all $x \in U$ and

$$\psi'(x) = -\frac{1}{\frac{\partial}{\partial \lambda}T(x, \psi(x))}\frac{\partial}{\partial x}T(x, \psi(x)) \in \mathcal{L}(X, \mathbb{K}) = X^*, \quad x \in U,$$

where again X^* denotes the dual space of X.

A.8   Convex Analysis

Definition A.67 *Let X be a normed space.*

(a) *A set $M \subset X$ is called* convex *if for all $x, y \in M$ and all $\lambda \in [0, 1]$ also $\lambda x + (1 - \lambda)y \in M$.*

(b) *Let $M \subset X$ be convex. A function $f : M \to \mathbb{R}$ is called* convex *if for all $x, y \in M$ and all $\lambda \in [0, 1]$*

$$f(\lambda x + (1 - \lambda)y) \leq \lambda f(x) + (1 - \lambda) f(y).$$

(c) A function $f : M \to \mathbb{R}$ is called concave *if $-f$ is convex; that is, if for all $x, y \in M$ and all $\lambda \in [0, 1]$*

$$f\big(\lambda x + (1 - \lambda)y\big) \geq \lambda f(x) + (1 - \lambda) f(y).$$

(d) f is called strictly convex *if*

$$f\big(\lambda x + (1 - \lambda)y\big) < \lambda f(x) + (1 - \lambda) f(y)$$

$x, y \in M$ and all $\lambda \in (0, 1)$ with $x \neq y$. The definition for a strictly concave function is formulated analogously.

The definition of convexity of a set or a function can be extended to more than two elements.

Lemma A.68 *Let X be a normed space.*

(a) A set $M \subset X$ is convex if, and only if, for any elements $x_j \in M$, $j = 1, \ldots, m$, and $\lambda_j \geq 0$ with $\sum_{j=1}^{m} \lambda_j = 1$ also the convex combination $\sum_{j=1}^{m} \lambda_j x_j$ belongs to M.

(b) Let $M \subset X$ be convex. A function $f : M \to \mathbb{R}$ is convex if, and only if,

$$f\left(\sum_{j=1}^{m} \lambda_j x_j\right) \leq \sum_{j=1}^{m} \lambda_j f(x_j)$$

for all $x_j \in M$, $j = 1, \ldots, m$, and $\lambda_j \geq 0$ with $\sum_{j=1}^{m} \lambda_j = 1$. For concave functions the characterization holds analogously.

For any set $A \subset X$ of a normed space A the set

$$\operatorname{conv} A = \left\{\sum_{j=1}^{m} \lambda_j a_j : a_j \in A, \ \lambda_j \geq 0, \ \sum_{j=1}^{m} \lambda_j = 1, \ m \in \mathbb{N}\right\} \qquad (A.53)$$

is convex by the previous lemma and is called the *convex hull* of A

The following separation theorem is one of the most important tools in the area of convex analysis.

Theorem A.69 *Let X be a normed space over $\mathbb{R}$ and $\mathcal{A}, \mathcal{B} \subset X$ two convex sets with $\mathcal{A} \cap \mathcal{B} = \emptyset$. Furthermore, let $\mathcal{A}$ be open. Then there exists a hyperplane which separates $\mathcal{A}$ and $\mathcal{B}$; that is, there exists $\ell \in X^*$ and $\gamma \in \mathbb{R}$ such that $\ell \neq 0$ and*

$$\langle \ell, a \rangle_{X^*, X} \geq \gamma \geq \langle \ell, b \rangle_{X^*, X} \quad \text{for all } a \in \mathcal{A} \text{ and } b \in \mathcal{B}.$$

Here, $\langle \ell, x \rangle_{X^*, X}$ denotes the dual pairing; that is, the application of $\ell \in X^*$ to $x \in X$.

For a proof we refer to, e.g., [139], Chapter II. The hyperplane is given by $\{x \in X : \langle \ell, x \rangle_{X^*, X} = \gamma\}$.

The convexity can be characterized easily for differentiable functions.

Lemma A.70 *Let X be a normed space and $M \subset X$ be an open convex set and $f : M \to \mathbb{R}$ be Fréchet differentiable on M. Then f is convex if, and only if,*

$$f(y) - f(x) - f'(x)(y - x) \geq 0 \quad \text{for all } x, y \in M.$$

f is strictly convex if, and only if, the inequality holds strictly for all $x \neq y$. For concave functions the characterizations hold analogously.

Proof: Let first f be convex, $x, y \in M$ and $t \in (0, 1]$. From the convexity of f we conclude that $f(x + t(y - x)) \leq f(x) + t[f(y) - f(y)]$, thus

$$\begin{aligned}
f(y) - f(x) &\geq \frac{1}{t}[f(x + t(y - x)) - f(x)] \\
&= \frac{1}{t}[f(x + t(y - x)) - f(x) - t\,f'(x)(y - x)] + f'(x)(y - x).
\end{aligned}$$

The first term on the right-hand side tends to zero as $t \to 0$. This proves $f(y) - f(x) \geq f'(x)(y - x)$.

Let now $f(u) - f(v) \geq f'(v)(u - v)$ for all $u, v \in M$. For $x, y \in M$ and $\lambda \in [0, 1]$ apply this twice to $v = \lambda x + (1 - \lambda)y$ and $u = y$ and $u = x$, respectively. With $y - v = -\lambda(x - y)$ and $x - v = (1 - \lambda)(x - y)$ this yields

$$\begin{aligned}
f(y) - f(v) &\geq f'(v)(y - v) = \lambda f'(v)(y - x), \\
f(x) - f(v) &\geq f'(v)(x - v) = (1 - \lambda)f'(v)(x - y).
\end{aligned}$$

Multiplying the first inequality by $1 - \lambda$ and the second by λ and adding these inequalities yields the assertion. □

Note that we could equally well write $\langle f'(x), y - x \rangle_{X^*, X}$ for $f'(x)(y - x)$. We use both notations synonymously.

This characterization motivates the definition of the subgradient of a convex function.

Definition A.71 *Let X be a normed space over $\mathbb{R}$ with dual X^*, $M \subset X$ be an open convex set, and $f : M \to \mathbb{R}$ be a convex function. For $x \in M$ the set*

$$\partial f(x) := \{\ell \in X^* : f(z) - f(x) - \langle \ell, z - x \rangle_{X^*, X} \geq 0 \text{ for all } z \in M\}$$

is called the subgradient of f at x.

As one sees from the function $f(x) = |x|$ for $x \in \mathbb{R}$ the subgradient ∂f is, in general, a multivalued function. It is not empty for continuous functions, and it consists of the derivative as the only element for differentiable functions.

Lemma A.72 *Let X be a normed space over $\mathbb{R}$ with dual X^*, $M \subset X$ be an open convex set, and $f : M \to \mathbb{R}$ be a convex and continuous function.*

(a) Then $\partial f(x) \neq \emptyset$ for all $x \in M$.

(b) If f is Fréchet differentiable at $x \in M$ then $\partial f(x) = \{f'(x)\}$. In particular, ∂f is single valued at x.

Proof: (a) Define the set $D \subset X \times \mathbb{R}$ by

$$D := \{(z,r) \in M \times \mathbb{R} : r > f(z)\}.$$

Then D is open because M is open and f is continuous and D is also convex (see Problem A.6). Fix $x \in M$. Then we observe that $(x, f(x)) \notin D$. The separation theorem for convex sets (see Theorem A.69) yields the existence of $(\ell, s) \in X^* \times \mathbb{R}$ with $(\ell, s) \neq (0,0)$ and $\gamma \in \mathbb{R}$ such that

$$\langle \ell, z \rangle_{X^*,X} + sr \leq \gamma \leq \langle \ell, x \rangle_{X^*,X} + s f(x) \quad \text{for all } r > f(z), \, z \in M.$$

Letting r tend to $+\infty$ implies that $s \leq 0$. Also $s \neq 0$ because otherwise $\langle \ell, z \rangle_{X^*,X} \leq \langle \ell, x \rangle_{X^*,X}$ for all $z \in M$ which would imply that also ℓ vanishes[2], a contradiction to $(\ell, s) \neq (0,0)$. Therefore, $s < 0$ and, without loss of generality (division by $|s|$), we can assume that $s = -1$. Letting r tend to $f(z)$ yields $\langle \ell, z \rangle_{X^*,X} - f(z) \leq \langle \ell, x \rangle_{X^*,X} - f(x)$ which shows that $\ell \in \partial f(x)$.

(b) $f'(x) \in \partial f(x)$ follows from Lemma A.70. Let $\ell \in \partial f(x)$ and $y \in X$ arbitrary with $y \neq 0$. For sufficiently small $t > 0$ we have that $x + ty \in M$ and thus

$$f(x + ty) - f(y) - tf'(x)y \geq t\left(\ell - f'(x)\right)y.$$

Division by $t > 0$ and letting t tend to zero implies that the left-hand side tends to zero by the definition of the derivative. Therefore, $(\ell - f'(x))y \leq 0$. Since this holds for all $y \in X$ we conclude that $\ell = f'(x)$. $\quad\square$

Lemma A.73 *Let $\psi : [0, a] \to \mathbb{R}$ be a continuous, concave, and monotonically increasing function with $\psi(0) = 0$.*

(a) Then $\psi(st) \leq \max\{1, s\} \, \psi(t)$ for all $s, t \geq 0$ with $st, t \in [0, a]$.

(b) The function $t \mapsto \left[\psi(\sqrt{t})\right]^2$ is concave on $[0, a^2]$.

(c) Let $K : X \to Y$ be a linear compact operator between Hilbert spaces and $a \geq \|K\|_{\mathcal{L}(X,Y)}$. Then

$$\left\|\psi\big((K^*K)^{1/2}\big)z\right\|_X \leq \psi\big(\|Kz\|_Y\big) \quad \text{for all } z \in X \text{ with } \|z\|_X \leq 1.$$

---

[2]The reader should verify this himself by using that M is an open set.

Proof: (a) If $s \leq 1$ the assertion follows from the monotonicity of ψ. If $s \geq 1$ then

$$\psi(t) = \psi\left(\frac{1}{s}(st) + \left(1 - \frac{1}{s}\right)0\right) \geq \frac{1}{s}\psi(st) + \left(1 - \frac{1}{s}\right)\psi(0) = \frac{1}{s}\psi(st).$$

(b) Set $\phi(t) = [\psi(\sqrt{t})]^2$ for $t \in [0, a^2]$. If ψ was twice differentiable then an elementary calculation shows that

$$\phi''(t) = \frac{\psi'(\sqrt{t})}{2t^{3/2}}\left[\sqrt{t}\,\psi'(\sqrt{t}) - \psi(\sqrt{t})\right] + \frac{1}{2t}\psi(\sqrt{t})\,\psi''(\sqrt{t}).$$

The first term is non-positive because $\psi' \geq 0$ and $s\,\psi'(s) - \psi(s) = \psi(0) - \psi(s) - (0-s)\psi'(s) \leq 0$ by Lemma A.70. The second term is also non-positive because $\psi \geq 0$ and $\psi'' \leq 0$. Therefore, $\phi''(t) \leq 0$ for all t which proves that ϕ is concave. If ψ is merely continuous then we approximate ψ by a sequence (ψ_ℓ) of smooth concave and monotonically increasing functions with $\psi_\ell(0) = 0$ and $\|\psi_\ell - \psi\|_\infty \to 0$ as ℓ tends to infinity. We sketch the proof but leave the details to the reader.

In the first step we approximate ψ by the interpolating polygonal function p_m with respect to $t_j = j\frac{a}{m}$, $j = 0, \ldots, m$, with values $\psi_j := \psi(t_j)$ at t_j. Then, for any $\ell \in \mathbb{N}$ there exists $m = m(\ell) \in \mathbb{N}$ with $\|p_m - \psi\|_\infty \leq 1/\ell$. Next, we extend p_m onto $\mathbb{R}$ by extending the first and the last segment linearly; that is, by setting $p_m(t) = \psi_1\frac{t}{t_1}$ for $t \leq 0$ and $p_m(t) = \psi(a) + \frac{\psi(a) - \psi_{m-1}}{a - t_{m-1}}(t - a)$ for $t \geq a$.

In the third step we smooth p_m by using a mollifier; that is, a non-negative function $\phi \in C^\infty(\mathbb{R})$ with $\phi(t) = 0$ for $|t| \geq 1$ and $\int_{-1}^1 \phi(t)dt = 1$. We set $\phi_\rho(t) = \frac{1}{\rho}\phi\left(\frac{t}{\rho}\right)$ and

$$\tilde{\psi}_\rho(t) = \int_{-\infty}^{\infty} \phi_\rho(t-s)\,p_m(s)\,ds = \int_{-\rho}^{\rho} \phi_\rho(s)\,p_m(t-s)\,ds, \quad t \in [0, a].$$

Then $\tilde{\psi}_\rho$ is in $C^\infty[0, a]$, concave, monotonically increasing and $\|\tilde{\psi}_\rho - p_m\|_{C[0,a]} \to 0$ as ρ tends to zero. Finally we set $\psi_\ell(t) = \tilde{\psi}_\rho(t) - \tilde{\psi}_\rho(0)$, where $\rho = \rho(\ell) > 0$ is such that $\|\psi_\ell - p_m\|_{C(0,a)} \leq 1/\ell$. This sketches the construction of the sequence ψ_ℓ.

From the first part we know that $t \mapsto [\psi_\ell(\sqrt{t})]^2$ is concave for every ℓ. Letting ℓ tend to infinity proves the assertion.

(c) We set again $\phi(t) = [\psi(\sqrt{t})]^2$ for $t \in [0, \|K\|^2_{\mathcal{L}(X,Y)}]$ and let $z \in X$ with $\|z\|_X \leq 1$. We decompose z in the form $z = z_0 + z^\perp$ with $z_0 \in \mathcal{N}(K)$ and $z^\perp \perp \mathcal{N}(K)$. Then $\psi((K^*K)^{1/2})z = \psi((K^*K)^{1/2})z^\perp$ and $Kz = Kz^\perp$. Therefore, it suffices to take $z \in X$ with $z \perp \mathcal{N}(K)$ and $\|z\|_X \leq 1$. With a singular system $\{\mu_j, x_j, y_j : j \in J\}$ of K we expand such a $z \in X$ as $z = \sum_{j \in J} z_j x_j$. We set $J_n = J$ if J is finite and $J_n = \{1, \ldots, n\}$ if $J = \mathbb{N}$ and $\hat{z}^{(n)} = \sum_{j \in J_n} \hat{z}_j x_j$ with

$\hat{z}_j = z_j / \sqrt{\sum_{\ell \in J_n} |z_\ell|^2}$. Then $\|\hat{z}^{(n)}\|_X^2 = \sum_{j \in J_n} |\hat{z}_j|^2 = 1$ and thus

$$
\begin{aligned}
\left\| \psi\big((K^*K)^{1/2}\big) \hat{z}^{(n)} \right\|_X^2 &= \sum_{j \in J_n} \big[\psi(\mu_j)\big]^2 |\hat{z}_j|^2 = \sum_{j \in J_n} \phi(\mu_j^2) |\hat{z}_j|^2 \\
&\leq \phi\bigg(\sum_{j \in J_n} \mu_j^2 |\hat{z}_j|^2 \bigg) = \left[\psi\bigg(\sqrt{\sum_{j \in J_n} \mu_j^2 |\hat{z}_j|^2} \bigg) \right]^2 \\
&= \Big[\psi\big(\|K\hat{z}^{(n)}\|_Y\big) \Big]^2
\end{aligned}
$$

where we used that ϕ is concave. Letting n tend to infinity yields $\hat{z}^{(n)} \to \hat{z} = z/\|z\|_X$ and thus $\left\| \psi\big((K^*K)^{1/2}\big) \hat{z} \right\|_X \leq \psi\big(\|K\hat{z}\|_Y\big)$. Therefore,

$$
\begin{aligned}
\left\| \psi\big((K^*K)^{1/2}\big) z \right\|_X &= \|z\|_X \left\| \psi\big((K^*K)^{1/2}\big) \hat{z} \right\|_X \\
&\leq \|z\|_X \, \psi\big(\|K\hat{z}\|_Y\big) = \|z\|_X \, \psi\bigg(\frac{1}{\|z\|_X} \|Kz\|_Y \bigg) \\
&\leq \psi\big(\|Kz\|_Y\big)
\end{aligned}
$$

where we used the estimate $\psi(st) \leq s\,\psi(t)$ from part (a) for $s = 1/\|z\|_X \geq 1$ and $t = \|Kz\|_Y$. $\qquad\square$

Lemma A.74 *There exist constants $c_+ > 0$ and $c_p \geq 0$ with $c_p = 0$ for $p \leq 2$ and $c_p > 0$ for $p > 2$ such that for all $z \in \mathbb{R}$*

$$
c_p |z|^p \leq |1 + z|^p - 1 - pz \leq
\begin{cases}
c_+ |z|^p & \text{if } p \leq 2 \text{ or } |z| \geq \frac{1}{2}, \\
c_+ |z|^2 & \text{if } p > 2 \text{ and } |z| \leq \frac{1}{2}.
\end{cases}
\tag{A.54}
$$

Proof: To show the upper estimate, let first $|z| \geq \frac{1}{2}$. Then $1 \leq 2|z|$ and thus

$$
|1+z|^p - 1 - pz \leq (1+|z|)^p + p \cdot 1 \cdot |z| \leq (3|z|)^p + p\,2^{p-1} |z|^{p-1}|z| = \big(3^p + p\,2^{p-1}\big)|z|^p.
$$

Let now $|z| \leq \frac{1}{2}$; that is, $1 \geq 2|z|$. Then $|1+z|^p - 1 - pz = (1+z)^p - 1 - pz := f(z)$. We compute $f'(z) = p\big[(1+z)^{p-1} - 1\big]$ and $f''(z) = p(p-1)(1+z)^{p-2}$. Taylor's formula yields

$$
(1+z)^p - 1 - pz = f(z) = f(0) + f'(0)z + \frac{1}{2} f''(\xi)z^2 = \frac{p(p-1)}{2}(1+\xi)^{p-2} z^2
$$

for some ξ with $|\xi| \leq |z|$.
Let first $p \geq 2$. Then $(1+\xi)^{p-2} \leq (3/2)^{p-2}$, and the estimate is shown.
Let now $p < 2$. Then $(1+\xi)^{p-2} = \frac{1}{(1+\xi)^{2-p}} \leq \frac{1}{|z|^{2-p}} = |z|^{p-2}$ because $1 + \xi \geq 1 - |\xi| \geq 2|z| - |z| = |z|$. Therefore, $f(z) \leq \frac{p(p-1)}{2}|z|^{p-2}z^2 = \frac{p(p-1)}{2}|z|^p$, and the upper estimate of (A.54) is shown.

To show the lower estimate, we choose $c_p \in (0,1)$ for $p > 2$ such that $c_p^{(p-1)/(p-2)} - c_p + \big(1 - c_p^{1/(p-2)}\big)^{p-1} \geq 0$ and set $c_p = 0$ for $p \leq 2$. We distinguish between several cases.

Case A: $z \geq 0$. Set $f(z) := |1+z|^p - 1 - pz - c_p|z|^p = (1+z)^p - 1 - pz - c_p z^p$ for $z \geq 0$. Then $f(0) = 0$ and $f'(z) = p[(1+z)^{p-1} - 1 - c_p z^{p-1}]$ and $f'(0) = 0$ and $f''(z) = p(p-1)[(1+z)^{p-2} - c_p z^{p-2}] \geq p(p-1)[z^{p-2} - c_p z^{p-2}] \geq 0$ because $c_p \leq 1$. Therefore, f' is monotonically increasing, thus positive. Therefore, f is monotonically increasing, thus positive.

Case B: $z \leq 0$. We replace z by $-z \geq 0$ and have to show that $f(z) := |1 - z|^p - 1 + pz - c_p z^p \geq 0$ for all $z \geq 0$.

Case B1: $0 \leq z \leq 1$. Then $f(z) = (1-z)^p - 1 + pz - c_p z^p$ and $f(0) = 0$ and $f'(z) = p[1 - (1-z)^{p-1} - c_p z^{p-1}]$ and $f'(0) = 0$ and $f'(1) = p(1 - c_p) > 0$.
If $p \leq 2$ then $c_p = 0$ and thus $f' \geq 0$, thus $f \geq 0$ on $[0, 1]$.
Let $p > 2$. Then $f''(z) = p(p-1)[(1-z)^{p-2} - c_p z^{p-2}]$ and $f''(0) > 0$ and $f''(1) < 0$. Since $f''' < 0$ on $(0, 1)$ there exists exactly one zero $\hat{z} \in (0, 1)$ of f''. Therefore, f' increases on $[0, \hat{z}]$ and decreases on $[\hat{z}, 1]$. Therefore, the minimal values of f' on $[0, 1]$ are obtained for $z = 0$ or $z = 1$ which are both non-negative. Therefore, $f' \geq 0$ on $[0, 1]$ which implies that also f is non-negative.

Case B2: $z \geq 1$. Then $f(z) = (z-1)^p - 1 + pz - c_p z^p$ and $f(1) = -1 + p - c_p > 0$ and $f'(z) = p[(z-1)^{p-1} + 1 - c_p z^{p-1}]$ and $f'(1) = p(1 - c_p) > 0$.
If $p \leq 2$ we conclude that $f'' \geq 0$ on $[1, \infty)$, thus $f' \geq 0$ on $[1, \infty)$ and thus also $f \geq 0$ on $[1, \infty)$.
Let finally $p > 2$. Then $f''(z) = p(p-1)[(z-1)^{p-2} - c_p z^{p-2}]$ and $f''(1) < 0$ and $f''(z) \to \infty$ as $z \to \infty$. The second derivative f'' vanishes at $\hat{z} = [1 - c_p^{1/(p-2)}]^{-1} > 1$, which is negative on $[1, \hat{z})$ and positive for $z > \hat{z}$. Therefore, $f'(z)$ decreases on $(1, \hat{z})$ and increases for $z > \hat{z}$. Its minimal value is attained at $\hat{z}$ which is computed as

$$\begin{aligned} f'(\hat{z}) &= p[(\hat{z}-1)^{p-1} - c_p \hat{z}^{p-1} + 1] \\ &= \frac{p}{(1 - c_p^{1/(p-2)})^{p-1}} [c_p^{(p-1)/(p-2)} - c_p + (1 - c_p^{1/(p-2)})^{p-1}] \geq 0 \end{aligned}$$

by the choice of c_p. Therefore, f' is positive for $z \geq 1$ and thus also f. $\square$

A.9 Weak Topologies

In this subsection, we recall the basics on weak topologies. We avoid the definition of the topology itself; that is, the definition of the family of open sets, but restrict ourselves to the concept of weak convergence (or weak∗ convergence).

Definition A.75 *Let X be a normed space and X^* its dual space with dual pairing $\langle \ell, x \rangle_{X^*, X}$ for $\ell \in X^*$ and $x \in X$.*

(a) A sequence (x_n) in X is said to converge weakly *to $x \in X$ if $\lim_{n \to \infty} \langle \ell, x_n \rangle_{X^*, X} = \langle \ell, x \rangle_{X^*, X}$ for all $\ell \in X^*$. We write $x_n \rightharpoonup x$ for the weak convergence.*

(b) A sequence (ℓ_n) in X^* is said to converge weak* to $\ell \in X^*$ if
$$\lim_{n\to\infty} \langle \ell_n, x \rangle_{X^*,X} = \langle \ell, x \rangle_{X^*,X} \text{ for all } x \in X. \text{ We write } \ell_n \xrightarrow{*} \ell \text{ for the}$$
weak* convergence.

First we note that weak convergence is indeed weaker than norm convergence. This follows directly from the continuity of the functional $\ell \in X^*$. Furthermore, we note that weak convergence in X^* means that $\langle T, \ell_n \rangle_{X^{**},X^*} \to \langle T, \ell \rangle_{X^{**},X^*}$ as $n \to \infty$ for all $T \in X^{**}$ where now $\langle \cdot, \cdot \rangle_{X^{**},X^*}$ denotes the dual pairing in (X^{**}, X^*). By the canonical imbedding $X \hookrightarrow X^{**}$ (see formula (A.14)) weak convergence in X^* is stronger than weak* convergence. In reflexive spaces (see Definition A.22) the bidual X^{**} can be identified with X and, therefore, weak and weak* convergence coincide. If X is a Hilbert space with inner product $(\cdot, \cdot)_X$ then, by the representation Theorem A.23 of Riesz, the dual space X^* is identified with X itself, and weak convergence $x_n \rightharpoonup x$ is equivalent to $(x_n, z)_X \to (x, z)_X$ for all $z \in X$.

We will need the following results which we cite without proof.

Theorem A.76 *Let X be a normed space and X^* its dual space and (x_n) a sequence in X which converges weakly to some $x \in X$. Then the following holds:*

(a) The weak limit is well-defined; that is, if $x_n \rightharpoonup x$ and $x_n \rightharpoonup y$ then $x = y$.

(b) The sequence (x_n) is bounded in norm; that is, $\|x_n\|_X$ is bounded in $\mathbb{R}$.

(c) If X is finite dimensional then (x_n) converges to x in norm. Therefore, in finite-dimensional spaces there is no difference between weak and norm convergence.

(d) Let $U \subset X$ be a convex and closed set. Then U is also weakly sequentially closed; that is, every weak limit point $x \in X$ of a weakly convergent sequence (x_n) in U also belongs to U.

(e) Let Y be another normed space and $K : X \to Y$ be a linear bounded operator. Then (Kx_n) converges weakly to Kx in Y. If K is compact then (Kx_n) converges in norm to Kx.

The following theorem of Alaoglu–Bourbaki (see, e.g., [139]) is the essential ingredient to assure the existence of global minima of the Tikhonov functional for nonlinear inverse problems.

Theorem A.77 *Let X be a Banach space with dual space X^*. Then every bounded sequence (ℓ_n) in X^* has a weak* convergent subsequence.*

For Hilbert spaces the theorem takes the following form.

Corollary A.78 *Let X be a Hilbert space. Every bounded sequence (x_n) in X contains a weak accumulation point; that is, a weakly convergent subsequence.*

A.10   Problems

A.1 (a) Show with Definition A.5 that a set $M \subset X$ is closed if, and only if, the limit of every convergent sequence $(x_k)_k$ in M also belongs to M.

(b) Show that a subspace $V \subset X$ is dense if, and only if, the orthogonal complement is trivial; that is, $V^\perp = \{0\}$.

A.2 Try to prove Theorem A.8.

A.3 Let $V \subset X \subset V'$ be a Gelfand triple (see Definition A.26) and $j : V \hookrightarrow X$ and $j' : X \hookrightarrow V'$ the corresponding embedding operators. Show that both of them are one-to-one with dense range.

A.4 Define the operator A from the sequence space ℓ^2 into itself by

$$(Ax)_k := \begin{cases} 0, & \text{if} \quad k = 1, \\ x_{k-1}, & \text{if} \quad k \geq 2, \end{cases}$$

for $x = (x_k) \in \ell^2$. Show that $\lambda = 1$ is not an eigenvalue of A but $I - A$ fails to be surjective. Therefore, 1 belongs to the spectrum of A.

A.5 Let $K : X \to Y$ a compact operator between Hilbert spaces X and Y, and let $f : [0, \|K\|^2_{\mathcal{L}(X,Y)}] \to \mathbb{R}$ be continuous. Define the operator $f(K^*K)$ from X into itself by (A.47). Show that this operator is well defined (that is, $f(K^*K)x$ defines an element in X for every $x \in X$), linear, and bounded and that it is compact if, and only if, $f(0) = 0$.

A.6 Let $A \subset X$ be an open set of a Hilbert space X and $f : A \to \mathbb{R}$ convex and continuous. Show that the set $D \subset X \times \mathbb{R}$, defined by

$$D := \{(z, r) \in A \times \mathbb{R} : r > f(z)\}.$$

is convex and open in $X \times \mathbb{R}$.

Appendix B

Proofs of the Results of Section 2.7

In this appendix, we give the complete proofs of the theorems and lemmas of Chapter 2, Section 2.7. For the convenience of the reader, we formulate the results again.

Theorem 2.20 *(Fletcher–Reeves)*
Let $K : X \to Y$ be a bounded, linear, and injective operator between Hilbert spaces X and Y. The conjugate gradient method is well-defined and either stops or produces sequences (x^m), (p^m) in X with the properties

$$\left(\nabla f(x^m), \nabla f(x^j)\right)_X \;=\; 0 \quad \text{for all } j \neq m, \tag{2.41a}$$

and

$$(Kp^m, Kp^j)_Y \;=\; 0 \quad \text{for all } j \neq m; \tag{2.41b}$$

that is, the gradients are orthogonal and the directions p^m are K-conjugate. Furthermore,

$$\left(\nabla f(x^j), K^*Kp^m\right)_X \;=\; 0 \quad \text{for all } j < m. \tag{2.41c}$$

Proof: First, we note the following identities:

(α) $\nabla f(x^{m+1}) \;=\; 2\,K^*(Kx^{m+1} - y) \;=\; 2\,K^*(Kx^m - y) - 2t_m K^*Kp^m \;=\; \nabla f(x^m) - 2t_m K^*Kp^m$.

(β) $\left(p^m, \nabla f(x^{m+1})\right)_X \;=\; \left(p^m, \nabla f(x^m)\right)_X - 2t_m(Kp^m, Kp^m)_Y \;=\; 0$ by the definition of t_m.

(γ) $t_m = \frac{1}{2}\left(\nabla f(x^m), p^m\right)_X \|Kp^m\|_Y^{-2} = \frac{1}{4}\|\nabla f(x^m)\|_X^2 / \|Kp^m\|_Y^2$ since $p^m = \frac{1}{2}\nabla f(x^m) + \gamma_{m-1}p^{m-1}$ and (β).

Now we prove the following identities by induction with respect to m:

© Springer Nature Switzerland AG 2021
A. Kirsch, *An Introduction to the Mathematical Theory of Inverse Problems*,
Applied Mathematical Sciences 120,
https://doi.org/10.1007/978-3-030-63343-1

(i) $\left(\nabla f(x^m), \nabla f(x^j)\right)_X = 0$ for $j = 0, \ldots, m - 1$,

(ii) $(Kp^m, Kp^j)_Y = 0$ for $j = 0, \ldots, m - 1$.

Let $m = 1$. Then, using (α),

(i) $\quad \left(\nabla f(x^1), \nabla f(x^0)\right)_X = \|\nabla f(x^0)\|_X^2 - 2t_0\left(Kp^0, K\nabla f(x^0)\right)_Y = 0,$

which vanishes by (γ) since $p^0 = \frac{1}{2}\nabla f(x^0)$.

(ii) By the definition of p^1 and identity (α), we conclude that

$$
\begin{aligned}
(Kp^1, Kp^0)_Y &= (p^1, K^*Kp^0)_X \\[2mm]
&= -\frac{1}{2t_0}\left[\frac{1}{2}\left(\nabla f(x^1) + \gamma_0 p^0, \nabla f(x^1) - \nabla f(x^0)\right)_X\right] \\[2mm]
&= -\frac{1}{2t_0}\left[\frac{1}{2}\|\nabla f(x^1)\|_X^2 - \gamma_0\left(p^0, \nabla f(x^0)\right)_X\right] = 0,
\end{aligned}
$$

where we have used (β), the definition of p^0, and the choice of γ_0.

Now we assume the validity of (i) and (ii) for m and show it for $m + 1$:

(i) For $j = 0, \ldots m - 1$ we conclude that (setting $\gamma_{-1} = 0$ in the case $j = 0$)

$$
\begin{aligned}
\left(\nabla f(x^{m+1}), \nabla f(x^j)\right)_X &= \left(\nabla f(x^m) - 2t_m K^*Kp^m, \nabla f(x^j)\right)_X \\[2mm]
&= -2t_m\left(\nabla f(x^j), K^*Kp^m\right)_X \\[2mm]
&= -4t_m\left(Kp^j - \gamma_{j-1}Kp^{j-1}, Kp^m\right)_Y = 0,
\end{aligned}
$$

where we have used $\frac{1}{2}\nabla f(x^j) + \gamma_{j-1}p^{j-1} = p^j$ and assertion (ii) for m.

For $j = m$, we conclude that

$$
\begin{aligned}
&\left(\nabla f(x^{m+1}), \nabla f(x^m)\right)_X \\[2mm]
&= \|\nabla f(x^m)\|_X^2 - 2t_m\left(\nabla f(x^m), K^*Kp^m\right)_X \\[2mm]
&= \|\nabla f(x^m)\|_X^2 - \frac{1}{2}\frac{\|\nabla f(x^m)\|_X^2}{\|Kp^m\|_Y^2}\left(\nabla f(x^m), K^*Kp^m\right)_X
\end{aligned}
$$

by (γ). Now we write

$$
\begin{aligned}
(Kp^m, Kp^m)_Y &= \left(Kp^m, K\left(\frac{1}{2}\nabla f(x^m) + \gamma_{m-1}p^{m-1}\right)\right)_Y \\[2mm]
&= \frac{1}{2}(Kp^m, K\nabla f(x^m))_Y,
\end{aligned}
$$

which implies that $\left(\nabla f(x^{m+1}), \nabla f(x^m)\right)_X$ vanishes.

(ii) For $j = 0, \ldots, m - 1$, we conclude that, using (α),

$$
\begin{aligned}
(Kp^{m+1}, Kp^j)_Y &= \left(\frac{1}{2}\nabla f(x^{m+1}) + \gamma_m p^m, K^*Kp^j\right)_X \\[2mm]
&= -\frac{1}{4t_j}\left(\nabla f(x^{m+1}), \nabla f(x^{j+1}) - \nabla f(x^j)\right)_X,
\end{aligned}
$$

which vanishes by (i).

For $j = m$ by (α) and the definition of p^{m+1}, we have

$$
\begin{aligned}
(Kp^{m+1}, Kp^m)_Y &= \frac{1}{2t_m} \left(\frac{1}{2} \nabla f(x^{m+1}) + \gamma_m p^m, \nabla f(x^m) - \nabla f(x^{m+1}) \right)_X \\
&= \frac{1}{2t_m} \left\{ \frac{1}{2} (\nabla f(x^{m+1}), \nabla f(x^m))_X - \frac{1}{2} \|\nabla f(x^{m+1})\|_X^2 \right. \\
&\qquad\qquad \left. + \gamma_m \underbrace{(p^m, \nabla f(x^m))_X}_{=\frac{1}{2}\|\nabla f(x^m)\|_X^2} - \gamma_m (p^m, \nabla f(x^{m+1}))_X \right\} \\
&= \frac{1}{4t_m} \left\{ \gamma_m \|\nabla f(x^m)\|_X^2 - \|\nabla f(x^{m+1})\|_X^2 \right\}
\end{aligned}
$$

by (i) and (β). This term vanishes by the definition of γ_m. Thus we have proven (i) and (ii) for $m + 1$ and thus for all $m = 1, 2, 3 \ldots$ To prove (2.41c) we write

$$
\left(\nabla f(x^j), K^* K p^m \right)_X = 2 \left(p^j - \gamma_{j-1} p^{j-1}, K^* K p^m \right)_X = 0 \quad \text{for } j < m,
$$

and note that we have already shown this in the proof. $\qquad\square$

Theorem 2.21 *Let (x^m) and (p^m) be the sequences of the conjugate gradient method. Define the space $V_m := \operatorname{span} \{p^0, \ldots, p^m\}$. Then we have the following equivalent characterizations of V_m:*

$$
\begin{aligned}
V_m &= \operatorname{span} \{\nabla f(x^0), \ldots, \nabla f(x^m)\} & (2.42a) \\
&= \operatorname{span} \{p^0, K^* K p^0, \ldots, (K^* K)^m p^0\} & (2.42b)
\end{aligned}
$$

for $m = 0, 1, \ldots$. Furthermore, x^m is the minimum of $f(x) = \|Kx - y\|_Y^2$ on V_{m-1} for every $m \geq 1$.

Proof: Let $\tilde{V}_m = \operatorname{span} \{\nabla f(x^0), \ldots, \nabla f(x^m)\}$. Then $V_0 = \tilde{V}_0$. Assume that we have already shown that $V_m = \tilde{V}_m$. Since $p^{m+1} = \frac{1}{2} \nabla f(x^{m+1}) + \gamma_m p^m$ we also have that $V_{m+1} = \tilde{V}_{m+1}$. Analogously, define the space $\hat{V}_m := \operatorname{span} \{p^0, \ldots, (K^* K)^m p^0\}$. Then $V_0 = \hat{V}_0$ and $V_1 = \hat{V}_1$. Assume that we have already shown that $V_j = \hat{V}_j$ for all $j = 0, \ldots, m$. Then we conclude that

$$
\begin{aligned}
p^{m+1} &= K^*(Kx^{m+1} - y) + \gamma_m p^m \\
&= K^*(Kx^m - y) - t_m K^* K p^m + \gamma_m p^m \\
&= p^m - \gamma_{m-1} p^{m-1} - t_m K^* K p^m + \gamma_m p^m \in \hat{V}_{m+1}. \qquad (B1)
\end{aligned}
$$

On the other hand, from

$$
(K^* K)^{m+1} p^0 = (K^* K) \left[(K^* K)^m p^0 \right] \in (K^* K)(V_m)
$$

and $K^* K p^j \in V_{j+1} \subset V_{m+1}$ by (B1) for $j = 0, \ldots, m$, we conclude also that $V_{m+1} = \hat{V}_{m+1}$.

Now every x^m lies in V_{m-1}. This is certainly true for $m = 1$, and if it holds for m then it holds also for $m + 1$ since $x^{m+1} = x^m - t_m p^m \in V_m$. x^m is the minimum of f on V_{m-1} if and only if $(Kx^m - y, Kz)_Y = 0$ for all $z \in V_{m-1}$. By (2.42a), this is the case if and only if $(\nabla f(x^m), \nabla f(x^j))_X = 0$ for all $j = 0, \ldots, m - 1$. This holds by the preceding theorem. $\quad\square$

Lemma 2.22 *(a) The polynomial $\mathbb{Q}_m$, defined by $\mathbb{Q}_m(t) = 1 - t\,\mathbb{P}_{m-1}(t)$ with $\mathbb{P}_{m-1}$ from (2.43), minimizes the functional*

$$H(\mathbb{Q}) := \|\mathbb{Q}(KK^*)y\|_Y^2 \quad \text{on} \quad \{\mathbb{Q} \in \mathcal{P}_m : \mathbb{Q}(0) = 1\}$$

and satisfies

$$H(\mathbb{Q}_m) = \|Kx^m - y\|_Y^2.$$

(b) For $k \neq \ell$, the following orthogonality relation holds:

$$\langle \mathbb{Q}_k, \mathbb{Q}_\ell \rangle := \sum_{j=1}^\infty \mu_j^2 \, \mathbb{Q}_k(\mu_j^2)\, \mathbb{Q}_\ell(\mu_j^2)\, |(y, y_j)_Y|^2 = 0. \tag{2.44}$$

If $y \notin \text{span}\{y_1, \ldots, y_N\}$ for any $N \in \mathbb{N}$, then $\langle \cdot, \cdot \rangle$ defines an inner product on the space $\mathcal{P}$ of all polynomials.

Proof: (a) Let $\mathbb{Q} \in \mathcal{P}_m$ be an arbitrary polynomial with $\mathbb{Q}(0) = 1$. Set $\mathbb{P}(t) := (1 - \mathbb{Q}(t))/t$ and $x := \mathbb{P}(K^*K)K^*y = -\mathbb{P}(K^*K)p^0 \in V_{m-1}$. Then

$$y - Kx = y - K\mathbb{P}(K^*K)K^*y = \mathbb{Q}(KK^*)y.$$

Thus

$$H(\mathbb{Q}) = \|Kx - y\|_Y^2 \geq \|Kx^m - y\|_Y^2 = H(\mathbb{Q}_m).$$

(b) Let $k \neq \ell$. From the identity

$$\frac{1}{2}\nabla f(x^k) = K^*(Kx^k - y) = -\sum_{j=1}^\infty \mu_j \mathbb{Q}_k(\mu_j^2)(y, y_j)_Y\, x_j,$$

we conclude that

$$0 = \frac{1}{4}\big(\nabla f(x^k), \nabla f(x^\ell)\big)_X = \langle \mathbb{Q}_k, \mathbb{Q}_\ell \rangle.$$

The properties of the inner product are obvious, except perhaps the definiteness. If $\langle \mathbb{Q}_k, \mathbb{Q}_k \rangle = 0$, then $\mathbb{Q}_k(\mu_j^2)\,(y, y_j)_Y$ vanishes for all $j \in \mathbb{N}$. The assumption on y implies that $(y, y_j)_Y \neq 0$ for infinitely many j. But then the polynomial $\mathbb{Q}_k$ has infinitely many zeros μ_j^2. This implies $\mathbb{Q}_k = 0$, which ends the proof. $\square$

The following lemma is needed for the proof of Theorem 2.24.

Lemma *Let $0 < m \leq m(\delta)$ where $m(\delta)$ satisfies (2.46) and define the space $X^\sigma = (K^*K)^{\sigma/2}$ for any $\sigma > 0$, equipped with the norm $\|x\|_{X^\sigma} := \|(K^*K)^{-\sigma/2}x\|_X$.*

Let $y^\delta \notin \text{span}\{y_1, \ldots, y_N\}$ *for any* $N \in \mathbb{N}$ *with* $\|y^\delta - Kx^*\|_Y \leq \delta$, *and let* $x^* \in X^\sigma$ *for some* $\sigma > 0$, *and* $\|x^*\|_{X^\sigma} \leq E$. *Then*

$$\|Kx^{m,\delta} - y^\delta\|_Y \leq \delta + (1+\sigma)^{(\sigma+1)/2} \frac{E}{\left|\frac{d}{dt}\mathbb{Q}_m^\delta(0)\right|^{(\sigma+1)/2}}$$

where $\mathbb{Q}_m^\delta$ *denotes the polynomial* $\mathbb{Q}_m$ *for* $y = y^\delta$.

Before we prove this lemma, we recall some properties of orthogonal functions (see [254]).

As we saw in Lemma 2.22, the polynomials $\mathbb{Q}_m$ are orthogonal with respect to the inner product $\langle \cdot, \cdot \rangle$. Therefore, the zeros $\lambda_{j,m}$, $j = 1, \ldots, m$, of $\mathbb{Q}_m$ are all real and positive and lie in the interval $(0, \|K\|^2_{\mathcal{L}(X,Y)})$. By their normalization, $\mathbb{Q}_m$ must have the form

$$\mathbb{Q}_m(t) = \prod_{j=1}^{m}\left(1 - \frac{t}{\lambda_{j,m}}\right).$$

Furthermore, the zeros of two subsequent polynomials interlace; that is,

$$0 < \lambda_{1,m} < \lambda_{1,m-1} < \lambda_{2,m} < \lambda_{2,m-1} < \cdots < \lambda_{m-1,m-1} < \lambda_{m,m} < \|K\|^2_{\mathcal{L}(X,Y)}.$$

Finally, from the factorization of $\mathbb{Q}_m$, we see that

$$\frac{d}{dt}\mathbb{Q}_m(t) = -\mathbb{Q}_m(t)\sum_{j=1}^{m}\frac{1}{\lambda_{j,m} - t} \quad \text{and}$$

$$\frac{d^2}{dt^2}\mathbb{Q}_m(t) = \mathbb{Q}_m(t)\left[\left(\sum_{j=1}^{m}\frac{1}{\lambda_{j,m} - t}\right)^2 - \sum_{j=1}^{m}\frac{1}{(\lambda_{j,m} - t)^2}\right]$$

$$= \mathbb{Q}_m(t)\sum_{\substack{j,\ell=1 \\ j\neq\ell}}^{m}\frac{1}{(\lambda_{j,m} - t)(\lambda_{\ell,m} - t)}.$$

For $0 \leq t \leq \lambda_{1,m}$, we conclude that $\frac{d}{dt}\mathbb{Q}_m(t) \leq 0$, $\frac{d^2}{dt^2}\mathbb{Q}_m(t) \geq 0$, and $0 \leq \mathbb{Q}_m(t) \leq 1$.

For the proof of the lemma and the following theorem, it is convenient to introduce two orthogonal projections. For any $\varepsilon > 0$, we denote by $L_\varepsilon : X \to X$ and $M_\varepsilon : Y \to Y$ the orthogonal projections

$$L_\varepsilon z := \sum_{\mu_n^2 \leq \varepsilon}(z, x_n)_X\, x_n, \quad z \in X,$$

$$M_\varepsilon z := \sum_{\mu_n^2 \leq \varepsilon}(z, y_n)_Y\, y_n, \quad z \in Y$$

where $\{\mu_n, x_n, y_n : n \in J\}$ denotes a singular system of K.

The following estimates are easily checked:

$$\|M_\varepsilon K x\|_Y \leq \sqrt{\varepsilon}\|L_\varepsilon x\|_X \quad \text{and} \quad \|(I - L_\varepsilon)x\|_X \leq \frac{1}{\sqrt{\varepsilon}}\|(I - M_\varepsilon)K x\|_Y$$

for all $x \in X$.

Proof of the lemma: Let $\lambda_{j,m}$ be the zeros of $\mathbb{Q}_m^\delta$. We suppress the dependence on δ. The orthogonality relation (2.44) implies that $\mathbb{Q}_m^\delta$ is orthogonal to the polynomial $t \mapsto \mathbb{Q}_m^\delta(t)/(\lambda_{1,m} - t)$ of degree $m - 1$; that is,

$$\sum_{n=1}^\infty \mu_n^2 \, \mathbb{Q}_m^\delta(\mu_n^2) \, \frac{\mathbb{Q}_m^\delta(\mu_n^2)}{\lambda_{1,m} - \mu_n^2} \, \left|(y^\delta, y_n)_Y\right|^2 = 0.$$

This implies that

$$\sum_{\mu_n^2 \leq \lambda_{1,m}} \mathbb{Q}_m^\delta(\mu_n^2)^2 \, \frac{\mu_n^2}{\lambda_{1,m} - \mu_n^2} \, \left|(y^\delta, y_n)_Y\right|^2$$

$$= \sum_{\mu_n^2 > \lambda_{1,m}} \mathbb{Q}_m^\delta(\mu_n^2)^2 \, \frac{\mu_n^2}{\mu_n^2 - \lambda_{1,m}} \, \left|(y^\delta, y_n)_Y\right|^2$$

$$\geq \sum_{\mu_n^2 > \lambda_{1,m}} \mathbb{Q}_m^\delta(\mu_n^2)^2 \left|(y^\delta, y_n)_Y\right|^2 .$$

From this, we see that

$$\|y^\delta - K x^{m,\delta}\|_Y^2 = \left(\sum_{\mu_n^2 \leq \lambda_{1,m}} + \sum_{\mu_n^2 > \lambda_{1,m}}\right) \mathbb{Q}_m^\delta(\mu_n^2)^2 \left|(y^\delta, y_n)_Y\right|^2$$

$$\leq \sum_{\mu_n^2 \leq \lambda_{1,m}} \underbrace{\mathbb{Q}_m^\delta(\mu_n^2)^2 \left\{1 + \frac{\mu_n^2}{\lambda_{1,m} - \mu_n^2}\right\}}_{=:\Phi_m(\mu_n^2)^2} \left|(y^\delta, y_n)_Y\right|^2$$

$$= \|M_{\lambda_{1,m}} \Phi_m(KK^*) y^\delta\|_Y^2 ,$$

where we have set

$$\Phi_m(t) := \mathbb{Q}_m^\delta(t) \sqrt{1 + \frac{t}{\lambda_{1,m} - t}} = \mathbb{Q}_m^\delta(t) \sqrt{\frac{\lambda_{1,m}}{\lambda_{1,m} - t}} .$$

Therefore,

$$\|y^\delta - K x^{m,\delta}\|_Y \leq \|M_{\lambda_{1,m}} \Phi_m(KK^*)(y^\delta - y^*)\|_Y + \|M_{\lambda_{1,m}} \Phi_m(KK^*) K x^*\|_Y .$$

We estimate both terms on the right-hand side separately:

$$\|M_{\lambda_{1,m}}\Phi_m(KK^*)(y^\delta - y^*)\|_Y^2 = \sum_{\mu_n^2 \leq \lambda_{1,m}} \Phi_m(\mu_n^2)^2 |(y^\delta - y^*, y_n)_Y|^2$$

$$\leq \max_{0 \leq t \leq \lambda_{1,m}} \Phi_m(t)^2 \|y^\delta - y^*\|_Y^2,$$

$$\|M_{\lambda_{1,m}}\Phi_m(KK^*)Kx^*\|_Y^2 = \sum_{\mu_n^2 \leq \lambda_{1,m}} \left[\Phi_m(\mu_n^2)^2 \mu_n^{2+2\sigma}\right] \mu_n^{-2\sigma} |(x^*, x_n)_X|^2$$

$$\leq \max_{0 \leq t \leq \lambda_{1,m}} \left[t^{1+\sigma}\Phi_m(t)^2\right] \|x^*\|_{X^\sigma}^2.$$

The proof is finished provided we can show that $0 \leq \Phi_m(t) \leq 1$ and

$$t^{1+\sigma}\Phi_m^2(t) \leq \left(\frac{1+\sigma}{|\frac{d}{dt}Q_m^\delta(0)|}\right)^{\sigma+1} \qquad \text{for all } 0 \leq t \leq \lambda_{1,m}.$$

The first assertion follows from $\Phi_m(0) = 1$, $\Phi_m(\lambda_{1,m}) = 0$, and

$$\frac{d}{dt}\left[\Phi_m(t)^2\right] = 2\,Q_m(t)\frac{d}{dt}Q_m(t)\frac{\lambda_{1,m}}{\lambda_{1,m} - t} + Q_m(t)^2\frac{\lambda_{1,m}}{(\lambda_{1,m} - t)^2}$$

$$= \Phi_m(t)^2 \left[\frac{1}{\lambda_{1,m} - t} - 2\sum_{j=1}^{m}\frac{1}{\lambda_{j,m} - t}\right] \leq 0.$$

Now we set $\psi(t) := t^{1+\sigma}\Phi_m(t)^2$. Then $\psi(0) = \psi(\lambda_{1,m}) = 0$. Let $\hat{t} \in (0, \lambda_{1,m})$ be the maximum of ψ in this interval. Then $\psi'(\hat{t}) = 0$, and thus by differentiation

$$(\sigma + 1)\hat{t}^\sigma \Phi_m(\hat{t})^2 + \hat{t}^{\sigma+1}\frac{d}{dt}\left[\Phi_m(\hat{t})^2\right] = 0;$$

that is,

$$\sigma + 1 = \hat{t}\left[2\sum_{j=1}^{m}\frac{1}{\lambda_{j,m} - \hat{t}} - \frac{1}{\lambda_{1,m} - \hat{t}}\right] \geq \hat{t}\sum_{j=1}^{m}\frac{1}{\lambda_{j,m} - \hat{t}}$$

$$\geq \hat{t}\sum_{j=1}^{m}\frac{1}{\lambda_{j,m}} = \hat{t}\left|\frac{d}{dt}Q_m^\delta(0)\right|.$$

This implies that $\hat{t} \leq (\sigma + 1)/|\frac{d}{dt}Q_m^\delta(0)|$. With $\psi(t) \leq \hat{t}^{\sigma+1}$ for all $t \in [0, \lambda_{1,m}]$, the assertion follows. $\qquad\square$

Theorem 2.24 *Assume that y^* and y^δ do not belong to the linear span of finitely many y_j. Let the sequence $x^{m(\delta),\delta}$ be constructed by the conjugate gradient method with stopping rule (2.46) for fixed parameter $r > 1$. Let $x^* = (K^*K)^{\sigma/2}z \in X^\sigma$ for some $\sigma > 0$ and $z \in X$. Then there exists $c > 0$ with*

$$\|x^* - x^{m(\delta),\delta}\|_X \leq c\,\delta^{\sigma/(\sigma+1)}\,E^{1/(\sigma+1)}, \tag{2.47}$$

where $E = \|z\|_X$.

Proof: Similar to the analysis of Landweber's method, we estimate the error by the sum of two terms: the first converges to zero as $\delta \to 0$ independently of m, and the second term tends to infinity as $m \to \infty$. The role of the norm $\|R_\alpha\|_{\mathcal{L}(Y,X)}$ here is played by $\left|\frac{d}{dt}\mathbb{Q}_m^\delta(0)\right|^{1/2}$.

First, let δ and $m := m(\delta)$ be fixed. Set for abbreviation

$$q := \left|\frac{d}{dt}\mathbb{Q}_m^\delta(0)\right|.$$

Choose $0 < \varepsilon \leq 1/q \leq \lambda_{1,m}$. With

$$\tilde{x} := x^* - \mathbb{Q}_m^\delta(K^*K)x^* = \mathbb{P}_{m-1}^\delta(K^*K)K^*y^*$$

we conclude that

$$
\begin{aligned}
\|x^* - x^{m,\delta}\|_X &\leq \|L_\varepsilon(x^* - x^{m,\delta})\|_X + \|(I - L_\varepsilon)(x^* - x^{m,\delta})\|_X \\
&\leq \|L_\varepsilon(x^* - \tilde{x})\|_X + \|L_\varepsilon(\tilde{x} - x^{m,\delta})\|_X \\
&\quad + \frac{1}{\sqrt{\varepsilon}}\|(I - M_\varepsilon)(y^* - Kx^{m,\delta})\|_Y \\
&\leq \|L_\varepsilon \mathbb{Q}_m^\delta(K^*K)x^*\|_X + \|L_\varepsilon \mathbb{P}_{m-1}^\delta(K^*K)K^*(y^* - y^\delta)\|_X \\
&\quad + \frac{1}{\sqrt{\varepsilon}}\|y^* - y^\delta\|_Y + \frac{1}{\sqrt{\varepsilon}}\|y^\delta - Kx^{m,\delta}\|_Y \\
&\leq E \max_{0 \leq t \leq \varepsilon}\left|t^{\sigma/2}\mathbb{Q}_m^\delta(t)\right| + \delta \max_{0 \leq t \leq \varepsilon}\left|\sqrt{t}\,\mathbb{P}_{m-1}^\delta(t)\right| + \frac{1+r}{\sqrt{\varepsilon}}\,\delta.
\end{aligned}
$$

From $\varepsilon \leq \lambda_{1,m}$ and $0 \leq \mathbb{Q}_m^\delta(t) \leq 1$ for $0 \leq t \leq \lambda_{1,m}$, we conclude that

$$0 \leq t^{\sigma/2}\mathbb{Q}_m^\delta(t) \leq \varepsilon^{\sigma/2} \quad \text{for } 0 \leq t \leq \varepsilon.$$

Furthermore,

$$0 \leq t\,\mathbb{P}_{m-1}^\delta(t)^2 = \underbrace{\left[1 - \mathbb{Q}_m^\delta(t)\right]}_{\leq 1}\underbrace{\frac{1 - \mathbb{Q}_m^\delta(t)}{t}}_{=-\frac{d}{dt}\mathbb{Q}_m^\delta(s)} \leq \left|\frac{d}{dt}\mathbb{Q}_m^\delta(0)\right|$$

for some $s \in [0, \varepsilon]$. Thus we have proven the basic estimate

$$\|x^* - x^{m,\delta}\|_X \leq E\,\varepsilon^{\sigma/2} + (1+r)\frac{\delta}{\sqrt{\varepsilon}} + \sqrt{q}\,\delta \quad \text{for} \quad 0 < \varepsilon \leq \frac{1}{q}. \qquad \text{(B2)}$$

$\varepsilon \in (0, 1/q)$ is a free parameter in this expression. We minimize the right-hand side with respect to ε. This gives

$$\varepsilon_*^{(\sigma+1)/2} = \frac{r+1}{\sigma}\frac{\delta}{E}.$$

Since we do not know if ε_* lies in the interval $(0, 1/q)$, we have to distinguish between two cases.

Case I: $\varepsilon_* \leq 1/q$. Then

$$\sqrt{q} \leq \frac{1}{\sqrt{\varepsilon_*}} = \left(\frac{\sigma}{r+1}\right)^{1/(\sigma+1)} \left(\frac{E}{\delta}\right)^{1/(\sigma+1)}$$

and thus

$$\|x^* - x^{m,\delta}\|_X \leq c\,\delta^{\sigma/(\sigma+1)}\,E^{1/(\sigma+1)}$$

with some constant $c > 0$, which depends only on σ and r. This case is finished.

Case II: $\varepsilon_* > 1/q$. In this case, we substitute $\varepsilon = 1/q$ in (B2) and conclude that

$$\|x^* - x^{m,\delta}\|_X \leq E q^{-\sigma/2} + (r+2)\sqrt{q}\,\delta \leq E\varepsilon_*^{\sigma/2} + (r+2)\sqrt{q}\,\delta$$
$$\leq \left(\frac{r+1}{\sigma}\right)^{\sigma/(\sigma+1)} \delta^{\sigma/(\sigma+1)}\,E^{1/(\sigma+1)} + (r+2)\sqrt{q}\,\delta.$$

It remains to estimate the quantity $q = q_m = \left|\frac{d}{dt}Q^\delta_{m(\delta)}(0)\right|$. Until now, we have not used the stopping rule. We will now use this rule to prove the estimate

$$q_m \leq c\left(\frac{E}{\delta}\right)^{2/(\sigma+1)} \qquad (B3)$$

for some $c > 0$, which depends only on σ and r. Analogously to q_m, we define $q_{m-1} := \left|\frac{d}{dt}Q^\delta_{m(\delta)-1}(0)\right|$. By the previous lemma, we already know that

$$r\delta < \|y^\delta - Kx^{m(\delta)-1,\delta}\|_Y \leq \delta + (1+\sigma)^{(\sigma+1)/2}\frac{E}{q_{m-1}^{(\sigma+1)/2}};$$

that is,

$$q_{m-1}^{(\sigma+1)/2} \leq \frac{(1+\sigma)^{(1+\sigma)/2}}{r-1}\frac{E}{\delta}. \qquad (B4)$$

We have to prove such an estimate for m instead of $m-1$.

Choose $T > 1$ and $\rho^* \in (0,1)$ with

$$\frac{T}{T-1} < r \quad \text{and} \quad T\frac{\rho^*}{1-\rho^*} \leq 2.$$

If $q_m \leq q_{m-1}/\rho^*$, then we are finished by (B4). Therefore, we assume that $q_m > q_{m-1}/\rho^*$. From $\lambda_{j,m} \geq \lambda_{j-1,m-1}$ for all $j = 2, \ldots, m$, we conclude that

$$q_m = \left|\frac{d}{dt}Q^\delta_m(0)\right| = \sum_{j=1}^m \frac{1}{\lambda_{j,m}} \leq \frac{1}{\lambda_{1,m}} + \sum_{j=1}^{m-1}\frac{1}{\lambda_{j,m-1}} = \frac{1}{\lambda_{1,m}} + q_{m-1}.$$

This implies that

$$q_{m-1} \leq \rho^* q_m \leq \frac{\rho^*}{\lambda_{1,m}} + \rho^* q_{m-1};$$

that is,

$$q_{m-1} \leq \frac{\rho^*}{(1-\rho^*)\,\lambda_{1,m}} \,.$$

Finally, we need

$$\frac{1}{\lambda_{2,m}} \leq \frac{1}{\lambda_{1,m-1}} \leq \sum_{j=1}^{m-1} \frac{1}{\lambda_{j,m-1}} = q_{m-1} \leq \frac{\rho^*}{1-\rho^*}\,\frac{1}{\lambda_{1,m}} \,.$$

Now we set $\varepsilon := T\,\lambda_{1,m}$. Then

$$\varepsilon \leq T\,\frac{\rho^*}{1-\rho^*}\,\lambda_{2,m} \leq 2\,\lambda_{2,m} \,.$$

Define the polynomial $\phi \in \mathcal{P}_{m-1}$ by

$$\phi(t) := \mathbb{Q}_m^\delta(t)\left(1 - \frac{t}{\lambda_{1,m}}\right)^{-1} = \prod_{j=2}^{m}\left(1 - \frac{t}{\lambda_{j,m}}\right).$$

For $t \leq \varepsilon$ and $j \geq 2$, we note that

$$1 \geq 1 - \frac{t}{\lambda_{j,m}} \geq 1 - \frac{\varepsilon}{\lambda_{2,m}} \geq -1;$$

that is,

$$|\phi(t)| \leq 1 \quad \text{for all } 0 \leq t \leq \varepsilon.$$

For $t \geq \varepsilon$, we conclude that

$$\left|1 - \frac{t}{\lambda_{1,m}}\right| = \frac{t - \lambda_{1,m}}{\lambda_{1,m}} \geq \frac{\varepsilon}{\lambda_{1,m}} - 1 = T - 1;$$

that is,

$$|\phi(t)| \leq \frac{1}{T-1}\,|\mathbb{Q}_m^\delta(t)| \quad \text{for all } t \geq \varepsilon.$$

Since $\phi(0) = 1$, we can apply Lemma 2.22. Using the projector M_ε, we conclude that

$$
\begin{aligned}
r\delta &< \|y^\delta - Kx_{m-1}^\delta\|_Y \leq \|\phi(KK^*)y^\delta\|_Y \\
&\leq \|M_\varepsilon\phi(KK^*)y^\delta\|_Y + \|(I - M_\varepsilon)\phi(KK^*)y^\delta\|_Y \\
&\leq \|M_\varepsilon(y^\delta - y^*)\|_Y + \|M_\varepsilon y^*\|_Y + \frac{1}{T-1}\underbrace{\|\mathbb{Q}_m^\delta(KK^*)y^\delta\|_Y}_{=\|y^\delta - Kx^{m,\delta}\|_Y} \\
&\leq \delta + \varepsilon^{(\sigma+1)/2}E + \frac{1}{T-1}\delta = \frac{T}{T-1}\delta + \left(T\,\lambda_{1,m}\right)^{(\sigma+1)/2}E,
\end{aligned}
$$

since $\|M_\varepsilon y^*\|_Y = \|M_\varepsilon Kx^*\|_Y \leq \varepsilon^{(\sigma+1)/2}\|x^*\|_{X^\sigma}$. Defining $c := r - \frac{T}{T-1}$, we conclude that $c\frac{\delta}{E} \leq \left(T\,\lambda_{1,m}\right)^{(\sigma+1)/2}$ and thus finally

$$q_m \leq \frac{1}{\lambda_{1,m}} + q_{m-1} \leq T\left(\frac{E}{c\,\delta}\right)^{2/(\sigma+1)} + q_{m-1}\,.$$

Combining this with (B4) proves (B3) and ends the proof. $\qquad\square$

Bibliography

[1] R.A. Adams, J. Fournier, *Sobolev Spaces*, 2nd repr. edn. (Academic Press, 2005)

[2] L.V. Ahlfors, *Complex Analysis* (McGraw Hill, New York, 1966)

[3] R.S. Anderssen, P. Bloomfield, Numerical differentiation procedures for non-exact data. Numer. Math. **22**, 157–182 (1973)

[4] L.E. Andersson, Algorithms for solving inverse eigenvalue problems for Sturm-Liouville equations, in *Inverse Methods in Action*, ed. by P.S. Sabatier (Springer, New York, 1990)

[5] G. Anger, *Inverse and Improperly Posed Problems in Partial Differential Equations* (Springer, New York, 1979)

[6] G. Anger, Uniquely determined mass distributions in inverse problems. Technical report, Veröffentlichungen des Zentralinstituts für Physik der Erde, 1979

[7] G. Anger, Einige Betrachtungen über inverse Probleme, Identifikationsprobleme und inkorrekt gestellte Probleme, in *Jahrbuch Überblicke Mathematik* (Springer, Berlin, 1982), pp. 55–71

[8] D.N. Arnold, W.L. Wendland, On the asymptotic convergence of collocation methods. Math. Comput. **41**, 349–381 (1983)

[9] D.N. Arnold, W.L. Wendland, The convergence of spline collocation for strongly elliptic equations on curves. Numer. Math. **47**, 310–341 (1985)

[10] K. Astala, L. Päivärinta, Calderón's inverse conductivity problem in the plane. Ann. Math. **163**, 265–299 (2006)

[11] K.E. Atkinson, A discrete Galerkin method for first kind integral equations. J. Integ. Equat. Appl. **1**, 343–363 (1988)

© Springer Nature Switzerland AG 2021
A. Kirsch, *An Introduction to the Mathematical Theory of Inverse Problems*,
Applied Mathematical Sciences 120,
https://doi.org/10.1007/978-3-030-63343-1

[12] K.E. Atkinson, I.H. Sloan, The numerical solution of first-kind logarithmic-kernel integral equations on smooth open arcs. Math. Comput. **56**, 119–139 (1991)

[13] G. Backus, F. Gilbert, The resolving power of gross earth data. Geophys. J. R. Astron. Soc. **16**, 169–205 (1968)

[14] G. Backus, F. Gilbert, Uniqueness in the inversion of inaccurate gross earth data. Philos. Trans. R. Soc. London **266**, 123–197 (1970)

[15] H.T. Banks, K. Kunisch, *Estimation Techniques for Distributed Parameter Systems* (Birkhäuser, Boston, 1989)

[16] V. Barcilon, Iterative solution of the inverse Sturm-Liouville problem. J. Math. Phys. **15**, 429–436 (1974)

[17] J. Baumeister, *Stable Solutions of Inverse Problems* (Vieweg, Braunschweig, 1987)

[18] H. Bialy, Iterative Behandlung linearer Funktionalgleichungen. Arch. Rat. Mech. Anal. **4**, 166 (1959)

[19] R.P. Boas, *Entire Functions* (Academic Press, 1954)

[20] L. Borcea, Electrical impedance tomography. Inverse Probl. **18**, R99–R136 (2002)

[21] H. Brakhage, On ill-posed problems and the method of conjugate gradients, in *Inverse and Ill-Posed Problems*, ed. by H.W. Engl, C.W. Groetsch (Academic Press, New York, 1987)

[22] G. Bruckner, On the regularization of the ill-posed logarithmic kernel integral equation of the first kind. Inverse Probl. **11**, 65–78 (1995)

[23] M. Brühl, *Gebietserkennung in der elektrischen Impedanztomographie.* PhD thesis, Universität Karlsruhe, Karlsruhe, Germany, 1999

[24] M. Brühl, Explicit characterization of inclusions in electrical impedance tomography. SIAM J. Math. Anal. **32**, 1327–1341 (2001)

[25] H. Brunner, Discretization of Volterra integral equations of the first kind (II). Numer. Math. **30**, 117–136 (1978)

[26] H. Brunner, A survey of recent advances in the numerical treatment of Volterra integral and integro-differential equations. J. Comp. Appl. Math. **8**, 213–229 (1982)

[27] L. Brynielson, On Fredholm integral equations of the first kind with convex constraints. SIAM J. Math. Anal. **5**, 955–962 (1974)

[28] A.L. Bukhgeim, Recovering a potential from Cauchy data in the two-dimensional case. J. Inverse Ill-Posed Probl. **15**, 1–15 (2007)

[29] K.E. Bullen, B. Bolt, *An Introduction to the Theory of Seismology* (Cambridge University Press, Cambridge, UK, 1985)

[30] J.P. Butler, J.A. Reeds, S.V. Dawson, Estimating solutions of first kind integral equations with nonnegative constraints and optimal smoothing. SIAM J. Numer. Anal. **18**, 381–397 (1981)

[31] B.L. Buzbee, A. Carasso, On the numerical computation of parabolic problems for preceding times. Math. Comput. **27**, 237–266 (1973)

[32] B. Caccin, C. Roberti, P. Russo, L.A. Smaldone, The Backus-Gilbert inversion method and the processing of sampled data. IEEE Trans. Signal Process. **40**, 2823–2825 (1992)

[33] F. Cakoni, D. Colton, H. Haddar, The interior transmission problem for regions with cavity. SIAM J. Math. Anal. **42**, 145–162 (2010)

[34] F. Cakoni, D. Colton, H. Haddar, *Inverse Scattering Theory and Transmission Eigenvalues*, Regional Conference Series in Applied Mathematics, vol. 88 (SIAM, Philadelphia, 2016)

[35] F. Cakoni, D. Gintides, New results on transmission eigenvalues. Inverse Probl. Imaging **4**, 39–48 (2010)

[36] F. Cakoni, D. Gintides, H. Haddar, The existence of an infinite discrete set of transmission eigenvalues. SIAM J. Math. Anal. **42**, 237–255 (2010)

[37] F. Cakoni, H. Haddar, Special issue on transmission eigenvalues. Inverse Probl. **29** (2013)

[38] A.P. Calderón, On an inverse boundary value problem, in *Seminar on Numerical Analysis and its Applications to Continuum Mechanics* (Sociedade Brasileira de Matemática, Rio de Janerio, 1980), pp. 65–73

[39] J.R. Cannon, *The One-Dimensional Heat Equation* (Addison-Wesley, Reading, MA, 1984)

[40] J.R. Cannon, C.D. Hill, Existence, uniqueness, stability and monotone dependence in a Stefan problem for the heat equation. J. Math. Mech. **17**, 1–19 (1967)

[41] J.R. Cannon, U. Hornung (eds.), *Inverse Problems* (Birkhäuser, Boston, 1986)

[42] C. Canuto, M. Hussani, A. Quarteroni, T. Zang, *Spectral Methods in Fluid Dynamics* (Springer, New York, 1987)

[43] A. Carasso, Error bounds in the final value problem for the heat equation. SIAM J. Math. Anal. **7**, 195–199 (1976)

[44] K. Chandrasekharan, *Classical Fourier Transforms* (Springer, New York, 1989)

[45] L.H. Chen, On the inverse spectral theory in a non-homogeneous interior transmission problem. *Complex Variables Elliptic Equations* (2015)

[46] L.H. Chen, An uniqueness result with some density theorems with interior transmission eigenvalues. Appl. Anal. (2015)

[47] M. Cheney, D. Isaacson, J.C. Newell, Electrical impedance tomography. SIAM Rev. **41**, 85–101 (1999)

[48] I. Cioranescu, *Geometry of Banach Spaces, Duality Mappings and Nonlinear Problems*, Mathematics and its Applications, vol. 62 (Kluwer Academic Publishers, Dordrecht, 1990)

[49] D. Colton, The approximation of solutions to the backwards heat equation in a nonhomogeneous medium. J. Math. Anal. Appl. **72**, 418–429 (1979)

[50] D. Colton, The inverse scattering problem for time-harmonic acoustic waves. SIAM Rev. **26**, 323–350 (1984)

[51] D. Colton, A. Kirsch, A simple method for solving inverse scattering problems in the resonance region. Inverse Probl. **12**, 383–393 (1996)

[52] D. Colton, A. Kirsch, L. Päivärinta, Far field patterns for acoustic waves in an inhomogeneous medium. SIAM J. Math. Anal. **20**, 1472–1483 (1989)

[53] D. Colton, R. Kress, *Integral Equation Methods in Scattering Theory* (Wiley-Interscience, New York, 1983)

[54] D. Colton, R. Kress, Eigenvalues of the far field operator and inverse scattering theory. SIAM J. Math. Anal. **26**, 601–615 (1995)

[55] D. Colton, R. Kress, *Inverse Acoustic and Electromagnetic Scattering Theory*, 4th edn. (Springer, New York, 2019)

[56] D. Colton, Y.-J. Leung, Complex transmission eigenvalues for spherically stratified media. Inverse Probl. **28** (2012)

[57] D. Colton, Y.-J. Leung, Complex eigenvalues and the inverse spectral problem for transmission eigenvalues. Inverse Probl. **29** (2013)

[58] D. Colton, Y.-J. Leung, The existence of complex transmission eigenvalues for spherically stratified media. Appl. Anal. **96**, 39–47 (2017)

[59] D. Colton, P. Monk, A novel method for solving the inverse scattering problem for time-harmonic acoustic waves in the resonance region. SIAM J. Appl. Math. **45**, 1039–1053 (1985)

[60] D. Colton, P. Monk, A novel method for solving the inverse scattering problem for time-harmonic acoustic waves in the resonance region II. SIAM J. Appl. Math. **46**, 506–523 (1986)

[61] D. Colton, P. Monk, The inverse scattering problem for time-harmonic acoustic waves in a penetrable medium. Quart. J. Mech. Appl. Math. **40**, 189–212 (1987)

[62] D. Colton, P. Monk, The inverse scattering problem for time-harmonic acoustic waves in an inhomogeneous medium. Quart. J. Mech. Appl. Math. **41**, 97–125 (1988)

[63] D. Colton, P. Monk, A new method for solving the inverse scattering problem for acoustic waves in an inhomogeneous medium. Inverse Probl. **5**, 1013–1026 (1989)

[64] D. Colton, L. Päivärinta, J. Sylvester, The interior transmission problem. Inverse Probl. Imaging **1**, 13–28 (2007)

[65] M. Costabel, Boundary integral operators on Lipschitz domains: elementary results. SIAM J. Math. Anal. **19**, 613–626 (1988)

[66] M. Costabel, V.J. Ervin, E.P. Stephan, On the convergence of collocation methods for Symm's integral equation on smooth open arcs. Math. Comput. **51**, 167–179 (1988)

[67] M. Costabel, E.P. Stephan, On the convergence of collocation methods for boundary integral equations on polygons. Math. Comput. **49**, 461–478 (1987)

[68] M. Costabel, W. Wendland, Strong ellipticity of boundary integral operators. J. Reine Angew. Math. **372**, 39–63 (1986)

[69] J. Cullum, Numerical differentiation and regularization. SIAM J. Numer. Anal. **8**, 254–265 (1971)

[70] J. Cullum, The effective choice of the smoothing norm in regularization. Math. Comput. **33**, 149–170 (1979)

[71] J.W. Daniel, The conjugate gradient method for linear and nonlinear operator equations. SIAM J. Numer. Anal. **4**, 10–26 (1967)

[72] P.J. Davis, *Interpolation and Approximation* (Blaisdell, New York, 1963)

[73] E. Deuflhard, E. Hairer (eds.), *Numerical Treatment of Inverse Problems in Differential and Integral Equations* (Springer, New York, 1983)

[74] T.F. Dolgopolova, V.K. Ivanov, Numerical differentiation. Comput. Math. Math. Phys. **6**, 570–576 (1966)

[75] R.G. Douglas, On majorization, factorization, and range inclusion of operators on Hilbert space. Proc. Am. Math. Soc. **17**, 413–415 (1966)

[76] P.P.B. Eggermont, Approximation properties of quadrature methods for Volterra equations of the first kind. Math. Comput. **43**, 455–471 (1984)

[77] P.P.B. Eggermont, Beyond superconvergence of collocation methods for Volterra equations of the first kind, in *Constructive Methods for the Practical Treatment of Integral Equations*, ed. by G. Hämmerlin, K.H. Hoffmann (Birkhäuser, Boston, 1985), pp. 110–119

[78] B. Eicke, A.K. Louis, R. Plato, The instability of some gradient methods for ill-posed problems. Numer. Math. **58**, 129–134 (1990)

[79] L. Eldén, Algorithms for the regularization of ill-conditioned least squares problems. BIT **17**, 134–145 (1977)

[80] L. Eldén, Regularization of the backwards solution of parabolic problems, in *Inverse and Improperly Posed Problems in Differential Equations*, ed. by G. Anger (Akademie Verlag, Berlin, 1979)

[81] L. Eldén, Time discretization in the backward solution of parabolic equations. Math. Comput. **39**, 53–84 (1982)

[82] L. Eldén, An algorithm for the regularization of ill-conditioned banded least squares problems. SIAM J. Sci. Stat. Comput. **5**, 237–254 (1984)

[83] J. Elschner, On spline approximation for a class of integral equations. I: Galerkin and collocation methods with piecewise polynomials. Math. Meth. Appl. Sci. **10**, 543–559 (1988)

[84] H. Engl, Necessary and sufficient conditions for convergence of regularization methods for solving linear operator equations of the first kind. Numer. Funct. Anal. Optim. **3**, 201–222 (1981)

[85] H. Engl, On least-squares collocation for solving linear integral equations of the first kind with noisy right-hand-side. Boll. Geodesia Sc. Aff. **41**, 291–313 (1982)

[86] H. Engl, Discrepancy principles for Tikhonov regularization of illposed problems leading to optimal convergence rates. J. Optim. Theory Appl. **52**, 209–215 (1987)

[87] H. Engl, On the choice of the regularization parameter for iterated Tikhonov-regularization of ill-posed problems. J. Approx. Theory **49**, 55–63 (1987)

[88] H. Engl, Regularization methods for the stable solution of inverse problems. Surv. Math. Ind. **3**, 71–143 (1993)

[89] H. Engl, H. Gfrerer, A posteriori parameter choice for general regularization methods for solving linear ill-posed problems. Appl. Numer. Math. **4**, 395–417 (1988)

[90] H. Engl, W. Grever, Using the L-curve for determining optimal regularization parameters. Numer. Math. **69**, 25–31 (1994)

[91] H. Engl, C.W. Groetsch (eds.), *Inverse and Ill-Posed Problems* (Academic Press, Boston, 1987)

[92] H. Engl, M. Hanke, A. Neubauer, *Regularization of Inverse Problems* (Kluwer Academic Publishers, Dordrecht, Netherlands, 1996)

[93] H. Engl, A. Neubauer, Optimal discrepancy principles for the Tikhonov regularization of integral equations of the first kind, in *Constructive Methods for the Practical Treatment of Integral Equations*, ed. by G. Hämmerlin and K.H. Hoffmann, vol. ISNM 73 (Birkhäuser Verlag, Basel, 1985), pp. 120–141

[94] R.E. Ewing, The approximation of certain parabolic equations backwards in time by Sobolev equations. SIAM J. Math. Anal. **6**, 283–294 (1975)

[95] A. Fasano, M. Primicerio, General free-boundary problems for the heat equation. Parts I, II. J. Math. Anal. Appl. **57, 58**, 694–723, 202–231 (1977)

[96] J. Flemming, *Variational Source Conditions, Quadratic Inverse Problems, Sparsity Promoting Regularization* (Birkhäuser, Cham, 2018)

[97] J.N. Franklin, On Tikhonov's method for ill-posed problems. Math. Comput. **28**, 889–907 (1974)

[98] V. Fridman, A method of successive approximations for Fredholm integral equations of the first kind. Uspeki Mat. Nauk. **11**, 233–234 (1956). in Russian

[99] B.G. Galerkin, Expansions in stability problems for elastic rods and plates. Vestnik Inzkenorov **19**, 897–908 (1915). in Russian

[100] I.M. Gelfand, N.J. Vilenkin, *Generalized Functions, Vol 4: Some Applications of Harmonic Analysis. Rigged Hilbert Spaces* (Academic Press, New York, 1964)

[101] I.M. Gel'fand, B.M. Levitan, On the determination of a differential operator from its spectral function. Am. Math. Soc. Trans. **1**, 253–304 (1951)

[102] H. Gfrerer, An a posteriori parameter choice for ordinary and iterated Tikhonov regularization of ill-posed problems leading to optimal convergence rates. Math. Comput. **49**, 507–522 (1987)

[103] D. Gilbarg, N.S. Trudinger, *Elliptic Partial Differential Equations of Second Order* (Springer, New York, 1983)

[104] S.F. Gilyazov, Regularizing algorithms based on the conjugate gradient method. USSR Comput. Math. Math. Phys. **26**, 9–13 (1986)

[105] V.B. Glasko, *Inverse Problems of Mathematical Physics* (American Institute of Physics, New York, 1984)

[106] G.H. Golub, D.P. O'Leary, Some history of the conjugate gradient method and Lanczos algorithms: 1948–1976. SIAM Rev. **31**, 50–102 (1989)

[107] G.H. Golub, C. Reinsch, Singular value decomposition and least squares solutions. Numer. Math. **14**, 403–420 (1970)

[108] R. Gorenflo, S. Vessella, *Abel Integral Equations, Analysis and Applications*, Lecture Notes in Mathematics, vol. 1461 (Springer, Berlin, 1991)

[109] J. Graves, P.M. Prenter, Numerical iterative filters applied to first kind Fredholm integral equations. Numer. Math. **30**, 281–299 (1978)

[110] C.W. Groetsch, *The Theory of Tikhonov Regularization for Fredholm Equations of the First Kind* (Pitman, Boston, 1984)

[111] C.W. Groetsch, *Inverse Problems in the Mathematical Sciences* (Vieweg, Braunschweig Wiesbaden, 1993)

[112] S. Gutman, M. Klibanov, Regularized quasi-Newton method for inverse scattering problems. Math. Comput. Model. **18**, 5–31 (1993)

[113] S. Gutman, M. Klibanov, Iterative method for multi-dimensional inverse scattering problems at fixed frequencies. Inverse Probl. **10**, 573–599 (1994)

[114] H. Haario, E. Somersalo, The Backus-Gilbert method revisited: background, implementation and examples. Numer. Funct. Anal. Optim. **9**, 917–943 (1985)

[115] J. Hadamard, *Lectures on the Cauchy Problem in Linear Partial Differential Equations* (Yale University Press, New Haven, 1923)

[116] P. Hähner, A periodic Faddeev-type solution operator. J. Diff. Equ. **128**, 300–308 (1996)

[117] G. Hämmerlin, K.H. Hoffmann (eds.), *Improperly Posed Problems and Their Numerical Treatment*, vol. ISNM 63 (Birkhäuser–Verlag, Basel, 1983)

[118] M. Hanke, Accelerated Landweber iterations for the solution of ill-posed equations. Numer. Math. **60**, 341–373 (1991)

[119] M. Hanke, Regularization with differential operators. An iterative approach. Numer. Funct. Anal. Optim. **13**, 523–540 (1992)

[120] M. Hanke, An ϵ-free a posteriori stopping rule for certain iterative regularization methods. SIAM J. Numer. Anal. **30**, 1208–1228 (1993)

[121] M. Hanke, Regularisierung schlecht gestellter Gleichungen. Technical report, University of Karlsruhe, Karlsruhe, 1993

[122] M. Hanke, *Conjugate Gradient Type Methods for Ill-Posed Problems*. Pitman Research Notes in Mathematics (Pitman, 1995)

[123] M. Hanke, M. Brühl, Recent progress in electrical impedance tomography. Inverse Probl. **19**, S65–S90 (2003)

[124] M. Hanke, H. Engl, An optimal stopping rule for the ν-method for solving ill-posed problems using Christoffel functions. J. Approx. Theor. **79**, 89–108 (1994)

[125] M. Hanke, C. Hansen, Regularization methods for large-scale problems. Surv. Math. Ind. **3**, 253–315 (1993)

[126] M. Hanke, A. Neubauer, O. Scherzer, A convergence analysis of the Landweber iteration for nonlinear ill-posed problems. Numer. Math. **72**, 21–37 (1995)

[127] C. Hansen, Analysis of discrete ill-posed problems by means of the L-curve. SIAM Rev. **34**, 561–580 (1992)

[128] S. Helgason, *The Radon Transform* (Birkhäuser-Verlag, Boston, 1980)

[129] G. Hellwig, *Partielle Differentialgleichungen* (Teubner Verlag, Stuttgart, 1960)

[130] G.T. Herman (ed.), *Image Reconstruction from Projections: The Fundamentals of Computerized Tomography* (Academic Press, New York, 1980)

[131] G.T. Herman, F. Natterer (eds.), *Mathematical Aspects of Computerized Tomography*. Lecture Notes in Medical Informatics, vol. 8 (Springer, New York, 1981)

[132] M.R. Hestenes, E. Stiefel, Methods of conjugate gradients for solving linear systems. J. Res. Nat. Bur. Stand. **49**, 409–436 (1952)

[133] J.W. Hilgers, On the equivalence of regularization and certain reproducing kernel Hilbert space approaches for solving first kind problems. SIAM J. Numer. Anal. **13**, 172–184 (1976)

[134] M. Hitrik, K. Krupchyk, P. Ola, L. Päivärinta, Transmission eigenvalues for operators with constant coefficients. Personal Communication (2010)

[135] H. Hochstadt, *The Functions of Mathematical Physics* (Wiley, New York, 1971)

[136] B. Hofmann, *Regularization for Applied Inverse and Ill-Posed Problems* (Teubner-Verlag, Leipzig, 1986)

[137] B. Hofmann, O. Scherzer, Local ill-posedness and source conditions of operator equations in Hilbert spaces. Inverse Probl. **14**, 1189–1206 (1998)

[138] T. Hohage, F. Weidling, Verification of a variational source condition for acoustic inverse medium scattering problems. Inverse Probl. **31**, 03 (2015)

[139] R.B. Holmes, *Geometric Functional Analysis and its Applications.* Graduate Texts in Mathematics, vol. 24 (Springer, New York, 1975)

[140] G.C. Hsiao, P. Kopp, W.L. Wendland, A Galerkin collocation method for some integral equations of the first kind. Computing **25**, 89–113 (1980)

[141] G.C. Hsiao, R.C. MacCamy, Solution of boundary value problems by integral equations of the first kind. SIAM Rev. **15**, 687–705 (1973)

[142] G.C. Hsiao, W.L. Wendland, A finite element method for some integral equations of the first kind. J. Math. Anal. Appl. **58**, 449–481 (1977)

[143] G.C. Hsiao, W.L. Wendland, The Aubin-Nitsche lemma for integral equations. J. Integ. Eq. **3**, 299–315 (1981)

[144] S.P. Huestis, The Backus-Gilbert problem for sampled band-limited functions. Inverse Probl. **8**, 873–887 (1992)

[145] D. Isaacson, M. Cheney, Effects of measurement precision and finite number of electrodes on linear impedance imaging algorithms. SIAM J. Appl. Math. **51**, 1705–1731 (1991)

[146] S. Järvenpää, E. Somersalo, Impedance imaging and electrode models, in *Inverse Problems in Medical Imaging and Nondestructive Testing.* Proceedings of the Conference in Oberwolfach (Springer, Vienna, 1996), pp. 65–74

[147] S. Joe, Y. Yan, A piecewise constant collocation method using cosine mesh grading for Symm's equation. Numer. Math. **65**, 423–433 (1993)

[148] M. Kac, Can one hear the shape of the drum? Am. Math. Month. **73**, 1–23 (1966)

[149] B. Kaltenbacher, A. Neubauer, O. Scherzer, *Iterative Regularization Methods for Nonlinear Ill-Posed Problems* (de Gruyter, Berlin, 2008)

[150] W.J. Kammerer, M.Z. Nashed, Iterative methods for best approximate solutions of integral equations of the first and second kinds. J. Math. Anal. Appl. **40**, 547–573 (1972)

[151] L.V. Kantorovic, G.P. Akilov, *Functional Analysis*, 2nd edn. (Pergamon Press, Oxford, UK, 1982)

[152] J.B. Keller, Inverse problems. Am. Math. Mon. **83**, 107–118 (1996)

[153] J.T. King, D. Chillingworth, Approximation of generalized inverses by iterated regularization. Numer. Funct. Anal. Optim. **1**, 499–513 (1979)

[154] A. Kirsch, The denseness of the far field patterns for the transmission problem. IMA J. Appl. Math. **37**, 213–225 (1986)

[155] A. Kirsch, Inverse problems, in *Trends in Mathematical Optimization*, ed. by K.-H. Hoffmann, J.-B. Hiriart-Urruty, C. Lemarechal, J. Zowe, vol. ISNM 84 (Birkhäuser–Verlag, Basel, 1988), pp. 117–137

[156] A. Kirsch, An inverse scattering problem for periodic structures, in *Methoden und Verfahren der mathematischen Physik*, ed. by E. Martensen, R.E. Kleinman, R. Kress (Peter Lang, Frankfurt, 1995), pp. 75–93

[157] A. Kirsch, Characterization of the shape of a scattering obstacle using the spectral data of the far field operator. Inverse Probl. **14**, 1489–1512 (1998)

[158] A. Kirsch, On the existence of transmission eigenvalues. Inverse Probl. Imaging **3**, 155–172 (2009)

[159] A. Kirsch, A note on Sylvester's proof of discreteness of interior transmission eigenvalues. C. R. Acad. Sci. Paris, Ser. I (2016)

[160] A. Kirsch, N. Grinberg, *The Factorization Method for Inverse Problems*. Oxford Lecture Series in Mathematics and its Applications 36 (Oxford University Press, Oxford, UK, 2008)

[161] A. Kirsch, F. Hettlich, *The Mathematical Theory of Time-Harmonic Maxwell's Equations* (Springer, 2015)

[162] A. Kirsch, B. Schomburg, G. Berendt, The Backus-Gilbert method. Inverse Probl. **4**, 771–783 (1988)

[163] A. Kirsch, B. Schomburg, G. Berendt, Mathematical aspects of the Backus–Gilbert method, in *Inverse Modeling in Exploration Geophysics*, ed. by B. Kummer, A. Vogel, R. Gorenflo, C.O. Ofoegbu (Vieweg–Verlag, Braunschweig, Wiesbaden, 1989)

[164] R.E. Kleinman, P.M. van den Berg, A modified gradient method for two-dimensional problems in tomography. J. Comput. Appl. Math. **42**, 17–36 (1992)

[165] I. Knowles, R. Wallace, A variational method for numerical differentiation. Numer. Math. **70**, 91–110 (1995)

[166] C. Kravaris, J.H. Seinfeld, Identification of parameters in distributed parameter systems by regularization. SIAM J. Control Optim. **23**, 217–241 (1985)

[167] C. Kravaris, J.H. Seinfeld, Identifiability of spatially-varying conductivity from point observation as an inverse Sturm-Liouville problem. SIAM J. Control Optim. **24**, 522–542 (1986)

[168] R. Kress, *Linear Integral Equations*, 3rd edn. (Springer, New York, 2013)

[169] R. Kress, I.H. Sloan, On the numerical solution of a logarithmic integral equation of the first kind for the Helmholtz equation. Numer. Math. **66**, 199–214 (1993)

[170] O.A. Ladyzenskaja, V.A. Solonnikov, N.N. Uralceva, *Linear and Quasi-Linear Equations of Parabolic Type* (American Mathematical Society, Providence, 1986)

[171] C. Lanczos, *Linear Differential Operators* (Van Nostrand, New York, 1961)

[172] L. Landweber, An iteration formula for Fredholm integral equations of the first kind. Am. J. Math. **73**, 615–624 (1951)

[173] M.M. Lavrentiev, *Some Improperly Posed Problems of Mathematical Physics* (Springer, New York, 1967)

[174] M.M. Lavrentiev, K.G. Reznitskaya, V.G. Yakhov, *One-Dimensional Inverse Problems of Mathematical Physics* (American Mathematical Society Translations, Providence, 1986)

[175] M.M. Lavrentiev, V.G. Romanov, V.G. Vasiliev, *Multidimensional Inverse Problems for Differential Equations*, Springer Lecture Notes, vol. 167 (Springer, New York, 1970)

[176] P.D. Lax, R.S. Phillips, *Scattering Theory* (Academic Press, New York, London, 1967)

[177] A. Lechleiter, A regularization technique for the factorization method. Inverse Probl. **22**, 1605–1625 (2006)

[178] P. Linz, Numerical methods for Volterra equations of the first kind. Comput. J. **12**, 393–397 (1969)

[179] P. Linz, *Analytical and Numerical Methods for Volterra Equations* (SIAM, Philadelphia, 1985)

[180] J. Locker, P.M. Prenter, Regularization with differential operators. I: General theory. J. Math. Anal. Appl. **74**, 504–529 (1980)

[181] A.K. Louis, Convergence of the conjugate gradient method for compact operators, in *Inverse and Ill-posed Problems*, ed. by H.W. Engl, C.W. Groetsch (Academic Press, Boston, 1987), pp. 177–183

[182] A.K. Louis, *Inverse und schlecht gestellte Probleme* (Teubner-Verlag, Stuttgart, 1989)

[183] A.K. Louis, Medical imaging, state of art and future developments. Inverse Probl. **8**, 709–738 (1992)

[184] A.K. Louis, P. Maass, Smoothed projection methods for the moment problem. Numer. Math. **59**, 277–294 (1991)

[185] A.K. Louis, F. Natterer, Mathematical problems in computerized tomography. Proc. IEEE **71**, 379–389 (1983)

[186] B.D. Lowe, M. Pilant, W. Rundell, The recovery of potentials from finite spectral data. SIAM J. Math. Anal. **23**, 482–504 (1992)

[187] B.D. Lowe, W. Rundell, The determination of a coefficient in a parabolic equation from input sources. IMA J. Appl. Math. **52**, 31–50 (1994)

[188] J.T. Marti, An algorithm for computing minimum norm solutions of Fredholm integral equations of the first kind. SIAM J. Numer. Anal. **15**, 1071–1076 (1978)

[189] J.T. Marti, On the convergence of an algorithm computing minimum-norm solutions to ill-posed problems. Math. Comput. **34**, 521–527 (1980)

[190] J. McLaughlin, P. Polyakov, On the uniqueness of a spherically symmetric speed of sound from transmission eigenvalues. J. Differ. Equ. **107**, 351–382 (1994)

[191] W. McLean, *Strongly Elliptic Systems and Boundary Integral Operators* (Cambridge University Press, Cambridge, UK, 2000)

[192] C.A. Miccelli, T.J. Rivlin, A survey of optimal recovery, in *Optimal Estimation in Approximation Theory*, ed. by C.A. Miccelli, T.J. Rivlin (Plenum Press, New York, 1977)

[193] K. Miller, Efficient numerical methods for backward solution of parabolic problems with variable coefficients, in *Improperly Posed Problems*, ed. by A. Carasso (Pitman, Boston, 1975)

[194] V.A. Morozov, Choice of parameter for the solution of functional equations by the regularization method. Sov. Math. Doklady **8**, 1000–1003 (1967)

[195] V.A. Morozov, The error principle in the solution of operational equations by the regularization method. USSR Comput. Math. Math. Phys. **8**, 63–87 (1968)

[196] V.A. Morozov, *Methods for Solving Incorrectly Posed Problems* (Springer, New York, 1984)

[197] V.A. Morozov, *Regularization Methods for Ill-Posed Problems* (CRC Press, Boca Raton, FL, 1993)

[198] D.A. Murio, *The Mollification Method and the Numerical Solution of Ill-Posed Problems* (Wiley, New York, 1993)

[199] A.I. Nachman, Reconstructions from boundary measurements. Ann. Math. **128**, 531–576 (1988)

[200] S.I. Nakagiri, Review of Japanese work of the last ten years on identifiability in distributed parameter systems. Inverse Probl. **9**, 143–191 (1993)

[201] M.Z. Nashed, On moment discretization and least-squares solution of linear integral equations of the first kind. J. Math. Anal. Appl. **53**, 359–366 (1976)

[202] M.Z. Nashed, G. Wahba, Convergence rates of approximate least squares solution of linear integral and operator equations of the first kind. Math. Comput. **28**, 69–80 (1974)

[203] I.P. Natanson, *Constructive Function Theory* (Frederick Ungar, New York, 1965)

[204] F. Natterer, Regularisierung schlecht gestellter Probleme durch Projektionsverfahren. Numer. Math. **28**, 329–341 (1977)

[205] F. Natterer, Error bounds for Tikhonov regularization in Hilbert scales. Appl. Anal. **18**, 29–37 (1984)

[206] F. Natterer, *The Mathematics of Computerized Tomography* (Teubner, Stuttgart, 1986)

[207] A.S. Nemirov'ski, B.T. Polyak, Iterative methods for solving linear ill-posed problems and precise information I. Eng. Cybern. **22**, 1–11 (1984)

[208] A.S. Nemirov'ski, B.T. Polyak, Iterative methods for solving linear ill-posed problems and precise information II. Eng. Cybern. **22**, 50–56 (1984)

[209] A. Neubauer, An a posteriori parameter choice for Tikhonov regularization in Hilbert scales leading to optimal convergence rates. SIAM J. Numer. Anal. **25**, 1313–1326 (1988)

[210] Y. Notay, On the convergence rate of the conjugate gradients in the presence of rounding errors. Numer. Math. **65**, 301–318 (1993)

[211] R.G. Novikov, Multidimensional inverse spectral problems. Funct. Anal. Appl. **22**, 263–272 (1988)

[212] L. Päivärinta, E. Somersalo (eds.), *Inverse Problems in Mathematical Physics*, Lecture Notes in Physics, vol. 422 (Springer, Berlin, 1993)

[213] L. Päivärinta, J. Sylvester, Transmission eigenvalues. SIAM J. Math. Anal. **40**, 738–753 (2008)

[214] R.L. Parker, Understanding inverse theory. Ann. Rev. Earth Planet. Sci. **5**, 35–64 (1977)

[215] L.E. Payne, *Improperly Posed Problems in Partial Differential Equations* (SIAM, Philadelphia, 1975)

[216] G.I. Petrov, Application of Galerkin's method to a problem of the stability of the flow of a viscous fluid. Priklad. Matem. Mekh. **4**, 3–12 (1940). (In Russian)

[217] D.L. Phillips, A technique for the numerical solution of certain integral equations of the first kind. J. Assoc. Comput. Mach. **9**, 84–97 (1962)

[218] J. Pöschel, E. Trubowitz, *Inverse Spectral Theory* (Academic Press, London, 1987)

[219] R. Ramlau, Morozov's discrepancy principle for Tikhonov regularization of non-linear operators. Numer. Funct. Anal. Opt **23**, 147–172 (2002)

[220] A.G. Ramm, Recovery of the potential from fixed energy scattering data. Inverse Probl. **4**, 877–886 (1988)

[221] Lord Rayleigh, On the dynamical theory of gratings. Proc. R. Soc. Lon. A **79**, 399–416 (1907)

[222] F. Rellich, Über das asymptotische Verhalten von Lösungen von $\Delta u + \lambda u = 0$ in unendlichen Gebieten. Jber. Deutsch. Math. Verein. **53**, 57–65 (1943)

[223] G.R. Richter, Numerical solution of integral equations of the first kind with nonsmooth kernels. SIAM J. Numer. Anal. **15**, 511–522 (1978)

[224] G.R. Richter, An inverse problem for the steady state diffusion equation. SIAMJAP **41**, 210–221 (1981)

[225] G.R. Richter, Numerical identification of a spatially varying diffusion coefficient. Math. Comput. **36**, 375–386 (1981)

[226] A. Rieder, *Keine Probleme mit Inversen Problemen* (Vieweg, Wiesbaden, 2003)

[227] J.R. Ringrose, *Compact Non-Self-Adjoint Operators* (Van Nostrand Reinhold, London, 1971)

[228] W. Ritz, Über lineare Funktionalgleichungen. Acta Math. **41**, 71–98 (1918)

[229] G. Rodriguez, S. Seatzu, Numerical solution of the finite moment problem in a reproducing kernel Hilbert space. J. Comput. Appl. Math. **33**, 233–244 (1990)

[230] W. Rudin, *Functional Analysis* (McGraw-Hill, New York, 1973)

[231] W. Rudin, *Principles of Mathematical Analysis*, 3rd edn. (McGraw-Hill, Auckland, 1976)

[232] W. Rundell, Inverse Sturm–Liouville problems. Technical report, University of Oulu, Oulu, 1996

[233] W. Rundell, P.E. Sacks, The reconstruction of Sturm-Liouville operators. Inverse Probl. **8**, 457–482 (1992)

[234] W. Rundell, P.E. Sacks, Reconstruction techniques for classical Sturm-Liouville problems. Math. Comput. **58**, 161–183 (1992)

[235] P.C. Sabatier, Positivity constraints in linear inverse problem - I. General theory. Geophys. J. R. Astron. Soc. **48**, 415–441 (1977)

[236] P.C. Sabatier, Positivity constraints in linear inverse problem - II. Applications. Geophys. J. R. Astron. Soc. **48**, 443–459 (1977)

[237] P.C. Sabatier (ed.), *Applied Inverse Problems*, Lecture Notes in Physics, vol. 85 (Springer, New York, 1978)

[238] G. Santhosh, M. Thamban Nair, A class of discrepancy principles for the simplified regularization of ill-posed problems. J. Austr. Math. Soc. Ser. B **36**, 242–248 (1995)

[239] J. Saranen, The modified quadrature method for logarithmic-kernel integral equations on closed curves. J. Integ. Eq. Appl. **3**, 575–600 (1991)

[240] J. Saranen, I.H. Sloan, Quadrature methods for logarithmic-kernel integral equations on closed curves. IMA J. Numer. Anal. **12**, 167–187 (1992)

[241] J. Saranen, W.L. Wendland, On the asymptotic convergence of collocation methods with spline functions of even degree. Math. Comput. **45**, 91–108 (1985)

[242] G. Schmidt, On spline collocation methods for boundary integral equations in the plane. Math. Meth. Appl. Sci. **7**, 74–89 (1985)

[243] E. Schock, On the asymptotic order of accuracy of Tikhonov regularization. J. Optim. Theory Appl. **44**, 95–104 (1984)

[244] B. Schomburg, G. Berendt, On the convergence of the Backus-Gilbert algorithm. Inverse Probl. **3**, 341–346 (1987)

[245] T. Schuster, B. Kaltenbacher, B. Hofmann, K.S. Kazimierski, *Regularization Methods in Banach Spaces*, Radon Series on Computational and Applied Mathematics, vol. 10 (De Gruyter, Berlin/Boston, 2012)

[246] T.I. Seidman, Nonconvergence results for the application of least squares estimation to ill-posed problems. J. Optim. Theory Appl. **30**, 535–547 (1980)

[247] R.E. Showalter, The final value problem for evolution equations. J. Math. Anal. Appl. **47**, 563–572 (1974)

[248] B.D. Sleeman, The inverse problem of acoustic scattering. IMA J. Appl. Math. **29**, 113–142 (1982)

[249] I.H. Sloan, Error analysis of boundary integral methods. Acta Numer. **1**, 287–339 (1992)

[250] I.H. Sloan, B.J. Burn, An unconventional quadrature method for logarithmic-kernel integral equations on closed curves. J. Integ. Eq. Appl. **4**, 117–151 (1992)

[251] I.H. Sloan, W.L. Wendland, A quadrature-based approach to improving the collocation method for splines of even degree. Z. Anal. Anw. **8**, 362–376 (1989)

[252] E. Somersalo, M. Cheney, D. Isaacson, Existence and uniqueness for electrode models for electric current computed tomography. SIAM J. Appl. Math. **52**, 1023–1040 (1992)

[253] J. Stefan, Über einige Probleme der Theorie der Wärmeleitung. S.-Ber. Wien Akad. Mat. Nat. **98**, 173, 616, 956, 1418 (1889)

[254] J. Stoer, R. Bulirsch, *Introduction to Numerical Analysis* (Springer, Heidelberg, 1980)

[255] T. Suzuki, R. Muayama, A uniqueness theorem in an identification problem for coefficients of parabolic equations. Proc. Jpn. Acad. Ser. A **56**, 259–263 (1980)

[256] J. Sylvester, Discreteness of transmission eigenvalues via upper triangular compact operators. SIAM J. Math. Anal. 341–354 (2011)

[257] J. Sylvester, Transmission eigenvalues in one-dimension. Inverse Probl. **29** (2013)

[258] J. Sylvester, G. Uhlmann, A global uniqueness theorem for an inverse boundary value problem. Ann. Math. **125**, 69–153 (1987)

[259] G. Talenti (ed.), *Inverse Problems* (Springer, Berlin, 1986)

[260] U. Tautenhahn, Q. Jin, Tikhonov regularization and a posteriori rules for solving nonlinear ill posed problems. Inverse Probl. **19**, 1–21 (2003)

[261] A.N. Tikhonov, Regularization of incorrectly posed problems. Sov. Doklady **4**, 1624–1627 (1963)

[262] A.N. Tikhonov, Solution of incorrectly formulated problems and the regularization method. Sov. Doklady **4**, 1035–1038 (1963)

[263] A.N. Tikhonov, V.Y. Arsenin, *Solutions of Ill-Posed Problems* (V.H. Winston & Sons, Washington DC, 1977)

[264] A.N. Tikhonov, A.V. Goncharsky, V.V. Stepanov, A.G. Yagola, *Numerical Methods for the Solution of Ill-Posed Problems* (Kluwer, Dordrecht, 1995)

[265] S. Twomey, The application of numerical filtering to the solution of integral equations encountered in indirect sensing measurements. J. Franklin Inst. **279**, 95–109 (1965)

[266] P.M. van den Berg, M.G. Coté, R.E. Kleinman, "Blind" shape reconstruction from experimental data. IEEE Trans. Ant. Prop. **43**, 1389–1396 (1995)

[267] J.M. Varah, Pitfalls in the numerical solution of linear ill-posed problems. SIAM J. Sci. Stat. Comput. **4**, 164–176 (1983)

[268] W. Wendland, On Galerkin collocation methods for integral equations of elliptic boundary value problems, in *Numerical Treatment of Integral Equations*, ed. by R. Leis, vol. ISNM 53 (Birkhäuser–Verlag, Basel, 1979), pp. 244–275

[269] J. Werner, *Optimization Theory and Applications* (Vieweg-Verlag, Braunschweig, Wiesbaden, 1984)

[270] G.M. Wing, *A Primer on Integral Equation of the First Kind. The Problem of Deconvolution and Unfolding* (SIAM, Philadelphia, 1992)

[271] J. Wloka, *Funktionalanalysis und Anwendungen* (de Gruyter, Berlin, New York, 1971)

[272] X.G. Xia, M.Z. Nashed, The Backus-Gilbert method for signals in reproducing kernel Hilbert spaces and wavelet subspaces. Inverse Probl. **10**, 785–804 (1994)

[273] X.G. Xia, Z. Zhang, A note on 'the Backus-Gilbert inversion method and the processing of sampled data'. IEEE Trans. Signal Process. **43**, 776–778 (1995)

[274] Y. Yan, I.H. Sloan, On integral equations of the first kind with logarithmic kernels. J. Integ. Eq. Appl. **1**, 549–579 (1988)

[275] Y. Yan, I.H. Sloan, Mesh grading for integral equations of the first kind with logarithmic kernel. SIAM J. Numer. Anal. **26**, 574–587 (1989)

[276] T. Yosida, *Lectures on Differential and Integral Equations* (Wiley Interscience, New York, 1960)

[277] D. Zidarov, *Inverse Gravimetric Problems in Geoprospecting and Geodesy* (Elsevier, Amsterdam, 1980)

Index

© Springer Nature Switzerland AG 2021
A. Kirsch, *An Introduction to the Mathematical Theory of Inverse Problems*,
Applied Mathematical Sciences 120,
https://doi.org/10.1007/978-3-030-63343-1

Printed in the United States
Baker & Taylor Publisher Services

Printed in the United States
by Baker & Taylor Publisher Services